HORMONES AND SIGNALING

ADVANCES IN
PHARMACOLOGY

VOLUME 47

HORMONES AND SIGNALING

ADVANCES IN
PHARMACOLOGY

HORMONES AND SIGNALING

Edited by

Bert W. O'Malley

Department of Cell Biology
Baylor College of Medicine
Houston, TX

ADVANCES IN
PHARMACOLOGY

VOLUME 47

ACADEMIC PRESS

San Diego   London   Boston   New York   Sydney   Tokyo   Toronto

Academic Press
A Harcourt Science and Technology Company
525 B Street, Suite 1900, San Diego, California 92101-4495, U.S.A.
http://www.apnet.com

Academic Press
24-28 Oval Road, London NW1 7DX, UK
http://www.hbuk.co.uk/ap/

International Standard Book Number: 0-12-032948-4

PRINTED IN THE UNITED STATES OF AMERICA
99 00 01 02 03 04   MM   9   8   7   6   5   4   3   2   1

Contents

New Insights into Glucocorticoid and Mineralocorticoid Signaling: Lessons from Gene Targeting

Holger M. Reichardt, François Tronche, Stefan Berger, Christoph Kellendonk, Günther Schütz

Orphan Nuclear Receptors: An Emerging Family of Metabolic Regulators

Robert Sladek, Vincent Giguère

Nuclear Receptor Coactivators

Stefan Westin, Michael G. Rosenfeld, Christopher K. Glass

Cytokines and STAT Signaling

Christian Schindler, Inga Strehlow

Coordination of cAMP Signaling Events through PKA Anchoring

John D. Scott, Mark L. Dell'Acqua, Iain D. C. Fraser, Steven J. Tavalin, Linda B. Lester

G Protein-Coupled Extracellular Ca^{2+} (Ca$^{2+}_o$)-Sensing Receptor (CaR): Roles in Cell Signaling and Control of Diverse Cellular Functions

Toru Yamaguchi, Naibedya Chattopadhyay, Edward M. Brown

Pancreatic Islet Development

Debra E. Bramblett, Hsiang-Po Huang, Ming-Jer Tsai

Genetic Analysis of Androgen Receptors in Development and Disease

A. O. Brinkmann, J. Trapman

An Antiprogestin Regulable Gene Switch for Induction of Gene Expression *in Vivo*

Yaolin Wang, Sophia Y. Tsai, Bert W. O'Malley

Steroid Receptor Knockout Models: Phenotypes and Responses Illustrate Interactions between Receptor Signaling Pathways *in Vivo*

Sylvia Hewitt Curtis, Kenneth S. Korach

Contributors

Numbers in parentheses indicate the pages on which the authors'contributions begin.

Stefan Berger (1) Division Molecular Biology of the Cell I, German Cancer Research Center Heidelberg, 69120 Heidelberg, Germany

Debra E. Bramblett (255) Department of Cell Biology, Baylor College of Medicine, Houston, Texas 77030

A. O. Brinkmann (317) Department of Endocrinology and Reproduction, Erasmus University Rotterdam

Edward M. Brown (209) Endocrine-Hypertension Division, Department of Medicine, Brigham and Women's Hospital and Harvard Medical School, Boston, MA

Naibedya Chattopadhyay (209) Endocrine-Hypertension Division, Department of Medicine, Brigham and Women's Hospital and Harvard Medical School, Boston, MA

Sylvia Hewitt Curtis (357) LRDT, Receptor Biology Section, National Institute of Environmental Health Sciences, NIH

Mark L. Dell'Acqua (175) Howard Hughes Medical Institute, Vollum Institute, Portland, Oregon 97201

Iain D. C. Fraser (175) Howard Hughes Medical Institute, Vollum Institute, Portland, Oregon 97201

Vincent Giguère (23) Molecular Oncology Group, McGill University Health Centre, Montréal, Québec, Canada H3A IAI

Christopher K. Glass (89) Howard Hughes Medical Institute, University of California, San Diego, La Jolla, California 92093-0651

Hsiang-Po Huang (255) Department of Cell Biology, Baylor College of Medicine, Houston, Texas 77030

Christoph Kellendonk (1) Division Molecular Biology of the Cell I, German Cancer Research Center Heidelberg, 69120 Heidelberg, Germany

Kenneth S. Korach (357) LRDT, Receptor Biology Section, National Institute of Environmental Health Sciences, NIH

Linda B. Lester (175) Division of Endocrinology, Oregon Health Sciences University, Portland, Oregon 97201

Bert W. O'Malley (343) Department of Cell Biology, Baylor College of Medicine, Houston, TX 77030

Holger M. Reichardt (1) Division Molecular Biology of the Cell I, German Cancer Research Center Heidelberg, 69120 Heidelberg, Germany

Michael G. Rosenfeld (89) Howard Hughes Medical Institute, University of California, San Diego, La Jolla California 92093-0651

Christian Schindler (113) Departments of Microbiology and Medicine, College of Physicians and Surgeons, Columbia University, New York, NY 10032

Günther Schütz (1) Division Molecular Biology of the Cell I, German Cancer Research Center Heidelberg, 69120 Heidelberg, Germany

John D. Scott (175) Howard Hughes Medical Institute, Vollum Institute, Portland, Oregon 97201

Robert Sladek (23) Molecular Oncology Group, McGill University Health Centre, Montréal, Québec, Canada H3A IAI

Inga Strehlow (113) Departments of Microbiology and Medicine, College of Physicians and Surgeons, Columbia University, New York, NY 10032

Steven J. Tavalin (175) Howard Hughes Medical Institute, Vollum Institute, Portland, Oregon 97201

J. Trapman (317) Department of Pathology, Erasmus University, Rotterdam

François Tronche (1) Division Molecular Biology of the Cell I, German Cancer Research Center Heidelberg, 69120 Heidelberg, Germany

Ming-Jer Tsai (255) Department of Cell Biology, Baylor College of Medicine, Houston, Texas 77030

Sophia Y. Tsai (343) Department of Cell Biology, Baylor College of Medicine, Houston, TX 77030

Yaolin Wang (343) Department of Cell Biology, Baylor College of Medicine, Houston, TX 77030

Stefan Westin (89) Division of Cellular and Molecular Medicine, Department of Medicine

Toru Yamaguchi (209) Endocrine-Hypertension Division, Department of Medicine, Brigham and Women's Hospital and Harvard Medical School, Boston, MA

Holger M. Reichardt
François Tronche
Stefan Berger
Christoph Kellendonk
Günther Schütz

Division Molecular Biology of the Cell I
German Cancer Research Center Heidelberg
69120 Heidelberg, Germany

New Insights into Glucocorticoid and Mineralocorticoid Signaling: Lessons from Gene Targeting

I. Generation of Mutant Mice by Gene Targeting Provides a Valuable Tool to Study Nuclear Receptor Action

During the past decade enormous progress has been achieved in the analysis of nuclear receptor function using newly established molecular genetic techniques. The introduction of gene targeting in embryonic stem cells (Thomas and Capecchi, 1987) has allowed ubiquitous inactivation of nuclear receptors in mice, providing important new information on the role of these molecules. During these studies already-known functions of nuclear receptors were confirmed, but highly unexpected findings have also suggested additional roles for some of the receptors in development, physiology, and pathology (Beato *et al.*, 1995; Couse *et al.*, 1995; Kastner *et al.*, 1995; Lydon *et al.*, 1995). Combination of mutants revealed redundancies in signaling pathways showing specific roles of isoforms

Hormones and Signaling

that only became apparent when several of them were absent (Subbarayan *et al.*, 1997).

Despite the power of classical gene targeting, this method suffers from the lack of spatial, temporal, and functional selectivity of the introduced mutations. This sometimes renders data interpretation difficult, as distinguishing between direct and indirect effects of gene inactivation is not always possible. Compensatory mechanisms and early lethality of the mutants can prevent identification of some physiological functions of the receptors. Therefore, it is desirable to generate mouse strains carrying more refined mutations in the genes of interest, thereby limiting the effect of the mutation to only certain cell types or functional properties of the receptor. This can be achieved by knock-in approaches using the Cre–loxP system (Gu *et al.*, 1993) leading to somatic mutations, gene replacements, or introduction of point mutations. To obtain somatic mutations, mice are generated in which an essential part of a gene is flanked by two loxP-sites. After crossing them with mice expressing Cre-recombinase in a cell-type specific manner, the gene can be selectively disrupted in a certain tissue. Another useful method for the analysis of nuclear receptors is the introduction of point mutations into a gene (Reichardt *et al.*, 1998). Very subtle mutations that give answers to many physiologically relevant questions can thereby be obtained. In the following, mouse mutants for the glucocorticoid receptor (GR) and the mineralocorticoid receptor (MR) will be discussed with emphasis on how these mice allow new insights into function and molecular mechanism of these genes.

II. Targeted Disruption of the Glucocorticoid Receptor Gene in Mice Reveals Essential Functions in Multiple Physiological Processes

A. The Role of Glucocorticoids in Physiology

Glucocorticoids are synthesized in the adrenal gland and participate in the regulation of physiological processes in a variety of organ systems (Miller and Blake Tyrrel, 1995). The majority of effects are mediated by the GR, although the MR is also able to bind glucocorticoids (Funder, 1992). In contrast to the MR, the affinity of the GR is much lower, such that receptor occupancy varies within the range of physiological glucocorticoid levels and thus ensures flexible responses to altered hormone concentrations in the blood.

One of the well-established functions assigned to glucocorticoids is the control of carbohydrate and lipid metabolism. For example, glucocorticoids induce glucose synthesis by activation of gluconeogenesis, and important genes involved in this process such as glucose-6-phosphatase, phosphoenolpyruvate carboxykinase (PEPCK), and tyrosine aminotransferase (TAT) have been shown to be direct targets of glucocorticoids (Ruppert *et al.*, 1990). In the lung glucocorticoids are able to promote maturation during development, and in the adrenal gland they are involved in the biosynthesis of catecholamines. Furthermore, glucocorticoids participate in the supression of inflammatory reactions, which is achieved by repressing mRNA expression of cytokines and regulation of lymphocyte migration (Barnes and Adcock, 1993). Additionally, glucocorticoids are able to induce apoptosis of thymocytes (Chapman *et al.*, 1996), and in the erythroid compartment they are involved in long-term proliferation of erythroblasts (Wessely *et al.*, 1997).

The regulation of glucocorticoid synthesis and release is tightly controlled via the hypothalamus–pituitary–adrenal (HPA) axis (Fink, 1997). Stress and other stimuli induce synthesis and release of CRH and probably also AVP from parvocellular neurons localized in the PVN of the hypothalamus. This leads to increased synthesis and secretion of ACTH from the anterior lobe of the pituitary and thereby stimulates glucocorticoid production and release in the adrenal gland. In a negative feedback loop, elevated glucocorticoid concentrations in the blood result in repression of CRH and ACTH on the level of transcription as well as secretion, leading to homeostasis of glucocorticoid levels. A scheme of the HPA axis and the major components involved in its regulation is depicted in Fig. 1.

Besides involvement in the negative feedback loop of the HPA-axis, glucocorticoids in the brain also influence behavior, learning, and memory (McEwen and Sapolsky, 1995; McGaugh *et al.*, 1996; Sapolsky, 1996). Cognitive processes, electrophysiological properties of hippocampal neurons, and anxiety clearly involve glucocorticoid action, and both the GR and the MR may contribute to these effects. Especially in the hippocampus, it could be shown that under basal conditions the MR is completely occupied and mediates tonic effects of glucocorticoids, whereas under conditions of elevated hormone levels the GR increasingly binds hormone, resulting in a dynamic response of the neurons (Joels and deKloet, 1994). Interestingly, different populations of hippocampal neurons show divergent requirements for glucocorticoids. In the pyramidal neurons of the CA regions, chronically elevated glucorticoid levels lead to damage of the cells (Magarinos *et al.*, 1996). Accordingly, memory impairment in elderly humans is often characterized by increased glucocorticoid levels (Lupien *et al.*, 1998). In contrast, granular neurons in the dentate gyrus require a certain level of glucocorticoids for survival (Sloviter *et al.*, 1989).

To investigate in which of these multiple functions of glucocorticoids the GR is involved and what consequences the lack of functional receptor might have in terms of viability, development and physiology, the GR-gene was disrupted in mice.

B. Glucocorticoid Receptor-Deficient Mice (GR[hypo]) Confirm a Major Role for the GR in Physiology

In order to study *in vivo* the role of the glucocorticoid receptor in physiology and development the GR-gene was disrupted by homologous

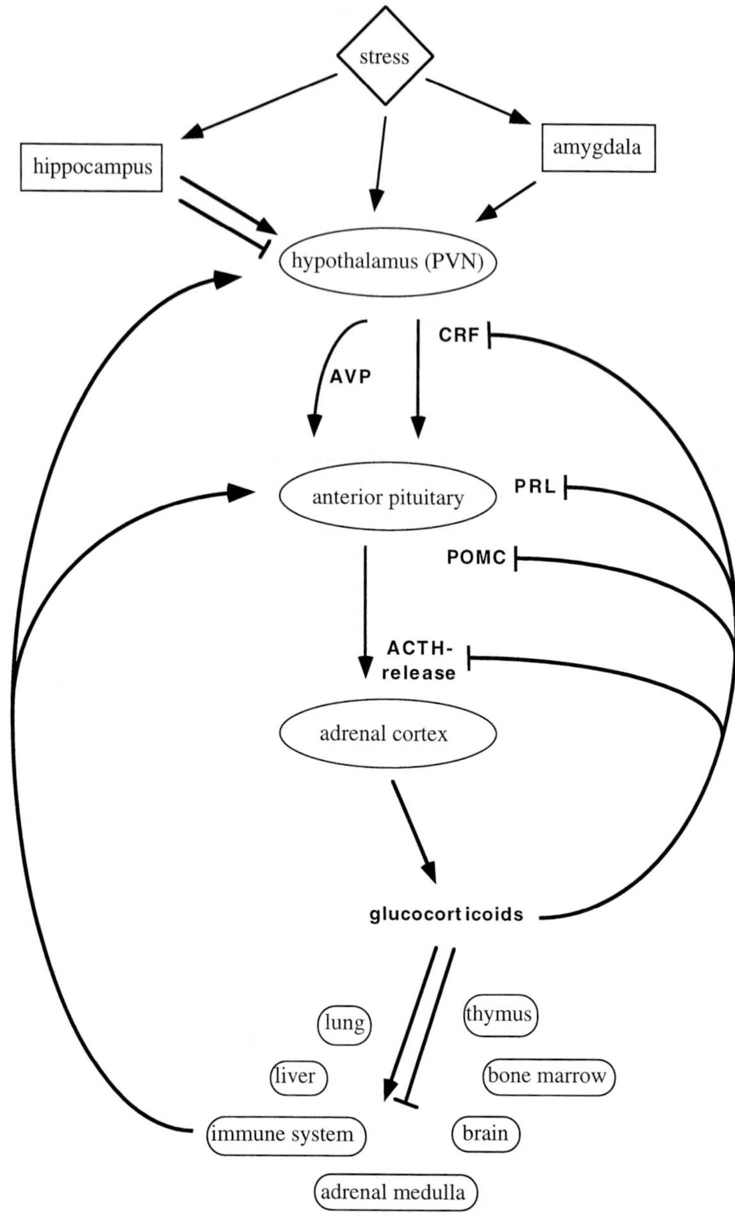

FIGURE I   The regulatory system of the hypothalamus–pituitary–adrenal (HPA) axis. Glucocorticoids released after stimuli such as stress lead to physiological responses in a variety of organs. The HPA axis is a complex regulatory circuit that controls this release of glucocorticoids from the adrenal gland. Homeostasis is achieved by feedback inhibition of several of the involved factors in the hypothalamus and the anterior pituitary. These include corticotropin-releasing factor (CRF), arginine vasopressin (AVP), proopiomelanocortin (POMC), adrenocorticotrope hormone (ACTH), and prolactin (PRL). Signals from the hippocampus and the amygdala as well as from the immune system exert further influences on the control of the HPA axis.

recombination in ES-cells using an insertion strategy (Cole *et al.*, 1995). We obtained mutant mice lacking functional GR-protein and found that the majority of homozygous mutants dies around birth due to atelectasis of the lungs. This demonstrates that transcriptional control of lung development and function crucially depends on glucocorticoid signalling via the GR. However, the exact cause of atelectasis and the primary targets of the GR in the lung still remain to be defined. As these mutants carry a hypomorphic allele of the glucocorticoid receptor gene the mouse strain is now called GR^{hypo} to distinguish it from the newly generated GR^{null} mice (see next section).

Because the majority of homozygous mutants do not reach adulthood, the main focus of the initial experiments was the analysis of development and gene expression during embryogenesis. In liver of newborn $GR^{hypo/hypo}$ mice, for example, impaired expression of gluconeogenic enzymes was observed. This was most prominent for mRNA-expression of tyrosine aminotransferase (TAT) and serine dehydrogenase (SDH), although expression was not completely abolished for any of the genes analyzed. The residual expression may result from activation of other signaling pathways, e.g., by glucagon, since expression of these genes is also dependent on CREB and related transcription factors (Ruppert *et al.*, 1990).

Regulation of glucocorticoid homeostasis is achieved by the HPA axis, which ensures fast adaption of corticosterone levels via a negative feedback circuit. Regulation is thought to be mainly mediated by GR localized in the anterior lobe of the pituitary and the paraventricular nucleus (PVN) of the hypothalamus. Thus, it was important to see how this system would respond to the lack of feedback control. As the HPA axis is fully established by day E16.5, the major components of the system could already be followed in embryos (Reichardt and Schütz, 1996). When corticosterone levels were measured in newborn $GR^{hypo/hypo}$ mice it became immediately evident that the system was out of equilibrium, as this hormone was nearly threefold elevated. This is probably due to an increase in the respective trophic hormone, namely ACTH, which was more than 10-fold elevated in $GR^{hypo/hypo}$ mice with respect to wild-type littermates. Next we analyzed mRNA and peptide expression of relevant components of the HPA axis to see if the changes in hormone concentrations were the result of altered gene transcription. In the anterior pituitary of $GR^{hypo/hypo}$ embryos we found a strong upregulation of proopiomelanocortin (POMC) mRNA starting around E16.5 of development, indicating that at least in part the high levels of ACTH are due to an increased transcriptional rate. However, CRH, the major releasing hormone for ACTH, also showed increased mRNA expression in the paraventricular nucleus of the hypothalamus, as well as elevated peptide levels in the median eminence. Interestingly, AVP, another releasing hormone for ACTH, was only moderately elevated compared to expectations from previous studies using adrenalectomy (Kretz *et al.*, 1999). This suggests that the major targets for transcriptional regulation by the GR are the genes encoding CRH and POMC.

From human pathology it is known that increased ACTH levels lead to alterations in the adrenal gland. For example, in Cushing's syndrome, pituitary tumors result in a massive overproduction of ACTH, and this is associated with adrenal hyperthrophy and hyperplasia (Miller and Blake Tyrrel, 1995). Interestingly, a comparable phenotype was found in the adrenals of GR[hypo/hypo] mice. The whole adrenal gland of the mutants is enlarged and the cortex shows massive signs of hyperplasia and hypertrophy. Because of the strong proliferation of the cortex, no defined central medulla but rather scattered patches of chromaffin cells are found in GR[hypo/hypo] mice. Additionally, mRNA expression of several steroidogenic enzymes is increased, which may also contribute to the elevated corticosterone levels measured in the serum of the mutants.

Approximately 20% of the GR[hypo/hypo] mice survive to adulthood (Cole *et al.*, 1995). This allowed study of at least some physiological functions of the GR in adult mice, although the number of animals available was limited. However, caution must be taken in the interpretation of these data, because the survivors represent only a distinct and small subgroup of the homozygous mutant with probably special characteristics (see next section). Initial studies on the role of the GR in the brain were undertaken to analyze the influence of the GR on the processing of spatial information. These experiments suggested that loss of the GR led to impairment of spatial learning along with increased motor activity (Oitzl *et al.*, 1997). In addition, the electrophysiological properties of hippocampal CA1 neurons, such as voltage-gated Ca currents and responses to serotonin and carbachol, were found to be altered (Hesen *et al.*, 1996). These studies illustrate the value of GR-deficient mice for this type of analysis, although it became clear from the sometimes conflicting results that brain-specific GR-mutants as well as mutants carrying point mutations in the GR are necessary to confirm and extend the results obtained in this initial approach.

C. Mice Carrying a Null Allele of the GR (GR[null/null]) Show Complete Penetrance of Perinatal Lethality

A finding that remained unexplained for a long time was the fact that about 20% of the homozygous GR[hypo/hypo] mutants survived to adulthood without any gross abnormalities. The discovery of two unusual GR-mRNA species in homozygous mutants provides a potential explanation for this observation (unpublished data). In one of these transcripts exon 1 is directly spliced onto exon 3 of the GR gene, thereby omitting exon 2, which codes for the entire NH_2-terminus of the protein. From studies on apoptosis it is known that a methionine in exon 3 can serve as a potential start site, giving rise to a GR protein of about half the normal size consisting only of the

DNA-binding domain and the ligand-binding domain (Dieken *et al.*, 1990). The second transcript is a fusion between the neomycin phosphotransferase-cDNA and exon 2 of the GR gene, possibly representing a polycistronic mRNA. Therefore, one cannot rule out that at least some transcriptional regulation remains in GR$^{hypo/hypo}$ mice, although at a strongly reduced level. In the light of these results a new GR mutant was generated (GR^{null}), this time by deleting exon3 and thus ensuring complete ablation of GR protein.

Interestingly, in contrast to GR$^{hypo/hypo}$ carrying the hypomorphic allele, all homozygous GR$^{null/null}$ mice die immediately after birth, and no exceptions have been observed to date. We hypothesize that survival of some of the GR$^{hypo/hypo}$ mice is indeed due to a remaining truncated form of the GR and that its transcriptional properties are at the limit of what is necessary for survival. In some cases this might be sufficient to overcome the critical first hours after birth, while in other cases it might not. In summary, we conclude that the GR is absolutely necessary for survival, although a strongly reduced level of GR activity might be sufficient to overcome perinatal lethality.

Analysis of GR$^{null/null}$ mice revealed only few differences compared to homozygous mutants carrying the hypomorphic allele. However, some of the phenotypes were more pronounced in GR$^{null/null}$ mice than in GR$^{hypo/hypo}$ mice. As discussed earlier, this was especially obvious in the case of postnatal survival, but also in the case of the adrenal medulla. A major difference was observed for the skin, which is only altered in GR$^{null/null}$ mice. Whereas GR$^{hypo/hypo}$ mice have normal skin at birth, the skin of GR$^{null/null}$ mice appears unmature at birth (unpublished data). In summary, from the analysis of GR$^{null/null}$ mice we conclude that the observations made in GR$^{hypo/hypo}$ mice are indeed characteristic for an absence of the GR during development. However, interpretation of effects observed in adult GR$^{hypo/hypo}$ mice requires caution, as it must be considered that a truncated GR protein may have some residual transcriptional properties.

In Table I the main phenotypes of GR^{hypo} and GR^{null} mice are summarized and the few minor differences between the two alleles are mentioned in each case.

III. Mice Carrying a DNA-Binding Defective GR (GR^{dim}) Demonstrate the Importance of Receptor Cross-talk with Other Transcription Factors

A. Transcriptional Regulation by the GR is Achieved via DNA-Binding-Dependent as Well as -Independent Mechanisms

The glucocorticoid receptor controls transcription by at least two distinct modes of action. It has long been known that binding of the GR to

TABLE I   Alterations of a Large Number of Physiological Functions Found in Mice Carrying a Targeted Disruption of the Glucocorticoid Receptor Gene (GR[hypo] and GR[null])

	Alterations found in glucocorticoid receptor deficient mice (GR[hypo] and GR[null])
Lung	Perinatal death due to atelectasis, low number of adult survivors in GR[hypo] mice, complete penetrance of lethal phenotype in GR[null] mice
Liver	Impaired expression of genes encoding gluconeogenic enzymes
Thymus	Loss of glucocorticoid-dependent thymocyte apoptosis
Bone marrow	Impaired proliferation of erythroid progenitor cells
Brain	Altered electrophysiological properties of hippocampal neurons in surviving adult GR[hypo] mice
Adrenal cortex	Hypertrophy and hyperplasia (more pronounced in GR[null] mice), increased mRNA expression of several steroidogenic enzymes
Adrenal medulla	Absence of adrenalin synthesis and reduced noradrenalin synthesis (more pronounced in GR[null] mice)
HPA axis	Deregulation of the negative feedback loop (upregulation of gene expression for POMC and CRH and increased hormone concentrations of corticosterone and ACTH in the serum)

regulatory elements in promoter and enhancer regions of certain genes called glucocorticoid response elements (GREs) leads to activation of transcription (Beato, 1989). These elements are present in a number of genes and and can function in an autonomous manner when put in front of a minimal promoter. By comparing GREs described in a variety of genes, a rather good consensus sequence was identified. Additionally, negative regulation dependent on DNA binding by the GR has been described. However, in this case the GR binds to so-called negative GREs (nGREs), which do not possess as good sequence conservation as positive GREs (Beato *et al.*, 1989). Also, this type of element has been described in several genes that are negatively regulated by glucocorticoids.

Only a few years ago, a second major mechanism by which the GR controls transcription was identified simultanously by several groups (Jonat *et al.*, 1990; Schüle *et al.*, 1990; Yang Yen *et al.*, 1990). Exemplified for AP-1-dependent activation, it was demonstrated that transcription can be repressed by the GR without itself binding to DNA. It is thought that this type of repression is mediated by protein–protein interaction of the GR with AP-1, possibly through an intermediary factor. Furthermore, cross-talk of the GR with other transcription factors has been described for a number of proteins, including NF-κB, CREB, GATA-1, and Stat-5 (Chang *et al.*, 1993; Imai *et al.*, 1993; Caldenhoven *et al.*, 1995; Stöcklin *et al.*, 1996). Interestingly, not only repression can be exerted by the GR via protein–protein interaction, but also activation. This has been shown *in vitro* for

Jun-homodimers as well as for Stat-5, where activation by the GR and another transcription factor acts synergistically (Pearce and Yamamoto, 1993; Stöcklin *et al.*, 1996).

Two further variants of cross-talk between the GR and other factors have been added to the constantly growing list of this type of transcriptional regulation. In both cases interference of the GR with the signal-transduction pathway involving JNK have been identified. One report describes that GR is able to prevent phosphorylation of Jun at Ser-63/73 by JNK and thereby recruitement of CBP that antgonizes AP-1 function (Caelles *et al.*, 1997). The second report presents data showing how transcriptional activation by the GR can be inhibited by activation of the JNK signal transduction pathway. These authors were able to show that *in vivo* JNK is able to phosphorylate GR on Ser-246, which leads to inhibition of transactivation (Rogatsky *et al.*, 1998). Taken together, these observations prove the existence of multiple modes of action of the GR and show that cross-talk between signalling pathways can be much more complex than previously thought.

Figure 2A illustrates the major elements of the present model for transcriptional regulation by the GR. After crossing the cell membrane, glucocorticoids bind to the GR, thereby releasing heat shock proteins from the receptor. After translocation into the nucleus, the GR either can activate transcription by binding to DNA as a dimer or repress transcription via protein–protein interaction.

B. A Point Mutation in the D-Loop of the GR Allows Separation of Different Modes of Action

In order to dissect DNA-binding-dependent and -independent functions of the GR *in vivo,* we took advantage of a point mutation that allows GRE-binding-dependent transactivation to be separated from transrepression via cross-talk with AP-1. This mutation A458T is located in the D-loop of the GR, a segment of five amino acids in the second zinc finger, which forms an important dimerization interface in the GR dimer (Fig. 2B). As demonstrated by studies in cell culture, the mutation strongly impairs dimerization and thereby DNA binding of the mutant receptor. However, transrepression of AP-1 induced expression is only slightly affected (Heck *et al.*, 1994).

To introduce this point mutation into the GR, a gene targeting strategy taking advantage of the Cre/loxP-system was employed. This allowed introduction of the mutation into the genome by homologous recombination without the selection cassette remaining in the genome, and thereby garanteed faithful expression of the mutated gene. Homozygous mice obtained by this strategy were fully viable, demonstrating that the DNA-binding function of the GR is dispensable for survival. As the introduced mutation is located in one of the dimerization domains of the GR, the mouse strain was called GR^{dim} (Reichardt *et al.*, 1998).

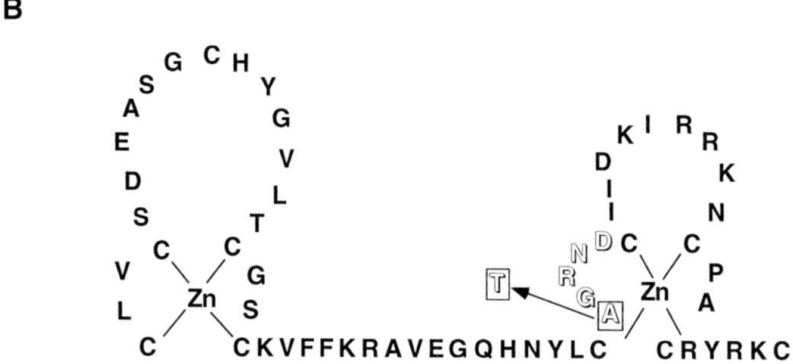

FIGURE 2   The glucocorticoid receptor (GR) can act via different mechanisms. (A) Model of the main pathways of transcriptional control by the GR. Glucocorticoids bind to the GR in the cytoplasm, heat shock proteins are released, and the receptor translocates into the nucleus. Here the receptor either binds as a dimer to glucocorticoid response elements (GREs) or represses, directly or indirectly, transcriptional activation by other factors such as AP-1 or NF-κB. (B) The DNA-binding domain of the GR. The five amino acids of the D-loop in the second zinc finger are highlighted. The mutation A458T, which was introduced in GR^dim mice, is depicted.

To examine whether the mutated glucocorticoid receptor really confers the desired characteristics on GR^{dim} mice, namely transcriptional control by crosstalk of the GR in the absence of DNA-binding-dependent regulation, several tests for transactivation and transrepression by the receptor were made. To analyze transcriptional activation via DNA binding of the GR, transient transfection studies with GRE-dependent reporters were performed in embryonic fibroblasts obtained from homozygous mutant and wild-type mice. We were able to show that activation of two different high-affinity

GRE reporters was less than 5% in mutant cells compared to wild-type cells. This demonstrates that transactivation is basically abolished. To confirm these results on a biochemical level, we performed bandshift experiments with liver nuclear extracts. Using two distinct GR antibodies we obtained supershifts with wild-type extracts, but hardly any with extracts derived from GR$^{dim/dim}$ mice, showing that DNA binding of the mutated GR was no longer possible, at least on the classical GREs examined. Finally, we wanted to demonstrate that also in the intact organism, GRE-dependent genes cannot be induced by glucocorticoids. Therefore, we analyzed induction of tyrosine aminotransferase (TAT) mRNA by dexamethasone in liver. In wild-type mice we could show that TAT mRNA is about fivefold induced by dexamethasone treatment, whereas in mutant mice no induction was observed. Taken together, these experiments prove that in GR$^{dim/dim}$ mice, DNA-binding-dependent transcriptional regulation is largely abolished.

The second main characteristic of the mutation besides the loss of DNA binding was normal transpression via protein–protein interaction. In the case of cross-talk between GR and AP-1, the best-studied paradigm is repression of TPA-induced collagenase mRNA expression. Using primary embryonic fibroblasts, it was possible to show that endogenous mRNA expression of collagenase-3 and gelatinase B could be induced by TPA treatment and efficiently repressed by dexamethasone. The degree of repression observed in mutant cells was in the same range as in wild-type cells. We conclude that repression by GR via cross-talk with AP-1 is hardly affected in GR$^{dim/dim}$ mice and propose that this most likely also accounts for other transcription factors with which the GR communicates by protein–protein interaction. In summary, GR$^{dim/dim}$ mice possess a GR that is no longer able to control transcription via binding to GREs in the regulatory regions of responsive genes, but still influences transcription by cross-talk with other protein factors. This allows *in vivo* effects dependent on DNA-binding to be distinguished from those that are not.

C. The HPA Axis Represents a Physiological System Involving Different Modes of GR Action

An important function of the GR is to control the level of glucocorticoids in the body via feedback regulation of the HPA axis (Fink, 1997). This is achieved by control of gene transcription and hormone release (Birnberg *et al.*, 1983). Because different modes of action have been proposed for some of these processes, it was interesting to see if regulation was indeed differently affected in GR$^{dim/dim}$ mice (Reichardt *et al.*, 1998). Thereby it should also be possible to assign a molecular mechanism to each process.

On the level of the hypothalamus, CRH is the main target for feedback repression by the GR. So far, no nGRE has been described in the promoter region, suggesting that repression might be mediated by cross-talk of the

GR with activating transcription factors such as CREB or Nur77. Consistent with this model we could not detect any difference in CRH immunoreactivity in the median eminence of GR$^{dim/dim}$ mice compared to wild-type animals. We conclude that indeed transrepression of CRH synthesis by the GR is independent of receptor DNA binding.

In the pituitary the situation is different. For two glucocorticoid-regulated genes in the anterior lobe, namely proopiomelanocortin (POMC) and prolactin (PRL), nGREs were described, although their functional importance remained controversial. When we analyzed mRNA expression of POMC and PRL, we found both strongly upregulated. This shows that DNA binding of the GR is required for repression of these genes and that in GR$^{dim/dim}$ mice absence of this receptor property leads to derepression. Obviously, in contrast to CRH expression in the hypothalamus, synthesis of POMC and PRL is controlled by the GR in a completely different manner, demonstrating that such different molecular mecanisms can be present even in the same physiological system.

Next we wanted to see how release of ACTH would be affected in GR$^{dim/dim}$ mice. From the analysis of GR$^{hypo/hypo}$ mice it was known that absence of the GR leads to a stong increase of ACTH in the serum. Because in GR$^{dim/dim}$ mice ACTH immunoreactivity was elevated in the anterior lobe of the pituitary, we were curious to see if this also led to an increased secretion of the hormone. However, no significant difference could be detected in the serum level. This suggests that in contrast to synthesis of ACTH, its release is differently affected in homozygous mutants, most likely because of a DNA-binding-independent mechanism responsible for repression of secretion.

GR$^{hypo/hypo}$ and GR$^{null/null}$ mice show severe hyperplasia and hypertrophy of the adrenal cortex. As this is explained by the elevated ACTH serum levels, it was interesting to analyze this parameter in GR$^{dim/dim}$ mice, as well. In accordance with not significantly altered ACTH levels, we found no severe abnormalites in the morphology of the adrenal gland. However, mRNA expression of side chain cleavage enzyme (SCC) in the cortex and serum corticosterone levels were increased.

D. Regulation of Catecholamine Synthesis in the Adrenal Medulla Is Achieved on Several Levels by the GR

Glucocorticoids play an important role in the adrenal medulla, but apart from a GRE identified in the rat PNMT promoter (Ross *et al.,* 1990), not much is known about the molecular mechanisms by which the GR participates in physiological processes in the adrenal medulla.

In GR$^{dim/dim}$ mice no obvious differences in size and morphology of the adrenal medulla were found. Interestingly, PNMT mRNA-expression was also unaltered, which was unexpected in the light of the description of a GRE in the rat promoter. However, when we analyzed hormone levels, we found a marked reduction of adrenalin and normal noradrenalin levels (unpublished data). Obviously, there must be a step in the control of adrenalin systhesis other than mRNA transcription that is dependent on DNA-binding of the receptor. This regulation could be at the level of PNMT enzyme activity, as it was reported that glucocorticoids are indeed involved in the control of this step (Wong *et al.*, 1995). Our results suggest that DNA-binding-dependent mechanisms account for regulation of adrenalin synthesis by the GR, although not by influencing PNMT mRNA expression. Additionally, in contrast to previous findings (Ross *et al.*, 1990), GRE binding to the PNMT promoter seems not to be crucial for basal expression of this gene.

E. GR^{dim} Mice Offer the Opportunity to Further Analyze Molecular Mechanisms of Glucocorticoid Physiology

One of the major advantages of GR^{dim} mice is the possibility of studying DNA-binding-independent regulation by the GR in adult mice. Among the most important functions which are considered to be mediated by cross-talk of the GR with various transcription factors such as AP-1 and NF-κB are all types of immunological processes (Gaillard, 1994). Therefore, we are presently analyzing various aspects of this type of control. First results suggest that repression of cytokines is widely intact in GR$^{dim/dim}$ mice and also that the immunosuppressive action of the GR is largely retained in *in vivo* models for inflammatory reactions (unpublished data). Therefore, some of these paradigms may allow use of GR^{dim} mice in the seach for pharmacological compounds acting preferentially through the DNA-binding-independent function of the GR.

In Table II the major observations made in GR^{dim} mice are summarized and subdivided by the presence or absence of a phenotype.

IV. Disruption of the Mineralocorticoid Receptor Gene Offers a Tool for the Molecular Analysis of Its Role in Kidney Physiology and Brain Functions

A. The MR Mediates Effects of Aldosterone and Corticosterone in a Limited Set of Target Organs

In epithelial cells, such as those that line the distal tubule in the kidney, the distal part of the colon, and the ducts of the salivary and the sweat glands, the mineralocorticoid receptor (MR) is involved in the regulation

TABLE II   Mice with a DNA-Binding Defective Glucocorticoid Receptor (GR[dim])
Show Alterations in Some Physiological Functions Also Observed in GR-Deficient
Mice, Whereas Other Functions Are Unaffected in These Mice

Conclusions from GR[dim] mice	
Differences observed	*No differences observed*
DNA-binding-dependent transcriptional regulation: Induction of GRE-dependent reporters in embryonic fibroblasts, GRE binding in bandshift experiments, induction of TAT-mRNA in liver by dexamethasone	Transcriptional regulation by protein–protein interaction: Repression of TPA-induced mRNA-expression of collagenase-3 and gelatinase B in embryonic fibroblasts
Regulation of some components of the HPA axis: POMC expression, PRL expression, corticosterone serum level	Regulation of some components of the HPA-axis: CRH-expression, ACTH serum level
Long-term proliferation of erythroid progenitor cells; glucocorticoid-dependent apoptosis of thymocytes	Survival of homozygous mutants and lung function
Expression of SCC in the adrenal cortex; adrenalin synthesis in the adrenal medulla	Adrenal size and morphology, PNMT expression in the adrenal medulla

of sodium reabsorption (Funder, 1993). In contrast, in the hippocampus the MR participates in the control of neuronal activities (Joels and deKloet, 1994). As the MR is able to bind both mineralocorticoids and glucocorticoids with approximately equal affinity, it is important to prevent access of corticosterone to the MR in aldosterone target tissues. In kidney and colon this is achieved through expression of the enzyme 11β-hydroxysteroid dehydrogenase type II (11βOHSDII), which enzymatically inactivates corticosterone (Funder *et al.*, 1988). Therefore, in these epithelial tissues the MR is predominantly activated by aldosterone. However, in the hippocampus no such protective mechanism exists. Consequently, because of the much higher serum concentration of corticosterone, the MR in the hippocampus is mainly occupied by this hormone.

The most prominent role of aldosterone is the control of sodium and potassium homeostasis, which is mainly maintained by regulated electrogenic Na^+ reabsorption in the kidney. This involves the apical amiloride-sensitive epithelial sodium channel (ENaC) and the basolateral (Na^+, K^+)-ATPase that provides the driving force for the transepithelial Na^+ transport (Rossier and Palmer, 1992). Both the ion channel and the ATPase have been proposed to be directly regulated in an MR-dependent manner, although it remains controversial whether this is achieved by a transcriptional or posttranscriptional mechanism (Verrey, 1995).

B. Impairment of Sodium Reabsorption Causes Postnatal Lethality of MR-Deficient Mice

In order to analyze the exact function of the MR in physiological processes such as the ones described above, we disrupted the MR gene by homologous recombination in ES cells (MR^{null}). Mice homozygous for the mutation are born at Mendelian frequency but die around day 10 after birth (Berger *et al.*, 1998).

As aldosterone-regulated sodium reabsorption is part of the renin–angiotensin–aldosterone system (RAAS), we analyzed how this regulatory circuit responds to the absence of MR. We found that major components of the system, namely renin, angiotensin II, and aldosterone, were elevated by one to two orders of magnitude. This indicates that the loss of negative feedback via the MR leads to a massive upregulation of the feedback system.

As aldosterone signalling is completely abolished in homozygous MR mutants, one should expect to observe characteristics similar to those known from human pseudohypoaldosteronism type I (Chang *et al.*, 1996). Indeed, a strongly increased fractional excretion of sodium along with hyperkalemia and hyponatremia was demonstrated. ENaC activity was dramatically reduced in kidney and colon, although residual activity was still measurable. Taken together, this indicates that sodium reabsorption is strongly impaired and raises the question of the responsible molecular mechanism.

With MR being a transcription factor, the most obvious interpretation of the data would have been that mRNA expression of at least some components of the involved sodium transport system would be impaired. Interestingly, none of the genes analyzed showed a significant change. mRNA expression of neither the subunits forming ENaC nor the ones that are part of the (Na^+, K^+)-ATPase was significantly altered. We conclude that control of different targets must be responsible for the marked effects observed for sodium reabsorption. It is especially likely that it is not transcription of these genes that is the primary site of regulation, but rather transcription of regulatory factors or even of posttranscriptional events such as translation or trafficking of ENaC subunits to the cell surface.

Unfortunately, because homozygous mutants die before weaning, questions on adult physiology could not be addressed. Therefore, as a first approach toward an answer to these questions, we intended to rescue the mutants. One possibility to achieve this is the application of betamethasone. It is known that the GR is present in the same cells as the MR; thus, it might be able to at least partially substitute for MR functions if sufficient concentrations of ligand reach the GR. Because of the presence of 11β-OHSDII, this is not the case for endogenous glucocorticoids. However, the synthetic steroid betamethasone overcomes the protective mechanism of 11βOHSDII when applied in high doses and therefore is able to activate transcription via the GR. Using daily injections of betamethasone, we were able to prolong survival of the homozygous mutants by approximately 2 weeks, but not longer. This indiates that the GR can indeed substitute the MR, but only to a limited extent. Most likely only a subset of the target genes for the MR can also be controlled by the GR. Another possibility is that the magnitude of transcriptional regulation by the GR is much lower because of differences in the N-terminal activation function of both receptors.

As this approach did not yield adult $MR^{-/-}$ mice, another possibility to rescue the homozygous mutants was tested. It is known from studies with adrenalectomized animals that application of salt solutions can compensate for the lack of sodium reabsorption. As it is likely that the inability to retain sodium is the main cause of death in MR mutants, one might thus be able to overcome the lethal phenotype by this treatment. Interestingly, it was indeed possible to obtain adult homozygous mutants by several injections of a salt solution per day until weaning and by afterwards keeping them on a salt diet (R. Greger; unpublished results). With this protocol, several-month-old mice have been obtained that now will allow questions on neuronal functions of the MR to be addressed.

However, this approach still has the drawback that an unphysiological salt diet is used, which might have side effects not directly related to the lack of the receptor itself. Therefore, conditional mutants of the MR are presently being generated that will allow the receptor to be selectively inactivated in the central nervous system or kidney and colon. This should enable us to study physiological processes in a specific cell type of adult mice without interference from outside effects.

Table III represents a summary of the phenotypic characteristics of MR-deficient mice and our main findings.

TABLE III   Impairment of Many Functions Related to Kidney and Colon Physiology Are Found in Mice Carrying a Targeted Disruption of the Mineralocorticoid Receptor Gene (MR^{null})

	Alterations found in mineralocorticoid receptor deficient mice
Survivability	Mice die around day 10 after birth, partial rescue with betamethasone, complete rescue with salt replacement
Ion balance	Increased fractional excretion of sodium, hyperkalemia, and hypernatremia
RAAS	Strong increase in renin, angiotensin II, and aldosterone in the serum
Na^+ transport system	mRNA expression of subunits for $(Na^+ + K^+)$-ATPase and ENaC unaltered in kidney and colon
ENaC activity	Strong but not complete reduction in kidney and colon

V. Outlook and Perspectives

As ubiquitous ablation of the GR as well as the MR leads to lethality before adulthood, analysis of most physiological functions of the receptors in the adult animal, such as challenge by stress and inflammation or behavioral studies, is impossible. Although the generation of GR^{dim} mice and the rescue of MR mutants already allows further insights into some of these questions, it is still desirable to obtain mice carrying cell-type-specific somatic mutations. Therefore, mouse mutations are presently being generated and analyzed that carry floxed alleles of the GR and MR genes. By homologous recombination in ES cells, mice in which an essential part of the gene is flanked by two loxP-sites can be obtained (GR^{flox} and MR^{flox}). By crossing these homozygous *floxed* mice with transgenic mice expressing Cre recombinase in a cell-type-specific manner, it is possible to disrupt the receptor gene only in the desired organ.

To address questions on the function of the GR in the central nervous system, a GR mutant was generated by the use of a nestin-Cre mouse (unpublished data). In these mice the GR is ablated in most of the cells in the CNS. First studies suggest impaired feedback regulation of the HPA axis along with increased corticosterone levels, which mimics to some extent the human Cushing's syndrome. As in those patients, abnormal fat distribution, signs of osteoporosis, and growth defects were observed in CNS-specific GR-mutants ($GR^{flox/flox}$; nestin-Cre).

With the generation of several distinct alleles of the glucocorticoid receptor gene, namely GR^{hypo}, GR^{null}, GR^{dim}, and GR^{flox}, it is now possible to compare certain phenotypes observed in these mouse strains. In Fig. 3, the four alleles of the GR-gene are shown along with their relative location within the GR locus. Whereas in GR^{hypo} the NH_2-terminal transactivation region was targeted, the other three alleles carry mutations in exon3 and 4 that code for the DNA-binding domain. Exemplified for the HPA axis, in Table IV components of this regulatory circuit are listed, as well as how they are differently affected in GR^{hypo}, GR^{null}, and GR^{dim} mice. This list is presently completed by the analysis of CNS-specific GR-mutants ($GR^{flox/flox}$; nestin-Cre) so that in the near future a comprehensive description on this system will be available.

During the past few years, generation of null alleles for hundreds of genes has increased tremendously our knowledge on the function of the respective proteins. The next couple of years will be reserved for the analysis of cell-type and function-specific mutations in these genes, which will enable us to dissect step by step how and by which molecular mechanisms the respective proteins control the physiology of their target organs. Based on the experience gained with the analysis of knockout mice, we can be sure that these studies will yield even more fascinating results.

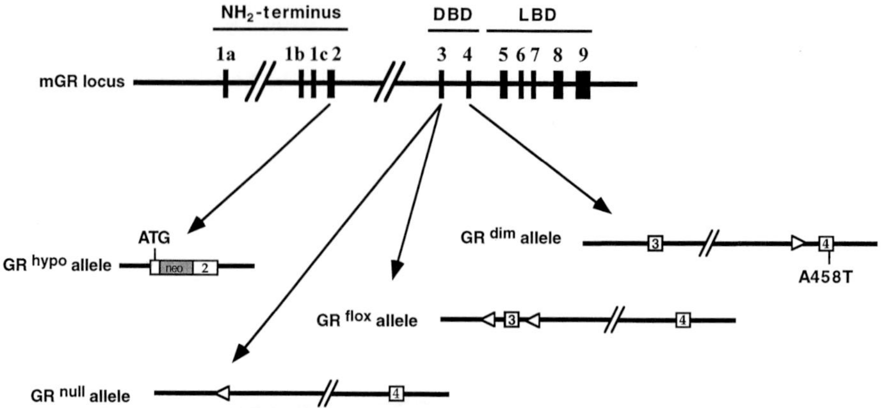

FIGURE 3   Generation of multiple GR alleles by gene targeting in ES cells. The mouse GR locus consists of 11 exons spanning more than 110 kb of genomic sequences. While the first exons code for the aminoterminal part of the receptor, exons 3 and 4 give rise to the DNA-binding domain (DBD) and exons 4–9 to the ligand-binding domain (LBD). The GR^{hypo} allele was generated by insertion of a PGK-neo cassette into exon 2 and the GR^{dim} allele by introduction of the point mutation A458T into exon 4 with a loxP-site (triangle) remaining in intron3. In the GR^{flox} allele exon 3 is flanked by two loxP-sites and the GR^{null} allele represents deletion of exon 3 after recombination of the two loxP-sites as shown for the GR^{flox} allele.

TABLE IV   Comparison of Some Major Components of the HPA Axis and How Their Regulation Is Differently Affected in GR^{hypo}, GR^{null}, and GR^{dim} Mice

	Regulation of the HPA-axis in mice carrying different GR-alleles	
Component analyzed	*GR^{hypo} or GR^{null}*	*GR^{dim}*
CRH expression in the PVN of the hypothalamus	Increased	n.d.
CRH immunoreactivity in the median emminence	5-fold increased	Unaltered
AVP expression in the hypothalamus	Slightly increased	n.d.
POMC mRNA expression in the anterior pituitary	Strongly increased	Strongly increased
ACTH serum level	>10-fold increased	Not significantly altered
Corticosterone serum level	3-fold increased	60% increased
Adrenal morphology	Hyperplasia/hypertrophy	No gross changes

References

Barnes, P. J., and Adcock, I. (1993). Anti-inflammatory actions of steroids: molecular mechanisms. *Trends Pharmacol. Sci.* **14**, 436–441.

Beato, M. (1989). Gene regulation by steroid hormones. *Cell* **56**, 335–344.

Beato, M., Chalepakis, G., Schauer, M., and Slater, E. P. (1989). DNA regulatory elements for steroid hormones. *J. Steroid. Biochem.* **32**, 737–747.

Beato, M., Herrlich, P., and Schütz, G. (1995). Steroid hormone receptors: many actors in search of a plot. *Cell* **83**, 851–857.

Berger, S., Bleich, M., Schmid, W., Cole, T. J., Peters, J., Watanabe, H., Kriz, W., Warth, R., Greger, R., and Schütz, G. (1998). Mineralocorticoid receptor knockout mice: Pathophysiology of Na$^+$ metabolism. *Proc. Natl. Acad. Sci. USA* **95**, 9424–9429.

Birnberg, N. C., Lissitzky, J. C., Hinman, M., and Herbert, E. (1983). Glucocorticoids regulate proopiomelanocortin gene expression *in vivo* at the levels of transcription and secretion. *Proc. Natl. Acad. Sci. USA* **80**, 6982–6986.

Caelles, C., Gonzalez Sancho, J. M., and Munoz, A. (1997). Nuclear hormone receptor antagonism with AP-1 by inhibition of the JNK pathway. *Genes Dev.* **11**, 3351–3364.

Caldenhoven, E., Liden, J., Wissink, S., Van de Stolpe, A., Raaijmakers, J., Koenderman, L., Okret, S., Gustafsson, J. A., and Van der Saag, P. T. (1995). Negative cross-talk between RelA and the glucocorticoid receptor: a possible mechanism for the antiinflammatory action of glucocorticoids. *Mol. Endocrinol.* **9**, 401–412.

Chang, T. J., Scher, B. M., Waxman, S., and Scher, W. (1993). Inhibition of mouse GATA-1 function by the glucocorticoid receptor: possible mechanism of steroid inhibition of erythroleukemia cell differentiation. *Mol. Endocrinol.* **7**, 528–542.

Chang, S. S., Grunder, S., Hanukoglu, A., Rosler, A., Mathew, P. M., Hanukoglu, I., Schild, L., Lu, Y., Shimkets, R. A., Nelson Williams, C., Rossier, B. C., and Lifton, R. P. (1996). Mutations in subunits of the epithelial sodium channel cause salt wasting with hyperkalaemic acidosis, pseudohypoaldosteronism type 1. *Nature Genetics* **12**, 248–253.

Chapman, M. S., Askew, D. J., Kuscuoglu, U., and Miesfeld, R. L. (1996). Transcriptional control of steroid-regulated apoptosis in murine thymoma cells. *Mol. Endocrinol.* **10**, 967–978.

Cole, T. J., Blendy, J. A., Monaghan, A. P., Krieglstein, K., Schmid, W., Aguzzi, A., Fantuzzi, G., Hummler, E., Unsicker, K., and Schütz, G. (1995). Targeted disruption of the glucocorticoid receptor gene blocks adrenergic chromaffin cell development and severely retards lung maturation. *Genes Dev.* **9**, 1608–1621.

Couse, J. F., Curtis, S. W., Washburn, T. F., Lindzey, J., Golding, T. S., Lubahn, D. B., Smithies, O., and Korach, K. S. (1995). Analysis of transcription and estrogen insensitivity in the female mouse after targeted disruption of the estrogen receptor gene. *Mol. Endocrinol.* **9**, 1441–1454.

Dieken, E. S., Meese, E. U., and Miesfeld, R. L. (1990). nti glucocorticoid receptor transcripts lack sequences encoding the amino-terminal transcriptional modulatory domain. *Mol. Cell. Biol.* **10**, 4574–4581.

Fink, G. (1997). Mechanism of negative and positive feedback of steroids in the hypothalamic–pituitary system. *In* "Principles of Medical Biology" (E. E. Bittar and N. Bittar, eds.), pp. 30–100. JAI-Press, London.

Funder, J. W. (1992). Glucocorticoid receptors. *J. Steroid. Biochem. Mol. Biol.* **43**, 389–394.

Funder, J. W. (1993). Mineralocorticoids, glucocorticoids, receptors and response elements. *Science* **259**, 1132–1133.

Funder, J. W., Pearce, P. T., Smith, R., and Smith, A. I. (1988). Mineralocorticoid action: target tissue specificity is enzyme, not receptor, mediated. *Science* **242**, 583–585.

Gaillard, R. C. (1994). Neuroendocrine–immune system Interactions. The immune–hypothalamo–pituitary–adrenal axis. *Trends Endo. Metab.* **5**, 303–309.

Gu, H., Zou, Y. R., and Rajewsky, K. (1993). Independent control of immunoglobulin switch recombination at individual switch regions evidenced through Cre-loxP-mediated gene targeting. *Cell* 73, 1155–1164.

Heck, S., Kullmann, M., Gast, A., Ponta, H., Rahmsdorf, H. J., Herrlich, P., and Cato, A. C. (1994). A distinct modulating domain in glucocorticoid receptor monomers in the repression of activity of the transcription factor AP-1. *Embo J.* 13, 4087–4095.

Hesen, W., Karst, H., Meijer, O., Cole, T. J., Schmid, W., de Kloet, E. R., Schutz, G., and Joels, M. (1996). Hippocampal cell responses in mice with a targeted glucocorticoid receptor gene disruption. *J. Neurosci.* 16, 6766–6774.

Imai, E., Miner, J. N., Mitchell, J. A., Yamamoto, K. R., and Granner, D. K. (1993). Glucocorticoid receptor-cAMP response element-binding protein interaction and the response of the phosphoenolpyruvate carboxykinase gene to glucocorticoids. *J. Biol. Chem.* 268, 5353–5356.

Joels, M., and deKloet, E. R. (1994). Mineralocorticoid and glucocorticoid receptors in the brain. Implications for ion permeability and transmitter systems. *Prog. Neurobiol.* 43, 1–36.

Jonat, C., Rahmsdorf, H. J., Park, K. K., Cato, A. C., Gebel, S., Ponta, H., and Herrlich, P. (1990). Antitumor promotion and antiinflammation: down-modulation of AP-1 (Fos/Jun) activity by glucocorticoid hormone. *Cell* 62, 1189–1204.

Kastner, P., Mark, M., and Chambon, P. (1995). Nonsteroid nuclear receptors: what are genetic studies telling us about their role in real life? *Cell* 83, 859–869.

Kretz, O., Reichardt, H. M., Schütz, G., and Bock, R. (1999). Corticotropin-releasing hormone expression is the major target for glucocorticoid feedback-control at the hypothalamic level. *Brain Research* 818, 488–491.

Lupien, S. J., de Leon, M., de Santi, S., Convit, A., Tarshish, C., Nair, N. P. V., Thakur, M., McEwen, B. S., Hauger, R. L., and Meaney, M. J. (1998). Cortisol level during human aging predict hippocampal atrophy and memory deficits. *Nature Neurosci.* 1, 69–73.

Lydon, J. P., DeMayo, F. J., Funk, C. R., Mani, S. K., Hughes, A. R., Montgomery, C. A., Jr., Shyamala, G., Conneely, O. M., and O'Malley, B. W. (1995). Mice lacking progesterone receptor exhibit pleiotropic reproductive abnormalities. *Genes Dev.* 9, 2266–2278.

Magarinos, A. M., McEwen, B. S., Flugge, G., and Fuchs, E. (1996). Chronic psychosocial stress causes apical dendritic atrophy of hippocampal CA3 pyramidal neurons in subordinate tree shrews. *J. Neurosci.* 16, 3534–3540.

McEwen, B. S., and Sapolsky, R. M. (1995). Stress and cognitive function. *Curr. Opin. Neurobiol.* 5, 205–216.

McGaugh, J. L., Cahill, L., and Roozendaal, B. (1996). Involvement of the amygdala in memory storage: interaction with other brain systems. *Proc. Natl. Acad. Sci. USA* 93, 13508–13514.

Miller, W. L., and Blake Tyrrel, J. (1995). The adrenal cortex. *In* "Endocrinology and Metabolism" (P. Felig, J. D. Baxter and L. A. Frohman, eds.), pp. 555–711. McGraw-Hill, New York.

Oitzl, M. S., de Kloet, E. R., Joels, M., Schmid, W., and Cole, T. J. (1997). Spatial learning deficits in mice with a targeted glucocorticoid receptor gene disruption. *Eur. J. Neurosci.* 9, 2284–2296.

Pearce, D., and Yamamoto, K. R. (1993). Mineralocorticoid and glucocorticoid receptor activities distinguished by nonreceptor factors at a composite response element. *Science* 259, 1161–1165.

Reichardt, H. M., and Schütz, G. (1996). Feedback control of glucocorticoid production is established during fetal development. *Mol. Med.* 2, 735–744.

Reichardt, H. M., Kaestner, K. H., Tuckermann, J., Kretz, O., Gass, P., Schmid, W., Herrlich, P., Angel, P., and Schütz, G. (1998). DNA binding of the glucocorticoid receptor is not essential for survival. *Cell* 93, 531–541.

Rogatsky, I., Logan, S. K., and Garabedian, M. J. (1998). Antagonism of glucocorticoid receptor transcriptional activation by the c-Jun N-terminal kinase. *Proc. Natl. Acad. Sci. USA* **95**, 2050–2055.

Ross, M. E., Evinger, M. J., Hyman, S. E., Carroll, J. M., Mucke, L., Comb, M., Reis, D. J., Joh, T. H., and Goodman, H. M. (1990). Identification of a functional glucocorticoid response element in the phenylethanolamine N-methyltransferase promoter using fusion genes introduced into chromaffin cells in primary culture. *J. Neurosci.* **10**, 520–530.

Rossier, C. R., and Palmer, L. G. (1992). Mechanisms of aldosterone action on sodium and potassium transport. *In* "The Kidney: Physiology and Pathophysiology" Seldin D. W., and Giebisch, G., eds., pp. 1373–1409. Raven, New York.

Ruppert, S., Boshart, M., Bosch, F. X., Schmid, W., Fournier, R. E., and Schütz, G. (1990). Two genetically defined trans-acting loci coordinately regulate overlapping sets of liver-specific genes. *Cell* **61**, 895–904.

Sapolsky, R. M. (1996). Why stress is bad for your brain. *Science* **273**, 749–750.

Schüle, R., Rangarajan, P., Kliewer, S., Ransone, L. J., Bolado, J., Yang, N., Verma, I. M., and Evans, R. M. (1990). Functional antagonism between oncoprotein c-Jun and the glucocorticoid receptor. *Cell* **62**, 1217–1226.

Sloviter, R. S., Valiquette, G., Abrams, G. M., Ronk, E. C., Sollas, A. L., Paul, L. A., and Neubort, S. (1989). Selective loss of hippocampal granule cells in the mature rat brain after adrenalectomy. *Science* **243**, 535–538.

Stöcklin, E., Wissler, M., Gouilleux, F., and Groner, B. (1996). Functional interactions between Stat5 and the glucocorticoid receptor. *Nature* **383**, 726–728.

Subbarayan, V., Kastner, P., Mark, M., Dierich, A., Gorry, P., and Chambon, P. (1997). Limited specificity and large overlap of the functions of the mouse RAR gamma 1 and RAR gamma 2 isoforms. *Mech. Dev.* **66**, 131–142.

Thomas, K. R., and Capecchi, M. R. (1987). Site-directed mutagenesis by gene targeting in mouse embryo-derived stem cells. *Cell* **51**, 503–512.

Verrey, F. (1995). Transcriptional control of sodium transport in tight epithelial by adrenal steroids. *J. Membr. Biol.* **144**, 93–110.

Wessely, O., Deiner, E., Beug, H., and von Lindern, M. (1997). The glucocorticoid receptor is a key regulator of the decision between self-renewal and differentiation in erythroid progenitors. *EMBO J.* **16**, 267–280.

Wong, D. L., Siddall, B., and Wang, W. (1995). Hormonal control of rat adrenal phenylethanolamine N-methyltransferase. Enzyme activity, the final critical pathway. *Neuropsychopharmacology* **13**, 223–234.

Yang Yen, H. F., Chambard, J. C., Sun, Y. L., Smeal, T., Schmidt, T. J., Drouin, J., and Karin, M. (1990). Transcriptional interference between c-Jun and the glucocorticoid receptor: mutual inhibition of DNA binding due to direct protein–protein interaction. *Cell* **62**, 1205–1215.

Robert Sladek
Vincent Giguère

Molecular Oncology Group
McGill University Health Centre
Montréal, Québec, Canada H3A IAI

Orphan Nuclear Receptors: An Emerging Family of Metabolic Regulators

The nuclear receptor superfamily includes more than 70 members sharing a conserved domain structure and general mode of action. Prototype members of the family function as transcription factors whose activity is modulated by classical endocrine hormones such as gonadal and adrenal steroids, vitamin D, and thyroid hormone. It was soon appreciated that the nuclear receptor family contained many members whose activity was not regulated by classical hormones: The search for ligands and functions associated with these putative receptors, referred to as orphan nuclear receptors, became an active and rewarding field of research. The discoveries that specific orphan nuclear receptors respond to nonclassical hormones such as vitamin A derivatives, prostanoids, sterols, and fatty acids, combined with the observation that nuclear receptor response elements are present in the promoter regions of genes encoding enzymes involved in cellular metabolism, suggest that orphan nuclear receptors may be important regulators of basic cellular function *in vivo*. This hypothesis has been corroborated by genetic linkage studies in patients with metabolic disorders as well as physiologic studies of mutant mice harboring defective orphan nuclear receptor genes. Members of the nuclear receptor family may perform key roles in balancing the metabolic demands of the whole organism, encoded by classical endocrine hormones, with those of individual organs and cells, encoded by lipophilic paracrine and intracrine signals.

I. Introduction

The concept that small lipophilic hormones could regulate cellular function by interacting with specific cellular proteins, called receptors, was first proposed more than 30 years ago (Jensen *et al.*, 1966). Formal proof of the nuclear receptor hypothesis was provided by the molecular cloning and functional dissection of the steroid hormone receptors. The subsequent isolation and characterization of receptors that were activated by nonsteroidal ligands such as vitamin A and thyroid hormone suggested that nuclear hormone receptors could potentially regulate gene expression in response to a wide variety of lipophilic ligands (Evans, 1988). Nuclear hormone receptors share considerable sequence and structural homology: to date, more than 70 gene products have been identified as members of the steroid receptor superfamily, primarily based on their similarity to known hormone receptors (Gronemeyer and Laudet, 1995). Nuclear receptors with no associated ligands at the time of their discovery are referred to as orphan nuclear receptors: Considerable effort has been devoted to identifying ligands and physiological functions for these gene products. Since developments in this field were last reviewed in this series (Willy and Mangelsdorf, 1998), candidate ligands have been identified for five nuclear receptors: In three cases, intracellular ligands formed as intermediate products of cell metabolism were shown to regulate the activity of specific family members. These findings suggest that many orphan nuclear receptors transduce intracellular signals essential for the regulation of basic cellular functions and modulate gene expression in response to environmental stimuli and metabolic demands. In the following paper, we provide a brief conceptual outline of the molecular mechanisms underlying nuclear receptor function, followed by more detailed examination of the specific orphan nuclear receptors implicated in metabolic regulation as well as the mechanisms used by these receptors to provide integrated control of metabolic target genes.

A. Structure and Action of Nuclear Receptors

Members of the nuclear receptor family have a well-conserved domain structure that parallels the functions of the receptor protein (Fig. 1). The nuclear receptor DNA binding domain (DBD), which displays the highest

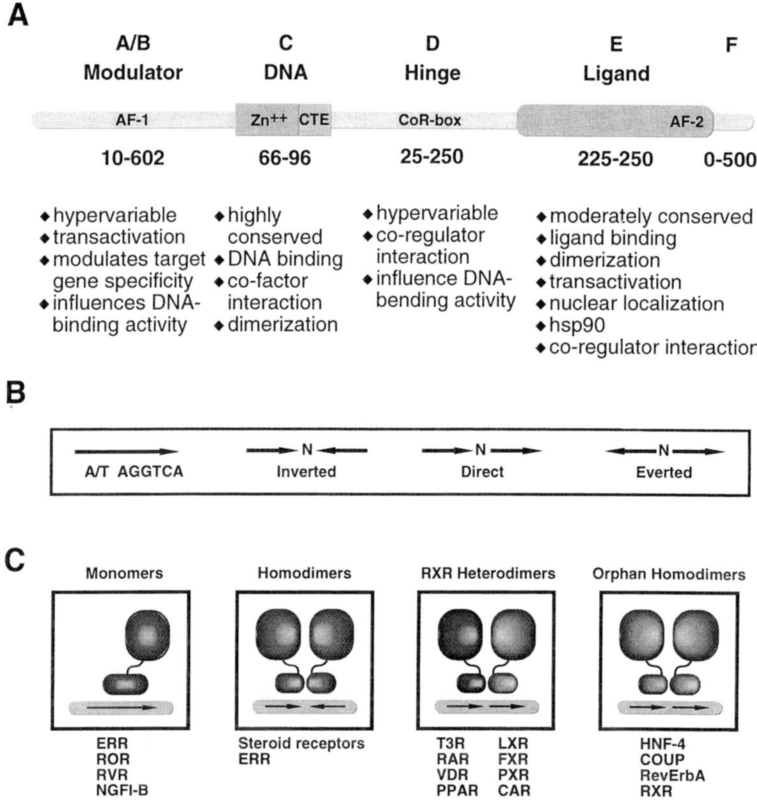

FIGURE I Nuclear receptor functional domains and DNA response elements. (A) Functional anatomy of a nuclear hormone receptor. The highly conserved DNA-binding motif is flanked by less well conserved N- and C-terminal domains. Ligand-dependent (AF2) and independent (AF1) transactivation functions are located in the receptor A/B and E domains, while dimerization functions are found in the C and E regions. DNA binding is regulated by residues in the core DNA binding domain (C) in cooperation with residues in the N-terminal (A/B) and hinge (D) regions. (B, C) Topology of nuclear receptor response elements and modes of orphan nuclear receptors DNA binding. Orphan receptors can bind to monomeric response elements containing a receptor-specific A/T rich flaking sequence. Alternatively, orphan receptor homodimers or RXR-heterodimers bind to response elements containing two copies of the AGGTCA core motif in inverted (IRn) direct (DRn) or everted (ERn) conformations. The affinity of receptor binding to dimeric response elements may be influenced by sequences contained 5′ to either half site. In some cases, RXR heterodimers may bind to monomeric response elements: This mode of binding is mediated by tethering of RXR and does not involve direct contact between the RXR DNA binding domain and DNA. Most orphan receptors display a narrow range of preferred DNA binding modes when studied using response elements containing consensus half sites (panel C). Binding rules are often less strictly followed when the receptors interact with natural HREs.

protein sequence conservation among family members, consists of a Cys-coordinated Zn-cluster occupying between 66 and 70 amino acids in the central region of the receptor. The ligand binding domain (LBD), which is less well conserved among family members, occupies the C-terminal region of the receptor. This region also contains a ligand-dependant transactivation function (AF-2), as well as protein surfaces that regulate dimerization among family members. The N-terminal region of the receptor, which displays the poorest sequence conservation among family members, contains a ligand-independent transactivation function (Berry *et al.*, 1990). For some family members, the N-terminal domain acts in concert with the central DNA-binding domain to modulate DNA binding specificity and affinity (Giguère *et al.*, 1994; Wong and Privalsky, 1995).

The classical model of steroid receptor function proposes that ligands interact with a cytosolic complex containing the receptor and chaperone proteins: This results in rapid translocation of the liganded receptor to the cell nucleus where it can bind to response elements contained in the enhancer and promoter regions of target genes. In practice, many nuclear hormone receptors are constitutively localized in the nucleus and bind to DNA in absence of ligand. In some cases, unliganded receptors bound to DNA can act as active repressors: this repression is mediated by corepressor proteins that interact with domains within the receptor protein and influence both the activity of the general transcription apparatus as well as the chromatin structure in the neighborhood of the promoter and transcription initiation site (reviewed in Horwitz *et al.*, 1996; Heinzel *et al.*, 1997; Nagy *et al.*, 1997). Ligand binding changes the protein conformation of receptor C-terminus, which displaces corepressor proteins from the transcription complex and facilitates binding of transcriptional coactivator proteins (reviewed in Glass *et al.*, 1997). Both ligand-induced conformational changes in receptor proteins and ligand-dependent binding of coactivator proteins can be detected *in vitro*: this provides a useful method to indirectly demonstrate interactions between ligands and receptors (Janowski *et al.*, 1996; Krey *et al.*, 1997; Kliewer *et al.*, 1998).

B. DNA-Binding and Hormone Response Elements

Nuclear receptors bind specifically and with high affinity to short DNA sequences called hormone response elements (HREs). Contacts between key residues in the receptor DNA binding domain and bases within the cognate HRE core motif are primarily responsible for recognizing the HRE core (reviewed in Glass, 1994). Additional binding site specificity can be determined by interactions between nucleotides upstream of the HRE core motif and residues contained within in the C-terminal extension (CTE) of the DBD or the N-terminus of the receptor (Wilson *et al.*, 1992, 1993a; Giguère *et al.*, 1995; Zhao *et al.*, 1998). Estrogen receptors, nonsteroidal hormone receptors, and orphan nuclear receptors bind to sites containing variants of the consensus core motif AGGTCA, whereas the mineralocorticoid, glucocorticoid, androgen and progesterone receptors bind as homodimers to sites containing palindromic repeats of the core element AGAACA (Fig. 1B). Nonsteroidal receptors commonly bind to a single core motif as monomers or to HREs containing direct repeats of the core motif either as homodimers or as heterodimers with the nuclear receptor RXR (reviewed in Glass, 1994). Receptors that bind as monomers usually stabilize their interaction with DNA through contacts between the CTE and bases upstream of the core motif: The sequence of the response element's upstream flank is usually characteristic for individual monomer binding receptors (Wilson *et al.*, 1991, 1993a; Giguère *et al.*, 1994). Most receptors that bind as dimers demonstrate highest affinity binding to HREs with specific spacing between the core motif half site: The preferred distance between half sites is determined by the dimerization surfaces present within the DNA-binding domain (Perlmann *et al.*, 1993; Rastinejad *et al.*, 1995). Gene regulation by certain nuclear receptors occurs independently of direct DNA binding and is mediated by protein–protein interactions involving other transcription factors or adapter proteins. The physiological importance of this mode of gene regulation has been best demonstrated for the glucocorticoid receptor (Karin, 1998; Reichardt *et al.*, 1998), but has yet to be fully investigated for orphan nuclear receptors.

Whereas most receptors bind with high affinity to a very limited number of response element configurations, natural response elements often contain multiple overlapping half sites that do not exactly match the half site consensus sequence. Modifications within the HRE core motifs may stabilize or destabilize binding of particular family members (Yu *et al.*, 1994), whereas the specific arrangement of half sites and the specific spacer sequences contained in complex HREs may provide an opportunity for direct interactions or competitive binding between multiple nuclear receptors, allowing the promoter activity to be influenced by convergence of multiple signals. The specific HRE configuration may also influence the conformation of the DNA-bound receptor and alter its ability to interact with coactivators and corepressors (Lefstin and Yamamoto, 1998). The ability of several nuclear receptors to bind in close proximity to complex promoter elements, together with their ability to differentially interact with a common set of coregulatory proteins, both appear to be key mechanisms underlying their ability to control metabolic target genes in response to competing input signals encoded by endocrine hormones, dietary components, and intracellular intermediary metabolites.

II. Orphan Nuclear Receptors Regulate Carbohydrate and Lipid Metabolism

Several lines of evidence have suggested that orphan nuclear receptors may play an essential role in regulating carbohydrate and lipid metabolism.

These include studies of adipocyte differentiation and physiology, detailed physiologic characterization of nuclear receptor "knockout" mice, and *in vivo* and *in vitro* identification of the promoter and enhancer elements regulating the expression of metabolic enzymes. In the following sections, we summarize current knowledge on a subset of orphan nuclear receptors that have been identified as important regulators of these metabolic pathways.

A. Hepatocyte Nuclear Factor 4

Hepatocyte nuclear factor 4 (HNF-4α) was initially identified as a hepatic protein that bound to enhancer elements required for transthyretin expression in hepatoma lines (Sladek *et al.*, 1990). The cloned HNF4α protein had similar DNA binding properties and antigenic determinants as LF-A1, a previously characterized liver-specific transcription factor. LF-A1 sites are present in several genes that were highly expressed in liver, suggesting that HNF-4α, together with other liver-enhanced transcription factors, might play a role in hepatic differentiation or in the establishment of the metabolic properties of hepatocytes.

I. HNF-4 Expression

Two genes encoding HNF-4 isoforms have been identified in humans: Each gene product differs significantly in its expression pattern and transactivation potential (Chartier *et al.*, 1994; Drewes *et al.*, 1996; Kritis *et al.*, 1996). In adult humans, HNF-4α is expressed at high levels in liver, kidney, intestine, and pancreas and at low levels in the testis (Drewes *et al.*, 1996), paralleling its expression pattern in adult rats (Sladek *et al.*, 1990). In contrast, the human HNF4γ isoform is not expressed in liver and is expressed at low levels in the kidney, intestine, and pancreas (Drewes *et al.*, 1996).

In the developing mouse, HNF-4α is expressed in primary endoderm at 4.5 dpc and in visceral endoderm between 5.5 dpc and 8.5 dpc (Duncan *et al.*, 1994). Although hepatic expression is not present at the earliest stages of liver differentiation (8.0 dpc), strong HNF-4α expression is detected in the liver primordia by 8.5 dpc and during all subsequent stages of development. By 9.5 dpc, low-level HNF-4α expression is present in the gut; by 10.5 dpc, expression is observed in the developing pancreas and the mesonephric tubules (Duncan *et al.*, 1994; Taraviras *et al.*, 1994). At later stages of development, high-level HNF-4α expression is present in the periphery of the liver and the kidney (Taraviras *et al.*, 1994).

2. HNF-4 Response Elements and Metabolic Target Genes

HNF-4 homodimers bind preferentially to DR-1 response elements (Jiang *et al.*, 1995). HNF-4 binding sites have been characterized in a variety of liver-specific genes including transthyretin (Costa *et al.*, 1989), serum coagulation factors (for example, see Reijnen *et al.*, 1992), urea cycle enzymes (Kimura *et al.*, 1993), and members of the cytochrome p450 family (Ström *et al.*, 1995). Under hypoxic conditions, HNF-4α has been shown to enhance activation of a synthetic promoter containing the erythropoietin enhancer and proximal promoter elements (Galson *et al.*, 1995). In addition, HNF-4α regulates the expression of genes involved in cholesterol, xenobiotic and amino acid metabolism, as well as all aspects of carbohydrate and lipid metabolism (Table I).

3. HNF-4 Activity Is Modulated by Fatty Acyl-CoA Thioesters and Phosphorylation

HNF-4α acts as a constitutive inducer of gene transcription, suggesting that its activity may be regulated by an ubiquitous intracellular ligand (Sladek *et al.*, 1990). Long-chain fatty acyl-CoA thioesters have been shown to modulate HNF-4α-mediated gene activation by binding directly to the HNF-4α ligand binding domain (Hertz *et al.*, 1998). Long chain fatty acyl-CoA thioesters are amphiphilic intracellular molecules that play important roles in the regulation of energy metabolism through direct interaction with a variety of cellular enzymes (reviewed in Faergeman and Knudsen, 1997). For example, acetyl-CoA carboxylase, a regulatory step in fatty acid synthesis, is inhibited by nanomolar concentrations of acyl-CoA thioesters (K_i 5.5 nM for palmitoyl-CoA) (Ogiwara *et al.*, 1978). Higher intracellular fatty acyl-CoA concentrations, which might result from prolonged fasting or diabetes mellitus, can inhibit the activity of glycolytic enzymes such as glucokinase and pyruvate dehydrogenase, resulting in fatty acid oxidation replacing glycolysis as the primary cellular energy source (Kruszynska *et al.*, 1990; Faergeman and Knudsen, 1997). In addition to their effects on enzyme activity, acyl-CoA thioesters may regulate gene expression by interfering with thyroid hormone receptor (T_3R) signaling (Li *et al.*, 1993). *In vitro*, submicromolar concentrations of oleoyl-CoA displace triiodothyronine (T_3) bound to T_3R. Although no studies were performed to determine if these agents altered T_3R activity, acyl-CoAs might play a physiologic role in modulating thyroid-hormone-induced lipogenesis by antagonizing T_3 binding to its receptor.

Acyl-CoA thioesters with side chains containing 12 or more backbone carbons bound with high affinity to HNF-4α (Hertz *et al.*, 1998). Palmitoyl-CoA showed saturable binding to HNF-4α with micromolar affinity (K_d 2.6 μM): other long-chain acyl-CoAs also bound to HNF-4α with affinities that were not significantly affected by the chain length or by the degree of side chain saturation. In contrast, acyl-CoAs displayed marked differences in their ability to regulate HNF-4α transcriptional activity: poly- and mononunsaturated acyl-CoAs inhibited HNF-4α, whereas different saturated acyl-CoAs activated (palmitoyl-CoA) or inhibited (stearoyl-CoA) HNF-4α. In no case did acyl-CoA treatment result in more than a twofold change in

TABLE I Orphan Nuclear Receptor Metabolic Target Genes

Target gene[a]	Receptor	HRE	Evidence		References
Malic enzyme (distal)	PPAR	DR1	*In vitro*	*In vivo*	Castelein *et al.*, 1994; Aoyama *et al.*, 1998
Malic enzyme (distal)	COUP-TF	DR1	*In vitro*		Baes *et al.*, 1995
Malic enzyme (proximal)	PPAR	DR1	*In vitro*	*In vivo*	Castelein *et al.*, 1994; Aoyama *et al.*, 1998
PEPCK (PEPCK1)	PPAR	DR1	*In vitro*		Tontonoz *et al.*, 1995
PEPCK (PEPCK2)	PPAR	DR1	*In vitro*		Tontonoz *et al.*, 1995
PEPCK	HNF-4	DR1	*In vitro*		Hall *et al.*, 1995
PEPCK	COUP-TF	DR1	*In vitro*		Hall *et al.*, 1995
Pyruvate kinase	HNF-4	DR1	*In vitro*	*In vivo*	Diaz Guerra *et al.*, 1993; Stoffel and Duncan, 1997
Pyruvate kinase	COUP-TF	DR1	*In vitro*		Diaz Guerra *et al.*, 1993
Acyl-CoA oxidase (ACOA)	PPAR	DR1	*In vitro*		Dreyer *et al.*, 1992; Tugwood *et al.*, 1992
Acyl-CoA oxidase (ACOB)	PPAR	DR1	*In vitro*		Krey *et al.*, 1995
Acyl-CoA Synthase	PPAR	DR1	*In vitro*		Schoonjans *et al.*, 1995
Beta-ketothiolase	PPAR	DR1	*In vitro*	*In vivo*	Kliewer *et al.*, 1992b; Lee *et al.*, 1995
CPT I	PPAR	DR1	*In vitro*		Mascaro *et al.*, 1998
HD	PPAR	Complex	*In vitro*	*In vivo*	Chu *et al.*, 1995; Lee *et al.*, 1995
HD	COUP-TF	DR1	*In vitro*		Miyata *et al.*, 1993
Lipoprotein lipase	PPAR	DR1 ?	*In vitro*		Schoonjans *et al.*, 1996a
MCAD	PPAR	Complex	*In vitro*		Gulick *et al.*, 1994
MCAD	ERR	ER14	*In vitro*		Sladek *et al.*, 1997; Vega and Kelly, 1997
MCAD	COUP-TF	ER14	*In vitro*		Carter *et al.*, 1994
MCAD	HNF-4	DR0	*In vitro*		Carter *et al.*, 1993
MCAD	SF-1	Complex	*In vitro*		Leone *et al.*, 1995
HMG CoA synthase	PPAR	DR1	*In vitro*		Rodriguez *et al.*, 1994
HMG CoA synthase	HNF-4	DR1	*In vitro*		Rodriguez *et al.*, 1998
HMG-CoA synthase	COUP-TF	DR1	*In vitro*		Rodriguez *et al.*, 1997
SCD-1	PPAR	DR1	*In vitro*		Miller and Ntambi, 1996; Kurebayashi *et al.*, 1997
Ornithine transcarbamylase	HNF-4	DR1	*In vitro*		Kimura *et al.*, 1993; Nishiyori *et al.*, 1994
Ornithine transcarbamylase	COUP-TF	DR1	*In vitro*		Kimura *et al.*, 1993; Nishiyori *et al.*, 1994

Gene	Factor	Element	Assay		Reference
ACBP	PPAR	DR1	*In vitro*		Elholm *et al.*, 1996
CETP	COUP-TF	DR7	*In vitro*		Gaudet and Ginsburg, 1995
CRBP-II	PPAR	DR1	*In vitro*		Kliewer *et al.*, 1992b
CRBP-II	HNF-4	DR1	*In vitro*	*In vivo*	Nakshatri and Chambon, 1994; Stoffel and Duncan, 1997
CRBP-II	COUP-TF	DR1	*In vitro*		Kliewer *et al.*, 1992a
CRBP-II	NGFI-B	DR1	*In vitro*		Wu *et al.*, 1997
L-FABP	PPAR	DR1	*In vitro*	*In vivo*	Issemann *et al.*, 1992; Lee *et al.*, 1995
I-FABP	HNF-4	DR1	*In vitro*	*In vivo*	Rottman and Gordon, 1993; Stoffel and Duncan, 1997
I-FABP	COUP-TF	DR1	*In vitro*		Rottman and Gordon, 1993
Transthyretin	HNF-4	DR1	*In vitro*		Costa *et al.*, 1989
FATP	PPAR	DR1	*In vitro*		Martin *et al.*, 1997
Ti-LPT	PPAR	DR1	*In vitro*		Simonson and Iwanij, 1995
ApoAI	PPAR	DR1	*In vitro*		Vu-Dac *et al.*, 1994
ApoAI	HNF-4	DR1 ?	*In vitro*		Chan *et al.*, 1993
ApoAI (A)	COUP-TF	DR2	*In vitro*		Ladias and Karathanasis, 1991
ApoAI (C)	COUP-TF	DR1	*In vitro*		Ladias and Karathanasis, 1991
ApoAI	NGFI-B	DR1 or DR2	*In vitro*		Wu *et al.*, 1997
ApoAII	HNF-4	DR1	*In vitro*		Ladias *et al.*, 1992
ApoAII	COUP-TF	DR1	*In vitro*		Ladias *et al.*, 1992
ApoAIV	HNF-4	DR4	*In vitro*		Ktistaki *et al.*, 1994
ApoAIV	COUP-TF	DR4	*In vitro*		Ktistaki *et al.*, 1994
ApoB	HNF-4	DR1	*In vitro*		Metzger *et al.*, 1993
ApoB	COUP-TF	DR1	*In vitro*		Ladias *et al.*, 1992
ApoCII	HNF-4	DR0	*In vitro*		Vorgia *et al.*, 1998
ApoCII	COUP-TF	DR0, DR4	*In vitro*		Vorgia *et al.*, 1998
ApoCIII	PPAR	DR1	*In vitro*		Krey *et al.*, 1993
ApoCIII	HNF-4	DR1	*In vitro*		Mietus-Snyder *et al.*, 1992
ApoCIII	COUP-TF	DR1	*In vitro*		Reue *et al.*, 1988; Ladias *et al.*, 1992
ApoVLDL II	HNF-4	DR1	*In vitro*		Wijnholds *et al.*, 1988; Beekman *et al.*, 1991
ApoVLDL II	COUP-TF	DR1	*In vitro*		Wijnholds *et al.*, 1988; Beekman *et al.*, 1991
CYP2A4	HNF-4	DR1	*In vitro*		Yokomori *et al.*, 1997

(continues)

TABLE I (continued)

Target gene[a]	Receptor	HRE	Evidence		References
CYP2C2	HNF-4	DR1	In vitro		Chen et al., 1994a
CYP2C13	HNF-4	DR1	In vitro		Legraverend et al., 1994
CYP2C13	COUP-TF	DR1	In vitro		Legraverend et al., 1994
CYP2C9, 2C18	HNF-4	DR1 ?	In vitro		Ibeanu and Goldstein, 1995
CYP2D6	HNF-4	DR1	In vitro		Cairns et al., 1996
CYP2D6	COUP-TF	DR1	In vitro		Cairns et al., 1996
CYP3A1, 3A2	PXR	DR3	In vitro		Kliewer et al., 1998
CYP4A1, 4A3	PPAR	DR1	In vitro	In vivo	Aldridge et al., 1995; Lee et al., 1995; Johnson et al., 1996
CYP4A6	PPAR	DR1	In vitro		Palmer et al., 1995
CYP4A6	COUP-TF	DR1	In vitro		Palmer et al., 1995
CYP7a (Distal)	HNF-4	DR1	In vitro		Cooper et al., 1997
CYP7a (Distal)	COUP-TF	DR1	In vitro		Stroup et al., 1997b
CYP7a (Proximal)	LXR	DR4	In vitro		Lehmann et al., 1997a
CYP7a (Proximal)	COUP-TF	DR4	In vitro		Stroup et al., 1997b
p450scc (CYP11A)	SF1	SFRE	In vitro		Rice et al., 1990; Clemens et al., 1994
p450c21 (CYP21)	SF1	SFRE	In vitro		Rice et al., 1990
p450c21 (CYP21)	NGFI-B	NBRE	In vitro		Wilson et al., 1993b; Davis and Lau, 1994
p45011β (CYP11B1)	SF1	SFRE	In vitro		Morohashi et al., 1994
p450aldo (CYP11B2)	SF1	SFRE	In vitro		Bogerd et al., 1990
p45017α (CYP17)	SF1	SFRE	In vitro		Bakke and Lund, 1995; Zhang and Mellon, 1996
p45017α (CYP17)	COUP-TF	DR9	In vitro		Zhang and Mellon, 1996
p45017α (CYP17)	NGFI-B	NBRE	In vitro		Zhang and Mellon, 1996
p450arom (CYP19)	SF1	SFRE	In vitro		Lynch et al., 1993

3β-HSD-2	SF1	SFRE	*In vitro*	Leers-Sucheta *et al.*, 1997
STAR	SF1	SFRE	*In vitro*	Sugawara *et al.*, 1996
SR-BI	SF1	SFRE	*In vitro*	Cao *et al.*, 1997
Lactoferrin	COUP-TF	DR1	*In vitro*	Liu and Teng, 1991
Lactoferrin	ERR	ERE	*In vitro*	Yang *et al.*, 1996
Lactoferrin	NGFI-B	DR1	*In vitro*	Wu *et al.*, 1997
Transferrin	PPAR	DR1	*In vitro*	Hertz *et al.*, 1996b
Transferrin	HNF-4	DR1	*In vitro*	Schaeffer *et al.*, 1993; Hertz *et al.*, 1996b
Transferrin	COUP-TF	DR1	*In vitro*	Schaeffer *et al.*, 1993
aP2 (ARE6)	PPAR	DR1	*In vitro*	Graves *et al.*, 1992
aP2 (ARE7)	PPAR	DR1	*In vitro*	Graves *et al.*, 1992
aP2 (ARE7)	COUP-TF	DR1	*In vitro*	Brodie *et al.*, 1996
AFP	COUP-TF	DR0	*In vitro*	Thomassin *et al.*, 1996
S14	PPAR	Indirect	*In vitro*	Ren *et al.*, 1996
ACTH receptor	SF1	SFRE	*In vitro*	Cammas *et al.*, 1997
Erythopoietin	HNF-4	DR2	*In vitro*	Galson *et al.*, 1995
Erythopoietin	COUP-TF	DR2	*In vitro*	Galson *et al.*, 1995
Leptin	PPAR	Unknown	*In vitro*	Hollenberg *et al.*, 1997
POMC	COUP-TF	DR0	*In vitro*	Drouin *et al.*, 1989
Rat insulin II	COUP-TF	DR1	*In vitro*	Hwung *et al.*, 1988; Ladias and Karathanasis, 1991

[a] Abbreviations for target genes: ACBP, acyl-CoA-binding protein; ACTH, adrenocorticotropin hormone; AFP, alpha-fetoprotein; Apo, apolipoprotein; CETP, cholesterol ester transport protein; CPT-I, carnitine palmitoyltransferase I; CRBP-II, cellular retinol binding protein; CYP, cytochrome p450; I-FABP, ileal fatty acid binding protein; L-FABP, liver fatty acid binding protein; FATP, fatty acid transport protein; HD, enoyl-CoA hydratase/3-hydroxyacyl-CoA dehydrogenase; HMG-CoA, 3-hydroxy-3methylglutaryl CoA; Ti-LPT, liver-specific sugar transporter; MCAD, medium chain acyl-coA dehydrogenase; PEPCK, phosphoenolpyruvate carboxykinase; POMC, proopiomelanocartin; SCD-1, stearoyl-CoA desaturase; STAR, steroid acute regulatory protein; SR-BI, scavenger receptor-B1.

HNF-4α activity measured in transient transfection studies (Hertz *et al.*, 1998). Since a mixture of acyl-CoAs may demonstrate complex and mutually antagonistic effects on HNF-4α activity, it may be difficult to demonstrate the importance of these ligands as modulators of HNF-4α *in vivo*. If acyl-CoAs were to act predominantly as inhibitors of HNF-4α function, they might play a role in the physiologic downregulation of glycolytic enzymes such as pyruvate kinase in fasting subjects. Conversely, certain saturated acyl-CoAs might increase pyruvate kinase expression in the same circumstances, opposing the direct inhibitory effects of these compounds on other glycolytic enzymes. Given the small effect that acyl-CoAs demonstrate in transient transfection studies, confirmation of their importance will require careful physiologic studies, perhaps including comparison of their effects on key HNF-4α targets in wild-type and knockout mice. Alternatively, identification of acyl-CoA thioesters as physiologic HNF-4α ligands could lead to the discovery of specific synthetic HNF-4α ligand, which could be used to differentiate the effects of these compounds mediated by HNF-4α from those effects mediated by direct enzyme inhibition.

HNF-4α transcriptional activity may also be regulated through phosphorylation-induced changes in HNF-4α DNA binding affinity. Overnight fasting (Viollet *et al.*, 1997) or long-term dietary protein restriction (Marten *et al.*, 1996) decreases hepatic HNF-4 DNA-binding activity, an effect that is reversed by treating liver extracts obtained from fasted animals with the threonine/serine protein phosphatase PP1. HNF-4α DNA binding affinity is also reduced in hepatic nuclear extracts obtained from fed animals treated with agents that increase intracellular cAMP concentration, such as glucagon and β-adrenergic agents (Viollet *et al.*, 1997). These effects are accompanied by changes in HNF-4α transcriptional activity *in vitro*: PKA-induced phosphorylation of the HNF-4α DNA binding domain decreases HNF-4α activity, either as a result of decreased HNF-4α binding to the target HRE, or as a result of competitive interactions between HNF-4α and other transcription factors. As similar physiologic states appear to regulate the efficiency with which HNF-4 binds its response element both intracellular concentrations of putative HNF-4 ligands, it would be interesting to determine how these stimuli are integrated control of HNF-4 activation potential.

4. HNF4 Involvement in Diabetes and Embryonic Development

Maturity onset diabetes of the young (MODY) is an uncommon form of non-insulin-dependent diabetes (NIDDM) that is characterized by obesity, early age of onset, and impaired insulin secretion. The disorder, which is inherited as an autosomal dominant trait, has been associated with mutations in the coding regions of the glucokinase (MODY2) and HNF-1 (MODY3) genes (Froguel *et al.*, 1993; Yamagata *et al.*, 1996b). A third MODY locus (MODY1) has been associated with mutations in the HNF-4α gene (Yamagata *et al.*, 1996a; Bulman *et al.*, 1997; Furuta *et al.*, 1997; Hani *et al.*,

1998). The link between MODY1 and HNF-4α appears specific to this form of diabetes, as HNF-4α mutations have not been identified in patients with other forms of NIDDM (Nakajima *et al.,* 1996). In addition, disruption of the HNF-4 binding site in the HNF-1 promoter has been identified in an Italian family with MODY, providing an unusual example of patients whose diabetes likely results from a combined impairment of HNF-4α and HNF-1 function (Gragnoli *et al.,* 1997). Interestingly, several of the common HNF-4α mutations identified in MODY1 patients can alter the subcellular distribution of HNF-4α or reduce its transcriptional activity *in vitro,* further illustrating the potential for a direct link between reduced HNF-4α function and the MODY phenotype (Stoffel and Duncan, 1997; Hani *et al.,* 1998; Sladek *et al.,* 1998).

Gene targeting experiments have shown that HNF-4α is a key regulator of murine gastrulation (Chen *et al.,* 1994b). Ablation of HNF-4α function results in apoptosis of embryonic ectoderm at 6.5 dpc, followed by abnormal mesoderm differentiation and embryonic death. Tetraploid rescue experiments, in which chimeras formed from $Hnf4\alpha^{+/+}$ tetraploid morulas with $Hnf4\alpha^{-/-}$ ES cells were allowed to develop until 9.5 dpc in surrogate pseudo-pregnant females, demonstrated that complementation of extraembryonic tissue defects allowed gastrulation to occur in $Hnf4\alpha^{-/-}$ embryos. As HNF-4α expression during early postimplantation development is restricted to primary and visceral endoderm, the HNF-4α phenotype might result from the failure of the visceral endoderm to secrete a trophic molecule required by the developing embryonic ectoderm (Chen *et al.,* 1994b). Although mice with a haploinsufficiency of HNF-4α do not develop glucose intolerance, complete $Hnf4\alpha$ ablation in either ES cells or 8.5 dpc embryos is associated with significantly reduced expression of glycolytic enzymes as well as glucose and fatty acid transport proteins (Stoffel and Duncan, 1997). Genetic ablation of HNF-4α function in mice did not significantly reduce expression of HNF-1 or known HNF-1 targets, suggesting that the two genes might act independently in causing the MODY phenotype (Stoffel and Duncan, 1997).

In summary, HNF-4 was originally identified as a liver-specific transcription factor that has now been shown to play a broader role in the regulation of metabolic enzymes and metabolite transport proteins. Although HNF-4 dysfunction has been linked to MODY1, the finding that *HNF4α* disruption may not be associated with other forms of diabetes suggests that modulation of HNF-4 activity by natural or synthetic ligands may have important therapeutic applications in the treatment of disorders of carbohydrate metabolism.

B. Peroxisomal Proliferator Activated Receptors

Peroxisome proliferators (PPs) are a group of structurally unrelated compounds that cause massive proliferation of hepatic peroxisomes, liver hyperplasia, and hepatic malignancies in rodents (reviewed in Lemberger *et al.*, 1996a; Schoonjans *et al.*, 1996b). This class of compounds includes many industrial chemicals and herbicides as well as drugs used to treat hypercholesterolemia and other lipid abnormalities. Peroxisomal proliferators potentially exert their short- and medium-term effects through two mechanisms: Genomic effects are mediated by members of the peroxisomal proliferator activated receptor family (PPARs), while nongenomic effects may be mediated by alterations in cellular energy metabolism and by the generation of peroxisomal metabolic by-products. The mediators of long-term peroxisomal proliferator effects are less clear, although their stimulation of peroxisomal hydrogen peroxide production has been suggested to cause their genotoxic effects.

Three PPAR isoforms have been identified in mammals: PPARα, PPARβ (also called FAAR, NUC1, and PPARδ), and PPARγ (reviewed in Lemberger *et al.*, 1996a). PPARα was initially isolated by screening a mouse liver cDNA library with a mixture of oligonucleotides directed against a conserved region of the nuclear receptor DNA binding domain (Issemann and Green, 1990). Chimeric receptors containing the ligand binding domains of the resulting cDNA clones together with the estrogen receptor DNA-binding domain were used to screen for potential ligands. PPs were included in the ligand pool based on their ability to rapidly activate transcription of specific target genes and to bind with high affinity to a hepatocellular protein (for references, see Issemann and Green, 1990). The remaining members of the family have been identified by techniques including homology screening, DNA affinity purification and microsequencing of nuclear proteins, and expression library screening using radiolabeled HREs.

I. PPAR Expression

PPAR isoforms display distinct expression patterns in adult animals and during development. In adult rats, PPARα is highly expressed in heart, liver, kidney, intestine, and brown fat, tissues that demonstrate high rates of fatty acid β oxidation (Issemann and Green, 1990; Braissant *et al.*, 1996). Interestingly, hepatic PPARα expression levels varied widely between animals (Braissant *et al.*, 1996), possibly because of hormonal modulation of PPARα expression by glucocorticoids (Lemberger *et al.*, 1994), physical stress (Lemberger *et al.*, 1996b), or changes in serum insulin levels (Steineger *et al.*, 1994). In contrast to PPARα, PPARβ is widely expressed in adult rodent tissues. High levels of PPARβ mRNA are detected in the brain, kidney, small intestine, and Sertoli cells (Amri *et al.*, 1995; Braissant *et al.*, 1996). Finally, PPARγ displays an isoform-specific pattern of expression: PPARγ1 transcripts are most abundant in the spleen, intestine, and white adipose tissue (Braissant *et al.*, 1996), whereas the PPARγ2 isoform is expressed at high levels in white and brown adipose tissue (Tontonoz *et al.*, 1994a).

2. PPAR Response Elements and Metabolic Gene Targets

PPARs bind preferentially to DR1 response elements as a heterodimer with RXR (Dreyer *et al.*, 1992; Kliewer *et al.*, 1992b; Tugwood *et al.*, 1992). Analysis of natural PPREs has demonstrated that the nucleotide sequence contained in the 5′ flank of the PPRE regulates the efficiency with which specific PPAR isoforms bind to DNA (Palmer *et al.*, 1995; Juge-Aubry *et al.*, 1997). PPAR binding sites have been identified in genes controlling all aspect of carbohydrate and lipid metabolism (Table I). In addition to directly binding to its cognate response element, PPAR may regulate gene expression by forming heterodimers with other nuclear receptors such as $T_3R\beta$ and LXRα (Bogazzi *et al.*, 1994; Miyata *et al.*, 1996) and may antagonize T_3R activity by titrating limiting amounts of the common heterodimeric partner, RXR (Juge-Aubry *et al.*, 1995).

3. PPARs Are Activated by a Wide Variety of Ligands

A group of structurally diverse synthetic compounds that could modulate PPARα activity was identified during its initial characterization; few of these agents were natural products and none were shown to bind the receptor directly (Issemann and Green, 1990). An early search for endogenous PPAR activators, performed using fractionated human serum, demonstrated that fatty acids could activate PPARα (Gottlicher *et al.*, 1992). Subsequently, PPARα activity was shown to be induced by eicosanoids (Yu *et al.*, 1995), carbaprostacyclin (Hertz *et al.*, 1996a), nonsteroidal anti-inflammatory drugs (NSAIDs) (Lehmann *et al.*, 1997b), and leukotriene $\beta4$ (LTB4) (Devchand *et al.*, 1996). Of these agents, LTB4 has been shown to directly bind to PPARα (K_d 90 nM), while ligand displacement assays have demonstrated direct interaction between PPARα and the synthetic PPAR activator WY 14,643, as well as NSAIDs, fatty acids, and eicosanoids (Devchand *et al.*, 1996; Krey *et al.*, 1997; Lehmann *et al.*, 1997b).

PPARβ and PPARγ are activated by common PPAR ligands such as DHAS and certain prostaglandins (Yu *et al.*, 1995), as well as by isoform-specific ligands (Kliewer *et al.*, 1994). The PPARγ2 isoform was identified by its ability to bind enhancer elements. The fact that the thiazolidinedione (TZD) class of antidiabetic drugs regulated expression of the adipocyte-specific protein aP2 through a PPAR response element (Harris and Kletzien, 1994) suggested that these drugs might also modulate PPARγ activity, a hypothesis that was confirmed *in vitro* (Forman *et al.*, 1995b; Lehmann *et al.*, 1995). Additional PPARγ ligands include PGJ2 (Forman *et al.*, 1995b; Kliewer *et al.*, 1995), PUFA (Kliewer *et al.*, 1997), and NSAIDs (Lehmann *et al.*, 1997a). Specific synthetic PPARδ ligands have been identified by screening biased chemical libraries; however, no natural high-affinity PPARδ ligand has been identified (Brown *et al.*, 1997). Ligand displacement experiments showed that the NSAID indomethacin as well as fatty acids interacted directly with PPARγ (Lehmann *et al.*, 1997a).

In addition to direct activation by PPAR ligands, PPAR-RXR hetero-dimers may also be activated by RXR-specific ligands. Transient transfection studies showed that maximal activation of the acyl-CoA oxidase (ACO) gene promoter was achieved by simultaneous treatment with the PPAR ligand WY14,643 and the RXR ligand 9-*cis*-retinoic acid: individual treatment with PPAR or RXR ligands resulted in lower levels of promoter activity (Keller *et al.*, 1993). The PP and retinoid pathways may act through a common response element, possibly by allowing both members of the PPAR:RXR heterodimer to assume an activated conformation following ligand binding. This synergism is also observed *in vivo,* where the efficacy of TZDs in reducing fasting hyperglycemia and hypertriglyceridemia in *db/ db* mice is potentiated by concurrent treatment with RXR-specific ligands (Mukherjee *et al.*, 1997).

Members of the PPAR family resemble other nuclear hormone receptors in that they can be activated in a ligand-independent manner (for a general review of ligand-independent activation of steroid hormone receptors, see O'Malley *et al.*, 1995). A ligand-independent activation domain (AF1) has been identified in the PPARγ N-terminus (Adams *et al.*, 1997) and has been shown to display isoform-specific transactivation potential (Werman *et al.*, 1997). Treatment of rat fibroblasts with insulin, epidermal growth factor, or 12-O-tetradecanoylphorbol 13-acetate (TPA) resulted in phosphorylation of PPARγ2 at a serine residue contained in a consensus MAP kinase within the PPARγ2 AF1 domain (Ser^{112}) (Hu *et al.*, 1996). Mutation of this residue decreases PPARγ ligand-independent activity, blocks insulin- and TPA-induced phosphorylation of PPARγ2, and decreases PPARγ2 sensitivity to transcriptional inhibition following TPA treatment (Adams *et al.*, 1997). *In vivo*, mutation of Ser^{112} potentiates PPARγ2 induction of adipogenesis: Adipocyte differentiation occurs more readily in fibroblasts stably expressing the mutant receptor when compared to fibroblasts expressing wild-type PPARγ2 and is not inhibited by TPA treatment (Hu *et al.*, 1996; Adams *et al.*, 1997).

4. PPARs Regulate Fat Metabolism, Adipocyte Differentiation, and Macrophage Function

Extensive *in vitro* studies support a crucial role for PPARγ in regulating fat synthesis and adipocyte differentiation (reviewed in Schoonjans *et al.*, 1996b; Hwang *et al.*, 1997; Spiegelman, 1998a). A link between PPARγ and adipocyte differentiation was initially established by studies of enhancer elements regulating expression of the adipocyte protein aP2. Promoter characterization identified a regulatory factor (ARF6) that bound to two enhancer elements (ARE6 and ARE7) required for high-level aP2 expression in adipocytes (Graves *et al.*, 1992). ARF6, purified by DNA affinity chromatography from a brown fat cell line, was shown to consist of the nuclear receptors RXRα and PPARγ (Tontonoz *et al.*, 1994a, 1994b). Subsequent studies demonstrated that PPARγ played an essential role in controlling adipocyte differentiation. Overexpression of CEBPβ/CEBPδ induces fibroblasts to undergo adipocyte differentiation, which is accompanied by increased expression of PPARγ (Wu *et al.*, 1995). Overexpression of PPARγ2 is sufficient to induce these cell lines to undergo adipocyte differentiation in the presence of PPAR-specific ligands, identifying PPARγ as a key downstream regulator of CEBP-induced adipocyte differentiation (Tontonoz *et al.*, 1994b). Although PPARγ2 activates an adipogenic program more efficiently that other PPAR isoforms, PPARα can also induce adipocyte differentiation in the presence of strong activating ligands (Brun *et al.*, 1996). In addition, coexpression of PPARγ2 and CEBPα or PPARγ2 and SREBP1 (Kim and Spiegelman, 1996) induces an adipogenic program in the absence of PPAR ligands.

In addition to its role in adipocyte differentiation, PPARγ has been implicated in the regulation of myeloid function (reviewed in Spiegelman, 1998b). The adherence and migration of activated macrophages into the subendothelial region of arteries characterize early lesions in atherosclerotic plaques. This is followed by the conversion of macrophages to foam cells, primarily as a result of scavenger receptor-mediated uptake of cholesterol from circulating lipoprotein particles. Oxidized LDL lipids potently stimulate macrophage lipid accumulation and accelerate the process of atherogenesis. Macrophage cholesterol metabolism can be studied *in vitro* using HL60 monocytic leukemia cells, a cell line that displays a three- to five fold increase in PPARγ expression when it undergoes monocytic differentiation (Tontonoz *et al.*, 1998). Following treatment with a combination of the PPARγ ligand PGJ2 and the RXR ligand LG268, HL60 cells express the macrophage markers CD11b and CD18 accompanied by increased levels of mRNA encoding PPARγ and the lipoprotein scavenger receptor CD36. A similar differentiation program could be induced by combined PGJ2/LG268 treatment of THP-1 monocytic leukemia cells, but not by treatment of non-PPARγ-expressing CDM-1 leukemia cells. Induction of CD36 scavenger receptor expression, which is mediated by direct PPAR activation of the CD36 gene promoter, would result in increased macrophage LDL and oxo-LDL uptake, promoting cholesterol deposition in atherosclerotic plaques (Tontonoz *et al.*, 1998).

Gene targeting experiments have demonstrated that PPARα is an essential mediator of the hepatic response to peroxisomal proliferators such as Wy-14,643, clofibrate, and DHEA-S (Lee *et al.*, 1995; Peters *et al.*, 1996). PPARα null mice develop normally and are externally indistinguishable from their wild-type littermates. Although normal numbers of hepatic peroxisomes were present in free-running PPAR$\alpha^{-/-}$ mice, the peroxisome activators Wy-14,643 and clofibrate did not result in hepatic enlargement, peroxisomal proliferation or upregulation of PP-inducible enzymes in the PPARα-deficient animals (Peters *et al.*, 1996). This demonstrated that PPARα was not essential for peroxisome biogenesis, but was the major PPAR isoform mediating the hepatic response to specific peroxisome proliferators. In addition, although knockout and wild-type mice displayed similar basal levels of most hepatic fatty-acid metabolizing enzymes, PPARα null mice showed significantly increased expression of short-chain 3-hydroxyacyl-CoA dehydrogenase, as well as decreased constitutive malic enzyme activity and decreased constitutive expression of a subset of fat-metabolizing enzymes (very long chain and long-chain acyl CoA dehydrogenase, long-chain acyl CoA synthase, short-chain-specific 3-ketoacyl-CoA thiolase, and peroxisomal D-type bifunctional protein) (Aoyama *et al.*, 1998). Differences in basal enzyme expression resulted in impaired liver metabolism of palmitic acid (C16:0), but not of lauric (C12:0) or lignoceric acid (C24:0). The lack of a more widespread defect of peroxisomal metabolism in PPARα null mice suggested either that functional redundancy exists among the PPARs or that other mechanisms might regulate the basal expression of the majority of peroxisomal enzymes (Lee *et al.*, 1995; Gonzalez, 1997).

Taken together, these results demonstrate that PPARs are important regulators of adipocyte differentiation, macrophage function and lipid metabolism. In performing these functions, PPARs act as integrators of endocrine and intracrine signaling pathways: regulation of PPAR phosphorylation by peptide hormones influences receptor responsiveness to intracellular and paracellular ligands. Pharmacologic modulators of PPARγ and PPARα activity are commonly used to treat diabetes mellitus and dyslipidemias. The demonstration that PPARγ may regulate macrophage cholesterol accumulation and consequently may promote atherogenesis requires further study in human patients to determine the safety of TZD treatment of atherosclerosis-prone diabetic patients. In addition, the potential role played by PPARγ in the regulation of the inflammatory response may provide novel therapies for the treatment of acute and chronic inflammatory disorders.

C. Estrogen-Receptor-Related Receptors

The estrogen-receptor-related (ERR) subfamily of orphan nuclear receptors contains two closely related members: ERRα and ERRβ (also known as ERR1 and ERR2) (Giguère *et al.*, 1988). ERRα was initially cloned using a low-stringency cross-hybridization strategy that identified gene products homologous to the estrogen receptor DNA binding domain. ERRβ was cloned from a human heart cDNA library based on its homology to ERRα (Giguère *et al.*, 1988). Subsequently, studies using protein micropurification and microsequencing techniques to identify mammalian proteins that bound to the SV40 major late promoter identified ERRα as a repressor of SV40 major late promoter activity, implicating ERRα in regulation of the early-to-late switch of SV40 gene expression (Wiley *et al.*, 1993).

I. ERR Expression

ERR isoforms are expressed in spatially and temporally distinct patterns during murine development. ERRα is widely expressed during murine development and may display strain-specific expression patterns (Bonnelye *et al.*, 1997; Sladek *et al.*, 1997). Embryonic ERRα is expression is first detected at 8.5 dpc: At that time, ERRα transcripts are present in the trophoblast, mesoderm cells of the visceral yolk sac, the primitive heart, and the neural tube. During subsequent stages of development, ERRα expression is detected in the brain and spinal cord, pituitary gland, heart, and intestinal mucosa, as well as the premuscular mass of the limb bud and brown adipose tissue. The developmental pattern of ERRα expression has been used to infer two potential roles for ERRα function *in vivo*. First, ERRα may play a role in regulating in cellular differentiation. During murine development, ERRα is expressed prominently in the intermediate zone of the developing spinal cord, a region that contains immediately postmitotic neurons, suggesting that ERRα might be upregulated as precursor cells cease proliferation and enter a program of neuronal differentiation. This hypothesis is supported by *in vivo* studies of ERRα expression in developing skin and muscle, as well as by *in vitro* experiments demonstrating early upregulation of ERRα during myogenesis (Bonnelye *et al.*, 1997) and adipogenesis (Sladek *et al.*, 1997). Second, ERRα expression during late fetal development and early postnatal life appears to be correlated with organ-specific preferences for metabolic substrates (Sladek *et al.*, 1997). ERRα, and to a lesser extent ERRβ, are most prominently expressed in organs demonstrating a high capacity for fatty acid β-oxidation or activation, suggesting that both ERR isoforms may play a role in regulating energy metabolism. This hypothesis is supported by *in vitro* studies that identify ERRα as a modulator of medium chain acyl-coenzyme A dehydrogenase (MCAD) expression, a key regulator of mitochondrial β-oxidation (Sladek *et al.*, 1997; Vega and Kelly, 1997).

In contrast to ERRα, ERRβ displays a more restricted developmental expression pattern (Pettersson *et al.*, 1996; Luo *et al.*, 1997). ERRβ transcripts are first detected in a subset of cells in extraembryonic ectoderm at 5.5 dpc. By 6.5 dpc, ERRβ is expressed in ectoderm contained in the amniotic fold: Fusion of the amniotic fold results in formation of the chorion, where ERRβ is expressed at 7.5 dpc. ERRβ expression diminishes coincident with fusion of the chorion and ectoplacental cone at 8.5 dpc. As fusion progresses, ERRβ expression is extinguished in all but the free margin of the chorion, whereas ERRα becomes upregulated in the remaining trophoblast cells (Luo *et al.*, 1997 and unpublished observations). This highly specific pattern of expression suggests that ERRβ might play a crucial role in regulating the development and function of the early placenta.

2. ERR Response Elements and Target Genes

ERRα and β bind as monomers to the extended half site TNAAGGTCA (ERRE) (Johnston *et al.*, 1997; Sladek *et al.*, 1997), as homodimers to the consensus estrogen responsive element (ERE) (Pettersson *et al.*, 1996), and to response elements containing direct repeats of the AGGTCA core motif (Sladek and Giguère, unpublished observations). Because of the configuration of the two types of ERRE, ERR targets potentially include all genes regulated by either SF-1 (see later discussion) or by the estrogen receptors. ERRα has been shown to regulate activity of the lactoferrin (Yang *et al.*, 1996), MCAD (Sladek *et al.*, 1997; Vega and Kelly, 1997), and osteopontin (Bonnelye *et al.*, 1997) promoters in cotransfection assays: Endogenous ERRα has been shown to bind to regulatory elements within the MCAD promoter (Table I). For most promoters studied, ERRα represses gene transcription as measured by transient transfection experiments (Sladek *et al.*, 1997) or in cell-free systems (Wiley *et al.*, 1993; Johnston *et al.*, 1997). Although the lack of ERR transcriptional activity may be due to the lack of adequate levels of its cognate ligand, a chimeric receptor containing the progesterone receptor DNA binding domain and the putative ERRα was constitutively active in yeast and CV-1 cells (Lydon *et al.*, 1992). In addition, both receptors display cell-type specific transcriptional activity: ERRα activates the osteopontin gene promoter in ROS 17.2/8 osteosarcoma cells, but not COS-1 cells (Bonnelye *et al.*, 1997), whereas ERRβ acts as a cell-type-specific repressor of glucocorticoid activation of the MMTV promoter (Trapp and Holsboer, 1996). ERRα transcriptional activity may also be influenced by the position of the ERRE within a complex response element, possibly mediated by protein–protein interactions between ERRα and other nuclear receptors allowed by the response element structure. For example, ERRα enhancement of estrogen responsiveness of the lactoferrin gene is mediated by ERRα interaction with an extended half-site in the complex lactoferrin ERE (Yang *et al.*, 1996) and may occur through direct interaction between ERRα and the estrogen receptor (Yang *et al.*, 1996; Klinge *et al.*, 1997).

3. ERRs: In Search of a Ligand

Although ERRα and ERRβ display significant homology to the estrogen receptors (ERα and ERβ), neither binds estrogens *in vitro*, and neither responds to estrogens in cotransfection assays (Giguère *et al.*, 1988; Yang *et al.*, 1996). The crystal structure of ligand-complexed ERα (Brzozowski *et al.*, 1997) indicates that most amino acid residues shown to be critical for recognition of estradiol are conserved between members of the ER and ERR families, including residues that form critical hydrogen bonds with the hydroxyl groups present in the A and D rings. This suggests that ER and ERR ligands should be structurally similar. The absence of known ERR ligands might explain in part the lack of ERR transcriptional activity observed in most cotransfection studies, since both ERR proteins possess a well-conserved AF2 domain.

4. Physiological Functions for ERRs

ERRβ expression during embryogenesis defines a subset of extraembryonic ectoderm that subsequently forms the dome of the chorion, suggesting that ERRβ may play a role in early placental development. Homozygous mutant embryos generated by targeted disruption of the *Estrrb* gene have severely impaired placental formation and die by 10.5 dpc. The mutants display abnormal chorion development associated with an overabundance of trophoblast giant cells and a severe deficiency of diploid trophoblast. The phenotype can be rescued by aggregation of *Estrrb* mutant embryos with tetraploid wild-type cells that contribute exclusively to extraembryonic tissues. The ERRβ phenotype resembles that seen in HNF-4α knockout embryos, in that the major histologic and developmental changes observed in the null mutants occur in tissues that do not normally express the targeted receptor during murine development (Chen *et al.*, 1994b; Luo *et al.*, 1997). These results indicate an important role for ERRβ in early placentation and suggest that an inductive signal originating from or modified by the chorion is required for normal trophoblast proliferation and differentiation (Luo *et al.*, 1997). ERRβ may therefore control key metabolic steps implicated in the synthesis of this putative factor.

The observations that ERRα is expressed in tissues that preferentially metabolize fatty acids and that ERRα can control the expression of MCAD *in vitro* suggest that ERRα may play an important role in regulating cellular energy balance *in vivo*. ERRα mutant mice display intrauterine growth deficiency and abnormal adult body composition, but otherwise develop normally and appear to have normal reproductive function. The animals will provide an excellent model for identifying possible physiologic processes regulated by ERRα as well as potential *in vivo* targets of ERRα action (Sladek, R., Luo, J. and Giguère, V., unpublished observations).

III. Orphan Nuclear Receptors Regulate Sterol and Steroid Hormone Metabolism

Study of steroid secreting cells provides a useful model of a complex pattern of differentiation. This biological system is particularly useful as there are many specific biochemical markers that can be conveniently assayed. The next section will examine SF-1 and NGFI-B, both of which have been implicated in regulation of steroidogenesis. In addition, we will also review the LXR family, which has been implicated in the regulation of cholesterol catabolism and bile salt synthesis, as well as the orphan receptor PXR, which has been identified as a regulator of steroid and xenobiotic catabolism.

A. Steroidogenic Factor I

Steroidogenic factor 1 (SF-1) was initially characterized as adrenal-gland specific factor that bound to conserved regulatory elements in the proximal promoter regions of steroid hydroxylases CYP11A, CYP11B2, and CYP21 (Parker and Schimmer, 1997). These regulatory elements contained a conserved AGGTCA consensus motif, suggesting that SF-1 was a member of the nuclear receptor superfamily. SF-1 was cloned from an adrenal gland library based on its homology to the RXRβ DNA-binding domain (Lala *et al.*, 1992). SF-1 and the nuclear receptor ELP (embryonic long terminal repeat binding protein) are produced from a common gene through alternative promoter and splice site utilization (Tsukiyama *et al.*, 1992; Ikeda *et al.*, 1993). In total, a single SF-1 transcript and three alternatively spliced ELP transcripts have been identified (Ninomiya *et al.*, 1995). In addition, SF-1 shares significant homology with the orphan receptor LRH-1 (also called PRH-1 and FRF) (Becker-André *et al.*, 1993; Galarneau *et al.*, 1996).

I. SF-I Expression

SF-1 expression is first detected in murine embryos at 9.0 dpc, when it is prominently expressed in the urogenital ridge (Ikeda *et al.*, 1994). At 10–10.5 dpc, two distinct populations of SF-1 expressing cells are present. The first lies adjacent to the aorta and represents precursors of adrenal steroidogenic tissue. The second lies adjacent to the coelomic epithelium and represents precursors of gonadal steroid producing cells. SF-1 expression precedes functional differentiation of the steroidogenic tissues as well as molecularly and histologically recognizable sexual differentiation of the embryonic gonad. In later stage embryos, gonadal SF-1 expression is sexually dimorphic: In female embryos, SF-1 expression declines following 12.5 dpc, whereas in male embryos, SF-1 expression persists in both the Leydig cells and the spermatogenic cords. SF-1 expression is also detected in the fetal ventromedial hypothalamic nucleus (VMH) after 11.5 dpc and in the fetal pituitary after 13.5 dpc. Pituitary SF-1 expression precedes the onset of FSH expression in gonadotropes, suggesting that SF-1 might either directly regulate gonadotropin production or regulate gonadotrope differentiation (Ingraham *et al.*, 1994). In adult mice, SF-1 expression is highest in steroid secreting cells of the adrenal gland and gonads; lower level expression is present in the spleen and pituitary gonadotropes (Ikeda *et al.*, 1993). Interestingly, prominent SF-1 expression is not seen in the placenta, suggesting that this tissue relies on other factors for the regulation of steroidogenesis.

2. SF-I Response Elements and Metabolic Target Genes

SF-1 binds to monomeric response elements (SFREs) with the consensus sequence TCAAGGTCA (for references, see Wilson *et al.*, 1993a). Potential and proven *in vivo* SF-1 targets include steroidogenic enzymes (reviewed in Parker and Schimmer, 1997), Mullerian inhibiting substance (Shen *et al.*, 1994), the pituitary glycoprotein alpha subunit promoter (Barnhart and Mellon, 1994), and the luteinizing hormone β subunit promoter (Halvorson *et al.*, 1996) (Table I). SF-1 response elements have also been characterized in the steroidogenic acute regulatory protein promoter (Sugawara *et al.*, 1996) and the oxytocin promoter (Wehrenberg *et al.*, 1994). In addition to its role in regulating steroid synthesis, SF-1 may also regulate cellular uptake of HDL and LDL cholesterol through a response element in the class B scavenger receptor SR-BI promoter (Cao *et al.*, 1997). For most promoters studied, SF-1 constitutively induces basal gene transcription as a result of interaction with nuclear receptor coactivators (Ito *et al.*, 1998). SF-1 activity may be regulated by phosphorylation of the receptor protein: *In vitro*, PKA-induced phosphorylation of SF-1 reduces the receptor's DNA-binding affinity, whereas *in vivo*, SF-1 phosphorylation may regulate cAMP-dependent gene induction (Honda *et al.*, 1993; Morohashi *et al.*, 1994; Zhang and Mellon, 1996).

3. SF-I Is Activated by Oxysterols

Although SF-1 acts as a constitutive transcriptional activator, studies have shown that specific oxysterol ligands may regulate SF-1 activity (Lala *et al.*, 1997). Micromolar concentrations of 25-hydroxycholesterol (25OHC) significantly increased SF-1 transcriptional activity (EC50 5 μM). The oxysterols 26-hydroxycholesterol (EC50 5 μM), 27-hydroxycholesterol (EC50 5 μM), and 21-hydroxypregnenolone (EC50 11 μM) are less efficient SF-1 activators, while the LXR activator 22(R)-hydroxycholesterol does not alter SF-1 activity. Although 25OHC treatment clearly results in increased SF-1 activity, a direct interaction between SF-1 and 25-OHC has not yet been demonstrated. Consequently, oxysterols may regulate SF-1 activity by indirect mechanisms, including altered intracellular concentrations of metabolic products involved in sterol synthesis, altered expression of nuclear receptor coregulatory proteins, and altered activity of 25-OHC regulated transcription factors such as SREBP-1 (Wang *et al.*, 1994). Regardless of mechanism, 25-OHC regulation of SF-1 activity illustrates an interesting feedforward control scheme in which increasing levels of a cholesterol metabolite act as an intracellular signal to induce the activity of cholesterol-degrading steroid hydroxylases. In addition to its effects on steroid synthesis, oxysterol regulation of SF-1 would also be predicted to increase cellular cholesterol and oxysterol uptake through upregulation of scavenger receptor SR-BI expression (Cao *et al.*, 1997). The forward gain of this control loop could be further increased by autoinduction of SF-1 expression (Nomura *et al.*, 1996). Elucidation of the physiologic function of a control scheme that is considerably more complex than simple feedforward regulation of downstream steroid hydroxylases will require extensive study of the regulation of intracellular oxysterol, SF-1 and coregulator levels.

4. SF-I Regulates Steroid Metabolism and Sexual Differentiation

Gene targeting experiments identify SF-1 as a key regulator of adrenal steroidogenesis and murine sex differentiation (Luo *et al.*, 1994, 1995; Sadovsky *et al.*, 1995; Shinoda *et al.*, 1995). SF-1 null mutants are viable at birth, but become rapidly volume depleted and die during the first 8 days of life. Adrenal cortical and medullary tissue is not detectable in the mutant embryos, and the mutant pups have decreased serum corticosterone and elevated serum ACTH levels. Treatment with corticosterone prevents death of the mutant pups, demonstrating that their mortality is caused by adrenal insufficiency. As suggested by SF-1 expression studies, gonadal development is markedly abnormal in the null mutant embryos. SF-1 null mice also display abnormal hypothalamic and pituitary development, which could reflect abnormal development of the VMH, decreased adrenal or gonadal steroid synthesis, or decreased SF-1 activity in developing pituitary cell lineages. SF-1 ablation does not affect placental development as assessed by histologic examination and studies of CYP11A and CYP17 expression.

Regulatory mechanisms underlying the role of SF-1 in adrenal and gonadal development have been studied extensively *in vitro*. Stable overexpression of SF-1 is sufficient to induce expression of the steroid hydroxylase CYP11A in embryonic stem cells, demonstrating that SF-1 is able to initiate a fate-determining program that converts pluripotent cells to steroid-synthesizing cells (Crawford *et al.*, 1997). In addition, a direct link between the sexually dimorphic pattern of SF-1 expression in the embryonic gonad and persistence of Mullerian duct structures in the knockout embryos is provided by the finding that SF-1 constitutively activates the Mullerian inhibitory substance (MIS) gene promoter *in vitro* (Shen *et al.*, 1994). The SF-1 knockout phenotype is similar to the phenotype of patients affected by X-linked adrenal hypoplasia congenita. This syndrome, which results from mutations within the nuclear orphan receptor DAX-1 locus, is characterized by adrenal hypoplasia, often associated with reduced serum gonadotropin levels and abnormal gonadal development (Muscatelli *et al.*, 1994; Zanaria *et al.*, 1994). A functional link between DAX-1 and SF-1 is suggested by the presence of a consensus SFRE (of undetermined function) in the 5′ flanking region of the DAX-1 gene (Burris *et al.*, 1995). In addition, DAX-1 can inhibit SF-1 mediated transactivation by a direct interaction between SF-1 and the DAX-1 carboxy-terminus domain (Ito *et al.*, 1997), resulting in downregulation of the MIS promoter activity (Nachtigal *et al.*, 1998). In contrast, isoforms of the Wilms' tumor 1 (WT-1) gene markedly increase SF-1 transactivation of the MIS promoter, also through direct interaction with SF-1 (Nachtigal *et al.*, 1998). WT-1 gene mutations are commonly associated with male genital ambiguity or male pseudohermaphroditism, suggesting that DAX-1 and WT-1 may reciprocally regulate SF-1 activity during the normal progress of mammalian sexual differentiation (Nachtigal *et al.*, 1998). Taken together, these studies demonstrate strong evidence for a direct role for SF-1 in regulating mammalian sexual development as well as the differentiation of steroidogenic tissues.

B. Nerve Growth Factor Induced Factor B

The nerve growth factor induced factor B (NGFI-B) subfamily of orphan nuclear receptors contains three members: NGFI-B (also known as Nur77, N10, TR3, NAK1, ST-59, and TIS-1), Nurr1 (also known as RNR-1, NOT, and HZF-3) and NOR-1 (also known as MINOR and TEC) (for references, see Willy and Mangelsdorf, 1998). NGFI-B was initially identified as a factor whose expression was upregulated in NGF-stimulated PC12 pheochromocytoma cells (Milbrandt, 1988).

1. NGFI-B Expression

NGFI-B and its related family members Nurr1 and NOR-1 are highly expressed in the adult nervous system, where they are induced as part of the immediate early response to stimuli such as growth factors, membrane depolarization, and seizures (Hazel *et al.*, 1988; Ryseck *et al.*, 1989; Watson and Milbrandt, 1990; Law *et al.*, 1992; Hedvat and Irving, 1995). Their expression outside the nervous system varies significantly among studies, which may reflect the short half-life of NGFI-B transcripts as well as the importance of growth factors in regulating NGFI-B expression (Hazel *et al.*, 1988; Ryseck *et al.*, 1989). In adult rodents, NGFI-B is expressed in the adrenal, thyroid, and pituitary glands, as well as the liver, testis, ovary, thymus, muscle, lung, and ventral prostate (Milbrandt, 1988; Ryseck *et al.*, 1989; Nakai *et al.*, 1990; Lim *et al.*, 1995; Bandoh *et al.*, 1997). Its expression is upregulated in T-cells undergoing apoptosis (Liu *et al.*, 1994; Woronicz *et al.*, 1994). Nurr1 is expressed in the adult liver (Scearce *et al.*, 1993), as well as the pituitary gland, thymus, and osteoblasts (Mages *et al.*, 1994; Bandoh *et al.*, 1997). NOR-1 is expressed at high levels in the pituitary gland and at intermediate or low levels in the adrenal glands, heart, skeletal muscle, thymus, kidney, epididymis, and submandibular glands (Ohkura *et al.*, 1994; Labelle *et al.*, 1995; Maruyama *et al.*, 1997). Renal expression of NOR-1, NGFI-B, and RNR-1 is upregulated during early stages of antigen-induced glomerulonephritis (Hayashi *et al.*, 1996), whereas hepatic NGFI-B and Nurr1 expression increases in liver as it regenerates following partial hepatectomy (Scearce *et al.*, 1993).

2. NGFI-B Response Elements and Gene Regulation

NGFI-B family members can bind DNA as monomers, as homodimers or as heterodimers with RXR. NGFI-B binds to monomeric response elements (NBREs) containing the 5′ extended core motif (AAAGGTCA) (Wilson *et al.*, 1991). NGFI-B site specificity is determined by DNA–protein contacts between nucleotides located 5′ to the core motif contained in the NGFI-B DNA response element and "A box" residues located adjacent to the C-terminal end of the receptor DNA binding domain (Wilson *et al.*, 1992, 1993a; Giguère *et al.*, 1995). NGFI-B and Nurr1 can also bind to DR5 response elements as heterodimers with RXR. On this element, the NGFI-B/Nurr1-RXR heterodimer displays little basal transcriptional activity, but is efficiently induced by RXR agonists (Perlmann and Jansson, 1995). RXR ligands can also induce transcription of Nurr1-RXR heterodimers when synthetic reporters containing multiple copies of the monomeric NBRE are used in cotransfection assays: Activation of the Nurr1-RXR heterodimer occurs in the absence of direct RXR DNA binding (Forman *et al.*, 1995c). NGFI-B has also been reported to bind as a homodimer to a nonconsensus response element (NurRE) contained within the POMC gene promoter (Philips *et al.*, 1997). Nonconsensus dimeric NGFI-B response elements may be essential for certain aspects of the function of the receptor *in vivo*. Physiologic processes that have been shown to depend on NGFI-B activation, such as the induction of T-cell apoptosis and the response to corticotrophin-releasing hormone, result in increased activity of NurRE-regulated but not NBRE-regulated synthetic promoters (Philips *et al.*, 1997).

3. NGFI-B Ligand-Dependent and Ligand-Independent Activation

No ligand has yet been identified for NGFI-B or other members of this subfamily. However, RXR ligands can activate both the NGFI-B-RXR and NURR1-RXR heterodimers, suggesting that retinoids could enhance the response to growth factors initiated by the rapid induction of expression of these orphan receptors (Forman *et al.*, 1995c; Perlmann and Jansson, 1995). As exemplified by the PPAR family, activation of the NGFI-B and Nurr1 heterodimeric complexes by retinoids does not exclude the existence of NGFI-B-specific ligands.

NGFI-B family members belong to the class of immediate-early genes that are induced by various growth factors, and as transcription factors their activity could well be regulated by posttranslational modification. NGFI-B nuclear localization, DNA-binding affinity, and transcriptional activity can be modulated by phosphorylation of the receptor protein (Fahrner *et al.*, 1990; Hazel *et al.*, 1991). Treatment of pheochromocytoma PC12 cells with differentiation-promoting agents, such as NGF or fibroblast growth factor, resulted in synthesis of a hyperphosphorylated form of NGFI-B, which is distributed in both the nucleus and cytosol (Fahrner *et al.*, 1990). In contrast, treatment with epidermal growth factor or phorbol esters stimulated the synthesis of a hypophosphorylated receptor protein that was found only in the nucleus. *In vivo*, different physiologic stimuli induce different patterns of NGFI-B phosphorylation, which may be linked to functional differences in NGFI-B transactivation potential (Hazel *et al.*, 1991). For example, membrane depolarization induces the synthesis and phosphorylation of a transcriptionally active receptor, whereas NGF treatment results in increased synthesis of NGFI-B protein with little or no transcriptional activity (Katagiri *et al.*, 1997).

4. NGFI-B Regulates Steroid Metabolism

NGFI-B has been identified as a regulator of adrenal steroidogenesis based on its ability to regulate the steroid 21-hydroxylase (CYP21) and the steroid 17-hydroxylase (CYP17) gene promoters (Wilson *et al.*, 1993b; Zhang and Mellon, 1997) (Table I). ACTH treatment strongly induces NGFI-B and Nurr1 expression in the adrenal gland and in Y-1 adrenocortical carcinoma cells (Davis and Lau, 1994). NGFI-B constitutively induces CYP21 promoter activity through a response element that is also recognized by the nuclear receptor SF-1 (Wilson *et al.*, 1993b). In addition, NGFI-B induces CYP17 promoter activity by interacting with factors bound to a complex regulatory element containing three imperfect core motifs. Activity of this promoter element is also regulated by SF-1, COUP-TF, and two unknown transcription factors designated StF-IT-1 and StF-IT-2 (the latter being required for NGFI-B induction of the CYP17 promoter) (Zhang and Mellon, 1997). A role for NGFI-B in regulating steroid synthesis has not yet been demonstrated *in vivo:* NGFI-B null mutants have no evidence of adrenal or gonadal dysfunction and display normal basal and stimulated CYP21 expression levels (Crawford *et al.*, 1995). The failure to observe any adrenal abnormalities in NGFI-B null mutants likely results from functional redundancy of NGFI-B family members: NGFI-B null mice show increased adrenal Nurr1 expression, which might compensate for NGFI-B ablation in this tissue (Crawford *et al.*, 1995).

C. LXR

The LXR subfamily of nuclear receptors contains two members: LXRα (Willy *et al.*, 1995) (initially cloned as RLD-1; Apfel *et al.*, 1994) and LXRβ (also known as NER, UR, RIP15 and OR-1) (Shinar *et al.*, 1994; Song *et al.*, 1994; Seol *et al.*, 1995; Teboul *et al.*, 1995).

I. LXR Expression

LXRα is expressed at high levels in rat pituitary, spleen, adipose tissue, lung, and liver, and at lower levels in testis, prostate, and skin, whereas LXRβ is ubiquitously expressed (Apfel *et al.*, 1994; Song *et al.*, 1994; Willy *et al.*, 1995).

2. LXR Response Elements and Gene Regulation

Both LXRα and LXRβ bind DNA as a heterodimer with RXR, showing preference for elements containing direct repeats of the consensus AGGTCA half-site separated by four base pairs (DR4) (Apfel *et al.*, 1994; Song *et al.*, 1994; Teboul *et al.*, 1995; Willy *et al.*, 1995). When natural DR4 response elements were studied, LXRα displayed strong site preference based on the sequence of the spacer region, demonstrating that nuclear receptor regulation of complex promoter elements could be determined by binding site characteristics other that half-site spacing (Apfel *et al.*, 1994). Both LXRα (Apfel *et al.*, 1994; Willy *et al.*, 1995) and LXRβ (Song *et al.*, 1994; Seol *et al.*, 1995; Teboul *et al.*, 1995) inconsistently demonstrate constitutive activity in transient transfection experiments. On DR4 elements, the LXRα-RXR heterodimer may be activated by RXR-specific ligands (Willy *et al.*, 1995).

3. LXR Is Activated by Oxysterols

LXR-RXR heterodimers can be activated by specific oxysterols that are intermediates in steroid and sterol metabolism: These include the cholesterol precursor FF-MAS; the bile acid precursors 27α- and 7α-hydroxycholesterol; and the steroid precursors 20(*S*)- and 22(*R*)-hydroxycholesterol (Janowski *et al.*, 1996). LXRα activation by 22(*R*)-hydroxycholesterol likely involves direct interaction between the receptor and its ligand. 22(*R*)-hydroxycholesterol alters the sensitivity of LXRα to chymotrypsin digestion and results in a novel protease-resistant digestion product, likely due to changes in the receptor's conformation. Oxysterols can also regulate LXRα- and LXRβ-dependent activation of the cholesterol 7α-hydroxylase gene promoter, suggesting that substrate-dependent modulation of Cyp7a transcriptional activation may regulate the conversion of cholesterol to bile acids (Lehmann *et al.*, 1997a). LXRα activity can also be regulated by other intermediary products involved in mevalonic acid metabolism: LXRα is induced by treatment with mevalonic acid, but is repressed by geranylgeraniol (Forman *et al.*, 1997). Whether these agents influence LXRα activity directly is unclear; however, they demonstrate that LXRα can be activated or repressed by related naturally occurring compounds and suggest that it may play a fundamental role in regulation of the isoprenoid pathway.

4. LXR Regulates Cholesterol and Bile Acid Metabolism

Gene targeting experiments have identified LXRα as an essential regulator of cholesterol and bile acid metabolism *in vivo* (Peet *et al.*, 1998). LXRα null mutants fed standard rodent chow develop normally and are externally and histologically identical to their wild-type littermates. When challenged by a diet supplemented with 2% cholesterol, LXRα null mutants become hypercholesterolemic and develop enlarged, pale livers as a result of the accumulation of large cholesterol deposits in the hepatic parenchyma. Hepatic cholesterol accumulation is associated with significant hepatocellular injury and necrosis, reflected both by histologic changes and by elevated serum transaminase level in the knockout animals. LXRα knockout mice are less able to convert excess dietary cholesterol to bile acids and have abnormal fecal bile salt composition. Degradation of cholesterol to form bile acids is regulated by the hepatic activity of cholesterol-7α-hydroxylase (Cyp7a): In normal animals, this enzyme is upregulated by cholesterol loading. LXRα null mutants and wild-type mice display similar Cyp7a expression levels when fed a standard diet; however, dietary cholesterol supplementation does not result in Cyp7a induction in LXRα knockout mice (Peet *et al.*, 1998). Taken together, these data suggest that LXRα mutants are unable to activate cholesterol-degrading enzymes in response to high cholesterol loads as a result of an impaired ability to detect elevated intracellular cholesterol levels. Given that LXRα activity is regulated by oxysterols, the LXRα knockout phenotype raises interesting questions about the relationship between oxysterol production and mechanisms of sensing intracellular cholesterol. In addition, characterization of hepatic gene expression in LXRα null mutants suggests that LXRα also may play a wider role in regulating fat metabolism: in particular, LXRα, together with its oxysterol ligands, may directly or indirectly influence the activity of enzymes and regulatory proteins involved in fatty acid and isoprenoid synthesis (Peet *et al.*, 1998).

D. Pregnane X Receptor

Pregnane X receptor (PXR) was cloned *in silico* by screening EST databases for sequences homologous to the ligand binding domains of previously cloned nuclear receptors (Kliewer *et al.*, 1998). Two PXR isoforms were identified: PXR-2 is an alternatively spliced gene product that lacks a 123-nucleotide region contained in the PXR-1 ligand-binding domain. PXR shares moderate homology with the *Xenopus* receptor ONR-1 and with the mammalian vitamin D receptor (VDR). PXR binds DR3 response elements as a heterodimer with RXR: DR3 elements are also recognized by VDR, suggesting that the two receptors may regulate common gene targets. In adult rodents, PXR is highly expressed in the intestine and liver; A similar pattern of PXR expression is seen in 19.5 dpc mouse embryos (Kliewer *et al.*, 1998).

A Gal4-PXR chimeric receptor was used in transient transfection assays that identified possible PXR ligands (Kliewer *et al.*, 1998). PXR-1 activity was strongly upregulated in response to synthetic glucocorticoid agonists and antagonists, as well as by pregnenolone derivatives. Naturally occurring pregnanes, including pregnenolone, 17α-hydroxypregnenolone, progesterone, 17α-hydroxyprogesterone, and 5β-pregnane-3,20-dione, activated PXR-1 (EC50 between 5 and 20 μM for individual compounds). A narrower range of PXR-1 activators also induced PXR-2, demonstrating that alternate splicing of PXR transcripts could alter the receptor ligand specificity. These agents likely exert their effects on PXR transcriptional activity by interacting directly with the ligand binding domain of the receptor. *In vitro* interaction between PXR and the nuclear receptor coactivator SRC-1 was significantly enhanced in the presence of pregnenolone 16α-carboxynitrile (PCN) and other synthetic steroids.

Synthetic glucocorticoids such as dexamethasone (DEX) and PCN have widespread effects on hepatic and intestinal cytochrome p450 expression (Elshourbagy and Guzelian, 1980; Miyata *et al.*, 1995; Quattrochi *et al.*,

1995). Putative PXR response elements identified in a DEX-inducible region of the CYP3A1 and CYP3A2 promoters have been shown to bind PXR-RXR heterodimers and to activate a synthetic promoter in response to PXR ligands (Kliewer *et al.*, 1998). CYP3A enzymes participate in the biotransformation and inactivation of a variety of xenobiotics and synthetic steroids. These results suggest that the intestinal and hepatic capacity for metabolism of certain synthetic steroids could be regulated by substrate-dependent modulation of PXR activity, as well as by the intracellular levels of natural PXR ligands. In addition, PXR may play a wider role in the regulation of hepatic sterol metabolism: The previously demonstrated role of PCN in the regulation of hepatic cholesterol metabolism might occur through PCN modulation of PXR transcriptional activity.

IV. Other Orphan Nuclear Receptors with Proposed Roles in Metabolic Control

Many of the characterized promoters involved in the regulation of cellular carbohydrate and lipid metabolism are activated by nuclear receptors binding as homodimers or RXR heterodimers to DR1-type response elements. DR1 response elements are also bound by COUP-TF, which has been shown to antagonize PPAR and HNF-4 activity in a variety of promoter contexts (reviewed in Tsai and Tsai, 1997). To date no definite role has been established for COUP-TF in the regulation of cellular metabolism; however, its frequent identification as an *in vitro* regulator of metabolic gene promoters justifies its inclusion in this review. In contrast to COUP-TF, the orphan nuclear receptor FXR has not been identified as a regulator of any cellular metabolic enzymes. However, rat FXR was initially shown to be activated by intermediate metabolites produced during terpenoid biosynthesis and was suggested to be a potential intracrine regulator of this pathway (Forman *et al.*, 1995a).

A. Chicken Ovalbumin Upstream Promoter Transcription Factor

The chicken ovalbumin upstream promoter transcription factor (COUP-TF) subfamily of orphan nuclear receptors contains two closely related members: COUP-TFI (also known as ear3) and COUP-TFII (also known as ARP-1) (reviewed in Tsai and Tsai, 1997). COUP-TFI was initially identified as a factor required for expression of the chicken ovalbumin gene, whereas COUP-TFII was characterized as a factor required for regulating expression of the apolipoprotein AI gene. During murine development, COUP-TFs are preferentially expressed in the central nervous system as well as in organs whose development depends on interactions between the mesenchyme and other epithelial layers. In adult animals, COUP-TFI and COUP-TFII display restricted and often complementary expression patterns in the brain and spinal cord, but are otherwise widely expressed.

COUP-TFI binds with highest affinity to DR1 type response elements (Hwung *et al.*, 1988). In contrast to other nuclear receptors, COUP-TFI does not display highly specific requirements for HRE recognition: In fact, COUP-TF binds to most response elements containing direct repeats of the AGGTCA core motif, as well as to elements containing everted and inverted repeats of the core motif (Cooney *et al.*, 1992). COUP-TFI homodimerizes in solution and can bind DNA as either a homodimer or as a heterodimer with the nuclear receptor RXR (Wang *et al.*, 1989; Kliewer *et al.*, 1992a). In most promoter contexts, COUP-TFI is a potent transcriptional repressor that efficiently antagonizes transcriptional activation mediated by nuclear receptors PPAR, HNF-4, RXR, ER, RAR, VDR, and T_3R (Kliewer *et al.*, 1992a; Mietus-Snyder *et al.*, 1992; Cooney *et al.*, 1993; Miyata *et al.*, 1993; Burbach *et al.*, 1994; Klinge *et al.*, 1997). Proposed mechanisms for COUP-mediated repression include competition for target response elements, competition for limiting amounts of RXR, formation of inactive receptor–receptor complexes, and active repression mediated by N- and C-terminal repression domains (see Tsai and Tsai, 1997, for references).

Given the wide range of response elements recognized by COUP-TFs, it is not surprising that these receptors have been identified *in vitro* as potential regulators of the expression of many metabolic enzymes (Table I). Although COUP-TFs may modulate enzyme expression indirectly through their involvement in neuron, muscle, and adipocyte differentiation (Lutz *et al.*, 1994; Muscat *et al.*, 1995; Brodie *et al.*, 1996), their importance in directly regulating metabolic enzymes in fully differentiated tissues is less clear. Although COUP-TF is a potent repressor of HNF-4 and PPAR stimulation of gene activity, a direct role for COUP-TF in the regulation of potential metabolic target genes has not yet been demonstrated *in vivo*, for example, by studies of gene expression in COUP-TFI or COUP-TFII null mice. A potential link between COUP-TF family members and metabolic gene regulation is suggested by studies of the COUP-TF homolog *svp* during *Drosophila* fat body differentiation (Hoshizaki *et al.*, 1994). *Svp* is transiently expressed in mesoderm cells undergoing early fat cell differentiation and is not expressed in terminally differentiating fat cells. Although *svp* mutants still form fat body cells, inactivation of *svp* results in altered expression of fat body markers, suggesting that the receptor plays a crucial role in regulating the maturation of early fat body precursor cells. It will be interesting to determine whether this aspect of COUP-TF function is conserved between *Drosophila* and mammals.

B. Farnesoid X-Activated Receptor

Farnesoid X-activated receptor (FXR) was cloned from a rat liver cDNA library using an oligonucleotide probe directed against conserved residues in the nuclear receptor DNA-binding domain (Forman *et al.*, 1995a). FXR had previously been cloned from mouse liver based on its ability to heterodimerize with RXR (Seol *et al.*, 1995). FXR shares high DNA binding domain homology with the *Drosophila* ecdysone receptor (EcR) and binds to ecdysone response elements (EcRE) as a heterodimer with RXR. In addition to binding to the EcRE, FXR-RXR heterodimers bind to response elements containing direct repeats of the AGGTCA core motif separated by two, four, or five bases, as well as elements containing an inverted repeat of the core motif separated by zero bases (Seol *et al.*, 1995). In adult rodents, FXR is highly expressed in the liver, adrenal cortex, and kidney, with the highest levels of renal expression being localized to the medullary rays and stripe (Forman *et al.*, 1995a; Seol *et al.*, 1995).

An initial ligand screen performed using a reporter gene controlled by multiple copies of the EcRE identified the insect juvenile hormone JHIII as a potential FXR ligand (Forman *et al.*, 1995a). Whether this effect occurs through direct interaction between JHIII and FXR is unclear: Previous studies had demonstrated that JHIII could increase RXR activity (Harmon *et al.*, 1995), suggesting that activation of the FXR:RXR heterodimer could be mediated through ligand interaction with RXR, with FXR merely tethering the active complex to DNA. FXR activity is also increased following treatment with micromolar concentrations of farnesol, a JHIII precursor. Interestingly, neither farnesol nor JHIII activate RIP14, the murine FXR homolog, in cotransfection studies using an EcRE-regulated reporter gene (Zavacki *et al.*, 1997). Direct binding of JHIII or farnesol to FXR has not been demonstrated (Forman *et al.*, 1995a). These studies are particularly important to validate the role of FXR as an intracrine regulator, given that the synthetic retinoid TTNPB activates RIP14 in the absence of direct receptor binding (Zavacki *et al.*, 1997). This suggests that FXR activation may be regulated by metabolites of the identified activators, or more likely by indirect cellular effects of ligand treatment, including alterations in the expression levels of specific nuclear receptor coactivators. Confirmation of FXR's putative role as a regulator of terpenoid metabolism will require further study of the receptor's function *in vivo,* perhaps including gene targeting experiments.

V. Models for Nuclear Receptor Control of Metabolic Target Genes

The following sections review basic mechanisms used by nuclear receptors to modulate expression of metabolic enzymes. The specific enzymes discussed are MCAD and CYP7a, which provide examples of complex response elements whose activity is modulated by direct and competitive interactions with nuclear receptors, as well as L-PK and PEPCK, which provide examples of nuclear receptors acting as promoter context-dependent accessory factors.

A. Medium Chain Acyl-CoA Dehydrogenase

Medium chain acyl-CoA dehydrogenase (MCAD) is a nuclearly encoded protein that catalyzes the initial step in mitochondrial fatty acid oxidation (reviewed in Schulz, 1991). The importance of MCAD as a regulator of cellular energy supply is emphasized by the significant morbidity caused by human MCAD deficiency: Affected individuals may experience hypoglycemia, coma, and sudden death, often precipitated by prolonged fasting or intercurrent illness (reviewed in Roe and Coates, 1995). Regulation of MCAD activity, through transcriptional control of MCAD gene expression, is an important control of the rate of cellular fatty acid β-oxidation (Nagao *et al.*, 1993). MCAD expression is closely linked to energy substrate supply as well as to tissue energy requirements both during development and in response to physiologic stresses in postnatal life (Carroll *et al.*, 1989; Hainline *et al.*, 1993; Nagao *et al.*, 1993). Organs that preferentially utilize lipids as a source of cellular energy, such as the heart, kidneys, and brown adipose tissue, possess the highest MCAD activity (Kelly *et al.*, 1989; Hainline *et al.*, 1993; Nagao *et al.*, 1993).

Initial characterization of the MCAD promoter by DNase I footprinting has identified six protected regions, three of which bind the ubiquitous transcription factor Sp1 (Leone *et al.*, 1995). The remaining protected sites, designated NRRE-1, NRRE-2, and NRRE-3, contain potential binding sites for members of the nuclear receptor family. Of these, the NRRE-1 element is thought to control developmental and organ-specific MCAD expression. Studies performed using transgenic mice have demonstrated that deletion of NRRE-1 element in the context of the MCAD promoter significantly decreases basal MCAD expression in heart, brown adipose tissue, and kidney and blocks the physiologic postnatal induction of MCAD expression (Disch *et al.*, 1996). MCAD expression may be regulated by intercellular levels of metabolic substrates: For example, fatty acids have been shown to control MCAD expression through PPAR regulation of NRRE-1 activity (Gulick *et al.*, 1994), while fatty acyl-CoA modulation of MCAD expression could theoretically be mediated by their effects on HNF-4 activity.

Several members of the nuclear receptor superfamily have been shown to interact specifically with response elements contained within NRRE-1 (Fig. 2A). In transient transfection studies, NRRE-1 activity can be upregulated by PPAR, HNF-4, RAR/RXR heterodimers, and RXR homodimers (Raisher *et al.*, 1992; Carter *et al.*, 1993; Gulick *et al.*, 1994), while basal NRRE-1 activity can be suppressed by COUP-TFI, COUP-TFII, and Ear3 (Carter *et al.*, 1993, 1994). ERRα blocks RAR/RXR responsiveness of the element, whereas the transcriptional effects of SF-1, which binds to

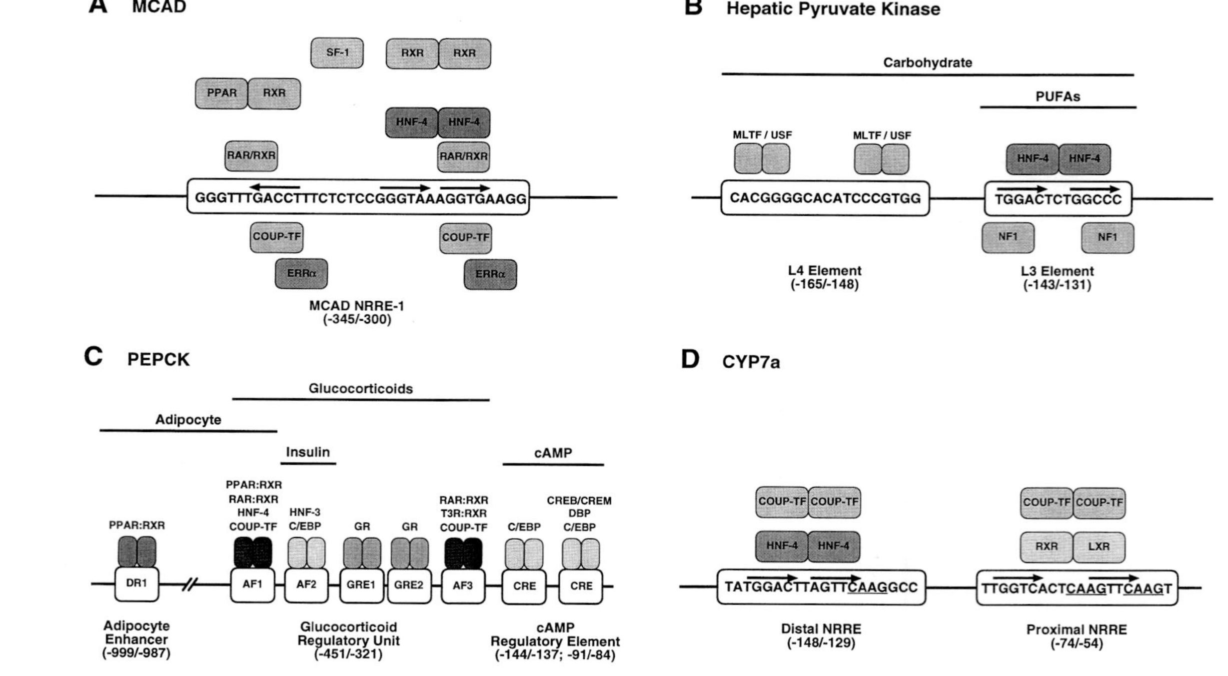

A   MCAD
SF-1
RXR   RXR
PPAR   RXR
HNF-4   HNF-4
RAR/RXR
RAR/RXR
GGGTTTGACCTTTCTCTCCGGGTAAAGGTGAAGG
COUP-TF
COUP-TF
ERRα
ERRα
MCAD NRRE-1
(-345/-300)

B   Hepatic Pyruvate Kinase
Carbohydrate
PUFAs
MLTF / USF
MLTF / USF
HNF-4   HNF-4
CACGGGGCACATCCCGTGG
TGGACTCTGGCCC
NF1
NF1
L4 Element
(-165/-148)
L3 Element
(-143/-131)

C   PEPCK
Glucocorticoids
Adipocyte
Insulin
cAMP
PPAR:RXR
PPAR:RXR
RAR:RXR
HNF-4
COUP-TF
HNF-3
C/EBP
GR
GR
RAR:RXR
T3R:RXR
COUP-TF
CREB/CREM
DBP
C/EBP
C/EBP
DR1
AF1
AF2
GRE1
GRE2
AF3
CRE
CRE
Adipocyte
Enhancer
(-999/-987)
Glucocorticoid
Regulatory Unit
(-451/-321)
cAMP
Regulatory Element
(-144/-137; -91/-84)

D   CYP7a
COUP-TF   COUP-TF
COUP-TF   COUP-TF
HNF-4   HNF-4
RXR   LXR
TATGGACTTAGTTCAAGGCC
TTGGTCACTCAAGTTCAAGT
Distal NRRE
(-148/-129)
Proximal NRRE
(-74/-54)

NRRE-1 *in vitro*, have not been determined (Leone *et al.*, 1995; Sladek *et al.*, 1997). Competitive binding of these and other unidentified factors to overlapping response elements contained within NRRE-1 determines the net contribution of the complex HRE to MCAD promoter activity. The ability of multiple factors to interact efficiently with NRRE-1 depends in part on the element's precise nucleotide sequence: For example, HNF-4 is unable to bind a DR0 element containing two consensus half sites, however it binds efficiently to the imperfect DR0 element contained in NRRE-1 (Carter *et al.*, 1993).

Several lines of evidence suggest that MCAD expression is regulated in response to cellular energy requirements of the competing activities of COUP-TFI and PPARα. MCAD expression in cultured cell lines is inversely correlated with COUP-TFI expression levels (Leone *et al.*, 1995), whereas pharmacologic inhibition of mitochondrial fatty acid import is associated with increased PPARα-mediated activation of NRRE-1 (Gulick *et al.*, 1994). In skeletal muscle, PPARα and MCAD expression parallels fiber-type-specific fatty acid β-oxidation capacity: Induction of MCAD expression in response to chronic muscle stimulation is preceded by increased PPARα protein levels (Cresci *et al.*, 1996). In heart muscle, cardiac failure results in the downregulation of MCAD expression associated with decreased levels PPARα protein and increased levels of COUP-TFI, Sp1, and Sp3 (Sack *et*

FIGURE 2 Promoter organization of the MCAD, L-PK, PEPCK, and Cyp7a genes. (A) Multiple nuclear receptor interact with consensus half sites contained within the MCAD NRRE-1. The receptor monomers are positioned to indicate their interactions with specific half sites, indicated by a horizontal arrow, contained within the complex response element. The half sites contacted by PPAR, SF-1, and RXR homodimers have not been determined. (B) Carbohydrate and PUFA regulation of L-PK is regulated by the promoter L3 and L4 elements. Full carbohydrate responsiveness of the promoter results from cooperative interactions between MLTF/USF factors and HNF-4. NF-1 consensus sites have been identified in other carbohydrate-regulated promoters; however, a physiologic role for the interaction between NF-1 and the L3 element has not been identified. (C) Glucocorticoid regulation of the PEPCK promoter is mediated by a complex glucocorticoid response unit ($-451/-321$) together with a proximal cAMP response element ($-91/-84$). HNF-3 factors binding to an accessory factor site (AF2) are responsible for insulin regulation of PEPCK expression as well as normal glucocorticoid responsiveness of the element. Nuclear receptors interacting with accessory factor sites (AF1 and AF3) induce PEPCK expression in response to their cognate ligand. Orphan nuclear receptors COUP-TF and HNF-4 are accessory factors that cooperate in regulating PEPCK promoter activity in response to glucocorticoids. (D) Two promoter elements are responsible for bile acid regulation of Cyp7a expression. A 57-kDa nuclear protein has been shown to interact with motifs (underlined) that overlap nuclear receptor consensus half sites (indicated by horizontal arrows) contained in the proximal and distal bile acid response elements: Bile acids may regulate Cyp7a expression by modulating expression of this factor. Cyp7a expression is activated by HNF-4 homodimers and LXR-RXR heterodimers interacting with the indicated elements. COUP-TF represses Cyp7a expression by interacting with either element. For references, please refer to the text.

al., 1997). The physiologic role played other nuclear factors in regulating MCAD expression is less clear. Although PPARs are highly expressed in brown fat, they do not play a significant role in MCAD induction in this tissue (Disch *et al.,* 1996); in this tissue, ERRα may be an important regulator of MCAD expression. Electrophoretic mobility shift experiments show that ERRα is the dominant factor binding NRRE-1 in primary cultures of brown adipocytes (Vega and Kelly, 1997) and forms part of a high-mobility retarded complex that is detected only in tissues expressing high levels of MCAD (Sladek *et al.,* 1997). It is hoped that further characterization of the nuclear receptors that interact with MCAD promoter elements will identify the factors that are important for mediating tissue-specific and developmental patterns of MCAD expression.

B. Hepatic Pyruvate Kinase

Pyruvate kinase (PK) catalyzes a key regulatory step in the glycolytic pathway by which energy is generated through the conversion of carbohydrates to pyruvate and lactate (reviewed in Pilkis and Granner, 1992). The hepatic pyruvate kinase (L-PK) isoenzyme is expressed primarily in the liver, where it forms the glycolytic half of a hepatic substrate cycle whose effect is opposed by the gluconeogenic enzymes PEPCK and pyruvate carboxylase. A complex regulatory network that integrates dietary, hormonal, and intracellular signals regulates substrate flux through the glycolytic pathway. Short-term regulation of PK enzyme activity is mediated by phosphorylation by cAMP-dependent protein kinase, as well as by allosteric activation by fructose 1,6-bisphosphate and allosteric inhibition by alanine and ATP. Long-term regulation results from changes in L-PK gene transcription, which is influenced by circulating hormone levels as well as by dietary stimuli. PK gene transcription is increased by insulin and glucose and is repressed by glucagon and PUFAs (Girard *et al.,* 1997; Towle *et al.,* 1997). In addition, normal constitutive and induced PK expression relies on the permissive effects of glucocorticoids and thyroid hormone (Vaulont *et al.,* 1986) and may also be influenced by products of glucose metabolism (Girard *et al.,* 1997).

Initial characterization of the L-PK gene promoter identified a 183-nucleotide promoter fragment that conferred appropriate cell-line-specific expression (Ginot *et al.,* 1989) as well as appropriate carbohydrate and hormonal responsiveness in cultured cells (Thompson and Towle, 1991). Promoter studies performed using transgenic mice demonstrated that the 183-nucleotide promoter fragment was sufficient to confer liver-specific PK expression (Cuif *et al.,* 1992). A complex response element between 150 nucleotides and 183 nucleotides upstream of the L-PK start site appropriately regulated hormonal responsiveness of the L-PK promoter *in vivo* (Cuif *et al.,* 1993). The PK proximal promoter region contains five elements that are protected in DNase I footprint experiments: Four elements contain consensus binding sites for the transcription factors HNF1 (L1), NF-1 (L2 and L3), HNF4 (L3), and MLTF (L4), whereas the fifth is bound by an unidentified factor (Vaulont *et al.*, 1989; Puzenat *et al.*, 1992). *In vitro* transcriptional activity of the L-PK promoter in hepatocyte nuclear extracts depends on the presence of the L1 and L3 elements (Vaulont *et al.*, 1989; Puzenat *et al.*, 1992), whereas carbohydrate, insulin, and glucagon responsiveness of the promoter is regulated by the L3 and L4 elements (Bergot *et al.*, 1992).

Although the L4 element can confer glucose responsiveness to a heterologous promoter, glucose regulation of the L-PK promoter depends on cooperative interactions among factors bound to the L3 and L4 elements (Bergot *et al.*, 1992; Liu *et al.*, 1993) (Fig. 2B). The L3-L4 hybrid element responds to glucose in the context of the S14 promoter, suggesting that glucose regulation of the L-PK promoter is not critically dependent on specific interactions between the hybrid element and other promoter regions (Liu and Towle, 1995). The precise mechanisms that allow HNF-4 and MLTF/USF to synergistically activate the hybrid element are unclear. HNF-4 binding to the L4 site does not seem to stabilize MLTF/USF binding to the L3 site (Liu and Towle, 1995). The hybrid element's ability to respond to glucose is impaired by altering the spacing between the L3 and L4 sites or by inverting the L4 site, suggesting that close proximity of the L3 and L4 sites allows HNF-4 and MLTF/USF to interact directly or through a bridging factor (Bergot *et al.*, 1992; Shih *et al.*, 1995). In addition, mutating the L3 site so that it binds other liver-enriched transcription factors, such as HNF-3 or CEB/P, abolishes the L-PK response to glucose (Liu and Towle, 1995). This suggests that a specific transactivation function mediated by HNF-4 is necessary for normal promoter activation. Alternatively, mutations within the L3 element may also destabilize binding of other transcription factors (such as NF-1 family members) to the HNF-4 site, resulting in decreased promoter responsiveness (Yamada *et al.*, 1997). This latter mechanism is unlikely, as HNF-4α knockout embryos display marked downregulation of L-PK expression (Stoffel and Duncan, 1997). Taken together, these data demonstrate a physiologic role for HNF-4 that is distinct from those of other liver-enriched transcription factors (Vaulont and Kahn, 1994): Synergistic interactions between HNF-4 and USF may provide a general mechanism of regulating a subset of liver-expressed targets, including the tyrosine aminotransferase (TAT) (Nitsch and Schutz, 1993) and PEPCK (Lucas *et al.*, 1991) gene promoters.

The L3-L4 element also mediates transcriptional repression of the L-PK gene in response to cAMP and PUFA (Bergot *et al.*, 1992; Liimatta *et al.*, 1994). CAMP regulation of L-PK transcription depends on normal L3-L4 contiguity: When the normal relationship between L3 and L4 is disrupted, cAMP activates rather than represses L-PK gene transcription (Bergot *et al.*, 1992). As HNF-4 DNA-binding activity can be reduced by increased levels of intracellular cAMP (Viollet *et al.*, 1997), it would be interesting to determine whether cAMP-mediated repression of L-PK activity results from alterations in the intracellular levels of an endogenous HNF-4 ligand, or from posttranslational modification of the receptor protein, such as phosphorylation-mediated changes in HNF-4 affinity for the L3 element. HNF-4 has also been implicated in the regulation of L-PK expression by linker scanning mutagenesis experiments that demonstrate that the L-PK PUFA-responsive region overlies the promoter L3 site (Liimatta *et al.*, 1994). PUFAs have no effect on the expression level of other HNF-4 regulated genes, such as TAT or PEPCK (Jump *et al.*, 1994), suggesting that their effect on L-PK expression may depend on the promoter context of the HNF-4 site (Liimatta *et al.*, 1994). As PUFA metabolites have been shown to directly modulate HNF-4 transcriptional activity, this data suggests that HNF-4 may play a direct role in regulating hepatic gene expression in response to dietary or hormonal factors (Hertz *et al.*, 1998).

C. Phosphoenolpyruvate Carboxykinase

Phosphoenolpyruvate carboxykinase (PEPCK), a key regulatory step in gluconeogenesis, catalyzes the conversion of oxaloacetate to phosphoenolpyruvate (reviewed in Pilkis and Granner, 1992; Hanson and Patel, 1994; Hanson and Reshef, 1997). Two forms of PEPCK are present in animal tissues: Cytosolic PEPCK is regulated closely in response to the physiologic state of the animal, whereas mitochondrial PEPCK is insensitive to dietary and hormonal signals. PEPCK is highly expressed in liver and kidney cortex, as well as in lung, jejunal mucosa, adipose tissue, and lactating breast; low-level PEPCK expression can be detected in many other tissues. Extrahepatic PEPCK fulfils metabolic roles other than gluconeogenesis: In adipose tissue, PEPCK plays an important role in providing glycerol for triglyceride synthesis, whereas in renal cortex, PEPCK forms part of a pathway that generates glucose as a by-product of glutamine-supported ammoniagenesis.

Hepatic PEPCK synthesis is controlled by the level of gene transcription as well as by changes in mRNA stability (Meyer *et al.*, 1991). PEPCK expression is induced by cyclic AMP, glucocorticoids, and thyroid hormone and is repressed by insulin (Hanson and Reshef, 1997). These hormones coordinately regulate PEPCK gene transcription in response to the physiologic state of the animal: Hepatic PEPCK gene transcription is decreased by carbohydrate ingestion and increased by fasting and diabetes mellitus. Although extrahepatic and hepatic PEPCK expression respond to some common stimuli, they are also regulated by distinct metabolic signals. Hepatic PEPCK expression levels are thought to be mainly determined by the antagonistic effects of insulin and cAMP (the intracellular mediator of glucagon signaling), whereas renal PEPCK expression is not altered by insulin, but is strongly induced by metabolic acidosis (reviewed in Hanson and Reshef, 1997).

Orphan nuclear receptors have been implicated in the tissue-specific expression of the PEPCK gene as well as in the modulation of its physiological response to glucocorticoids (Fig. 2C). A 460-nucleotide promoter fragment is sufficient to target expression of a PEPCK/GH transgene to the liver and kidney and to regulate appropriate hormonal and dietary responsiveness of the fusion gene in transgenic mice (McGrane *et al.*, 1988). Adipocyte expression of the transgene is not detected unless an additional 1500 nucleotides of 5′ flanking sequence is included in the construct, suggesting that upstream enhancer elements regulate PEPCK expression in adipocytes (McGrane *et al.*, 1990). An adipocyte-specific enhancer has been characterized in this region by transient transfection of cultured adipocytes: Normal adipocyte-specific function of the enhancer element was shown to depend on the presence of a functional PPRE, directly implicating PPARs in the regulation of the PEPCK gene in differentiated adipocytes (Tontonoz *et al.*, 1995).

Hormonal regulation of PEPCK expression is mediated by a set of transcription factors that bind to contiguous elements in the PEPCK proximal promoter (Fig. 2C). Glucocorticoids induce PEPCK activity through a complex response unit (GRU) that contains two weak glucocorticoid receptor binding sites (GRE1 and GRE2) (Imai *et al.*, 1990), together with three accessory factor binding sites (designated AF1, AF2, and AF3) (Imai *et al.*, 1990; Scott *et al.*, 1996). Independent mutation of any single accessory factor binding site impairs glucocorticoid induction of the GRU, whereas combined mutation of any two accessory factor sites effectively ablates the glucocorticoid response (Imai *et al.*, 1990; Scott *et al.*, 1998). The AF2 element potentiates glucocorticoid induction of PEPCK by stabilizing GR binding to the GRU (Wang *et al.*, 1996): This element also mediates insulin repression of PEPCK, likely by modulating the activity of the liver-enriched transcription factor HNF-3 (O'Brien *et al.*, 1995). Orphan nuclear receptors modulate PEPCK glucocorticoid responsiveness by interacting with the AF1 and AF3 elements. AF1 contains two half sites that can bind the nuclear receptor heterodimer RAR-RXR (Hall *et al.*, 1992) as well as HNF4 and COUP-TF (Hall *et al.*, 1995). Simultaneous mutation of both AF1 half sites blocks glucocorticoid responsiveness of the element, whereas cotransfection of either HNF-4 or COUP-TF potentiates the glucocorticoid response (Hall *et al.*, 1995). COUP-TF can also enhance glucocorticoid induction PEPCK by interacting with the AF3 element (Scott *et al.*, 1996). The molecular mechanisms underlying orphan receptor modulation of PEPCK-GRU function have not yet been determined.

These results identify potential roles for COUP-TF and HNF-4 in the modulation of glucocorticoid induction of hepatic PEPCK expression and for PPARγ in the regulation of adipocyte PEPCK expression. Normal function of the PEPCK GRU relies on cooperative interactions between orphan nuclear receptors, HNF3, C/EBP, and nuclear hormone receptors (Imai *et al.*, 1990;

Hall *et al.*, 1995). Interestingly, HNF-4α has been shown to regulate the expression of enzymes in opposing limbs of the L-PK/PEPCK substrate cycle: Modulation of intracellular concentrations of HNF-4 ligand or of the receptor's phosphorylation status might allow HNF-4 to coordinately regulate the expression of the PEPCK and L-PK gene products.

D. Cholesterol 7α-Hydroxylase

Cholesterol 7α-hydroxyase (CYP7a) is a microsomal enzyme that catalyzes the rate-limiting step in the hepatic conversion of cholesterol to bile acids (reviewed in Myant and Mitropoulos, 1977; Vlahcevic *et al.*, 1991). Cyp7a plays an important role in regulating hepatic cholesterol metabolism, both by directly controlling the rate of cholesterol catabolism to form bile acids and by indirectly controlling the rate of cholesterol secretion into the systemic circulation (reviewed in Russell and Setchell, 1992; Bjorkhem *et al.*, 1997). Cyp7a activity may also indirectly regulate hepatic cholesterol uptake by facilitating the intestinal absorption of dietary sterols. Increased Cyp7a activity, produced by transient overexpression of Cyp7a in Syrian hamsters, decreases hepatic LDL secretion and plasma LDL cholesterol concentrations (Spady *et al.*, 1995). In contrast, ablation of Cyp7a activity results in death of up to 85% of the affected offspring within 18 days of birth. Homozygous null Cyp7a mutant pups display a biphasic pattern of mortality: Early deaths (between 1 and 4 days of life) are prevented by treatment with fat-soluble vitamins, whereas late deaths (between 11 and 18 days of life) are prevented by treatment with oral bile salts (Ishibashi *et al.*, 1996). Those animals that do not die during the postnatal period display spontaneous amelioration of their symptoms by 3 weeks of age, as a result of induction of oxysterol-7α-hydroxylase (Cyp7b1), the initial step of a non-liver-specific pathway that catabolizes oxysterols to form bile acids (Schwarz *et al.*, 1996).

Cyp7a expression is primarily regulated at the level of gene transcription: While the importance of individual factors in controlling Cyp7a transcription varies considerably among species, rodent Cyp7a expression has been shown to be regulated by cholesterol (Jelinek *et al.*, 1990; Li *et al.*, 1990), mevalonate (Sundseth and Waxman, 1990), bile acids (Pandak *et al.*, 1991), steroid and thyroid hormones (Ness *et al.*, 1990; Hylemon *et al.*, 1992), diurnal rhythm (Li *et al.*, 1990; Noshiro *et al.*, 1990; Lavery and Schibler, 1993), insulin (Twisk *et al.*, 1995), glucagon (Hylemon *et al.*, 1992), growth hormone (Rudling *et al.*, 1997), cytokines (Feingold *et al.*, 1996), and drugs such as dexamethasone and cholestyramine (Jelinek *et al.*, 1990; Li *et al.*, 1990). The mechanisms by which Cyp7a expression is regulated by bile acids and dietary factors such as cholesterol and fatty acids have considerable clinical and biological significance (reviewed in Myant and Mitropoulos, 1977; Vlahcevic *et al.*, 1991).

Control of Cyp7a expression is coordinated with the expression of other enzymes involved in hepatic cholesterol metabolism (reviewed in Goldstein and Brown, 1990). Increases in the rate of cholesterol catabolism resulting from induction of Cyp7a expression are accompanied by reduction in the rate of cholesterol biosynthesis resulting from repression of HMG-CoA reductase and HMG-CoA synthase expression. The metabolic products involved in regulating this gene network have not been clearly identified. Studies in which primary hepatocyte cultures or adult rats are treated with chemical inhibitors of cholesterol synthesis have demonstrated a role for cholesterol in the regulation of Cyp7a activity (Jones *et al.*, 1993; Doerner *et al.*, 1995). In these studies, decreased intracellular cholesterol concentrations were associated with decreased Cyp7a expression: Cholesterol or mevalonate treatment increased Cyp7a expression. In contrast, HMG-CoA reductase is more potently repressed by oxysterols than by cholesterol (Kandutsch *et al.*, 1978). This discrepancy may result from the interaction among multiple transcription factors mediating cholesterol, oxysterol, and fatty acid signaling pathways. For example, 25-OHC can alter the activity of genes regulated by the nuclear receptor SF-1 (Lala *et al.*, 1997), as well as by the basic helix–loop–helix factors SREBP-1 and SREBP-2 (Hua *et al.*, 1993; Yokoyama *et al.*, 1993; Wang *et al.*, 1994).

In vivo evidence for bile salt regulation of Cyp7a expression has been provided by rodent studies in which the normal enterohepatic recirculation of bile acids has been disrupted by surgical biliary tract diversion (Pandak *et al.*, 1991). Bile acid diversion induces hepatic Cyp7a expression, which can be reversed by intraduodenal bile salt infusions. Cyp7a expression is also repressed by bile salt treatment of cultured cells, suggesting that bile acid repression of Cyp7a activity is not mediated by altered intestinal absorption of cholesterol or dietary lipids (Taniguchi *et al.*, 1994). Bile acids influence Cyp7a gene transcription through response elements identified in the Cyp7a proximal promoter (Hoekman *et al.*, 1993; Chiang and Stroup, 1994; Crestani *et al.*, 1994; Ramirez *et al.*, 1994): Their action requires intact PKC-mediated signaling pathways and may not be mediated by direct interaction between bile acids and a nuclear protein (Stravitz *et al.*, 1995, 1996). In addition, elements within the Cyp7a 3′-UTR may mediate post-transcriptional regulation of Cyp7a gene expression in response to bile acids (Agellon and Cheema, 1997).

DNA footprinting studies show that the proximal Cyp7a promoter contains protected elements containing consensus binding sites for the liver-enriched transcription factors CEBP, HNF-1, HNF-3, and HNF-4, as well as the orphan nuclear receptor COUP-TFII (Cooper *et al.*, 1997) (Fig. 2D). Functional studies have shown that two of these elements repress Cyp7a expression in response to bile acids (Chiang and Stroup, 1994; Stroup *et al.*, 1997a): A distal site contains degenerate DR1 and DR5 response elements that have been shown to interact with the nuclear receptors HNF-4 and COUP-TF (Chiang and Stroup, 1994; Cooper *et al.*, 1997; Stroup *et al.*, 1997b), whereas a proximal site contains a nonfunctional DR4 consensus thyroid hormone response element that interacts with COUP-TF (Crestani *et al.*, 1995; Stroup *et al.*, 1997b). Neither HNF-4 nor COUP-TF directly represses Cyp7a expression in response to bile acids: Rather, bile acids appear to regulate Cyp7a by altering expression of a 57-kDa protein that interacts with conserved sequence motifs overlapping but distinct from the nuclear receptor binding site identified in the bile-acid responsive element (Chiang and Stroup, 1994).

The proximal bile acid responsive element is also a target for LXR, a nuclear receptor whose activity is regulated by oxysterols (see earlier discussion) (Lehmann *et al.*, 1997a). Transient transfection studies show that LXR activates this element in response to oxysterols in the context of a heterologous promoter. The importance of LXR regulation of Cyp7a gene expression has also been demonstrated *in vivo*. LXRα-deficient mice have abnormalities in bile acid metabolism and are unable to increase Cyp7a expression in response to dietary cholesterol (Peet *et al.*, 1998). Given these findings, it would be interesting to determine whether bile acid regulation of the Cyp7a promoter might also be modulated by oxysterols: Changes in hepatic bile acid flux could alter modulate LXR transcriptional activity by changing intracellular oxysterol concentrations; in addition, bile acids might regulate posttranslational modification of the receptor protein, resulting in decreased LXRα DNA-binding activity.

Taken together, these results demonstrate an important role for oxysterol-activated nuclear receptors in regulating tissue-specific cholesterol metabolism (Peet *et al.*, 1998). On the one hand, 25-hydroxycholesterol and other specific oxysterols increase SF-1 activity in the adrenal gland, ovary, and testis, increasing the conversion of cholesterol to steroid hormones as a result of increased expression of cholesterol side-chain cleavage enzyme and other steroid hydroxylases (Lala *et al.*, 1997). On the other hand, a distinct group of oxysterols, including 22(R)-hydroxycholesterol, increase LXR activity, increasing the conversion of cholesterol to bile acids as a result of increased hepatic Cyp7a expression (Lehmann *et al.*, 1997a). In the case of Cyp7a expression, the activity of these regulatory pathways can be modified by dietary factors: Cholesterol efficiently induces Cyp7a expression in animals fed a diet supplemented with polyunsaturated fatty acids, but reduces Cyp7a activity in animals fed a diet supplemented with monounsaturated or saturated fatty acids (Cheema *et al.*, 1997). It would be interesting to determine whether these modulatory effects result from diet-induced changes in oxysterol metabolism or whether they may be mediated by modulation of the activity of HNF-4 or other nuclear receptors.

VI. Summary and Perspective

Orphan nuclear receptors are increasingly being identified as important regulators of genes encoding enzymes involved in cellular metabolism (Fig. 3). Members of the nuclear receptor frequently bind to similar HREs, suggesting that many novel receptors will be linked to metabolic functions through characterization of factors regulating receptor expression, identification of receptor target and ligands, and careful characterization of metabolic pathways in receptor knockout mice. In addition, many of the orphan nuclear receptors implicated in cellular differentiation, such as those regulating adipogenesis and myogenesis, are excellent candidates for factors regulating the tissue-specific expression of networks of metabolic enzymes. Such tissue-specific control strategies would be facilitated by the high degree of convergence of nuclear receptor signaling pathways as well as by the ability of nuclear receptors to modulate gene expression in cooperation with transcription factors whose activity is controlled by signals such as cAMP, oxysterols, carbohydrates, and insulin. Further characterization of metabolic gene regulatory programs will likely identify common strategies that are used to recruit coactivator proteins and general transcription factors to activate gene expression.

Because *in vitro* experiments are not well suited to study the regulation of cellular metabolism by hormones and energy substrates, the identification of orphan receptor roles in metabolic gene regulation will likely be revealed through careful physiologic studies of genetically altered mice, often using animals that have been physiologically stressed. The high degree of functional redundancy among family members has often resulted in subtle or limited phenotypes in knockout animals, suggesting that ablation of multiple orphan receptors may be required to further identify their physiologic functions. In addition, several orphan receptors have proven to be essential for fetal viability: The application of more sophisticated transgenic technologies, including tissue-specific gene targeting and inducible gene expression systems, may be required to determine if they also play a role in maintaining metabolic homeostasis.

In the long term, the development of specific pharmacologic agents may provide a fruitful approach to the study of orphan receptor function. Synthetic receptor-specific ligands may prove to be metabolically stable regulators of receptor activity and also provide a means of separating the cellular responses mediated by competitive or allosteric modulation of enzyme activity from the responses mediated by direct receptor binding. Identification of receptor ligands will help to determine whether similar metabolites mediate short-term regulation of enzyme activity as well as long-term regulation of metabolic gene transcription. The identification of acyl-CoA thioesters as HNF-4 ligands provides evidence that the expression and specific activity of metabolic enzymes may be coordinately controlled by meta-

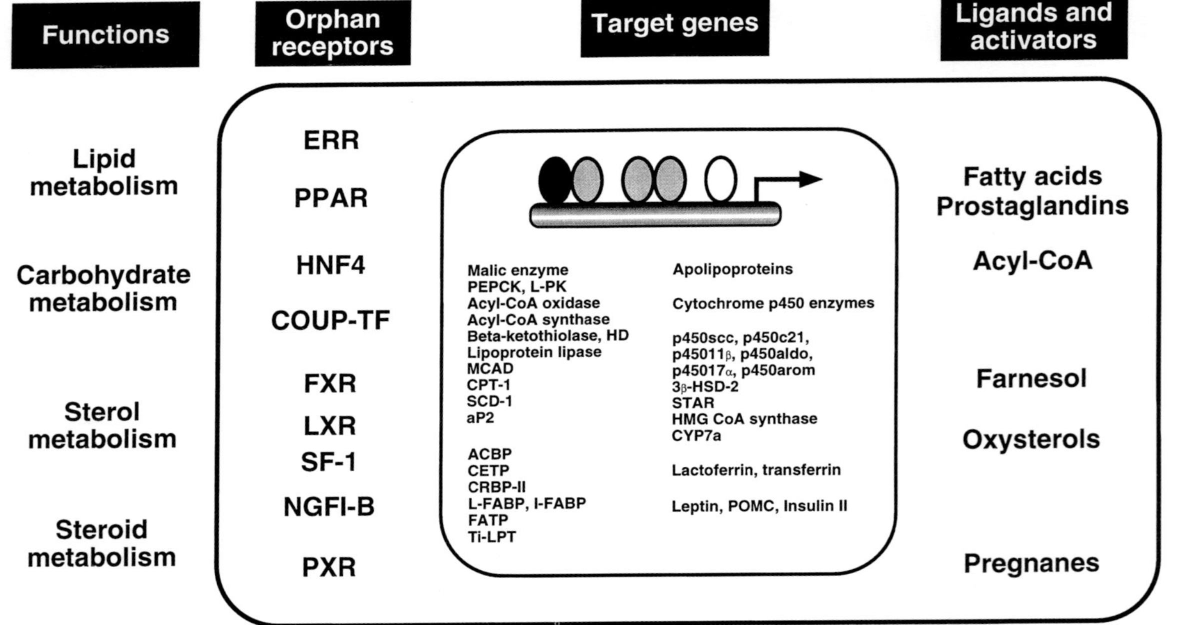

FIGURE 3 Nuclear orphan receptors display diverse effects in the regulation of cellular metabolism. Orphan nuclear receptors integrate intracrine, paracrine, and endocrine signals to regulate the expression of enzymes involved in many aspects of basic cellular metabolism. Pharmacologic modulation of orphan receptor activity may provide novel treatment strategies for diabetes mellitus, atherosclerosis, and other metabolic disorders.

bolic products that are intermediates in the regulated pathway. In contrast, regulation of SF-1 and LXR activity by a wide range of oxysterols, as well as regulation of PPAR activity by prostaglandins suggests that other metabolic pathways may be regulated by a distinct class of signaling metabolites whose production may be controlled by intracellular and paracrine effects.

Studies have revealed the existence of unexpected links between intracellular metabolic products and control of gene expression by ligand-inducible transcription factors. The characterization of orphan nuclear receptors and their ligands may identify useful and novel strategies for the treatment of metabolic disorders, including atherosclerotic heart disease, obesity, and diabetes mellitus.

Acknowledgments

Work performed in the authors' laboratory is supported by the Medical Research Council of Canada (MRCC), the National Cancer Institute of Canada, and the Cancer Research Society Inc. V. G. is a Scientist of the MRCC.

References

Adams, M., Reginato, M. J., Shao, D., Lazar, M. A., and Chatterjee, V. K. (1997). Transcriptional activation by peroxisome proliferator-activated receptor γ is inhibited by phosphorylation at a consensus mitogen-activated protein kinase site. *J. Biol. Chem.* **272**, 5128–5132.

Agellon, L. B., and Cheema, S. K. (1997). The 3′-untranslated region of the mouse cholesterol 7α-hydroxylase mRNA contains elements responsive to post-transcriptional regulation by bile acids. *Biochem. J.* **328**, 393–399.

Aldridge, T. C., Tugwood, J. D., and Green, S. (1995). Identification and characterization of DNA elements implicated in the regulation of CYP4A1 transcription. *Biochem. J.* **306**, 473–479.

Amri, E. Z., Bonino, F., Ailhaud, G., Abumrad, N. A., and Grimaldi, P. A. (1995). Cloning of a protein that mediates transcriptional effects of fatty acids in preadipocytes. Homology to peroxisome proliferator-activated receptors. *J. Biol. Chem.* **270**, 2367–2371.

Aoyama, T., Peters, J. M., Iritani, N., Nakajima, T., Furihata, K., Hashimoto, T., and Gonzalez, F. J. (1998). Altered constitutive expression of fatty acid-metabolizing enzymes in mice lacking the peroxisome proliferator-activated receptor α (PPARα). *J. Biol. Chem.* **273**, 5678–5684.

Apfel, R., Benbrook, D., Lerhardt, E., Ortiz, M. A., Salbert, G., and Pfahl, M. (1994). A novel orphan receptor specific for a subset of thyroid hormone-responsive elements and its interaction with the retinoid/thyroid hormone receptor subfamily. *Mol. Cell. Biol.* **14**, 7025–7035.

Baes, M., Castelein, H., Desmet, L., and Declercq, P. E. (1995). Antagonism of COUP-TF and PPAR alpha/RXR alpha on the activation of the malic enzyme gene promoter: modulation by 9-cis RA. *Biochem. Biophys. Res. Com.* **215**, 338–345.

Bakke, M., and Lund, J. (1995). Mutually exclusive interactions of two nuclear orphan receptors determine activity of a cyclic adenosine $3',5'$-monophosphate-responsive sequence in the bovine CYP17 gene. *Mol. Endocrinol.* **9**, 327–339.

Bandoh, S., Tsukada, T., Maruyama, K., Ohkura, N., and Yamaguchi, K. (1997). Differential expression of NGFI-B and RNR-1 genes in various tissues and developing brain of the rat: comparative study by quantitative reverse transcription-polymerase chain reaction. *J. Neuroendocrinol.* **9**, 3–8.

Barnhart, K. M., and Mellon, P. L. (1994). The orphan nuclear receptor, steroidogenic factor-1, regulates the glycoprotein hormone α-subunit gene in pituitary gonadotropes. *Mol. Endocrinol.* **8**, 878–885.

Becker-André, M., André, E., and DeLamarter, J. F. (1993). Identification of nuclear receptor mRNAs by RT-PCR amplification of conserved zinc-finger motif sequences. *Biochem. Biophys. Res. Commun.* **194**, 1371–1379.

Beekman, J. M., Wijnholds, J., Schippers, I. J., Pot, W., Gruber, M., and Ab, G. (1991). Regulatory elements and DNA-binding proteins mediating transcription from the chicken very-low-density apolipoprotein II gene. *Nucleic Acids Res.* **19**, 5371–5377.

Bergot, M. O., Diaz-Guerra, M. J., Puzenat, N., Raymondjean, M., and Kahn, A. (1992). Cis-regulation of the L-type pyruvate kinase gene promoter by glucose, insulin and cyclic AMP. *Nucleic Acids Res.* **20**, 1871–1877.

Berry, M., Metzger, D., and Chambon, P. (1990). Role of the two activating domains of the oestrogen receptor in the cell-type and promoter-context dependent agonistic activity of the anti-oestrogen 4-hydroxytamoxifen. *EMBO J.* **9**, 2811–2818.

Bjorkhem, I., Lund, E., and Rudling, M. (1997). Coordinate regulation of cholesterol 7α-hydroxylase and HMG-CoA reductase in the liver. *Subcellular Biochemistry* **28**, 23–55.

Bogazzi, F., Hudson, L. D., and Nikodem, V. M. (1994). A novel heterodimerization partner for thyroid hormone receptor. Peroxisome proliferator-activated receptor. *J. Biol. Chem.* **269**, 11683–11686.

Bogerd, A. M., Franklin, A., Rice, D. A., Schimmer, B. P., and Parker, K. L. (1990). Identification and characterization of two upstream elements that regulate adrenocortical expression of steroid 11 beta-hydroxylase. *Mol. Endocrinol.* **4**, 845–850.

Bonnelye, E., Vanacker, J. M., Dittmar, T., Begue, A., Desbiens, X., Denhardt, D. T., Aubin, J. E., Laudet, V., and Fournier, B. (1997). The ERR-1 orphan receptor is a transcriptional activator expressed during bone development. *Mol. Endocrinol.* **11**, 905–916.

Braissant, O., Foufelle, F., Scotto, C., Dauca, M., and Wahli, W. (1996). Differential expression of peroxisome proliferator-activated receptors (PPARs): tissue distribution of PPAR-α, -β, and -γ in the adult rat. *Endocrinology* **137**, 354–366.

Brodie, A. E., Manning, V. A., and Hu, C. Y. (1996). Inhibitors of preadipocyte differentiation induce COUP-TF binding to a PPAR/RXR binding sequence. *Biochem. Biophys. Res. Com.* **228**, 655–661.

Brown, P. J., Smith-Oliver, T. A., Charifson, P. S., Tomkinson, N. C., Fivush, A. M., Sternbach, D. D., Wade, L. E., Orband-Miller, L., Parks, D. J., Blanchard, S. G., Kliewer, S. A., Lehmann, J. M., and Willson, T. M. (1997). Identification of peroxisome proliferator-activated receptor ligands from a biased chemical library. *Chem. Biol.* **4**, 909–918.

Brun, R. P., Tontonoz, P., Forman, B. M., Ellis, R., Chen, J., Evans, R. M., and Spiegelman, B. M. (1996). Differential activation of adipogenesis by multiple PPAR isoforms. *Genes Dev.* **10**, 974–984.

Brzozowski, A. M., Pike, A. C. W., Dauter, Z., Hubbard, R. E., Bonn, T., Engström, L., Greene, G. L., Gustafsson, J.-Å., and Carlquist, M. (1997). Molecular basis of agonism and antagonism in the oestrogen receptor. *Nature* **389**, 753–758.

Bulman, M. P., Dronsfield, M. J., Frayling, T., Appleton, M., Bain, S. C., Ellard, S., and Hattersley, A. T. (1997). A missense mutation in the hepatocyte nuclear factor 4 α gene in a UK pedigree with maturity-onset diabetes of the young. *Diabetologia* **40**, 859–862.

Burbach, J. P. H., Lopez da Silva, S., Cox, J. J., Adan, R. A. H., Cooney, A. J., Tsai, M.-J., and Tsai, S. Y. (1994). Repression of estrogen-dependent stimulation of the oxytocin gene by chicken ovalbumin upstream promoter transcription factor 1. *J. Biol. Chem.* **269**, 15046–15053.

Burris, T. P., Guo, W. W., Le, T., and McCabe, E. R. B. (1995). Identification of a putative steroidogenic factor-1 response element in the Dax-1 promoter. *Biochem. Biophys. Res. Com.* **214**, 576–581.

Cairns, W., Smith, C. A. D., McLaren, A. W., and Wolf, C. R. (1996). Characterization of the human cytochrome P4502D6 promoter. A potential role for antagonistic interactions between members of the nuclear receptor family. *J. Biol. Chem.* **271**, 25269–25276.

Cammas, F. M., Pullinger, G. D., Barker, S., and Clark, A. J. (1997). The mouse adrenocorticotropin receptor gene: cloning and characterization of its promoter and evidence for a role for the orphan nuclear receptor steroidogenic factor 1. *Mol. Endocrinol.* **11**, 867–876.

Cao, G., Garcia, C. K., Wyne, K. L., Schultz, R. A., Parker, K. L., and Hobbs, H. H. (1997). Structure and localization of the human gene encoding SR-BI/CLA-1. Evidence for transcriptional control by steroidogenic factor 1. *J. Biol. Chem.* **272**, 33068–33076.

Carroll, J. E., McGuire, B. S., Chancey, V. F., and Harrison, K. B. (1989). Acyl-CoA dehydrogenase enzymes during early postnatal development in the rat. *Biol. Neonate* **55**, 185–190.

Carter, M. E., Gulick, T., Raisher, B. D., Caira, T., Ladias, J. A. A., Moore, D. D., and Kelly, D. P. (1993). Hepatocyte nuclear factor-4 activates medium chain acyl-CoA dehydrogenase gene transcription by interacting with a complex regulatory element. *J. Biol. Chem.* **268**, 13805–13810.

Carter, M. E., Gulick, T., Moore, D. D., and Kelly, D. P. (1994). A pleiotropic element in the medium-chain acyl coenzyme A dehydrogenase gene promoter mediates transcriptional regulation by multiple nuclear receptor transcription factors and defines novel receptor-DNA binding motifs. *Mol. Cell. Biol.* **14**, 4360–4372.

Castelein, H., Gulick, T., Declercq, P. E., Mannaerts, G. P., Moore, D. D., and Baes, M. I. (1994). The peroxisome proliferator activated receptor regulates malic enzyme gene expression. *J. Biol. Chem.* **269**, 26754–26758.

Chan, J., Nakabayashi, H., and Wong, N. C. (1993). HNF-4 increases activity of the rat Apo A1 gene. *Nucleic Acids Res.* **21**, 1205–1211.

Chartier, F. L., Bossu, J. P., Laudet, V., Fruchart, J. C., and Laine, B. (1994). Cloning and sequencing of cDNAs encoding the human hepatocyte nuclear factor 4 indicate the presence of two isoforms in human liver. *Gene* **147**, 269–272.

Cheema, S. K., Cikaluk, D., and Agellon, L. B. (1997). Dietary fats modulate the regulatory potential of dietary cholesterol on cholesterol 7α-hydroxylase gene expression. *J. Lipid Res.* **38**, 315–323.

Chen, D., Lepar, G., and Kemper, B. (1994a). A transcriptional regulatory element common to a large family of hepatic cytochrome P450 genes is a functional binding site of the orphan receptor HNF-4. *J. Biol. Chem.* **269**, 5420–5427.

Chen, W. S., Manova, K., Weinstein, D. C., Duncan, S. A., Plump, A. S., Prezioso, V. R., Bachvarova, R. F., and Darnell, J. E., Jr. (1994b). Disruption of the HNF-4 gene, expressed in visceral endoderm, leads to cell death in embryonic ectoderm and impaired gastrulation of mouse embryos. *Genes Dev.* **8**, 2466–2477.

Chiang, J. Y., and Stroup, D. (1994). Identification and characterization of a putative bile acid-responsive element in cholesterol 7α-hydroxylase gene promoter. *J. Biol. Chem.* **269**, 17502–17507.

Chu, R., Lin, Y., Rao, M. S., and Reddy, J. K. (1995). Cooperative formation of higher order peroxisome proliferator-activated receptor and retinoid X receptor complexes on the peroxisome proliferator responsive element of the rat hydratase–dehydrogenase gene. *J. Biol. Chem.* **270**, 29636–29639.

Clemens, J. W., Lala, D. S., Parker, K. L., and Richards, J. S. (1994). Steroidogenic factor-1 binding and transcriptional activity of the cholesterol side-chain cleavage promoter in rat granulosa cells. *Endocrinology* **134**, 1499–1508.

Cooney, A. J., Tsai, S. Y., O'Malley, B. W., and Tsai, M.-J. (1992). Chicken ovalbumin upstream promoter transcription factor (COUP-TF) dimers bind to different GGTCA response elements, allowing COUP-TF to repress hormonal induction of the vitamin D3, thyroid hormone, and retinoic acid receptors. *Mol. Cell. Biol.* **12**, 4153–4163.

Cooney, A. J., Leng, X., Tsai, S. Y., O'Malley, B. W., and Tsai, M.-J. (1993). Multiple mechanisms of chicken ovalbumin upstream promoter transcription factor-dependent repression of transactivation by the vitamin D, thyroid hormone, and retinoid acid receptors. *J. Biol. Chem.* **268**, 4152–4160.

Cooper, A. D., Chen, J., Botelho-Yetkinler, M. J., Cao, Y., Taniguchi, T., and Levy-Wilson, B. (1997). Characterization of hepatic-specific regulatory elements in the promoter region of the human cholesterol 7α-hydroxylase gene. *J. Biol. Chem.* **272**, 3444–3452.

Costa, R. H., Grayson, D. R., and Darnell, J. E., Jr. (1989). Multiple hepatocyte-enriched nuclear factors function in the regulation of transthyretin and α1-antitrypsin genes. *Mol. Cell. Biol.* **9**, 1415–1425.

Crawford, P. A., Sadovsky, Y., Woodson, K., Lee, S. L., and Milbrandt, J. (1995). Adrenocortical function and regulation of the steroid 21-hydroxylase gene In NGFI-B-deficient mice. *Mol. Cell. Biol.* **15**, 4331–4336.

Crawford, P. A., Sadovsky, Y., and Milbrandt, J. (1997). Nuclear receptor steroidogenic factor 1 directs embryonic stem cells toward the steroidogenic lineage. *Mol. Cell. Biol.* **17**, 3997–4006.

Cresci, S., Wright, L. D., Spratt, J. A., Briggs, F. N., and Kelly, D. P. (1996). Activation of a novel metabolic gene regulatory pathway by chronic stimulation of skeletal muscle. *Am. J. Physiol.* **270**, C1413–1420.

Crestani, M., Karam, W. G., and Chiang, J. Y. (1994). Effects of bile acids and steroid/thyroid hormones on the expression of cholesterol 7α-hydroxylase mRNA and the CYP7 gene in HepG2 cells. *Biochem. Biophys. Res. Com.* **198**, 546–553.

Crestani, M., Stroup, D., and Chiang, J. Y. (1995). Hormonal regulation of the cholesterol 7α-hydroxylase gene (CYP7). *J. Lipid Res.* **36**, 2419–2432.

Crowe, D. T., Hwung, Y. P., Tsai, S. Y., and Tsai, M. J. (1988). Characterization of the cis and trans elements essential for rat insulin II gene expression. *Prog. Clin. Biol. Res.* **284**, 211–224.

Cuif, M. H., Cognet, M., Boquet, D., Tremp, G., Kahn, A., and Vaulont, S. (1992). Elements responsible for hormonal control and tissue specificity of L-type pyruvate kinase gene expression in transgenic mice. *Mol. Cell. Biol.* **12**, 4852–4861.

Cuif, M. H., Porteu, A., Kahn, A., and Vaulont, S. (1993). Exploration of a liver-specific, glucose/insulin-responsive promoter in transgenic mice. *J. Biol. Chem.* **268**, 13769–13772.

Davis, I. J., and Lau, L. F. (1994). Endocrine and neurocrine regulation of the orphan receptors Nur77 and Nurr-1 in the adrenal glands. *Mol. Cell. Biol.* **14**, 3469–3483.

Devchand, P. R., Keller, H., Peters, J. M., Vazquez, M., Gonzalez, F. J., and Wahli, W. (1996). The PPARα-leukotriene B4 pathway to inflammation control. *Nature* **384**, 39–43.

Diaz Guerra, M. J., Bergot, M. O., Martinez, A., Cuif, M. H., Kahn, A., and Raymondjean, M. (1993). Functional characterization of the L-type pyruvate kinase gene glucose response complex. *Mol. Cell. Biol.* **13**, 7725–7733.

Disch, D. L., Rader, T. A., Cresci, S., Leone, T. C., Barger, P. M., Vega, R., Wood, P. A., and Kelly, D. P. (1996). Transcriptional control of a nuclear gene encoding a mitochondrial fatty acid oxidation enzyme in transgenic mice: role for nuclear receptors in cardiac and brown adipose expression. *Mol. Cell. Biol.* **16**, 4043–4051.

Doerner, K. C., Gurley, E. C., Vlahcevic, Z. R., and Hylemon, P. B. (1995). Regulation of cholesterol 7α-hydroxylase expression by sterols in primary rat hepatocyte cultures. *J. Lipid Res.* **36**, 1168–1177.

Drewes, T., Senkel, S., Holewa, B., and Ryffel, G. U. (1996). Human hepatocyte nuclear factor 4 isoforms are encoded by distinct and differentially expressed genes. *Mol. Cell. Biol.* **16**, 925–931.

Dreyer, C., Krey, G., Keller, H., Givel, F., Helftenbein, G., and Wahli, W. (1992). Control of the peroxisomal β-oxidation pathway by a novel family of nuclear hormone receptors. *Cell* **68**, 879–887.

Drouin, J., Nemer, M., Charron, J., Gagner, J. P., Jeannotte, L., Sun, Y. L., Therrien, M., and Tremblay, Y. (1989). Tissue-specific activity of the pro-opiomelanocortin (POMC) gene and repression by glucocorticoids. *Genome* **31**, 510–519.

Duncan, S. A., Manova, W. S., Chen, W. S., Hoodless, P., Weinstein, D., Bachvarova, R. F., and Darnell, J. E. (1994). Expression of transcription factor HNF-4 in the extraembryonic endoderm, gut, and nephrogenic tissue of the developing mouse embryo: HNF-4 is a marker for primary endoderm in the implanting blastocyst. *Proc. Natl. Acad. Sci. USA* **91**, 7598–7602.

Elholm, M., Bjerking, G., Knudsen, J., Kristiansen, K., and Mandrup, S. (1996). Regulatory elements in the promoter region of the rat gene encoding the acyl-CoA-binding protein. *Gene* **173**, 233–238.

Elshourbagy, N. A., and Guzelian, P. S. (1980). Separation, purification, and characterization of a novel form of cytochrome P-450 from rats treated with pregenolone-16α-carbonitrile. *J. Biol. Chem.* **255**, 1279–1285.

Evans, R. M. (1988). The steroid and thyroid hormone receptor superfamily. *Science* **240**, 889–895.

Faergeman, N. J., and Knudsen, J. (1997). Role of long-chain fatty acyl-CoA esters in the regulation of metabolism and in cell signalling. *Biochem. J.* **323**, 1–12.

Fahrner, T. J., Carroll, S. L., and Milbrandt, J. (1990). The NGFI-B protein, an inducible member of the thyroid/steroid receptor family, is rapidly modified posttranscriptionally. *Mol. Cell. Biol.* **10**, 6454–6459.

Feingold, K. R., Spady, D. K., Pollock, A. S., Moser, A. H., and Grunfeld, C. (1996). Endotoxin, TNF, and IL-1 decrease cholesterol 7α-hydroxylase mRNA levels and activity. *J. Lipid Res.* **37**, 223–228.

Forman, B. M., Goode, E., Chen, J., Oro, A. E., Bradley, D. J., Perlmann, T., Noonan, D. J., Burka, L. T., McMorris, T., Lamph, W. W., Evans, R. M., and Weinberberg, C. (1995a). Identification of a nuclear receptor that is activated by farnesol metabolites. *Cell* **81**, 687–693.

Forman, B. M., Tontonoz, P., Chen, J., Brun, R. P., Spiegelman, B. M., and Evans, R. M. (1995b). 15-deoxy-Δ12,14-prostaglandin J2 is a ligand for the adipocyte determination factor PPARγ. *Cell* **83**, 803–812.

Forman, B. M., Umesono, K., Chen, J., and Evans, R. M. (1995c). Unique response pathway are established by allosteric interactions among nuclear hormone receptors. *Cell* **81**, 541–550.

Forman, B. M., Ruan, B., Chen, J., Schroepfer, G. J., and Evans, R. M. (1997). The orphan nuclear receptor LXRα is positively and negatively regulated by distinct products of mevalonate metabolism. *Proc. Natl. Acad. Sci. USA* **94**, 10588–10593.

Froguel, P., Zouali, H., Vionnet, N., Velho, G., Vaxillaire, M., Sun, F., Lesage, S., Stoffel, M., Takeda, J., Passa, P., Permutt, M. A., Beckmann, J. S., Bell, G. I., and Cohen, D. (1993). Familial hyperglycemia due to mutations in glucokinase. Definition of a subtype of diabetes mellitus. *N. Engl. J. Med.* **328**, 697–702.

Furuta, H., Iwasaki, N., Oda, N., Hinokio, Y., Horikawa, Y., Yamagata, K., Yano, N., Sugahiro, J., Ogata, M., Ohgawara, H., Omori, Y., Iwamoto, Y., and Bell, G. I. (1997). Organization and partial sequence of the hepatocyte nuclear factor-4 α/MODY1 gene and identification of a missense mutation, R127W, in a Japanese family with MODY. *Diabetes* **46**, 1652–1657.

Galarneau, L., Pare, J. F., Allard, D., Hamel, D., Levesque, L., Tugwood, J. D., Green, S., and Belanger, L. (1996). The alpha(1)-fetoprotein locus is activated by a nuclear receptor of the *Drosophila* Ftz-F1 family. *Mol. Cell. Biol.* **16**, 3853–3865.

Galson, D. L., Tsuchiya, T., Tendler, D. S., Huang, L. R., Ren, Y., Ogura, T., and Bunn, H. F. (1995). The orphan receptor hepatic nuclear factor 4 functions as a transcriptional activator for tissue-specific and hypoxia-specific erythropoietin gene expression and is antagonized by EAR3/COUP-TF1. *Mol. Cell. Biol.* **15**, 2135–2144.

Gaudet, F., and Ginsburg, G. S. (1995). Transcriptional regulation of the cholesteryl ester transfer protein gene by the orphan nuclear hormone receptor apolipoprotein AI regulatory protein-1. *J. Biol. Chem.* **270**, 29916–29922.

Giguère, V., Yang, N., Segui, P., and Evans, R. M. (1988). Identification of a new class of steroid hormone receptors. *Nature* **331**, 91–94.

Giguère, V., Tini, M., Flock, G., Ong, E. S., Evans, R. M., and Otulakowski, G. (1994). Isoform-specific amino-terminal domains dictate DNA-binding properties of RORα, a novel family of orphan nuclear receptors. *Genes Dev.* **8**, 538–553.

Giguère, V., McBroom, L. D. B., and Flock, G. (1995). Determinants of target gene specificity for RORα1: monomeric DNA-binding by an orphan nuclear receptor. *Mol. Cell. Biol.* **15**, 2517–2526.

Ginot, F., Decaux, J. F., Cognet, M., Berbar, T., Levrat, F., Kahn, A., and Weber, A. (1989). Transfection of hepatic genes into adult rat hepatocytes in primary culture and their tissue-specific expression. *Eur. J. Biochem.* **180**, 289–294.

Girard, J., Ferre, P., and Foufelle, F. (1997). Mechanisms by which carbohydrates regulate expression of genes for glycolytic and lipogenic enzymes. *Ann. Rev. Nutr.* **17**, 325–352.

Glass, C. K. (1994). Differential recognition of target genes by nuclear receptors monomers, dimers, and heterodimers. *Endocr. Rev.* **15**, 391–407.

Glass, C. K., Rose, D. W., and Rosenfeld, M. G. (1997). Nuclear receptor coactivators. *Curr. Opin. Cell Biol.* **9**, 222–232.

Goldstein, J. L., and Brown, M. S. (1990). Regulation of the mevalonate pathway. *Nature* **343**, 425–430.

Gonzalez, F. J. (1997). Recent update on the PPARα-null mouse. *Biochimie* **79**, 139–144.

Gottlicher, M., Widmark, E., Li, Q., and Gustafsson, J. A. (1992). Fatty acids activate a chimera of the clofibric acid-activated receptor and the glucocorticoid receptor. *Proc. Natl. Acad. Sci. USA* **89**, 4653–4657.

Gragnoli, C., Lindner, T., Cockburn, B. N., Kaisaki, P. J., Gragnoli, F., Marozzi, G., and Bell, G. I. (1997). Maturity-onset diabetes of the young due to a mutation in the hepatocyte nuclear factor-4 α binding site in the promoter of the hepatocyte nuclear factor-1α gene. *Diabetes* **46**, 1648–1651.

Graves, R. A., Tontonoz, P., and Spiegelman, B. M. (1992). Analysis of a tissue-specific enhancer: ARF6 regulates adipogenic gene expression. *Mol. Cell. Biol.* **12**, 1202–1208.

Gronemeyer, H., and Laudet, V. (1995). Transcription factors 3: nuclear receptors. *Protein Profile* **2**, 1173–1308.

Gulick, T., Cresci, S., Caira, T., Moore, D. D., and Kelly, D. P. (1994). The peroxisome proliferator activated receptor regulates mitochondrial fatty acid oxidative enzyme gene expression. *Proc. Natl. Acad. Sci. USA* **91**, 11012–11016.

Hainline, B. E., Kahlenbeck, D. J., Grant, J., and Strauss, A. W. (1993). Tissue specific and developmental expression of rat long- and medium-chain acyl-CoA dehydrogenases. *Biochem. Biophys. Acta* **1216**, 460–468.

Hall, R. K., Scott, D. K., Noisin, E. L., Lucas, P. C., and Granner, D. K. (1992). Activation of the phosphoenolpyruvate carboxykinase gene retinoic acid response element is dependent on a retinoic acid receptor/coregulator complex. *Mol. Cell. Biol.* **12**, 5527–5535.

Hall, R. K., Sladek, F. M., and Granner, D. K. (1995). The orphan receptors COUP-TF and HNF-4 serve as accessory factors required for induction of phosphoenolpyruvate carboxykinase gene transcription by glucocorticoids. *Proc. Natl. Acad. Sci. USA* **92**, 412–416.

Halvorson, L. M., Kaiser, U. B., and Chin, W. W. (1996). Stimulation of luteinizing hormone beta gene promoter activity by the orphan nuclear receptor, steroidogenic factor-1. *J. Biol. Chem.* **271**, 6645–6650.

Hani, E., Suaud, L., Boutin, P., Chevre, J. C., Durand, E., Philippi, A., Demenais, F., Vionnet, N., Furuta, H., Velho, G., Bell, G. I., Laine, B., and Froguel, P. (1998). A missense mutation in hepatocyte nuclear factor-4α, resulting in a reduced transactivation activity, in human late-onset non-insulin-dependent diabetes mellitus. *J. Clin. Invest.* **101**, 521–526.

Hanson, R. W., and Patel, Y. M. (1994). Phosphoenolpyruvate carboxykinase (GTP): the gene and the enzyme. *Adv. Enzymol. Rel. Ar. Mol. Biol.* **69**, 203–281.

Hanson, R. W., and Reshef, L. (1997). Regulation of phosphoenolpyruvate carboxykinase (GTP) gene expression. *Ann. Rev. Biochem.* **66**, 581–611.

Harmon, M. A., Boehm, M. F., Heyman, R. A., and Mangelsdorf, D. J. (1995). Activation of mammalian retinoid X receptors by the insect growth regulator methoprene. *Proc. Natl. Acad. Sci. USA* **92**, 6157–6160.

Harris, P. K., and Kletzien, R. F. (1994). Localization of a pioglitazone response element in the adipocyte fatty acid-binding protein gene. *Mol. Pharmacol.* **45**, 439–445.

Hayashi, K., Ohkura, N., Miki, K., Osada, S., and Tomino, Y. (1996). Early induction of the NGFI-B/Nur77 family genes in nephritis induced by anti-glomerular basement membrane antibody. *Mol. Cell. Endocrinol.* **123**, 205–209.

Hazel, T. G., Nathans, D., and Lau, L. F. (1988). A gene inducible by serum growth factors encodes a member of the steroid and thyroid hormone receptor superfamily. *Proc. Natl. Acad. Sci. USA* **85**, 8444–8448.

Hazel, T. G., Misra, R., Davis, I. J., Greenberg, M. E., and Lau, L. E. (1991). Nur77 is differentially modified in PC12 cells upon membrane depolarization and growth factor treatment. *Mol. Cell. Biol.* **11**, 3239–3246.

Hedvat, C. V., and Irving, S. G. (1995). The isolation and characterization of MINOR, a novel mitogen-inducible nuclear orphan receptor. *Mol. Endocrinol.* **9**, 1692–1700.

Heinzel, T., Lavinsky, R. M., Mullen, T. M., Soderstrom, M., Laherty, C. D., Torchia, J., Yang, W. M., Brard, G., Ngo, S. D., Davie, J. R., Seto, E., Eisenman, R. N., Rose, D. W., Glass, C. K., and Rosenfeld, M. G. (1997). A complex containing N-CoR, mSin3 and histone deacetylase mediates transcriptional repression. *Nature* **387**, 43–48.

Hertz, R., Berman, I., Keppler, D., and Bar-Tana, J. (1996a). Activation of gene transcription by prostacyclin analogues is mediated by the peroxisome-proliferator-activated receptor (PPAR). *Eur. J. Biochem.* **235**, 242–247.

Hertz, R., Seckbach, M., Zakin, M. M., and Bar-Tana, J. (1996b). Transcriptional suppression of the transferrin gene by hypolipidemic peroxisome proliferators. *J. Biol. Chem.* **271**, 218–224.

Hertz, R., Magenheim, J., Berman, I., and Bar-Tana, J. (1998). Fatty acyl-CoA thioesters are ligands of hepatic nuclear factor-4-α. *Nature* **392**, 512–516.

Hoekman, M. F., Rientjes, J. M., Twisk, J., Planta, R. J., Princen, H. M., and Mager, W. H. (1993). Transcriptional regulation of the gene encoding cholesterol 7α-hydroxylase in the rat. *Gene* **130**, 217–223.

Hollenberg, A. N., Susulic, V. S., Madura, J. P., Zhang, B., Moller, D. E., Tontonoz, P., Sarraf, P., Spiegelman, B. M., and Lowell, B. B. (1997). Functional antagonism between CCAAT/Enhancer binding protein-α and peroxisome proliferator-activated receptor-γ on the leptin promoter. *J. Biol. Chem.* **272**, 5283–5290.

Honda, S.-I., Morohashi, K.-I., Nomura, M., Takeya, H., Kitajima, M., and Omura, T. (1993). Ad4BP regulating steroidogenic P-450 gene is a member of steroid hormone receptor superfamily. *J. Biol. Chem.* **268**, 7494–7502.

Horwitz, K. B., Jackson, T. A., Rain, D. L., Richer, J. K., Takimoto, G. S., and Tung, L. (1996). Nuclear receptor coactivators and corepressors. *Mol. Endocrinol.* **10**, 1167–1177.

Hoshizaki, D. K., Blackburn, T., Price, C., Ghosh, M., Miles, K., Ragucci, M., and Sweis, R. (1994). Embryonic fat-cell lineage in *Drosophila melanogaster*. *Development* **120**, 2489–2499.

Hu, E., Kim, J. B., Sarraf, P., and Spiegelman, B. M. (1996). Inhibition of adipogenesis through MAP kinase-mediated phosphorylation of PPARγ. *Science* **274**, 2100–2103.

Hua, X., Yokoyama, C., Wu, J., Briggs, M. R., Brown, M. S., Goldstein, J. L., and Wang, X. (1993). SREBP-2, a second basic–helix–loop–helix–leucine zipper protein that stimulates transcription by binding to a sterol regulatory element. *Proc. Natl. Acad. Sci. USA* **90**, 11603–11607.

Hwang, C. S., Loftus, T. M., Mandrup, S., and Lane, M. D. (1997). Adipocyte differentiation and leptin expression. *Ann. Rev. Cell Dev. Biol.* **13**, 231–259.

Hwung, Y.-P., Crowe, D. T., Wang, L.-H., Tsai, S. Y., and Tsai, M.-J. (1988). The COUP transcription factor binds to an upstream promoter element of the rat insulin II gene. *Mol. Cell. Biol.* **8**, 2070–2077.

Hylemon, P. B., Gurley, E. C., Stravitz, R. T., Litz, J. S., Pandak, W. M., Chiang, J. Y., and Vlahcevic, Z. R. (1992). Hormonal regulation of cholesterol 7α-hydroxylase mRNA levels and transcriptional activity in primary rat hepatocyte cultures. *J. Biol. Chem.* **267**, 16866–16871.

Ibeanu, G. C., and Goldstein, J. A. (1995). Transcriptional regulation of human CYP2C genes: functional comparison of CYP2C9 and CYP2C18 promoter regions. *Biochemistry* **34**, 8028–8036.

Ikeda, Y., Lala, D. S., Luo, X., Kim, E., Moisan, M.-P., and Parker, K. L. (1993). Characterization of the mouse FTZ-F1 gene, which encodes a key regulator of steroid hydroxylase gene expression. *Mol. Endocrinol.* **7**, 852–860.

Ikeda, Y., Shen, W.-H., Ingraham, H. A., and Parker, K. L. (1994). Developmental expression of mouse steroidogenic factor-1, an essential regulator of the steroid hydroxylases. *Mol. Endocrinol.* **8**, 654–662.

Imai, E., Stromstedt, P. E., Quinn, P. G., Carlstedt-Duke, J., Gustafsson, J. A., and Granner, D. K. (1990). Characterization of a complex glucocorticoid response unit in the phosphoenolpyruvate carboxykinase gene. *Mol. Cell. Biol.* **10**, 4712–4719.

Ingraham, H. A., Lala, D. S., Ikeda, Y., Luo, X., Shen, W.-H., Nachtigal, M. W., Abbud, R., Nilson, J. H., and Parker, K. L. (1994). The nuclear receptor steroidogenic factor 1 acts at multiple levels of the reproductive axis. *Genes Dev.* **8**, 2302–2312.

Ishibashi, S., Schwarz, M., Frykman, P. K., Herz, J., and Russell, D. W. (1996). Disruption of cholesterol 7α-hydroxylase gene in mice. I. Postnatal lethality reversed by bile acid and vitamin supplementation. *J. Biol. Chem.* **271**, 18017–18023.

Issemann, I., and Green, S. (1990). Activation of a member of the steroid hormone receptor superfamily by peroxisome proliferators. *Nature* **347**, 645–650.

Issemann, I., Prince, R., Tugwood, J., and Green, S. (1992). A role for fatty acids and liver fatty acid binding protein in peroxisome proliferation? *Biochem. Soc. Trans.* **20**, 824–827.

Ito, M., Yu, R., and Jameson, J. L. (1997). Dax-1 inhibits SF-1-mediated transactivation via a carboxy-terminal domain that is deleted in adrenal hypoplasia congenita. *Mol. Cell. Biol.* **17**, 1476–1483.

Ito, M., Yu, R. N., and Jameson, J. L. (1998). Steroidogenic factor-1 contains a carboxy-terminal transcriptional activation domain that interacts with steroid receptor coactivator-1. *Mol. Endocrinol.* **12**, 290–301.

Janowski, B. A., Willy, P. J., Rama Devi, T., Falck, J. R., and Mangelsdorf, D. J. (1996). An oxysterol signalling pathway mediated by the nuclear receptor LXRα. *Nature* **383**, 728–731.

Jelinek, D. F., Andersson, S., Slaughter, C. A., and Russell, D. W. (1990). Cloning and regulation of cholesterol 7α-hydroxylase, the rate-limiting enzyme in bile acid biosynthesis. *J. Biol. Chem.* **265**, 8190–8197.

Jensen, E. V., Jacobson, H. I., Flesher, J. W., Saha, N. N., Gupta, G. N., Smith, S., Colucci, V., Shiplacoff, D., Neuman, H. G., Desombre, E. R., and Jungblut, P. W. (1966). Estrogen receptors in target tissues. *In* "Steroid Dynamics" (G. Pincus, T. Nakao, and J. F. Tait, eds.), pp. 133–156. Academic Press, New York.

Jiang, G., Nepomuceno, L., Hopkins, K., and Sladek, F. M. (1995). Exclusive homodimerization of the orphan receptor hepatocyte nuclear factor 4 defines a new subclass of nuclear receptors. *Mol. Cell. Biol.* **15**, 5131–5143.

Johnson, E. F., Palmer, C. N., Griffin, K. J., and Hsu, M. H. (1996). Role of the peroxisome proliferator-activated receptor in cytochrome P450 4A gene regulation. *FASEB J.* **10**, 1241–1248.

Johnston, S. D., Liu, X., Zuo, F., Eisenbraum, T. L., Wiley, S. R., Kraus, R. J., and Mertz, J. E. (1997). Estrogen-related receptor α1 functionally binds as a monomer to extended half-site sequences including ones contained within estrogen-response elements. *Mol. Endocrinol.* **11**, 342–352.

Jones, M. P., Pandak, W. M., Heuman, D. M., Chiang, J. Y., Hylemon, P. B., and Vlahcevic, Z. R. (1993). Cholesterol 7α-hydroxylase: evidence for transcriptional regulation by cholesterol or metabolic products of cholesterol in the rat. *J. Lipid Res.* **34**, 885–892.

Juge-Aubry, C. E., Gorla-Bajszczak, A., Pernin, A., Lemberger, T., Wahli, W., Burger, A. G., and Meier, C. A. (1995). Peroxisome proliferator-activated receptor mediates cross-talk with thyroid hormone receptor by competition for retinoid X receptor. Possible role of a leucine zipper-like heptad repeat. *J. Biol. Chem.* **270**, 18117–18122.

Juge-Aubry, C., Pernin, A., Favez, T., Burger, A. G., Wahli, W., Meier, C. A., and Desvergne, B. (1997). DNA binding properties of peroxisome proliferator-activated receptor subtypes on various natural peroxisome proliferator response elements. Importance of the 5′-flanking region. *J. Biol. Chem.* **272**, 25252–25259.

Jump, D. B., Clarke, S. D., Thelen, A., and Liimatta, M. (1994). Coordinate regulation of glycolytic and lipogenic gene expression by polyunsaturated fatty acids. *J. Lipid Res.* **35**, 1076–1084.

Kandutsch, A. A., Chen, H. W., and Heiniger, H. J. (1978). Biological activity of some oxygenated sterols. *Science* **201**, 498–501.

Karin, M. (1998). New twists in gene regulation by glucocorticoid receptor: is DNA binding dispensable? *Cell* **93**, 487–490.

Katagiri, Y., Hirata, Y., Milbrandt, J., and Guroff, G. (1997). Differential regulation of the transcriptional activity of the orphan nuclear receptor NGFI-B by membrane depolarization and nerve growth factor. *J. Biol. Chem.* **272**, 31278–31284.

Keller, H., Dreyer, C., Medin, J., Mahfoudi, A., Ozato, K., and Wahli, W. (1993). Fatty acids and retinoids control lipid metabolism through activation of peroxisome proliferator-activated receptor-retinoid X receptor heterodimers. *Proc. Natl. Acad. Sci. USA* **90**, 2160–2164.

Kelly, D. P., Gordon, J. L., and Strauss, A. W. (1989). The tissue-specific expression and developmental regulation of two nuclear genes encoding rat mitochondrial proteins: medium-chain acyl CoA dehydrogenase and mitochondrial malate dehydrogenase. *J. Biol. Chem.* **264**, 18921–18925.

Kim, J. B., and Spiegelman, B. M. (1996). ADD1/SREBP1 promotes adipocyte differentiation and gene expression linked to fatty acid metabolism. *Genes Dev.* **10**, 1096–1107.

Kimura, A., Nishiyori, A., Murakami, T., Tsukamoto, T., Hata, S., Osumi, T., Okamura, R., Mori, M., and Takiguchi, M. (1993). Chicken ovalbumin upstream promoter-transcription factor (COUP-TF) represses transcription from the promoter of the gene for ornithine transcarbamylase in a manner antagonistic to hepatocyte nuclear factor-4 (HNF-4). *J. Biol. Chem.* **268**, 11125–11133.

Kliewer, S. A., Umesono, K., Heyman, R. H., Mangelsdorf, D. J., Dyck, J. A., and Evans, R. M. (1992a). Retinoid X receptor-COUP-TF interactions modulate retinoic acid signaling. *Proc. Natl. Acad. Sci. USA* **89**, 1448–1452.

Kliewer, S. A., Umesono, K., Noonan, D. J., Heyman, R. A., and Evans, R. M. (1992b). Convergence of 9-cis retinoic acid and peroxisome proliferator signalling pathways through heterodimer formation of their receptors. *Nature* **358**, 771–774.

Kliewer, S. A., Forman, B. M., Blumberg, B., Ong, E. S., Borgmeyer, U., Mangelsdorf, D. J., Umesono, K., and Evans, R. M. (1994). Differential expression and activation of a family of murine peroxisome proliferator-activated receptors. *Proc. Natl. Acad. Sci. USA* **91**, 7355–7359.

Kliewer, S. A., Lenhard, J. M., Willson, T. M., Patel, I., Morris, D. C., and Lehmann, J. M. (1995). A prostaglandin J2 metabolite binds peroxisome proliferator-activated receptor γ and promotes adipocyte differentiation. *Cell* **83**, 813–819.

Kliewer, S. A., Sundseth, S. S., Jones, S. A., Brown, P. J., Wisely, G. B., Koble, C. S., Devchand, P., Wahli, W., Willson, T. M., Lenhard, J. M., and Lehmann, J. M. (1997). Fatty acids and eicosanoids regulate gene expression through direct interactions with peroxisome proliferator-activated receptors alpha and gamma. *Proc. Natl. Acad. Sci. USA* **94**, 4318–4323.

Kliewer, S. A., Moore, J. T., Wade, L., Staudinger, J. L., Watson, M. A., Jones, S. A., McKee, D. D., Oliver, B. B., Wilson, T. M., Zetterström, R. H., Perlmann, T., and Lehmann, J. M. (1998). An orphan nuclear receptor activated by pregnanes defines a novel steroid signaling pathway. *Cell* **92**, 73–82.

Klinge, C. M., Silver, B. F., Driscoll, M. D., Sathya, G., Bambara, R. A., and Hilf, R. (1997). Chicken ovalbumin upstream promoter-transcription factor interacts with estrogen receptor, binds to estrogen response elements and half-sites, and inhibits estrogen-induced gene expression. *J. Biol. Chem.* **272**, 31465–31474.

Krey, G., Keller, H., Mahfoudi, A., Medin, J., Ozato, K., Dreyer, C., and Wahli, W. (1993). Xenopus peroxisome proliferator activated receptors: genomic organization, response element recognition, heterodimer formation with retinoid X receptor and activation by fatty acids. *J. Steroid Biochem. Molec. Biol.* **47**, 65–73.

Krey, G., Mahfoudi, A., and Wahli, W. (1995). Functional interactions of peroxisome proliferator-activated receptor, retinoid-X-receptor, and Sp1 in the transcriptional regulation of the acyl-coenzyme-A oxidase promoter. *Mol. Endocrinol.* **9**, 219–231.

Krey, G., Braissant, O., Lhorset, F., Kalkhoven, E., Perroud, M., Parker, M. G., and Wahli, W. (1997). Fatty acids, eicosanoids, and hypolipidemic agents identified as ligands of peroxisome proliferator-activated receptors by coactivator-dependent receptor ligand assay. *Mol. Endocrinol.* **11**, 779–791.

Kritis, A. A., Argyrokastritis, A., Moschonas, N. K., Power, S., Katrakili, N., Zannis, V. I., Cereghini, S., and Talianidis, I. (1996). Isolation and characterization of a third isoform of human hepatocyte nuclear factor 4. *Gene* **173**, 275–280.

Kruszynska, Y. T., McCormack, J. G., and McIntyre, N. (1990). Effects of non-esterified fatty acid availability on insulin stimulated glucose utilization and tissue pyruvate dehydrogenase activity in the rat. *Diabetologia* **33**, 396–402.

Ktistaki, E., Lacorte, J.-M., Katrakilli, N., Zannis, V. I., and Talianidis, I. (1994). Transcriptional regulation of the apolipoprotein A-IV gene involves synergism between a proximal orphan receptor element and a distant enhancer located in the upstream promoter region of the apolipoprotein C-III gene. *Nucleic Acids Res.* **22**, 4689–4696.

Kurebayashi, S., Hirose, T., Miyashita, Y., Kasayama, S., and Kishimoto, T. (1997). Thiazolidinediones downregulate stearoyl-CoA desaturase 1 gene expression in 3T3-L1 adipocytes. *Diabetes* **46**, 2115–2118.

Labelle, Y., Zucman, J., Stenman, G., Kindblom, L. G., Knight, J., Turccarel, C., Dockhorndworniczak, B., Mandahl, N., Desmaze, C., Peter, M., Aurias, A., Delattre, O., and Thomas, G. (1995). Oncogenic conversion of a novel orphan nuclear receptor by chromosome translocation. *Hum. Mol. Genet.* **4**, 2219–2226.

Ladias, J. A. A., and Karathanasis, S. K. (1991). Regulation of the apolipoprotein AI gene by ARP-1, a novel member of the steroid receptor superfamily. *Science* **251**, 561–565.

Ladias, J. A., Hadzopoulou-Cladaras, M., Kardassis, D., Cardot, P., Cheng, J., Zannis, V., and Cladaras, C. (1992). Transcriptional regulation of human apolipoprotein genes ApoB, ApoCIII, and ApoAII by members of the steroid hormone receptor superfamily HNF-4, ARP-1, EAR-2, and EAR-3. *J. Biol. Chem.* **267**, 15849–15860.

Lala, D. S., Rice, D. A., and Parker, K. L. (1992). Steroidogenic factor I, a key regulator of steroidogenic enzyme expression, is the mouse homolog of *fushi tarazu*-factor I. *Mol. Endocrinol.* **6**, 1249–1258.

Lala, D. S., Syka, P. M., Lazarchik, S. B., Mangelsdorf, D. J., Parker, K. L., and Heyman, R. A. (1997). Activation of the orphan nuclear receptor steroidogenic factor 1 by oxysterols. *Proc. Natl. Acad. Sci. USA* **94**, 4895–4900.

Lavery, D. J., and Schibler, U. (1993). Circadian transcription of the cholesterol 7α-hydroxylase gene may involve the liver-enriched bZIP protein DBP. *Genes Dev.* **7**, 1871–1884.

Law, S. W., Conneely, O. M., DeMayo, F. J., and O'Malley, B. W. (1992). Identification of a new brain-specific transcription factor, NURR1. *Mol. Endocrinol.* **6**, 2129–2135.

Lee, S. S.-T., Pineau, T., Drago, J., Lee, E. J., Owens, J. W., Kroetz, D. L., Fernandez-Salguero, P. M., Westphal, H., and Gonzalez, F. J. (1995). Targeted disruption of the α isoform of the peroxisome proliferator-activated receptor gene in mice results in abolishment of the pleiotropic effects of peroxisome proliferators. *Mol. Cell. Biol.* **15**, 3012–3022.

Leers-Sucheta, S., Morohashi, K., Mason, J. I., and Melner, M. H. (1997). Synergistic activation of the human type II 3β-hydroxysteroid dehydrogenase/δ5-δ4 isomerase promoter by the transcription factor steroidogenic factor-1/adrenal 4-binding protein and phorbol ester. *J. Biol. Chem.* **272**, 7960–7967.

Lefstin, J. A., and Yamamoto, K. R. (1998). Allosteric effects of DNA on transcriptional regulators. *Nature* **392**, 885–888.

Legraverend, C., Eguchi, H., Ström, A., Lahuna, O., Mode, A., Tollet, P., Westin, S., and Gustafsson, J.-Å. (1994). Transactivation of the rat CYP2C13 gene promoter involves HNF-1, HNF-3, and members of the orphan receptor subfamily. *Biochemistry* **33**, 9889–9897.

Lehmann, J. M., Moore, L. B., Smith-Oliver, T. A., Wilkinson, W. O., Wilson, T. M., and Kliewer, S. A. (1995). An antidiabetic thiazolidinedione is a high affinity ligand for peroxisome-activated receptor gamma. *J. Biol. Chem.* **270**, 12953–12956.

Lehmann, J. M., Kliewer, S. A., Moore, L. B., Smith-Oliver, T. A., Oliver, B. B., Su, J. L., Sundseth, S. S., Winegar, D. A., Blanchard, D. E., Spencer, T. A., and Willson, T. M. (1997a). Activation of the nuclear receptor LXR by oxysterols defines a new hormone response pathway. *J. Biol. Chem.* **272**, 3137–3140.

Lehmann, J. M., Lenhard, J. M., Oliver, B. B., Ringold, G. M., and Kliewer, S. A. (1997b). Peroxisome proliferator-activated receptors alpha and gamma are activated by indomethacin and other non-steroidal anti-inflammatory drugs. *J. Biol. Chem.* **272**, 3406–3410.

Lemberger, T., Staels, B., Saladin, R., Desvergne, B., Auwerx, J., and Wahli, W. (1994). Regulation of the peroxisome proliferator-activated receptor alpha gene by glucocorticoids. *J. Biol. Chem.* **269**, 24527–24530.

Lemberger, T., Desvergne, B., and Wahli, W. (1996a). Peroxisome proliferator-activated receptors: a nuclear receptor signaling pathway in lipid physiology. *Ann. Rev. Cell Dev. Biol.* **12**, 335–363.

Lemberger, T., Saladin, R., Vazquez, M., Assimacopoulos, F., Staels, B., Desvergne, B., Wahli, W., and Auwerx, J. (1996b). Expression of the peroxisome proliferator-activated receptor alpha gene is stimulated by stress and follows a diurnal rhythm. *J. Biol. Chem.* **271**, 1764–1769.

Leone, T. C., Cresci, S., Carter, M. E., Zhang, Z. F., Lala, D. S., Strauss, A. W., and Kelly, D. P. (1995). The human medium chain acyl-CoA dehydrogenase gene promoter consists of a complex arrangement of nuclear receptor response elements and Sp1 binding sites. *J. Biol. Chem.* **270**, 16308–16314.

Li, Y. C., Wang, D. P., and Chiang, J. Y. (1990). Regulation of cholesterol 7α-hydroxylase in the liver. Cloning, sequencing, and regulation of cholesterol 7α-hydroxylase mRNA. *J. Biol. Chem.* **265**, 12012–12019.

Li, Q., Yamamoto, N., Morisawa, S., and Inoue, A. (1993). Fatty acyl-CoA binding activity of the nuclear thyroid hormone receptor. *J. Cell. Biochem.* **51**, 458–464.

Liimatta, M., Towle, H. C., Clarke, S., and Jump, D. B. (1994). Dietary polyunsaturated fatty acids interfere with the insulin/glucose activation of L-type pyruvate kinase gene transcription. *Mol. Endocrinol.* **8**, 1147–1153.

Lim, R. W., Yang, W. L., and Yu, H. (1995). Signal-transduction-pathway-specific desensitization of expression of orphan nuclear receptor TIS1. *Biochem. J.* **308**, 785–789.

Liu, Y. H., and Teng, C. T. (1991). Characterization of estrogen-responsive mouse lactoferrin promoter. *J. Biol. Chem.* **266**, 21880–21885.

Liu, Z., and Towle, H. C. (1995). Functional synergism in the carbohydrate-induced activation of liver-type pyruvate kinase gene expression. *Biochem. J.* **308**, 105–111.

Liu, Z., Thompson, K. S., and Towle, H. C. (1993). Carbohydrate regulation of the rat L-type pyruvate kinase gene requires two nuclear factors: LF-A1 and a member of the c-myc family. *J. Biol. Chem.* **268**, 12787–12795.

Liu, Z.-G., Smith, S. W., McLaughlin, K. A., Schwartz, L. M., and Osborne, B. A. (1994). Apoptotic signals delivered through the T-cell receptor of a T-cell hybrid require the immediate-early gene nur77. *Nature* **367**, 281–284.

Lucas, P. C., O'Brien, R. M., Mitchell, J. A., Davis, C. M., Imai, E., Forman, B. C., Samuels, H. H., and Granner, D. K. (1991). A retinoic acid response element is part of a pleiotropic domain in the phosphoenolpyruvate carboxykinase gene. *Proc. Natl. Acad. Sci. USA* **88**, 2184–2188.

Luo, X., Ikeda, Y., and Parker, K. L. (1994). A cell-specific nuclear receptor is essential for adrenal and gonadal development and sexual differentiation. *Cell* **77**, 481–490.

Luo, X., Ikeda, Y., Schlosser, D. A., and Parker, K. L. (1995). Steroidogenic factor 1 is the essential transcript of the mouse Ftz-F1 gene. *Mol. Endocrinol.* **9**, 1233–1239.

Luo, J., Sladek, R., Bader, J.-A., Rossant, J., and Giguère, V. (1997). Placental abnormalities in mouse embryos lacking orphan nuclear receptor ERRβ. *Nature* **388**, 778–782.

Lutz, B., Kuratani, S., Cooney, A. J., Wawersik, S., Tsai, S. Y., Eichele, G., and Tsai, M.-J. (1994). Developmental regulation of the orphan receptor COUP-TF II gene in spinal motor neurons. *Development* **120**, 25–36.

Lydon, J. P., Power, R. F., and Conneely, O. M. (1992). Differential modes of activation define orphan subclasses within the steroid/thyroid receptor superfamily. *Gene Expression* **2**, 273–283.

Lynch, J. P., Lala, D. S., Peluso, J. J., Luo, W., Parker, K. L., and White, B. A. (1993). Steroidogenic factor 1, an orphan nuclear receptor, regulates the expression of the rat aromatase gene in gonadal tissues. *Mol. Endocrinol.* **7**, 776–786.

Mages, H. W., Rilke, O., Bravo, R., Senger, G., and Kroczek, R. A. (1994). NOT, a human immediate-early response gene closely related to the steroid/thyroid hormone receptor NAK1/TR3. *Mol. Endocrinol.* **8**, 1583–1591.

Marten, N. W., Sladek, F. M., and Straus, D. S. (1996). Effect of dietary protein restriction on liver transcription factors. *Biochem. J.* **317**, 361–370.

Martin, G., Schoonjans, K., Lefebvre, A. M., Staels, B., and Auwerx, J. (1997). Coordinate regulation of the expression of the fatty acid transport protein and acyl-CoA synthetase genes by PPARα and PPARγ activators. *J. Biol. Chem.* **272**, 28210–28217.

Maruyama, K., Tsukada, T., Bandoh, S., Sasaki, K., Ohkura, N., and Yamaguchi, K. (1997). Expression of the putative transcription factor NOR-1 in the nervous, the endocrine and the immune systems and the developing brain of the rat. *Neuroendocrinology* **65**, 2–8.

Mascaro, C., Acosta, E., Ortiz, J. A., Marrero, P. F., Hegardt, F. G., and Haro, D. (1998). Control of human muscle-type carnitine palmitoyltransferase I gene transcription by peroxisome proliferator-activated receptor. *J. Biol. Chem.* **273**, 8560–8563.

McGrane, M. M., de Vente, J., Yun, J., Bloom, J., Park, E., Wynshaw-Boris, A., Wagner, T., Rottman, F. M., and Hanson, R. W. (1988). Tissue-specific expression and dietary regulation of a chimeric phosphoenolpyruvate carboxykinase/bovine growth hormone gene in transgenic mice. *J. Biol. Chem.* **263**, 11443–11451.

McGrane, M. M., Yun, J. S., Moorman, A. F., Lamers, W. H., Hendrick, G. K., Arafah, B. M., Park, E. A., Wagner, T. E., and Hanson, R. W. (1990). Metabolic effects of developmental, tissue-, and cell-specific expression of a chimeric phosphoenolpyruvate carboxykinase (GTP)/bovine growth hormone gene in transgenic mice. *J. Biol. Chem.* **265**, 22371–22379.

Metzger, S., Halaas, J. L., Breslow, J. L., and Sladek, F. M. (1993). Orphan receptor HNF-4 and bZip protein C/EBP alpha bind to overlapping regions of the apolipoprotein B gene promoter and synergistically activate transcription. *J. Biol. Chem.* **268**, 16831–16838.

Meyer, S., Hoppner, W., and Seitz, H. J. (1991). Transcriptional and post-transcriptional effects of glucose on liver phosphoenolpyruvate-carboxykinase gene expression. *Eur. J. Biochem.* **202**, 985–991.

Mietus-Snyder, M., Sladek, F. M., Ginsburg, G. S., Kuo, C. F., Ladias, J. A. A., Darnell, J. E., and Karathanasis, S. K. (1992). Antagonism between Apolipoprotein AI regulatory protein 1, Ear3/COUP-TF, and hepatocyte nuclear factor 4 modulates Apolipoprotein CIII gene expression in liver and intestinal cells. *Mol. Cell. Biol.* **12**, 1708–1718.

Milbrandt, J. (1988). Nerve growth factor induces a gene homologous to the glucocorticoid receptor gene. *Neuron* **1**, 183–188.

Miller, C. W., and Ntambi, J. M. (1996). Peroxisome proliferators induce mouse liver stearoyl-CoA desaturase 1 gene expression. *Proc. Natl. Acad. Sci. USA* **93**, 9443–9448.

Miyata, K. S., Zhang, B., Marcus, S. L., Capone, J. P., and Rachubinski, R. A. (1993). Chicken ovalbumin upstream promoter transcription factor (COUP-TF) binds to a peroxisome proliferator-responsive element and antagonizes peroxisome proliferator-mediated signaling. *J. Biol. Chem.* **268**, 19169–19172.

Miyata, M., Nagata, K., Yamazoe, Y., and Kato, R. (1995). Transcriptional elements directing a liver-specific expression of P450/6βA (CYP3A2) gene-encoding testosterone 6β-hydroxylase. *Arch. Biochem. Biophys.* **318**, 71–79.

Miyata, K. S., McCaw, S. E., Patel, H. V., Rachubinski, R. A., and Capone, J. P. (1996). The orphan nuclear hormone receptor LXRα interacts with the peroxisome proliferator-activated receptor and inhibits peroxisome proliferator signaling. *J. Biol. Chem.* **271**, 9189–9192.

Morohashi, K., Iida, H., Nomura, M., Hatano, O., Honda, S., Tsukiyama, T., Niwa, O., Hara, T., Takakusu, A., Shibata, Y., and Omura, T. (1994). Functional difference between Ad4BP and ELP, and their distributions in steroidogenic tissues. *Mol. Endocrinol.* **8**(5), 643–653.

Mukherjee, R., Davies, P. J. A., Crombie, D. L., Bischoff, E. D., Cesario, R. M., Jow, L., Hamann, L. G., Boehm, M. F., Mondon, C. E., Nadzan, A. M., Paterniti, J. R., and Heyman, R. A. (1997). Sensitization of diabetic and obese mice to insulin by retinoid X receptor agonists. *Nature* **386**, 407–410.

Muscat, G. E. O., Rea, S., and Downes, M. (1995). Identification of a regulatory function for an orphan receptor in muscle: COUP-TF II affects the expression of the myoD gene family during myogenesis. *Nucleic Acids Res.* **23**, 1311–1318.

Muscatelli, F., Strom, T. M., Walker, A. P., Zanaria, E., Récan, D., Meindl, A., Bardoni, B., Guioli, S., Zehetner, G., Rabi, W., Schwarz, H. P., Kaplan, J.-C., Camerino, G., Meitinger, T., and Monaco, A. P. (1994). Mutations in the DAX-1 gene give rise to both X-linked adrenal hypoplasia congenita and hypogonadotropic hypogonadism. *Nature* **372**, 672–676.

Myant, N. B., and Mitropoulos, K. A. (1977). Cholesterol 7α-hydroxylase. *J. Lipid Res.* **18**, 135–153.

Nachtigal, M. W., Hirokawa, Y., Enyeart-VanHouten, D. L., Flanagan, J. N., Hammer, G. D., and Ingraham, H. A. (1998). Wilms' tumor 1 and Dax-1 modulate the orphan nuclear receptor SF-1 in sex-specific gene expression. *Cell* **93**, 445–454.

Nagao, M., Parimoo, B., and Tanaka, K. (1993). Developmental, nutritional, and hormonal regulation of tissue-specific expression of the genes encoding various acyl-CoA dehydrogenases and α-subunit of electron transfer flavoprotein in rat. *J. Biol. Chem.* **268**, 24114–24124.

Nagy, L., Kao, H. Y., Chakravarti, D., Lin, R. J., Hassig, C. A., Ayer, D. E., Schreiber, S. L., and Evans, R. M. (1997). Nuclear receptor repression mediated by a complex containing SMRT, mSin3a, and histone deacetylase. *Cell* **89**, 373–380.

Nakai, A., Kartha, S., Sakurai, A., Toback, F. G., and DeGroot, L. J. (1990). A human early response gene homologous to murine nur77 and rat NGFI-B, and related to the nuclear receptor superfamily. *Mol. Endocrinol.* **4**, 1438–1443.

Nakajima, H., Yoshiuchi, I., Hamaguchi, T., Tomita, K., Yamasaki, T., Iizuka, K., Okita, K., Moriwaki, M., Ono, A., Oue, T., Horikawa, Y., Shingu, R., Miyagawa, J., Namba, M., Hanafusa, T., and Matsuzawa, Y. (1996). Hepatocyte nuclear factor-4 α gene mutations in Japanese non-insulin dependent diabetes mellitus (NIDDM) patients. *Res. Comm. Mol. Pathol. Pharmacol.* **94**, 327–330.

Nakshatri, H., and Chambon, P. (1994). The directly repeated RG(G/T)TCA motifs of the rat and mouse cellular retinol-binding protein II genes are promiscuous binding sites for RAR, RXR, HNF-4, and ARP-1 homo- and heterodimers. *J. Biol. Chem.* **269**, 890–902.

Ness, G. C., Pendleton, L. C., Li, Y. C., and Chiang, J. Y. (1990). Effect of thyroid hormone on hepatic cholesterol 7α-hydroxylase, LDL receptor, HMG-CoA reductase, farnesyl pyrophosphate synthetase and apolipoprotein A-I mRNA levels in hypophysectomized rats. *Biochem. Biophys. Res. Commun.* **172**, 1150–1156.

Ninomiya, Y., Okada, M., Kotomura, N., Suzuki, K., Tsukiyama, T., and Niwa, O. (1995). Genomic organization and isoforms of the mouse ELP gene. *J. Biochem. (Tokyo)* **118**, 380–389.

Nishiyori, A., Tashiro, H., Kimura, A., Akagi, K., Yamamura, K., Mori, M., and Takiguchi, M. (1994). Determination of tissue specificity of the enhancer by combinatorial operation of tissue-enriched transcription factors. Both HNF-4 and C/EBP beta are required for liver-specific activity of the ornithine transcarbamylase enhancer. *J. Biol. Chem.* **269**, 1323–1331.

Nitsch, D., and Schutz, G. (1993). The distal enhancer implicated in the developmental regulation of the tyrosine aminotransferase gene is bound by liver-specific and ubiquitous factors. *Mol. Cell. Biol.* **13**, 4494–4504.

Nomura, M., Nawata, H., and Morohashi, K. (1996). Autoregulatory loop in the regulation of the mammalian ftz-f1 gene. *J. Biol. Chem.* **271**, 8243–8249.

Noshiro, M., Nishimoto, M., and Okuda, K. (1990). Rat liver cholesterol 7α-hydroxylase. Pretranslational regulation for circadian rhythm. *J. Biol. Chem.* **265**, 10036–10041.

O'Brien, R. M., Noisin, E. L., Suwanichkul, A., Yamasaki, T., Lucas, P. C., Wang, J. C., Powell, D. R., and Granner, D. K. (1995). Hepatic nuclear factor 3- and hormone-regulated expression of the phosphoenolpyruvate carboxykinase and insulin-like growth factor- binding protein 1 genes. *Mol. Cell. Biol.* **15**, 1747–1758.

O'Malley, B. W., Schrader, W. T., Mani, S., Smith, C., Weigel, N. L., Connelley, O. M., and Clark, J. H. (1995). An alternative ligand-independent pathway for activation of steroid receptors. *Rev. Prog. Horm. Res.* **50**, 333–347.

Ogiwara, H., Tanabe, T., Nikawa, J., and Numa, S. (1978). Inhibition of rat-liver acetyl-coenzyme-A carboxylase by palmitoyl- coenzyme A. Formation of equimolar enzyme–inhibitor complex. *Eur. J. Biochem.* **89**, 33–41.

Ohkura, N., Hijikuro, M., Yamamoto, A., and Miki, K. (1994). Molecular cloning of a novel thyroid/steroid receptor superfamily gene from cultured rat neuronal cells. *Biochem. Biophys. Res. Commun.* **205**, 1959–1965.

Palmer, C. N. A., Hsu, M. H., Griffin, K. J., and Johnson, E. F. (1995). Novel sequence determinants in peroxisome proliferator signaling. *J. Biol. Chem.* **270**, 16114–16121.

Pandak, W. M., Li, Y. C., Chiang, J. Y., Studer, E. J., Gurley, E. C., Heuman, D. M., Vlahcevic, Z. R., and Hylemon, P. B. (1991). Regulation of cholesterol 7α-hydroxylase mRNA and transcriptional activity by taurocholate and cholesterol in the chronic biliary diverted rat. *J. Biol. Chem.* **266**, 3416–3421.

Parker, K. L., and Schimmer, B. P. (1997). Steroidogenic factor 1: a key determinant of endocrine development and function. *Endocr. Rev.* **18**, 361–377.

Peet, D. J., Turley, S. D., Ma, A., Janowski, B. A., Lobaccaro, J.-M. A., Hammer, R. E., and Mangelsdorf, D. J. (1998). Cholesterol and bile acid metabolism are impaired in mice lacking the nuclear oxysterol receptor LXRα. *Cell* **93**, 693–704.

Perlmann, T., and Jansson, L. (1995). A novel pathway for vitamin A signaling mediated by RXR heterodimerization with NGFI-B and NURR1. *Genes Dev.* **9**, 769–782.

Perlmann, T., Rangarajan, P. N., Umesono, K., and Evans, R. M. (1993). Determinants for selective RAR and TR recognition of direct repeat HREs. *Genes Dev.* **7**, 1411–1422.

Peters, J. M., Zhou, Y. C., Ram, P. A., Lee, S. S., Gonzalez, F. J., and Waxman, D. J. (1996). Peroxisome proliferator-activated receptor α required for gene induction by dehydroepian-drosterone-3 β-sulfate. *Mol. Pharmacol.* **50**, 67–74.

Pettersson, K., Svensson, K., Mattsson, R., Carlsson, B., Ohlsson, R., and Berkenstam, A. (1996). Expression of a novel member of estrogen response element-binding nuclear receptors is restricted to the early stages of chorion formation during mouse embryogenesis. *Mech. Dev.* **54**, 211–223.

Philips, A., Lesage, S., Gingras, R., Maira, M. H., Gauthier, Y., Hugo, P., and Drouin, J. (1997). Novel dimeric Nur77 signaling mechanism in endocrine and lymphoid cells. *Mol. Cell. Biol.* **17**, 5946–5951.

Pilkis, S. J., and Granner, D. K. (1992). Molecular physiology of the regulation of hepatic gluconeogenesis and glycolysis. *Ann. Rev. Phys.* **54**, 885–909.

Puzenat, N., Vaulont, S., Kahn, A., and Raymondjean, M. (1992). Combinatorial crosstalk of transacting factors binding to the L-type pyruvate kinase promoter elements analysed in vitro. *Biochem. Biophys. Res. Com.* **189**, 1119–1128.

Quattrochi, L. C., Mills, A. S., Barwick, J. L., Yockey, C. B., and Guzelian, P. S. (1995). A novel cis-acting element in a liver cytochrome P-450 3A gene confers synergistic induction by glucocorticoids plus antiglucocorticoids. *J. Biol. Chem.* **270**, 2007–2012.

Raisher, B. D., Gulick, T., Zhang, Z., Strauss, A. W., Moore, D. D., and Kelly, D. P. (1992). Identification of a novel retinoid-responsive element in the promoter region of the medium chain acyl-coenzyme A dehydrogenase gene. *J. Biol. Chem.* **267**, 20264–20269.

Ramirez, M. I., Karaoglu, D., Haro, D., Barillas, C., Bashirzadeh, R., and Gil, G. (1994). Cholesterol and bile acids regulate cholesterol 7α-hydroxylase expression at the transcriptional level in culture and in transgenic mice. *Mol. Cell. Biol.* **14**, 2809–2821.

Rastinejad, F., Perlmann, T., Evans, R. M., and Sigler, P. B. (1995). Structural determinants of nuclear receptor assembly on DNA direct repeats. *Nature* **375**, 203–211.

Reichardt, H. M., Kaestner, K. H., Tuckermann, J., Kretz, O., Wessely, O., Bock, R., Gass, P., Schmid, W., Herrlich, P., Angel, P., and Schütz, G. (1998). DNA binding of the glucocorticoid receptor is not essential for survival. *Cell* **93**, 531–541.

Reijnen, M. J., Sladek, F. M., Bertina, R. M., and Reitsma, P. H. (1992). Disruption of a binding site for hepatocyte nuclear factor 4 results in hemophilia B Leyden. *Proc. Natl. Acad. Sci. USA* **89**, 6300–6303.

Ren, B., Thelen, A., and Jump, D. B. (1996). Peroxisome proliferator-activated receptor α inhibits hepatic S14 gene transcription. Evidence against the peroxisome proliferator-activated receptor α as the mediator of polyunsaturated fatty acid regulation of S14 gene transcription. *J. Biol. Chem.* **271**, 17167–17173.

Reue, K., Leff, T., and Breslow, J. L. (1988). Human apolipoprotein CIII gene expression is regulated by positive and negative cis-acting elements and tissue-specific protein factors. *J. Biol. Chem.* **263**, 6857–6864.

Rice, D. A., Kirkman, M. S., Aitken, L. D., Mouw, A. R., Schimmer, B. P., and Parker, K. L. (1990). Analysis of the promoter region of the gene encoding mouse cholesterol side-chain cleavage enzyme. *J. Biol. Chem.* **265**, 11713–11720.

Rodriguez, J. C., Gil-Gomez, G., Hegardt, F. G., and Haro, D. (1994). Peroxisome proliferator-activated receptor mediates induction of the mitochondrial 3-hydroxy-3-methylglutaryl-CoA synthase gene by fatty acids. *J. Biol. Chem.* **269**, 18767–18772.

Rodriguez, J. C., Ortiz, J. A., Hegardt, F. G., and Haro, D. (1997). Chicken ovalbumin upstream-promoter transcription factor (COUP-TF) could act as a transcriptional activator or repressor of the mitochondrial 3-hydroxy-3-methylglutaryl-CoA synthase gene. *Biochem. J.* **326**, 587–592.

Rodriguez, J. C., Ortiz, J. A., Hegardt, F. G., and Haro, D. (1998). The hepatocyte nuclear factor 4 (HNF-4) represses the mitochondrial HMG-CoA synthase gene. *Biochem. Biophys. Res. Com.* **242**, 692–696.

Roe, C. R., and Coates, P. M. (1995). Mitochondrial fatty acid oxidation disorders. *In* "The Metabolic and Molecular Bases of Inherited Disease" (C. R. Scriver, ed.). McGraw-Hill, New York.

Rottman, J. N., and Gordon, J. I. (1993). Comparison of the patterns of expression of rat intestinal fatty acid binding protein/human growth hormone fusion genes in cultured intestinal epithelial cell lines and in the gut epithelium of trangenic mice. *J. Biol. Chem.* **268**, 11994–12002.

Rudling, M., Parini, P., and Angelin, B. (1997). Growth hormone and bile acid synthesis. Key role for the activity of hepatic microsomal cholesterol 7α-hydroxylase in the rat. *J. Clin. Invest.* **99**, 2239–2245.

Russell, D. W., and Setchell, K. D. (1992). Bile acid biosynthesis. *Biochemistry* **31**, 4737–4749.

Ryseck, R.-P., Macdonald-Bravo, H., Mattéi, M.-G., Ruppert, S., and Bravo, R. (1989). Structure, mapping and expression of a growth factor inducible gene encoding a putative nuclear hormonal binding receptor. *EMBO J.* **11**, 3327–3335.

Sack, M. N., Disch, D. L., Rockman, H. A., and Kelly, D. P. (1997). A role for Sp1 and nuclear receptor transcription factors in a cardiac hypertrophic growth program. *Proc. Natl. Acad. Sci. USA* **94**, 6438–6443.

Sadovsky, Y., Crawford, P. A., Woodson, K. G., Polish, J. A., Clements, M. A., Tourtellotte, L. M., Simburger, K., and Milbrandt, J. (1995). Mice deficient in the orphan receptor steroidogenic factor I lack adrenal glands and gonads but express P450 side-chain-cleavage enzyme in the placenta and have normal embryonic serum levels of corticosteroids. *Proc. Natl. Acad. Sci. USA* **92**, 10939–10943.

Scearce, L. M., Laz, T. M., Hazel, T. G., Lau, L. F., and Taub, R. (1993). RNR-1, a nuclear receptor in the NGFI-B/Nurr77 family that is rapidly induced in regenerating liver. *J. Biol. Chem.* **268**, 8855–8861.

Schaeffer, E., Guillou, F., Part, D., and Zakin, M. M. (1993). A different combination of transcription factors modulates the expression of the human transferrin promoter in liver and Sertoli cells. *J. Biol. Chem.* **268**, 23399–23408.

Schoonjans, K., Watanabe, M., Suzuki, H., Mahfoudi, A., Krey, G., Wahli, W., Grimaldi, P., Staels, B., Yamamoto, T., and Auwerx, J. (1995). Induction of the acyl-coenzyme A synthetase gene by fibrates and fatty acids is mediated by a peroxisome proliferator response element in the C promoter. *J. Biol. Chem.* **270**, 19269–19276.

Schoonjans, K., Peinado-Onsurbe, J., Lefebvre, A.-M., Heyman, R. A., Briggs, M., Deeb, S., Staels, B., and Auwerx, J. (1996a). PPARα and PPARγ activators direct a distinct tissue-specific transcriptional response via a PPRE in the lipoprotein lipase gene. *EMBO J.* **15**, 5336–5348.

Schoonjans, K., Staels, B., and Auwerx, J. (1996b). The peroxisome proliferator activated receptors (PPARs) and their effects on lipid metabolism and adipocyte differentiation. *Biochim. Biophys. Acta* **1302**, 93–109.

Schulz, H. (1991). Beta oxidation of fatty acids. *Biochim. Biophys. Acta* **1081**, 109–120.

Schwarz, M., Lund, E. G., Setchell, K. D. R., Kayden, H. J., Zerwekh, J. E., Bjorkhem, I., Herz, J., and Russell, D. W. (1996). Disruption of cholesterol 7α-hydroxylase gene in mice. II. Bile acid deficiency is overcome by induction of oxysterol 7α-hydroxylase. *J. Biol. Chem.* **271**, 18024–18031.

Scott, D. K., Mitchell, J. A., and Granner, D. K. (1996). The orphan receptor COUP-TF binds to a third glucocorticoid accessory factor element within the phosphoenolpyruvate carboxykinase gene promoter. *J. Biol. Chem.* **271**, 31909–31914.

Scott, D. K., Stromstedt, P. E., Wang, J. C., and Granner, D. K. (1998). Further characterization of the glucocorticoid response unit in the phosphoenolpyruvate carboxykinase gene. The role of the glucocorticoid receptor-binding sites. *Mol. Endocrinol.* **12**, 482–491.

Seol, W., Choi, H.-S., and Moore, D. D. (1995). Isolation of proteins that interact specifically with the retinoid X receptor: two novel orphan receptors. *Mol. Endocrinol.* **9**, 72–85.

Shen, W.-H., Moore, C. C. D., Ikeda, Y., Parker, K. L., and Ingraham, H. A. (1994). Nuclear receptor steroidogenic factor 1 regulates the Müllerian inhibiting substance gene: a link to the sex determination cascade. *Cell* **77**, 651–661.

Shih, H. M., Liu, Z., and Towle, H. C. (1995). Two CACGTG motifs with proper spacing dictate the carbohydrate regulation of hepatic gene transcription. *J. Biol. Chem.* **270**, 21991–21997.

Shinar, D. M., Endo, N., Rutledge, S. J., Vogel, R., Rodan, G. A., and Schmidt, A. (1994). NER, a new member of the gene family encoding the human steroid hormone nuclear receptor. *Gene* **147**, 273–276.

Shinoda, K., Lei, H., Yoshii, H., Nomura, M., Nagano, M., Shiba, H., Sasaki, H., Osawa, Y., Ninomiya, Y., Niwa, O., Morohashi, K., and Li, E. (1995). Developmental defects of the ventromedial hypothalamic nucleus and pituitary gonadotroph in the Ftz-F1 disrupted mice. *Dev. Dynamics* **204**, 22–29.

Simonson, G. D., and Iwanij, V. (1995). Genomic organization and promoter sequence of a gene encoding a rat liver-specific type-I transport protein. *Gene* **154**, 243–247.

Sladek, F. M., Zhong, W., Lai, E., and Darnell, J. E., Jr. (1990). Liver-enriched transcription factor HNF-4 is a novel member of the steroid hormone receptor superfamily. *Genes Dev.* **4**, 2353–2365.

Sladek, R., Bader, J.-A., and Giguère, V. (1997). The orphan nuclear receptor estrogen-related receptor α is a transcriptional regulator of the human medium-chain acyl coenzyme A dehydrogenase gene. *Mol. Cell. Biol.* **17**, 5400–5409.

Sladek, F. M., Dallas-Yang, Q., and Nepomuceno, L. (1998). MODY1 mutation Q268X in hepatocyte nuclear factor 4α allows for dimerization in solution but causes abnormal subcellular localization. *Diabetes* **47**, 985–90.

Song, C., Kokontis, J. M., Hiipakka, R. A., and Liao, S. (1994). Ubiquitous receptor: a receptor that modulates gene activation by retinoic acid and thyroid hormone receptors. *Proc. Natl. Acad. Sci. USA* **91**, 10809–10813.

Spady, D. K., Cuthbert, J. A., Willard, M. N., and Meidell, R. S. (1995). Adenovirus-mediated transfer of a gene encoding cholesterol 7α-hydroxylase into hamsters increases hepatic enzyme activity and reduces plasma total and low density lipoprotein cholesterol. *J. Clin. Invest.* **96**, 700–709.

Spiegelman, B. M. (1998a). PPAR-γ: Adipogenic regulator and thiazolidinedione receptor. *Diabetes* **47**, 507–514.

Spiegelman, B. M. (1998b). PPARγ in monocytes: less pain, any gain? *Cell* **93**, 153–155.

Steineger, H. H., Sorensen, H. N., Tugwood, J. D., Skrede, S., Spydevold, O., and Gautvik, K. M. (1994). Dexamethasone and insulin demonstrate marked and opposite regulation of the steady-state mRNA level of the peroxisomal proliferator-activated receptor (PPAR) in hepatic cells. Hormonal modulation of fatty-acid-induced transcription. *Eur. J. Biochem.* **225**, 967–974.

Stoffel, M., and Duncan, S. A. (1997). The maturity-onset diabetes of the young (MODY1) transcription factor HNF4α regulates expression of genes required for glucose transport and metabolism. *Proc. Natl. Acad. Sci. USA* **94**, 13209–13214.

Stravitz, R. T., Vlahcevic, Z. R., Gurley, E. C., and Hylemon, P. B. (1995). Repression of cholesterol 7α-hydroxylase transcription by bile acids is mediated through protein kinase C in primary cultures of rat hepatocytes. *J. Lipid Res.* **36**, 1359–1369.

Stravitz, R. T., Rao, Y. P., Vlahcevic, Z. R., Gurley, E. C., Jarvis, W. D., and Hylemon, P. B. (1996). Hepatocellular protein kinase C activation by bile acids: implications for regulation of cholesterol 7α-hydroxylase. *Am. J. Physiol.* **271**, G293–303.

Ström, A., Westin, S., Eguchi, H., Gustafsson, J.-Å., and Mode, A. (1995). Characterization of orphan nuclear receptor binding elements in sex-differentiated members of the CYP2C gene family expressed in rat liver. *J. Biol. Chem.* **270**, 11276–11281.

Stroup, D., Crestani, M., and Chiang, J. Y. (1997a). Identification of a bile acid response element in the cholesterol 7α-hydroxylase gene CYP7A. *Am. J. Physiol.* **273**, G508–517.

Stroup, D., Crestani, M., and Chiang, J. Y. (1997b). Orphan receptors chicken ovalbumin upstream promoter transcription factor II (COUP-TFII) and retinoid X receptor (RXR) activate and bind the rat cholesterol 7alpha-hydroxylase gene (CYP7A). *J. Biol. Chem.* **272**, 9833–9839.

Sugawara, T., Holt, J. A., Kiriakidou, M., and Strauss, J. F., 3rd (1996). Steroidogenic factor 1-dependent promoter activity of the human steroidogenic acute regulatory protein (StAR) gene. *Biochemistry* **35**, 9052–9059.

Sundseth, S. S., and Waxman, D. J. (1990). Hepatic P-450 cholesterol 7α-hydroxylase. Regulation in vivo at the protein and mRNA level in response to mevalonate, diurnal rhythm, and bile acid feedback. *J. Biol. Chem.* **265**, 15090–15095.

Taniguchi, T., Chen, J., and Cooper, A. D. (1994). Regulation of cholesterol 7α-hydroxylase gene expression in Hep-G2 cells. Effect of serum, bile salts, and coordinate and noncoordinate regulation with other sterol-responsive genes. *J. Biol. Chem.* **269**, 10071–10078.

Taraviras, S., Monaghan, A. P., Schutz, G., and Kelsey, G. (1994). Characterization of the mouse HNF-4 gene and its expression during mouse embryogenesis. *Mech. Dev.* **48**, 67–79.

Teboul, M., Enmark, E., Li, Q., Wikstrom, A. C., Pelto-Huikko, M., and Gustafsson, J. A. (1995). OR-1, a member of the nuclear receptor superfamily that interacts with the 9-cis-retinoic acid receptor. *Proc. Natl. Acad. Sci. USA* **92**, 2096–2100.

Thomassin, H., Bois-Joyeux, B., Delille, R., Ikonomova, R., and Danan, J. L. (1996). Chicken ovalbumin upstream promoter-transcription factor, hepatocyte nuclear factor 3, and CCAAT/enhancer binding protein control the far-upstream enhancer of the rat alpha-fetoprotein gene. *DNA & Cell Biology* **15**, 1063–1074.

Thompson, K. S., and Towle, H. C. (1991). Localization of the carbohydrate response element of the rat L-type pyruvate kinase gene. *J. Biol. Chem.* **266**, 8679–8682.

Tontonoz, P., Hu, E., Graves, R. A., Budavari, A. J., and Spiegelman, B. M. (1994a). mPPARγ2: tissue-specific regulator of an adipocyte enhancer. *Genes Dev.* **8**, 1224–1234.

Tontonoz, P., Hu, E., and Spiegelman, B. M. (1994b). Stimulation of adipogenesis in fibroblasts by PPARγ2, a lipid-activated transcription factor. *Cell* **79**, 1147–1156.

Tontonoz, P., Hu, E., Devine, J., Beale, E. G., and Spiegelman, B. M. (1995). PPARγ2 regulates adipose expression of the phosphoenolpyruvate carboxykinase gene. *Mol. Cell. Biol.* **15**, 351–357.

Tontonoz, P., Nagy, L., Alvarez, J. G. A., Thomazy, V. A., and Evans, R. M. (1998). PPARγ promotes monocyte/macrophage differentiation and uptake of oxidized LDL. *Cell* **93**, 241–252.

Towle, H. C., Kaytor, E. N., and Shih, H. M. (1997). Regulation of the expression of lipogenic enzyme genes by carbohydrate. *Ann. Rev. Nutr.* **17**, 405–433.

Trapp, T., and Holsboer, F. (1996). Nuclear orphan receptor as a repressor of glucocorticoid receptor transcriptional activity. *J. Biol. Chem.* **271**, 9879–9882.

Tsai, S. Y., and Tsai, M.-J. (1997). Chick ovalbumin upstream promoter-transcription factors (COUP-TFs): coming of age. *Endocr. Rev.* **18**, 229–240.

Tsukiyama, T., Ueda, H., Hirose, S., and Niwa, O. (1992). Embryonal long terminal repeat-binding protein is a murine homolog of FTZ-F1, a member of the steroid receptor superfamily. *Mol. Cell. Biol.* **12**, 1286–1291.

Tugwood, J. D., Issemann, I., Anderson, R. G., Bundell, K. R., McPheat, W. L., and Green, S. (1992). The mouse peroxisome proliferator activated receptor recognizes a response element in the 5′ flanking sequence of the rat acyl CoA oxidase gene. *EMBO J.* **11**, 433–449.

Twisk, J., Hoekman, M. F., Lehmann, E. M., Meijer, P., Mager, W. H., and Princen, H. M. (1995). Insulin suppresses bile acid synthesis in cultured rat hepatocytes by down-regulation of cholesterol 7α-hydroxylase and sterol 27-hydroxylase gene transcription. *Hepatology* **21**, 501–510.

Vaulont, S., and Kahn, A. (1994). Transcriptional control of metabolic regulation genes by carbohydrates. *FASEB J.* **8**, 28–35.

Vaulont, S., Munnich, A., Decaux, J. F., and Kahn, A. (1986). Transcriptional and post-transcriptional regulation of L-type pyruvate kinase gene expression in rat liver. *J. Biol. Chem.* **261**, 7621–7625.

Vaulont, S., Puzenat, N., Levrat, F., Cognet, M., Kahn, A., and Raymondjean, M. (1989). Proteins binding to the liver-specific pyruvate kinase gene promoter. A unique combination of known factors. *J. Mol. Biol.* **209**, 205–219.

Vega, R. B., and Kelly, D. P. (1997). A role for estrogen-related receptor α in the control of mitochondrial fatty acid β-oxidation during brown adipocyte differentiation. *J. Biol. Chem.* **272**, 31693–31699.

Viollet, B., Kahn, A., and Raymondjean, M. (1997). Protein kinase A-dependent phosphorylation modulates DNA-binding activity of hepatocyte nuclear factor 4. *Mol. Cell. Biol.* **17**, 4208–4219.

Vlahcevic, Z. R., Heuman, D. M., and Hylemon, P. B. (1991). Regulation of bile acid synthesis. *Hepatology* **13**, 590–600.

Vorgia, P., Zannis, V. I., and Kardassis, D. (1998). A short proximal promoter and the distal hepatic control region-1 (HCR-1) contribute to the liver specificity of the human apolipoprotein C-II gene. Hepatic enhancement by HCR-1 requires two proximal hormone response elements which have different binding specificities for orphan receptors HNF-4, ARP-1, and EAR-2. *J. Biol. Chem.* **273**, 4188–4196.

Vu-Dac, N., Schoonjans, K., Laine, B., Fruchart, J. C., Auwerx, J., and Staels, B. (1994). Negative regulation of the human apolipoprotein A-I promoter by fibrates can be attenuated by the interaction of the peroxisome proliferator-activated receptor with its response element. *J. Biol. Chem.* **269**, 31012–31018.

Wang, L.-H., Tsai, S. Y., Cook, R. G., Beattie, W. G., Tsai, M.-J., and O'Malley, B. W. (1989). COUP transcription factor is a member of the steroid receptor superfamily. *Nature* **340**, 163–166.

Wang, X., Sato, R., Brown, M. S., Hua, X., and Goldstein, J. L. (1994). SREBP-1, a membrane-bound transcription factor released by sterol-regulated proteolysis. *Cell* **77**, 53–62.

Wang, J. C., Stromstedt, P. E., O'Brien, R. M., and Granner, D. K. (1996). Hepatic nuclear factor 3 is an accessory factor required for the stimulation of phosphoenolpyruvate carboxykinase gene transcription by glucocorticoids. *Mol. Endocrinol.* **10**, 794–800.

Watson, M. A., and Milbrandt, J. (1990). Expression of the nerve growth factor-regulated NGFI-A and NGFI-B genes in the developing rat. *Development* **110**, 173–183.

Wehrenberg, U., von Goedecke, S., Ivell, R., and Walther, N. (1994). The orphan receptor SF-1 binds to the COUP-like element in the promoter of the actively transcribed oxytocin gene. *J. Neuroendocrinol.* **6**, 1–4.

Werman, A., Hollenberg, A., Solanes, G., Bjorbaek, C., Vidal-Puig, A. J., and Flier, J. S. (1997). Ligand-independent activation domain in the N terminus of peroxisome proliferator-activated receptor gamma (PPARγ). Differential activity of PPARγ1 and -2 isoforms and influence of insulin. *J. Biol. Chem.* **272**, 20230–20235.

Wijnholds, J., Philipsen, J. N., and Ab, G. (1988). Tissue-specific and steroid-dependent interaction of transcription factors with the oestrogen-inducible apoVLDL II promoter in vivo. *EMBO J.* **7**, 2757–2763.

Wiley, S. R., Kraus, R. J., Zuo, F., Murray, E. E., Loritz, K., and Mertz, J. E. (1993). SV40 early-to-late switch involves titration of cellular transcriptional repressors. *Genes Dev.* **7**, 2206–2219.

Willy, P. J., and Mangelsdorf, D. J. (1998). Nuclear orphan receptors: the search for novel ligands and signaling pathways. *In* "Hormones and Signaling" (B. W. O'Malley, ed.), pp. 307–358. Academic Press, San Diego.

Willy, P. J., Umesono, K., Ong, E. S., Evans, R. M., Heyman, R. A., and Mangelsdorf, D. J. (1995). LXR, a nuclear receptor that defines a distinct retinoid response pathway. *Genes Dev.* **9**, 1033–1045.

Wilson, T. E., Fahrner, T. J., Johnson, M., and Milbrandt, J. (1991). Identification of the DNA binding site for NGFI-B by genetic selection in yeast. *Science* **252**, 1296–1300.

Wilson, T. E., Paulsen, R. E., Padgett, K. A., and Milbrandt, J. (1992). Participation of non-zinc finger residues in DNA binding by two nuclear orphan receptors. *Science* **256**, 107–110.

Wilson, T. E., Fahrner, T. J., and Milbrandt, J. (1993a). The orphan receptors NGFI-B and steroidogenic factor 1 establish monomer binding as a third paradigm of nuclear receptor-DNA interaction. *Mol. Cell. Biol.* **13**, 5794–5804.

Wilson, T. E., Mouw, A. R., Weaver, C. A., Milbrandt, J., and Parker, K. L. (1993b). The orphan nuclear receptor NGFI-B regulates expression of the gene encoding steroid 21-hydroxylase. *Mol. Cell. Biol.* **13**, 861–868.

Wong, C.-W., and Privalsky, M. L. (1995). Role of the N terminus in DNA recognition by the v-erb A protein, an oncogenic derivative of a thyroid hormone receptor. *Mol. Endocrinol.* **9**, 551–562.

Woronicz, J. D., Calnan, B., Ngo, V., and Winoto, A. (1994). Requirement for the orphan steroid receptor nur77 in apoptosis of T-cell hybridomas. *Nature* **367**, 277–280.

Wu, Z., Xie, Y., Bucher, N. L. R., and Farmer, S. R. (1995). Conditional ectopic expression of C/EBPβ in NIH-3T3 cells induces PPARγ and stimulates adipogenesis. *Genes Dev.* **9**, 2350–2363.

Wu, Q., Li, Y., Liu, R., Agadir, A., Lee, M. O., Liu, Y., and Zhang, X. K. (1997). Modulation of retinoic acid sensitivity in lung cancer cells through dynamic balance of orphan receptors Nur77 and COUP-TF and their heterodimerization. *EMBO J.* **16**, 1656–1669.

Yamada, K., Tanaka, T., and Noguchi, T. (1997). Members of the nuclear factor 1 family and hepatocyte nuclear factor 4 bind to overlapping sequences of the L-II element on the rat pyruvate kinase L gene promoter and regulate its expression. *Biochem. J.* **324**, 917–925.

Yamagata, K., Furuta, H., Oda, N., Kaisaki, P. J., Menzel, S., Cox, N. J., Fajans, S. S., Signorini, S., Stoffel, M., and Bell, G. I. (1996a). Mutations in the hepatocyte nuclear factor-4α gene in maturity-onset diabetes of the young (MODY1). *Nature* **384**, 458–460.

Yamagata, K., Oda, N., Kaisaki, P. J., Menzel, S., Furuta, H., Vaxillaire, M., Southam, L., Cox, R. D., Lathrop, G. M., Boriraj, V. V., Chen, X., Cox, N. J., Oda, Y., Yano, H., Le Beau, M. M., Yamada, S., Nishigori, H., Takeda, J., Fajans, S. S., Hattersley, A. T., Iwasaki, N., Hansen, T., Pedersen, O., Polonsky, K. S., Turner, R. C., Velho, G., Chèvre, J.-C., Froguel, P., and Bell, G. I. (1996b). Mutations in the hepatocyte nuclear factor-1α gene in maturity-onset diabetes of the young (MODY3). *Nature* **384**, 455–458.

Yang, N., Shigeta, H., Shi, H. P., and Teng, C. T. (1996). Estrogen-related receptor, hERR1, modulates estrogen receptor-mediated response of human lactoferrin gene promoter. *J. Biol. Chem.* **271**, 5795–5804.

Yokomori, N., Nishio, K., Aida, K., and Negishi, M. (1997). Transcriptional regulation by HNF-4 of the steroid 15α-hydroxylase P450 (Cyp2a-4) gene in mouse liver. *J. Steroid Biochem. Molec. Biol.* **62**, 307–314.

Yokoyama, C., Wang, X., Briggs, M. R., Admon, A., Wu, J., Hua, X., Goldstein, J. L., and Brown, M. S. (1993). SREBP-1, a basic–helix–loop–helix–leucine zipper protein that controls transcription of the low density lipoprotein receptor gene. *Cell* **75**, 187–197.

Yu, R. T., McKeown, M., Evans, R. M., and Umesono, K. (1994). Relationship between *Drosophila* gap gene *tailless* and a vertebrate nuclear receptor Tlx. *Nature* **370**, 375–379.

Yu, K., Bayona, W., Kallen, C. B., Harding, H. P., Ravera, C. P., McMahon, G., Brown, M., and Lazar, M. A. (1995). Differential activation of peroxisome proliferator-activated receptors by eicosanoids. *J. Biol. Chem.* **270**, 23975–23983.

Zanaria, E., Muscatelli, F., Bardoni, B., Strom, T. M., Guioli, S., Guo, W., Lalli, E., Moser, C., Walker, A. P., McCabe, E. R. B., Meitinger, T., Monaco, A. P., Sassone-Corsi, P., and Camerino, G. (1994). An unusual member of the nuclear hormone receptor superfamily responsible for X-linked adrenal hypoplasia congenita. *Nature* **372**, 635–641.

Zavacki, A. M., Lehmann, J. M., Seol, W., Willson, T. M., Kliewer, S. A., and Moore, D. D. (1997). Activation of the orphan receptor RIP14 by retinoids. *Proc. Natl. Acad. Sci. USA* **94**, 7909–7914.

Zhang, P., and Mellon, S. H. (1996). The orphan nuclear receptor steroidogenic factor-1 regulates the cyclic adenosine 3′-5′-monophosphate-mediated transcriptional activation of rat cytochrome P450c17 (17α-hydroxylase/17-20 lyase). *Mol. Endocrinol.* **10**, 147–158.

Zhang, P., and Mellon, S. H. (1997). Multiple organ nuclear receptors converge to regulate rat p450c17 gene transcription: novel mechanisms of orphan nuclear receptor action. *Mol. Endocrinol.* **11**, 891–904.

Zhao, Q., Khorasanizadeh, S., Miyoshi, Y., Lazar, M. A., and Rastinejad, F. (1998). Structural elements of an orphan nuclear receptor-DNA complex. *Mol. Cell* **1**, 849–861.

Stefan Westin*
Michael G. Rosenfeld†
Christopher K. Glass*

Division of Endocrinology and Metabolism
*Division of Cellular and Molecular Medicine
Department of Medicine

†Howard Hughes Medical Institute
University of California, San Diego
La Jolla California 92093-0651

Nuclear Receptor Coactivators

I. Nuclear Receptor Structure and Function

Members of the nuclear receptor superfamily regulate gene expression by binding to cis-active elements in target genes and either activating or repressing transcription. In addition to being the largest known family of eukaryotic transcription factors, nuclear receptors have become particularly attractive as model systems for the study of regulated transcription because their activities are directly controlled by the binding of small molecular weight ligands. This mechanism of action has greatly facilitated biochemical approaches to the study of transcriptional activation by members of the nuclear receptor family and has allowed the identification of several novel proteins that appear to play important and general roles in transcriptional coactivation and corepression.

Hormones and Signaling

Nuclear receptors are considered to consist of six domains (A–F) based on regions of conserved sequence and function (Fig.1) (reviewed in Beato *et al.*, 1995; Evans, 1988; Mangelsdorf *et al.*, 1995). The N-terminal A and B domains are highly variable among members of the nuclear receptor superfamily. These regions often contain a ligand-induced activation function, referred to as AF-1, that may synergize with the C-terminal activation function (AF-2, see later discussion) in a promoter- and cell-specific manner. The C domain, which encodes two zinc finger modules involved in DNA binding, is the most conserved domain and defines the nuclear receptor

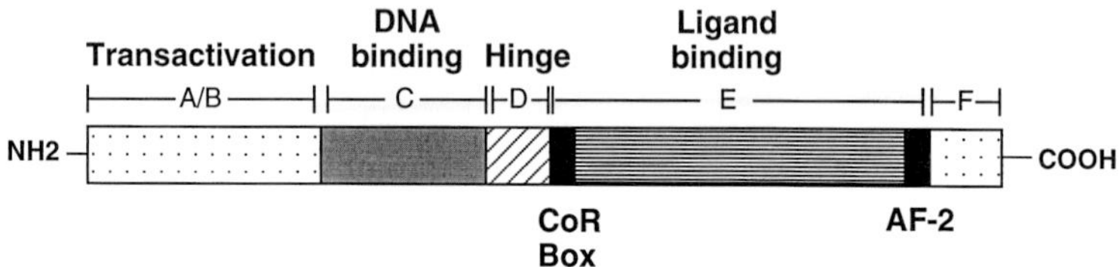

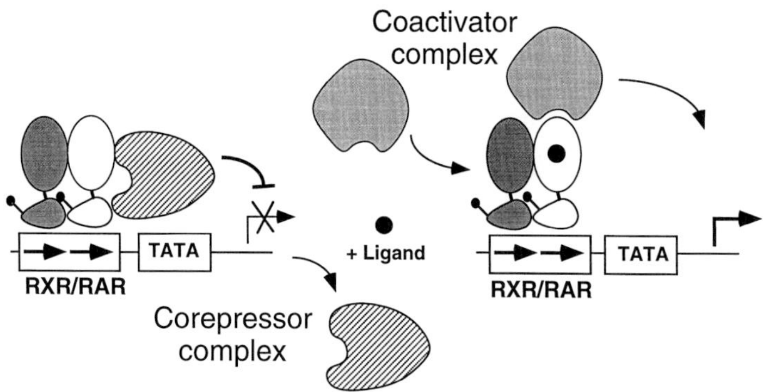

FIGURE 1 Nuclear receptor structure and function. (a) Functional domains of nuclear receptors, indicating N-terminal A/B domain that often contains a ligand-independent transactivation function, the highly conserved C domain that mediates DNA binding, the hinge region D, the conserved ligand binding domain E, and a highly variable C-terminal sequence F. The ligand binding domain also mediates dimerization, ligand-independent repression, and ligand-dependent transactivation functions. Ligand-dependent transactivation depends on a C-terminal sequence, termed AF-2, that undergoes a conformational change on binding of ligand. Ligand-independent repression is a specialized function of a subset of nuclear receptors, including retinoic acid and thyroid hormone receptors, and depends on a sequence at the N-terminal end of the ligand binding domain, termed the CoR box. (b) Model for mechanisms of repression and activation by RAR/RXR heterodimers. In the absence of ligand, RAR/RXR heterodimers interact with a co-repressor complex that represses gene transcription. Upon binding of retinoic acid, the corepressor complex dissociates and is replaced by a coactivator complex, resulting in transcriptional activation.

superfamily (reviewed in Glass, 1996). The D domain, which is poorly conserved across the nuclear receptor family, contains amino acids that also participate in DNA recognition by a subset of nuclear receptors that bind to DNA as monomers and heterodimers. The E domain of nuclear receptors represents a complex domain (referred to as the ligand binding domain, LBD), involved in determining the ligand binding, dimerization, and trans-activation properties of a particular nuclear receptor. The N-terminal end of the E domain contains a region that in the retinoic acid receptor (RAR) and thyroid hormone receptor (TR) is termed the CoR box and has been shown to be important for interaction of the co-repressors N-CoR and SMRT (Chen and Evans, 1995; Hörlein *et al.*, 1995; Kurokawa *et al.*, 1995). The C-terminal end of the E domain contains a highly conserved sequence that is required for ligand-dependent transactivation, termed AF-2. The AF-2 domain was initially delineated by mutational analysis of the estrogen receptor (Danielian *et al.*, 1992). Point mutations were identified that had little or no effect on the binding of regulatory ligands, dimerization, or DNA binding, but that abolished transcriptional activation (Barettino *et al.*, 1994; Danielian *et al.*, 1992; Durand *et al.*, 1994; Tone *et al.*, 1994). This region was therefore predicted to serve a critical function for interacting with putative coactivator proteins necessary for ligand-dependent transcriptional activation (Fig. 1).

Crystal structures of the apo retinoid X receptor α (RXRα) and holo RARγ, TRα, estrogen receptor α (ERα), and progesterone receptor (PR) ligand binding domains have been solved (Bourguet *et al.*, 1995; Brzozowski *et al.*, 1997; Renaud *et al.*, 1995; Wagner *et al.*, 1995; Williams and Sigler, 1998). The overall structures are very similar, consisting of 12 conserved alpha helices. A conserved β turn resides between helix 5 and helix 6, which is sandwiched between antiparallel bundles of helices 1–5 and 6–12 and contributes to the hydrophobic core of the LBD. In the liganded RAR, TR, ER, and PR structures, the ligand is buried deep within the hydrophobic core of the ligand-binding domain. Several notable differences are apparent in comparing the apo RXR structure with the liganded structures of the RAR, TR, ER, and PR structures. The most striking difference is the position of helix 12, which contains the conserved AF-2 core. This helix projects away from the LBD in the RXR structure, but is tightly folded against helix 4 and makes direct contact with the bound ligands in the RAR, TR, ER, and PR structures. These findings support the notion that the AF-2 helix adopts a distinct conformation upon the binding of ligand that allows the recruitment of coactivators. Intriguingly, the AF-2 helix adopts a more extended conformation in the estrogen receptor bound to the antagonist raloxifene, consistent with the idea that antagonists work in part by inducing a nonproductive conformation of the LBD. A second striking difference between the unliganded and liganded LBD structures is that the loop region between helices 2 and 3 in RXR, which also extends away from the LBD, is tucked under helix 6 in the RAR and TR structures. In addition, the liganded structures of the LBDs are more compact than that of the unliganded RXR LBD, indicating that ligands may play a structural role in reconfiguring other surface features of the LBD.

II. Nuclear Receptor Coactivators

A. Biochemically Defined Factors

The observation that estrogen, progesterone, and glucocorticoid receptors could interfere with each other's function (Meyer *et al.*, 1989) and the identification of a conserved region required for ligand-dependent transcriptional activity provided initial evidence for the existence of coactivator proteins that mediate AF-2 function. Several biochemical approaches have led to the identification of a relatively large number of potential coactivator proteins (Table I). One biochemical approach has been to fuse the ligand binding domain of a nuclear receptor to glutathione-S-transferase (GST). The fusion proteins can then be bound to a glutathione affinity matrix and incubated with whole-cell extracts prepared from metabolically labeled cells in the presence or absence of ligand. Using this approach, several proteins were observed that bound specifically to the ER-LBD in a ligand- and AF-2-dependent manner. The most abundant of these were proteins of 140 and 160 kDa molecular weight (p140 and p160) (Cavailles *et al.*, 1994; Halachmi *et al.*, 1994). A minor component of 300 kDa molecular weight has also been observed and has been demonstrated to correspond to CBP

TABLE I   Cellular Proteins That Associate with Nuclear Receptors in a Ligand-Dependent Manner

Protein	Receptors	References
p160 (ERAP 160, RIP 160)	ER, RAR	Halachmi *et al.*, 1994
p140 (ERAP 140, RIP 140)	ER, RAR	Cavailles *et al.*, 1994
		Kurokawa *et al.*, 1995
GRIP 95, GRIP 120, GRIP 170	GR	Eggert *et al.*, 1995
p300	ER	Hanstein *et al.*, 1996
TRAP complex	TR	Fondell *et al.*, 1996
TRAP 80, TRAP 93, TRAP 95	VDR RAR RXR	
TRAP 97, TRAP 100, TRAP 150	PPARα, PPARγ, ER	
TRAP 170, TRAP 220, TRAP 230	(Only TRAP 220)	
DRIP complex	VDR	Rachez *et al.*, 1998
DRIP70, DRIP78, DRIP88	TR	
DRIP95, DRIP100, DRIP125	RXR	
DRIP140, DRIP160, DRIP180	RAR	
DRIP230, DRIP240, DRIP250		

and the adenovirus E1A-associated p300 protein (Hanstein *et al.*, 1996). The p140 and p160 proteins interacted directly with the estrogen receptor ligand binding domain, because they could also be detected using a ^{32}P-ER-LBD probe following resolution of unlabeled ER-associated proteins by SDS polyacrylamide gel electrophoresis and transfer to nitrocellulose membranes (far Western blotting). The p140 and p160 proteins were also found to interact with the RAR and RXR in a ligand- and AF-2-dependent manner, suggesting that they recognized conserved structural features in the LBD (Kurokawa *et al.*, 1995).

Intriguingly, the use of a GST–vitamin D receptor fusion protein to isolate interacting proteins led to the identification of a distinct complex that appears to consist of at least 10 proteins ranging in size from 65 to 250 kDa (Table I) (Rachez *et al.*, 1998). These Vitamin D receptor interacting proteins (DRIPs) also interacted with other nuclear receptors in an AF-2-dependent manner. DRIPs enhanced VDR transcriptional activity on a naked DNA template in a cell-free *in vitro* transcription assay. It is not yet clear why such different sets of proteins were obtained using different GST–nuclear receptor fusion proteins. It is possible that different receptors have distinct preferences for specific complexes and/or that coactivator complexes may vary in a cell type-specific manner. These observations underscore the possibility that there may be several distinct coactivator complexes that mediate nuclear receptor function.

As a second approach, the DNA-bound glucocorticoid receptor was used as a probe to detect interacting proteins. These studies resulted in the identification of glucocorticoid receptor interacting proteins (GRIPs) of 95, 120, and 170 kDa (Eggert *et al.*, 1995). The requirement for ligand or the AF-2 function of the GR for interaction were not addressed in these studies, but purified GRIP170 stimulated ligand-independent transcriptional activation by the glucocorticoid receptor *in vitro,* raising the possibility that GRIP 170 may interact with the N-terminal activation domain of the GR rather than the ligand binding domain.

As a third approach, an epitope (FLAG)-tagged version of the thyroid hormone receptor was stably introduced into HeLa cells and used to isolate thyroid hormone receptor associated proteins (TRAPs) (Table I) (Fondell *et al.*, 1996). Immunoprecipitation of the FLAG-tagged TR from T3-stimulated cells resulted in the specific recovery of nine polypeptides, ranging in molecular weight from 80 to 230 kDa. Because of difficulties in accurately determining molecular weights of large proteins by SDS-PAGE, it is not clear whether any of the TRAPs correspond to the proteins demonstrated to interact with the ER, but several appear to correspond to the proteins that are associated with the vitamin D receptor. These studies did not address which of the several TRAPs interact with the TR directly, or whether they are present within a single complex. However, the epitope-tagged TR–TRAP complex isolated from thyroid hormone-treated cells was much more effective in stimulating transcription *in vitro* than the epitope-tagged TR isolated from unstimulated cells. In concert, these biochemical studies suggest that transcriptional activation by nuclear receptors may involve the concerted actions of a number of associated proteins. Cognate cDNAs have been described for 2 of the 10 TRAPs (TRAP 220 and TRAP 100) (Yuan *et al.,* 1998). Both of these appear to reside in a single complex with other TRAPs, but only TRAP 220 shows a direct ligand-dependent interaction with TRα. TRAP 220 also interacts with other nuclear receptors, such as the VDR, RAR, RXR, PPARα, PPARγ, and to a lesser extent with the ER, in a ligand-dependent manner and appears to be identical to DRIP 230 (Rachez *et al.,* 1998).

B. Structure–Function Analysis of Cloned Coactivators

To clone cDNAs encoding potential coactivator proteins, strategies based on protein purification, genetic screens, the yeast two hybrid system, and direct expression screening of bacteriophage cDNA libraries have been employed successfully. These efforts have resulted in the identification of a large number of cDNAs that encode interacting proteins, many of which are novel (Table II).

Several interacting proteins have been identified that exhibit sequence motifs suggesting transcriptional roles. These include TIF1 (Le Douarin *et al.,* 1995), which was identified on the basis of a genetic screen in yeast for proteins that could potentiate RXR- and RAR-dependent transcription, and RIP140 (Cavailles *et al.,* 1995), identified by screening bacteriophage-based expression libraries with the ER-LBD in the presence of estrogen. TIF1 contains several conserved domains found in transcriptional regulatory proteins, including a RING finger domain, a coiled-coil domain, and a bromodomain. TIF1 associates with several nuclear receptors in a ligand-dependent manner both *in vivo* and *in vitro*. Intriguingly, overexpression of TIF1 in mammalian cells strongly inhibits ligand-dependent activation by RXR. Although this observation might argue against a role as a coactivator, it is possible that overexpression of TIF1 alters the stoichiometry of a TIF1-containing complex required for activation. It is not clear whether TIF1, which has a predicted molecular weight of 112 kDa, corresponds to any of the proteins identified to interact with nuclear receptor in biochemical assays. TIF1 has been documented to possess a protein kinase activity (Fraser *et al.,* 1998) and undergoes a ligand-dependent hyperphosphorylation as a consequence of nuclear receptor binding. In addition to autophosphorylation, TIF1 selectively phosphorylates the transcription factors TF$_{II}$Eα, TAF$_{II}$28, and TAF$_{II}$55 *in vitro*. These studies raise the possibility that TIF1 may act by phosphorylating and modifying the activity of components of the transcriptional machinery.

TABLE II Cloned Proteins That Interact with Nuclear Receptors in a Ligand-Dependent Manner

Protein	Reported receptor interactions	Comments	References
RIP 140	ER	Weakly stimulates ER function at low levels of overexpression. High levels repress ER activity. May correspond to p140.	Cavailles *et al.*, 1995
TIF1	ER, RAR, RXR	Potentiates RAR and RXR activity in yeast, but inhibits RXR function when overexpressed in mammalian cells. Contains a protein kinase activity.	Le Douarin *et al.*, 1995
SRC-1/NCoA-1	PR, ER, RAR, RXR, TR, GR	Potentiates activities of several nuclear receptors. A component of p160. Related to TIF2/GRIP1 and p/CIP.	Oñate *et al.*, 1995 Kamei *et al.*, 1996 Yao *et al.*, 1996
TIF2/GRIP1	PR, ER, RAR, RXR	Potentiates activities of several nuclear receptors. A component of p160. Related to SRC-1/NCoA-1.	Voegel *et al.*, 1996 Hong *et al.*, 1996
p/CIP/ACTR/AIB1/ RAC3/TRAM1	ER, RAR, RXR, TR, PPAR	Potentiates activities of several nuclear receptors. A component of p160. Related to SRC-1/NCoA-1.Takeshita *et al.*, 1997	Torchia *et al.*, 1997 Chen *et al.*, 1997 Anzick *et al.*, 1997 Li *et al.*, 1997
ARA$_{70}$	AR	Specifically interacts with and potentiates the activity of the androgen receptor.	Yeh and Chang, 1996
CBP/p300	RAR, RXR, ER, TR	Essential coactivators for nuclear receptors and several other classes of regulated transcription factors. Interacts with SRC-1, p/CIP, and TIF2.	Hanstein *et al.*, 1996 Kamei *et al.*, 1996 Chakravarti *et al.*, 1996
PBP/TRIP-2/ DRIP230	TR, RAR, RXR, VDR, PPARα, PPARγ	A component of the TRAP and DRIP complexes that stimulate T$_3$receptor and VDR-dependent transcription *in vitro*.	Zhu *et al.*, 1997 Lee *et al.*, 1995 Rachez *et al.*, 1998
PGC-1		Interacts with PPARγ, TRβ, RARα, and ERα and functions as a cell-type and promoter-specific coactivator.	Puigserver *et al.*, 1998

The predicted amino acid sequence of RIP140 is less informative than that of TIF1, although it contains a serine/threonine-rich region in the middle of the protein, flanked on each side by acidic and basic domains. Although the predicted molecular mass of RIP140 is 127 kDa, it migrates equivalently with the 140 kDa protein that interacts with the GST-ER LBD in biochemical assays, suggesting that it may encode the p140 protein (Cavailles *et al.*, 1995). When RIP140 levels are moderately increased in cells, ER-dependent transcription is modestly increased (two- to three-fold), whereas at higher levels of RIP140 expression, ER-dependent transcription is repressed. As suggested for TIF1, these findings may reflect a role of RIP140 as a component of a complex involved in transcriptional activation in which the stoichiometry of RIP140 and additional interacting proteins is critical.

Expression cloning strategies have also led to the cloning of human and murine proteins that encode the biochemically defined p160 factors. These include SRC-1/N-CoA1, TIF2/GRIP1/NCoA-2, and p/CIP/ACTR/AIB1/RAC3/TRAM1 (Anzick *et al.*, 1997; Chen *et al.*, 1997; Hong *et al.*, 1996; Kamei *et al.*, 1996; Li *et al.*, 1997; Oñate *et al.*, 1995; Takeshita *et al.*, 1997; Torchia *et al.*, 1997; Voegel *et al.*, 1996). Alternative splicing has been observed to result in several different C-terminal ends of SRC-1/N-CoA1 and p/CIP/ACTR/AIB1/RAC3/TRAM1. Intriguingly, AIB1 was identified by characterizing a genomic region that is amplified in breast cancer, suggesting that AIB1 may contribute to the development of steroid-dependent cancers (Anzick *et al.*, 1997). Although the p160 proteins are only 30% identical overall, their alignment reveals several areas of 50–65% homology (Fig. 2). Interestingly, these proteins are most highly related in an N-terminal region that contains a PAS-A-bHLH homology domain. PAS domains have been shown to function as dimerization motifs and were initially identified in several nuclear proteins, including Period (Per), the Aryl hydrocarbon receptor (AHR), and its heterodimeric partner ARNT, and Single minded (Sim) (Huang, *et al.*, 1993). Also, similarly to a subgroup of PAS family members, the p160 proteins have a conserved bHLH domain at the N-terminus of the PAS domain, but they have not as yet been reported to have a DNA binding activity. The central region of the p160 protein contain the nuclear receptor interaction domain. A third region of increased homology is observed in the C-terminus that has been found to mediate interactions with CBP and p300. Remarkably, SRC-1 and TIF2 contain two autonomous transcription activation functions in their C-termini, referred to as AD1 and AD2 (Kalkhoven *et al.*, 1998; Voegel *et al.*, 1998). The AD1 coincides with the CBP interaction domain (see later discussion), and it has been shown to be a CBP-dependent transactivation function. The AD2 does not interact with CBP, suggesting that the activity is mediated by a factor or factors distinct from CBP.

Evidence that the p160 factors serve as functional nuclear receptor coactivators derive from several types of experiments. First, overexpression of SRC-1/NCoA-1, GRIP-1/TIF2, and p/CIP/ACTR/AIB-1 has been shown to potentiate ligand-dependent transcription by several nuclear receptors (Anzick *et al.*, 1997; Chen *et al.*, 1997; Hong *et al.*, 1996; Kamei *et al.*, 1996; Oñate *et al.*, 1995; Torchia *et al.*, 1997; Voegel *et al.*, 1996). These experiments suggest that the p160 factors are functionally limiting in cells and that they are capable of serving as coactivators when recruited to nuclear receptors in response to ligand. To address the question of whether p160 factors are required for ligand-dependent transcription, a second experimental approach has been to assess the effects of nuclear microinjection of specific blocking antibodies raised against each of the p160 factors on ligand-dependent transcription. Using this approach, microinjection of anti SRC-1/NCoA-1 antibody into Rat 1 fibroblasts abolished transcriptional activation by the retinoic acid receptor and TR, and reduced estrogen dependent transcription by about 60% (Torchia *et al.*, 1997). The effects of anti-SRC-1 antibody could be reversed by coinjection of a cDNA directing the expression of TIF2, suggesting that SRC-1/NCoA-1 and TIF2 are functionally redundant. Microinjection of anti-SRC-1 antibody did not block STAT-1-dependent transactivation in response to IFNγ or CREB-dependent transactivation by cAMP, indicating that it does not play a general role as a transcriptional coactivator (Torchia *et al.*, 1997). Intriguingly, microinjection of anti p/CIP antibodies abolished the transcriptional activities not only of the retinoic acid and estrogen receptors, but also of STAT1 and CREB (Torchia *et al.*, 1997). These inhibitory effects could not be overcome by microinjection of SRC-1 expression plasmids, suggesting that p/CIP is not functionally redundant with SRC-1 or TIF2. This is a surprising result given the sequence similarity of p/CIP, SRC-1, and TIF2, and the basis for these functional differences is not yet clear.

The *in vivo* biological function of SRC-1 has been assessed in mice in which the SRC-1 gene was inactivated by gene targeting (Xu *et al.*, 1998). Although SRC-1 null mutants showed no obvious external phenotype, target organs such as uterus, prostate, testis, and mammary gland exhibited decreased growth and development in response to steroid hormones. Thus, the loss of SRC-1 function results in partial resistance to hormone. Interestingly, expression of TIF2 was increased in the homozygotes, perhaps indicating a compensatory mechanism for the loss of SRC-1 function in target tissues. It will therefore be of importance to evaluate the consequences of disrupting TIF2 and p/CIP, alone and in combination.

C. Roles of CBP and p300 as Nuclear Receptor Coactivators

In addition to the p160 proteins, several lines of evidence suggest that CREB binding protein (CBP) and the adenovirus E1A-associated protein p300 also function as essential coactivators for several nuclear receptors

(Chakravarti et al., 1996; Kamei et al., 1996; Yao et al., 1996). CBP and p300 are large, structurally and functionally conserved proteins that have been demonstrated to serve coactivator roles for several classes of transcription factors (Fig. 3). The idea that CBP and/or p300 might be involved in nuclear receptor function was initially suggested by the observation that several nuclear receptors can antagonize transcriptional activation by AP-1 (reviewed in Beato, 1989). Although this antagonism could result from any one of several different mechanisms, one possibility is that nuclear receptors and AP-1 proteins ultimately require a common and rate-limiting set of coactivator proteins, and that antagonism results from competition for these factors. As CBP and p300 have previously been demonstrated to function as essential coactivators for AP-1 proteins (Arias et al., 1994), experiments were performed to assess their ability to also serve as nuclear receptor coactivators. These studies revealed a ligand-dependent interaction between CBP and the estrogen, retinoic acid, thyroid hormone, and retinoid X receptors (Chakravarti et al., 1996; Kamei et al., 1996). Furthermore, immunoprecipitation experiments documented that liganded RAR and CBP interact in cells. Using the retinoic acid and estrogen receptors, this interaction was mapped in a yeast two hybrid system to the extreme N-terminus of CBP (Chakravarti et al., 1996; Kamei et al., 1996). This region is conserved in p300, which also interacts with the same set of nuclear receptors (Chakravarti et al., 1996; Hanstein et al., 1996; Kamei et al., 1996). (See Fig. 3.)

Evidence that interactions of CBP/p300 with nuclear receptors are of functional importance derive from several types of experiments. First, overexpression of CBP/p300 potentiates ligand-dependent transcriptional responses of nuclear receptors (Chakravarti et al., 1996; Hanstein et al., 1996; Kamei et al., 1996). Second, forced expression of adenovirus E1A, which inhibits CBP/p300 function, inhibits retinoic acid receptor-dependent transcription (Kamei et al., 1996). Third, microinjection of antibodies directed against CBP/p300 blocks ligand-dependent transcriptional activation by RAR, RXR, and GR (Chakravarti et al., 1996; Kamei et al., 1996). The blockade of retinoic acid responses by injection of anti-CBP/p300 antibodies suggests that these proteins serve as essential coactivators.

Interestingly, a study has demonstrated that p300 and ER cooperatively activate transcription via differential enhancement of initiation and reinitiation. Using chromatin templates in vitro, it was shown that in the absence of ligand-activated ER, p300 has little effect on transcription, whereas p300 was observed to act synergistically with ligand-activated ER to enhance transcription. The study furthermore suggested a two-stroke mechanism for transcriptional activation by ligand-activated ER and p300. In the first stroke, ER and p300 function cooperatively to increase the efficiency of productive transcription initiation. In the second stroke, ER promotes the reassembly of the transcription preinitiation complex (Kraus and Kadonaga, 1998).

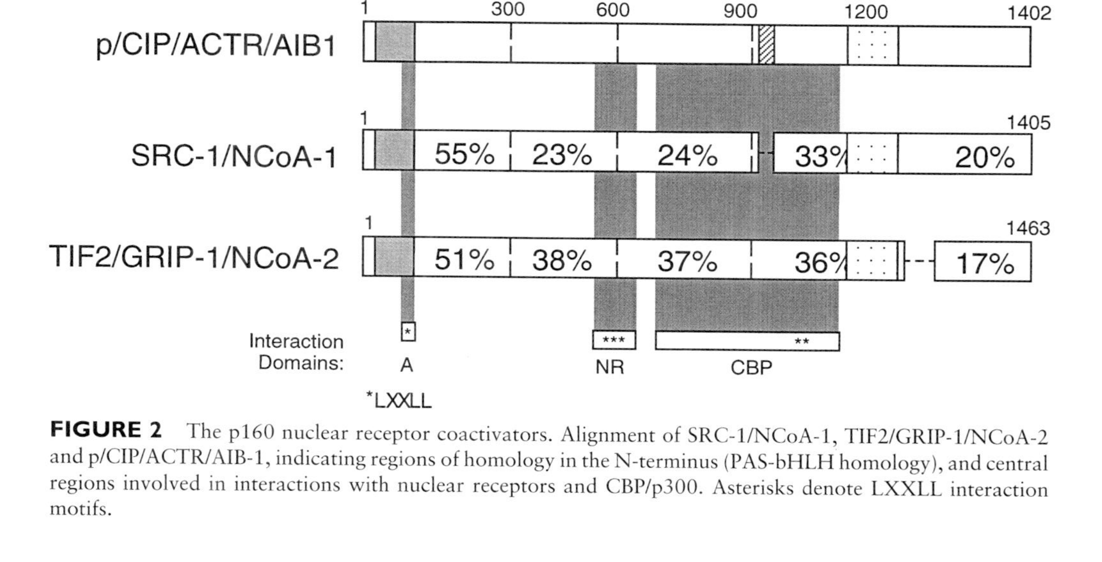

FIGURE 2 The p160 nuclear receptor coactivators. Alignment of SRC-1/NCoA-1, TIF2/GRIP-1/NCoA-2 and p/CIP/ACTR/AIB-1, indicating regions of homology in the N-terminus (PAS-bHLH homology), and central regions involved in interactions with nuclear receptors and CBP/p300. Asterisks denote LXXLL interaction motifs.

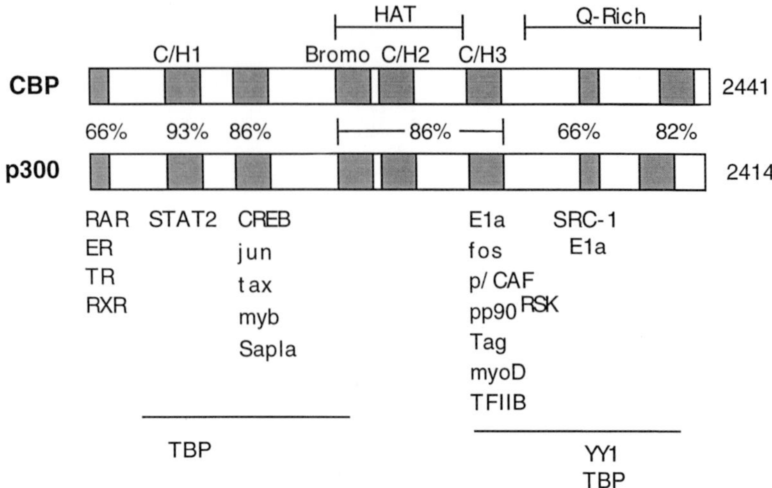

FIGURE 3   Alignment of mouse CBP and human p300. Regions involved in interaction with STAT2, nuclear receptors, jun, CREB, YY1, fos, E1A, p/CAF, pp90[RSK,] and SRC-1 are indicated, as are the three zinc fingers (C/Hx), histone acetyltransferase (HAT) domain, and the bromodomain.

The biological function of CBP and p300 have been assessed in mice in which the p300 gene was inactivated by gene targeting (Yao *et al.*, 1998). Animals with p300 null mutant died between days 9 and 11.5 of gestation, exhibiting defects in neurulation, cell proliferation, and heart development. Cells derived from p300-deficient embryos displayed specific transcriptional defects and proliferated poorly. Most strikingly, transcriptional activity of retinoic acid but not CREB was impaired in cells derived from the p300 null mutant. Surprisingly, animals heterozygous for p300 also exhibited considerable embryonic lethality. In addition, double heterozygosity for p300 and CBP was also associated with embryonic death. These results suggest that mouse development is sensitive to the overall gene dosage of p300 and CBP. In a related study, by using hammerhead ribozymes capable of cleaving either p300 or CBP mRNA, it was shown that these coactivators exhibit distinct roles in retinoic acid–induced F9 cell differentiation (Kawasaki *et al.*, 1998). F9 cells expressing a p300-specific ribozyme became resistant to RA-induced differentiation, whereas cells expressing a CBP-specific ribozyme were still able to express neuron-specific markers in response to RA. RA induction of the cell cycle inhibitor p21[Cip1] required normal levels of p300 but not CBP, whereas the reverse was seen for p27[Kip1]. In contrast, both ribozymes blocked RA-induced apoptosis. These findings suggest that, despite their similarities, p300 and CBP have distinct functions during RA-induced differentiation of F9 cells.

III. Mechanisms of Coactivator Interaction

Nearly all proteins that have been identified to interact with nuclear receptors contain a short leucine-rich motif of consensus sequence LXXLL (where L denotes leucine and X denotes any amino acid) (Table III). The LXXLL motif was first described to mediate interaction of TIF1α with the retinoic acid receptor and was subsequently demonstrated to play a similar role in RIP140 and the p160 coactivators (Heery *et al.*, 1997; Le Douarin *et al.*, 1996; Torchia *et al.*, 1997). The nuclear receptor interaction domain of SRC-1/NCoA1, TIF2/GRIP-1/NCoA2, and p/CIP/ACTR/AIB1/RAC3/ TRAM-1, spanning approximately 130 amino acids, contains three LXXLL motifs, each separated by approximately 50 amino acids. Individual mutations of each motif have been analyzed for their role in interaction and coactivation of nuclear receptors. Biochemically, mutation of individual LXXLL helical domains (HD) of SRC-1/NCoA1 and TIF2 does not appear to significantly alter the binding to ER, RAR, and RXR (Kalkhoven *et al.*,

TABLE III   AF-2 and HD Helix Alignments

NR/NCoA coordinate motif		*Amino acid sequence alignment* *-432112345678* *LXXLL*
SRC-1,	HD1	TSHKLVOLLTTT
NCoA-2,	HD1	GQTKLLOLLTTK
p/CIP,	HD1	GHKKLLOLLTCS
SRC-1,	HD2	RHKILHRLLQEG
NCoA-2,	HD2	KHKILHRLLQDS
p/CIP,	HD2	KHRILHKLLQNG
SRC-1,	HD3	DHQLLRYLLDKD
NCoA-2,	HD3	ENALLRYLLDKD
p/CIP,	HD3	NNALLRYLLDRD
RIP140		VLTYLEGLLMH1
		DSTLLASLLQSF
		ASSHLKTLLKKS
		ACSQLALLLSS
		NSLLLHLLKSQN
		KVTLLOLLLGHK
		RRTVLOLLLGNP
		KNGLLSRLLRQN
		SFNVLKOLLLSE
RXRα	AF-2	IDTFLMEMLEAP

1998; Voegel *et al.*, 1998), or to RAR/RXR heterodimers bound to DNA (Westin). Disruption of two out the three HDs has a more dramatic effect on the binding to heterodimeric nuclear receptors. When all three LXXLL motifs of SRC-1 are mutated (LXXLL to LXXAA), interaction with ER and coactivation of ER-dependent transcription was abolished (Kalkhoven *et al.*, 1998). However, an intact HD2 appears to be sufficient for interaction to ER, RAR, and RXR, suggesting that the HD2 is the preferred motif for these receptors. There is a good correlation of binding of TIF2 and SRC-1/NCoA1 to nuclear receptors and coactivation of transcription. For ER, only HD2 is sufficient for stimulatory activity. This is in line with microinjection experiments where an HD2 mutant of SRC-1/NCoA1 is unable to rescue inhibition by microinjected anti-NCoA1 IgG of estrogen-dependent transcription (Torchia *et al.*, 1997). In contrast an HD3 mutant completely rescued ER-dependent transcription. Conversely, a 50% rescue was seen with an HD2 mutant on RAR-dependent transcription, whereas an HD3 mutant was unable to rescue RAR-dependent transcription. These findings suggest that although some redundancy exists as to which HD can bind to any given receptor, some receptor specificity does exist in terms of stimulatory effects on transcription.

The structure of a ternary complex of a fragment of SRC-1 containing LXXLL helical domains 1 and 2 bound to a dimer of the liganded PPARγ LBD has been solved (Nolte *et al.*, 1998). This structure illustrates the mechanisms of ligand-dependent recruitment of coactivators containing LXXLL motifs to nuclear receptors and suggests a structural basis for cooperative assembly of coactivator complexes. Although PPARγ is thought to activate transcription as a heterodimer with RXR, the interface of the PPARγ homodimer observed in the crystal structure appears to utilize the same amino acids that are predicted to make up the heterodimer interface with RXR. Thus, the ternary complex structure is probably representative of naturally occurring dimers and heterodimers. Both subunits of the PPARγ dimer were occupied by ligand, with each of the two AF-2 domains folded against their respective LBDs in an "active" conformation and forming part of the ligand binding pocket. One molecule of SRC-1 interaction domain cocrystalized with the PPARγ dimer, such that the first LXXLL motif interacted with the AF-2 domain of one subunit of the dimer, and the second LXXLL motif interacted with the other subunit of the dimer. The interactions of the two LXXLL motifs with their respective LBD subunits were nearly identical. In each case, the LXXLL motif formed a short alpha helix. A backbone amide at the N-terminal end of the helix formed an ionic contact with an AF-2 glutamate residue that is highly conserved in many ligand-dependent nuclear receptors. A backbone carbonyl at the C-terminal end of the LXXLL helix formed an ionic contact with a conserved lysine residue present in helix 3 of the ligand binding domain. Together, the glutamate and lysine residues of the LBD constitute a ligand-activated charge clamp that grips the two ends of the LXXLL helix, allowing the leucine residues to pack tightly into an intervening hydrophobic pocket. The nonconserved X amino acids of the LXXLL motif are solvent exposed and do not contribute to binding interaction. This mechanism of binding therefore does not discriminate between different LXXLL motifs. Amino acids N-terminal of the LXXLL motif were not observed to be in contact with the PPARγ LBD, and thus are not likely to be involved in coactivator discrimination, at least for PPARγ. In contrast, amino acids C-terminal to the LXXLL motif were ordered and in close proximity to the LBD, suggesting that they could be involved in determining specificity of interaction of different coactivators containing LXXLL motifs. The linking amino acids between the first and second LXXLL motifs were disordered, suggested that they do not contribute to specificity.

The observation that two LXXLL motifs from a single SRC-1 molecule were used to interact with both components of a PPARγ dimer raises the question of whether similar interactions might be relevant to the assembly of coactivator complexes on naturally occurring nuclear receptor dimers or heterodimers. Although a single LXXLL motif in SRC-1 appears to be sufficient for coactivation of the estrogen receptor, studies suggest that the utilization of two LXXLL motifs underlies cooperative effects of RXR ligands on the transcriptional activities of RAR/RXR and PPAR/RXR heterodimers (Westin *et al.*, 1998). PPAR/RXR heterodimers represent a subset of RXR heterodimers that are permissive to RXR ligands. Thus, the PPAR/RXR heterodimer can be independently activated by PPAR-specific or RXR-specific ligands, which together can have synergistic transcriptional effects (Kliewer *et al.*, 1992). Biochemical studies of PPAR/RXR heterodimers indicated that RXR or PPAR-specific ligands could independently induce binding of SRC-1, and that together they exerted cooperative effects (Westin *et al.*, 1998). Although SRC-1 binding in response to a single ligand required only a single LXXLL motif, cooperative effects of two ligands required two LXXLL motifs. These results are thus consistent with a model in which a single molecule of SRC-1 docks onto a heterodimer of PPAR/RXR, with one LXXLL motif contacting PPAR, and a second LXXLL motif contacting RXR.

In contrast to PPAR/RXR heterodimers, RAR/RXR heterodimers represent a subset of RXR heterodimers that are nonpermissive to RXR ligands (Forman *et al.*, 1995; Kurokawa *et al.*, 1994). Thus, RAR/RXR heterodimers activate transcription in response to RAR-specific ligands, such as all-*trans* retinoic acid, but not in response to RXR-specific ligands. The failure of RXR ligands to activate RAR/RXR heterodimers has been demonstrated to result from allosteric inhibition of the binding of ligands to RXR by heterodimerization with RAR (Forman *et al.*, 1995; Kurokawa *et al.*, 1994). Intriguingly, cell-based studies have demonstrated that RXR-specific ligands can potentiate the transcriptional effects of limiting concentrations of RAR-specific ligands (Chen *et al.*, 1996). These observations imply relief of allosteric inhibition of RXR in cells in response to the binding of RAR-specific ligands. Consistent with this, RXR-specific ligands, although inactive alone, have been demonstrated to potentiate the binding of SRC-1 to RAR/RXR heterodimers in the presence of a limiting concentration of an RAR-specific ligand (Westin *et al.*, 1998). This effect results from relief of allosteric inhibition of RXR by binding of SRC-1 to RAR. The ability of RXR ligands to potentiate SRC-1 binding depends on the presence of two LXXLL motifs in SRC-1 and the AF-2 domains of both RAR and RXR. Thus, it is likely that a single molecule of SRC-1 binds to the heterodimer with one LXXLL motif docked into RAR, and a second LXXLL motif docked into RXR.

The solution of the structure of the SRC-1/PPARγ ternary complex and the structure of the unliganded PPARγ ligand binding domain has suggested a structural basis for the mechanism of allosteric inhibition of RXR by RAR (Nolte *et al.*, 1998; Westin *et al.*, 1998). Examination of the unliganded PPARγ LBD structure indicated that the AF-2 domains of the two members of the dimer were in two different conformations. One AF-2 domain was folded against the ligand binding domain in an "active" configuration, while the second AF-2 domain was extended away from the body of the ligand binding domain. Upon solution of the SRC-1/PPARγ ternary complex, it became evident that the position of the extended AF-2 domain in the unliganded PPARγ crystal was docked into the SRC-1 binding site of the "active" PPARγ subunit of a neighboring dimer. This observation is consistent with the fact that the AF-2 domains of many nuclear receptors bear a striking resemblance to the LXXLL motifs present in nuclear receptor coactivators (e.g., Table III), and it raises the possibility that allosteric interactions between RXR heterodimers could be accounted for by interactions between the AF-2 domain of RXR and the coactivator binding site of its dimeric partner. Molecular modeling of an RAR/RXR heterodimer suggested that it would be possible for the AF-2 domain of RXR to rotate around a relatively disordered loop and dock into the SRC-1 binding site of RAR (Westin *et al.*, 1998). This would be predicted to prevent closure of the RXR ligand binding pocket, accounting for allosteric inhibition of the binding of ligand to RXR. Consistent with this model, the AF-2 helix of RXR was found to bind to RAR in the absence of ligand and to be displaced from RAR in the presence of an RAR-specific ligand and SRC-1. In contrast, the RXR/AF-2 helix did not interact with PPARγ, consistent with the observation that PPARγ is a permissive partner of RXR-dependent transcription. In concert, these studies suggest a model for allosteric inhibition and coactivator assembly on RAR/RXR heterodimers in which the AF-2 domain of RXR is docked to the RAR coactivator interaction site in the absence of RAR ligands, preventing the binding of RXR ligands. Recruitment of SRC-1 via one of three LXXLL recognition motifs in response to an RAR-specific ligand displaces the RXR-AF-2 domain from RAR, relieving allosteric inhibition and allowing ligands to bind to RXR. The binding of an RXR ligand can then promote the interaction of a second LXXLL motif from the same SRC-1 molecule with RXR, stabilizing the complex. This model explains the selective responsiveness of RAR/RXR heterodimers to RAR ligands and resolves the apparent paradox of why RXR ligands potentiate the effects of RAR-specific ligands. The differential affinities of RAR and PPAR for the RXR-AF-2 domain also explain why RAR allosterically inhibits RXR, but PPAR does not. It is likely that this mechanism of allosteric regulation is involved in establishing receptor-specific responses of other RXR heterodimers. The finding that two LXXLL motifs within a single SRC-1 molecule are utilized for cooperative binding to a dimeric or heterodimeric nuclear receptor is supported by the crystal structure of SRC-1 complexed to a liganded PPARγ dimer and is likely to be prototypic for the assembly of other nuclear receptor coactivator complexes.

IV. Coactivator Complexes

The finding that the p160 factors CBP and p300 can each serve as coactivators of several nuclear receptors raises a number of intriguing questions regarding the mechanisms of transcriptional activation. On the one hand, SRC-1/TIF2/p/CIP and CBP/p300 could serve as the basis for independent pathways for transcriptional activation, perhaps acting in concert with other receptor-associated factors. However, studies suggest the p160 proteins and CBP/p300 function cooperatively by participation in the formation of a coactivator complex (Fig. 4) (Hanstein *et al.*, 1996; Kamei *et al.*, 1996; Korzus *et al.*, 1998; Torchia *et al.*, 1997; Westin *et al.*, 1998; Yao *et al.*,

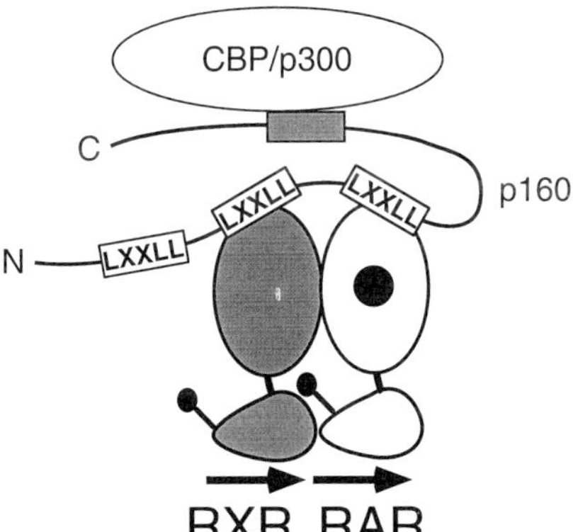

FIGURE 4 Mechanism of p160/CBP coactivator assembly on RAR/RXR heterodimers. p160 proteins are proposed to interact with RAR/RXR heterodimers via two LXXLL motifs that dock into the activation surfaces of RAR and RXR. CBP is recruited to the complex via the p160 interaction domain.

1996). Coexpression of SRC-1 and CBP has been found to result in synergistic, rather than additive, transcriptional responses to the progesterone and estrogen receptors (Smith *et al.*, 1996). Further, as alluded to previously, each of the p160 proteins have been demonstrated to interact with a conserved region in the C-terminus of CBP and p300 (Torchia *et al.*, 1997; Yao *et al.*, 1996), and immunoprecipitation experiments indicate that these interactions with CBP occur in cells (Torchia *et al.*, 1997). This interaction could thus provide nuclear receptors with two independent ways to interact with CBP/p300: directly with the N-terminus, and indirectly with the C-terminus through SRC-1. However, CBP and the p160 proteins appear to interact with a common coactivator binding site on the retinoic acid receptor. This is consistent with the observation that the interaction of CBP with nuclear receptors is mediated by an N-terminal region that contains an LXXLL motif. In the case of the retinoic acid receptor, SRC-1 binds with a much higher affinity than CBP, whereas in the case of PPARγ, the hierarchy of affinity is reversed (Schulman *et al.*, 1998). These observations are consistent with the prediction that the specific amino acids surrounding the LXXLL motif may affect receptor specificity. In studies of RAR/RXR heterodimers bound to DNA, CBP by itself bound very weakly, but was effectively recruited to the heterodimer in the presence of SRC-1 or p/CIP. This interaction depended on the CBP interaction domain of SRC-1 and p/CIP (Westin *et al.*, 1998). Consistent with these findings, mutation of the CBP interaction domain of SRC-1 abolished its ability to function as a coactivator of the retinoic acid receptor in single-cell microinjection assays.

An independent line of evidence supporting a critical role for CBP-p/160 coactivator complexes in retinoic acid receptor-dependent transcription derives from studies of the mechanisms by which the adenovirus E1A oncoprotein inhibits RA-dependent transcription (Kurokawa *et al.*, 1998). In addition to interaction with the conserved C/H3 domain of CBP and p300, E1A was also found to interact with the p160 binding domain of CBP/p300. Biochemical studies demonstrated that p/CIP and E1A competed for interaction with CBP, and overexpression of E1A was found to prevent the formation of CBP–p/CIP coactivator complexes in cells. Furthermore, mutant forms of E1A that were unable to interact with the C/H3 domains of CBP, but retained the ability to interact with the p160 interaction domain, were capable of blocking RA-dependent differentiation of p19 cells. In concert, these findings support an important role of CBP/p300–p160 coactivator complexes in mediating the transcriptional effects of retinoic acid receptors on endogenous target genes.

Because CBP and p300 appear to be expressed at functionally limiting levels, these studies suggest that one possible role of the p160 factors is to increase the affinity, and therefore probability, of CBP/p300 interactions. As already mentioned, SRC-1, TIF2, and p/CIP possess an intrinsic activator function (Kalkhoven *et al.*, 1998; Torchia *et al.*, 1997; Voegel *et al.*, 1998), and this function is in part CBP/p300-dependent. It will certainly be of interest to determine whether SRC-1 and TIF2 participate directly in additional interactions with core transcriptional machinery or proteins involved in regulating chromatin structure.

The precise roles of CBP/p300 in transactivation by nuclear receptors remain unclear. Interaction with components of core transcriptional machinery have been described, but their importance remains to be established. In the case of CREB, CBP has been suggested to serve as the mechanism for recruitment of RNA polymerase II through the associated factor RNA helicase A (Nakajima *et al.*, 1997). RNA helicase A binds to the C/H3 domain of CBP/p300, and this interaction appears to be required for CREB function. However, the C/H3 domain of CBP is not required for coactivation of the retinoic acid receptor, indicating that CREB and RAR utilize CBP in fundamentally different ways (Kurokawa *et al.*, 1998). This difference is further exemplified by the respective requirement of the HAT function of CBP by CREB and RAR. Point mutations that abolish the HAT activity of CBP have no effect on its ability to serve as a coactivator of RAR in the single cell microinjection assay, but abolish its ability to serve as a coactivator of CREB (Korzus *et al.*, 1998).

CBP and p300 have also been reported to be associated with p/CAF, which is related to the yeast transcriptional coactivator GCN5 and also contains a histone acetyltransferase function (Yang *et al.*, 1996). Microinjection of anti p/CAF antibodies blocked RAR and CREB-dependent transcription, but not transactivation by STAT1 (Korzus *et al.*, 1998). Intriguingly, the HAT activity of p/CAF was required for RAR function, but not for function of CREB (Korzus *et al.*, 1998). The mechanisms by which p/CAF is recruited into the CBP–nuclear receptor coactivator complex is unclear. Although initially discovered by virtue of its interaction with the C/H3 domain of CBP and p300, the purification of a p/CAF complex did not result in the co-purification of CBP or p300 (Ogryzko *et al.*, 1998). Instead, the p/CAF complex contained a number of proteins previously identified as TAFs containing histone-like motifs. p/CAF has also been reported to interact with SRC-1 and pCIP/ACTR, as well as with nuclear receptors (Chen *et al.*, 1997; Korzus *et al.*, 1998). The recruitment of p/CAF and associated factors into a nuclear receptor coactivator complex may thus result from the concerted effects of multiple weak interactions.

V. Conclusions

Several lines of evidence indicate that the p160 factors and CBP/p300 function as components of a nuclear receptor coactivator complex and, at least in some contexts, are required for ligand-dependent transcription. It remains likely, however, that additional interacting proteins will prove to have important functional roles, either by participating in the formation of complexes that contain SRC-1 and/or CBP family members, or by providing distinct pathways for transcriptional activation. Candidates for proteins that could serve such modulatory roles include TIF1 and RIP140, which interact with several nuclear receptors, but do not potentiate their transcriptional activities to the extent observed for SRC-1, TIF2, and CBP/p300 when overexpressed.

In addition, studies indicate the existence of receptor- and cell-type-specific coactivator proteins that modulate nuclear receptor function. For example, PGC-1 (for PPAR gamma coactivator-1), was isolated on the basis of a search for a brown fat–specific coactivator necessary for transcriptional activation of the UCP-1 promoter by PPARγ (Puigserver *et al.*, 1998). PGC-1 shows no homology to the p160 family of coactivators and may thus represent a novel family of nuclear receptor coactivators. PGC-1 is selectively expressed in brown fat and was shown to coactivate PPARγ and TR function on the uncoupling protein (UCP-1) promoter. Ectopic expression of PGC-1 in white adipose cells activated expression of UCP-1 and key mitochondrial enzymes of the respiratory chain, and increased the cellular content of mitochondrial DNA. Remarkably, PGC-1 mRNA expression is elevated upon cold exposure of mice in both brown fat and skeletal muscle, which are key thermogenic tissues. These results indicate that PGC-1 plays a key role in linking nuclear receptors to the transcriptional program of adaptive thermogenesis, and suggest that additional coactivators exist that are required for specific physiologic responses.

Of particular interest are the discoveries of receptor-associated complexes that interact with the thyroid hormone receptor (TRAPs) and the vitamin D receptor (DRIPs) that are clearly distinct from the CBP/p160 coactivator complexes and stimulate transcription *in vitro*. A critical question is whether these complexes provide alternative routes to transcriptional activation or work in concert, perhaps sequentially, with the CBP/p300 complexes. Indirect evidence suggesting that CBP is required for initial, but not subsequent, rounds of transcription has been provided by elegant *in vitro* transcription studies on chromatinized templates (Kraus and Kadonaga, 1998). The ability to evaluate these proteins in *in vitro* transcription assays should provide the most direct means of testing the roles of specific complexes of coactivator complexes in ligand-dependent transcription.

References

Anzick, S. L., Dononen, J., Walker, R. L., Azorsa, D. O., Tanner, M. M., Guan, X.-Y., Sauter, G., Kallioniemi, O.-P., Trent, J. M., and Meltzer, P. S. (1997). AIB1, a steroid receptor coactivator amplified in breast and ovarian cancer. *Science* 277(5328), 965–968.

Arias, J., Alberts, A. S., Brindle, P., Claret, F. X., Smeal, T., Karin, M., Feramisco, J., and Montminy, M. (1994). Activation of cAMP and mitogen responsive genes relies on a common nuclear factor. *Nature* 370, 226–229.

Barettino, D., Vivanco Ruiz, M. M., and Stunnenberg, H. G. (1994). Characterization of the ligand-dependent transactivation domain of thyroid hormone receptor. *EMBO J.* 13, 3039–3049.

Beato, M. (1989). Gene regulation by steroid hormones. *Cell* 56, 335–344.

Beato, M., Herrlich, P., and Schütz, G. (1995). Steroid hormone receptors: many actors in search of a plot. *Cell* 83, 851–857.

Bourguet, W., Ruff, M., Chambon, P., Gronemeyer, H., and Moras, D. (1995). Crystal structure of the ligand-binding domain of the human nuclear receptor RXR-α. *Nature* 375, 377–382.

Brzozowski, A. M., Pike, A. C. W., Dauter, Z., Hubbard, R. E., Bonn, T., Engström, O., Öhman, L., Greene, G. L., Gustafsson, J.-Å., and Carlquist, M. (1997). Molecular basis of agonism and antagonism in the oestrogen receptor. *Nature* 389(6652), 753–758.

Cavailles, V., Dauvois, S., Danielian, P. S., and Parker, M. G. (1994). Interaction of proteins with transcriptionally active estrogen receptors. *Proc. Natl. Acad. Sci. USA* 91, 10009–10013.

Cavailles, V., Dauvois, S., L'Horset, F., Lopez, G., Hoarc, S., Kushner, P. J., and Parker, M. G. (1995). Nuclear factor RIP140 modulates transcriptional activation by the estrogen receptor. *EMBO J.* 14, 3741–3751.

Chakravarti, D., LaMorte, V. J., Nelson, M. C., Nakajima, T., Schulman, I. G., Juguilon, H., Montminy, M., and Evans, R. M. (1996). Role of CBP/p300 in nuclear receptor signalling. *Nature* 383, 99–103.

Chen, J. D., and Evans, R. M. (1995). A transcriptional co-repressor that interacts with nuclear hormone receptors. *Nature* 377, 454–457.

Chen, J.-Y., Clifford, J., Zusi, C., Starrett, J., Tortolani, D., Ostrowski, J., Reczek, P. R., Chambon, P., and Gronemeyer, H. (1996). Two distinct actions of retinoid-receptor ligands. *Nature* 382, 819–822.

Chen, H., Lin, R. J., Schiltz, R. L., Chakravarti, D., Nash, A., Nagy, L., Privalsky, M. L., Nakatani, Y., and Evans, R. M. (1997). Nuclear receptor coactivator ACTR is a novel histone acetyltransferase and forms a multimeric activation complex with P/CAF and CBP/p300.*Cell* 90(3), 569–580.

Danielian, P. S., White, R., Lees, J. A., and Parker, M. G. (1992). Identification of a conserved region required for hormone-dependent transcriptional activation by steroid hormone receptors. *EMBO J.* 11, 1025–1033.

Durand, B., Saunders, M., Gaudon, C., Roy, B., Losson, R., and Chambon, P. (1994). Activation function 2 (AF-2) of retinoic acid receptor and 9-cis retinoic acid receptor: presence of a conserved autonomous constitutive activating domain and influence of the nature of the response element on AF-2 activity. *EMBO J.* 13, 5370–5382.

Eggert, M., Möws, C. C., Tripier, D., Arnold, R., Michel, J., Nickel, J., Schmidt, S., Beato, M., and Renkawitz, R. (1995). A fraction enriched in a novel glucocorticoid receptor-interacting protein stimulates receptor-dependent transcription *in vitro*. *J. Biol. Chem.* 270(51), 30755–30759.

Evans, R. M. (1988). The steroid and thyroid hormone receptor superfamily. *Science* 240, 889–895.

Fondell, J. D., Ge, H., and Roeder, R. G. (1996). Ligand induction of a transcriptionally active thyroid hormone receptor coactivator complex. *Proc. Natl. Acad. Sci. USA* 93, 8329–8333.

Forman, B. M., Umesono, K., Chen, J., and Evans, R. M. (1995). Unique response pathways are established by allosteric interactions among nuclear hormone receptors. *Cell* 81, 541–550.

Fraser, R. A., Heard, D. J., Adam, S., Lavigne, A. C., Le Douarin, B., Tora, L., Losson, R., Rochette-Egly, C., and Chambon, P. (1998). The putative cofactor TIFα is a protein kinase that is hyperphosphorylated upon interaction with liganded nuclear receptors. *J. Biol. Chem.* [273](26), 16199–16204.

Glass, C. K. (1996). Some new twists in the regulation of gene expression by thyroid hormone and retinoic acid receptors. *J. Endocrinol* **150**, 349–357.

Halachmi, S., Marden, E., Martin, G., MacKay, H., Abbondanza, C., and Brown, M. (1994). Estrogen receptor-associated proteins: possible mediators of hormone-induced transcription. *Science* **264**, 1455–1458.

Hanstein, B., Eckner, R., DiRenzo, J., Halachmi, S., Liu, H., Searcy, B., Kurokawa, R., and Brown, M. (1996). p300 is a component of an estrogen receptor coactivator complex. *Proc. Natl. Acad. Sci. USA* **93**, 11540–11545.

Heery, D. M., Kalkhoven, E., Hoare, S., and Parker, M. G. (1997). A signature motif in transcriptional co-activators mediates binding to nuclear receptors. *Nature* **387**(6634), 733–736.

Hong, H., Kohli, K., Trivedi, A., Johnson, D. L., and Stallcup, M. R. (1996). GRIP1, a novel mouse protein that serves as a transcriptional coactivator in yeast for the hormone binding domains of steroid receptors. *Proc. Natl. Acad. Sci. USA* **93**, 4948–4952.

Hörlein, A., Näär, A., Heinzel, T., Torchia, J., Gloss, B., Kurokawa, R., Kamei, Y., Ryan, A., Söderström, M., Glass, C. K., and Rosenfeld, M. G. (1995). Ligand-independent repression by the thyroid hormone receptor is mediated by a nuclear receptor co-repressor, N-CoR. *Nature* **377**, 397–404.

Huang, Z. J., Edery, I., and Rosbach, M. (1993). PAS is a dimerization domain common to *Drosophila* period and several transcription factors. *Nature* **364**, 259–262.

Kalkhoven, E., Valentine, J. E., Heery, D. M., and Parker, M. G. (1998). Isoforms of steroid receptor co-activator 1 differ in their ability to potentiate transcription by the oestrogen receptor. *EMBO J.* **17**(1), 232–243.

Kamei, Y., Xu, L., Heinzel, T., Torchia, J., Kurokawa, R., Gloss, B., Lin, S.-C., Heyman, R., Rose, D., Glass, C., and Rosenfeld, M. (1996). A CBP integrator complex mediates transcriptional activation and AP-1 inhibition by nuclear receptors. *Cell* **85**, 403–414.

Kawasaki, H., Eckner, R., Yao, T.-P., Taira, K., Chiu, R., Livingston, D. M., and Yokoyama, K. K. (1998). Distinct roles of the co-activators p300 and CBP in retinoic-acid-induced F9-cell differentiation. *Nature* **393**, 284–289.

Kliewer, S. A., Umesono, K., Noonan, D. J., Heyman, R. A., and Evans, R. M. (1992). Convergence of 9-cis retinoic acid and peroxisome proliferator signalling pathways through heterodimer formation of their receptors. *Nature* **358**, 771–774.

Korzus, E., Torchia, J., Rose, D. W., Xu, L., Kurokawa, R., McInerney, E. M., Mullen, T.-M., Glass, C. K., and Rosenfeld, M. G. (1998). Transcription factor-specific requirements for coactivators and their acetyltransferase functions. *Science* **279**, 703–707.

Kraus, W. L., and Kadonaga, J. T. (1998). p300 and estrogen receptor cooperatively activate transcription via differential enhancement of initiation and reinitiation. *Genes & Dev.* **12**(3), 331–342.

Kurokawa, R., DiRenzo, J., Boehm, M., Sugarman, J., Gloss, B., Rosenfeld, M. G., Heyman, R. A., and Glass, C. K. (1994). Regulation of retinoid signaling by receptor polarity and allosteric control of ligand binding. *Nature* **371**, 528–531.

Kurokawa, R., Söderström, M., Hörlein, A., Halachmi, S., Brown, M., Rosenfeld, M. G., and Glass, C. K. (1995). Polarity-specific activities of retinoic acid receptors determined by a co-repressor. *Nature* **377**, 451–454.

Kurokawa, R., Kalafus, D., Ogliastro, M.-H., Kioussi, C., Xu, L., Torchia, J., Rosenfeld, M. G., and Glass, C. K. (1998). Differential use of CREB binding protein–coactivator complexes. *Science* **279**, 700–703.

Le Douarin, B., Zechel, C., Garnier, J.-M., Lutz, Y., Tora, L., Pierrat, B., Heery, D., Gronemeyer, H., Chambon, P., and Losson, R. (1995). The N-terminal part of TIF1, a putative mediator of the ligand-dependent activation function (AF-2) of nuclear receptors, is fused to B-raf in the oncogenic protein T18. *EMBO J.* **14**, 2020–2033.

Le Douarin, B., Nielson, A. L., Garnier, J.-M., Ichinose, H., Jeanmougin, F., Losson, R., and Chambon, P. (1996). A possible involvement of TIF1α and TIF1β in the epigenetic control of transcription by nuclear receptors. *EMBO J.* 15(23), 6701–6715.

Lee, J. W., Ryan, F., Swaffield, J. C., Johnston, S. A., and Moore, D. D. (1995). Interaction of thyroid-hormone receptor with a conserved transcriptional mediator. *Nature* 374, 91–94.

Li, H., Gomes, P. J., and Chen, J. D. (1997). RAC3, a steroid/nuclear receptor-associated coactivator that is related to SRC-1 and TIF2. *Proc. Natl. Acad. Sci. USA* 94, 8479–8484.

Mangelsdorf, D. J., Thummel, C., Beato, M., Herrlich, P., Schütz, G., Umesono, K., Blumberg, B., Kastner, P., Mark, M., Chambon, P., and Evans, R. M. (1995). The nuclear receptor superfamily: the second decade. *Cell* 83, 835–839.

Meyer, M., Gronemeyer, H., Turcotte, B., Bocquel, M., Tasset, D., and Chambon, P. (1989). Steroid hormone receptors compete for factors that mediate their enhancer function. *Cell* 57, 433–442.

Nakajima, T., Uchida, C., Anderson, S. F., Lee, C.-G., Hurwitz, J., Parvin, J. D., and Montminy, M. (1997). RNA helicase A mediates association of CBP with RNA polymerase II. *Cell* 90, 1107–1112.

Nolte, R. T., Wisely, G. B., Westin, S., Cobb, J. E., Lambert, M. H., Kurokawa, R., Rosenfeld, M. G., Willson, T. M., Glass, C. K., and Milburn, M. V. (1998). Ligand binding and coactivator assembly of the peroxisome proliferator-activated receptor γ. *Nature*, in press.

Ogryzko, V. V., Kotani, T., Zhang, X., Schiltz, R. L., Howard, T., Yang, X.-J., Howard, B. H., Qin, J., and Nakatani, Y. (1998). Histone-like TAFs within the PCAF histone acetylase complex. *Cell* 94, 35–44.

Oñate, S. A., Tsai, S. Y., Tsai, M.-J., and O'Malley, B. W. (1995). Sequence and characterization of a coactivator for the steroid hormone receptor superfamily. *Science* 270, 1354–1357.

Puigserver, P., Wu, Z., Park, C. W., Graves, R., Wright, M., and Spiegelman, B. M. (1998). A cold-inducible coactivator of nuclear receptors linked to adaptive thermogenesis. *Cell* 92, 829–839.

Rachez, C., Suldan, Z., Ward, J., Chang, C.-P. B., Burakov, D., Erdjument-Bromage, H., Tempst, P., and Freedman, L. P. (1998). A novel protein complex that interacts with the vitamin D³ receptor in a ligand-dependent manner and enhances VDR transactivation in a cell-free system. *Genes & Dev.* 12, 1787–1800.

Renaud, J.-P., Rochel, N., Ruff, M., Vivat, V., Chambon, P., Gronemeyer, H., and Moras, D. (1995). Crystal structure of the RAR-γ ligand-binding domain bound to all-*trans* retinoic acid. *Nature* 378, 681–689.

Schulman, I. G., Shao, G., and Heyman, R. A. (1998). Transactivation by retinoid X receptor–peroxisome proliferator-activated receptor γ (PPARγ) heterodimers: intermolecular synergy requires only the PPARγ hormone-dependent activation function. *Mol. Cell. Biol.* 18, 3483–3494.

Smith, C. L., Oñate, S. A., Tsai, M.-J., and O'Malley, B. W. (1996). CREB binding protein acts synergistically with steroid receptor coactivator-1 to enhance steroid receptor-dependent transcription. *Proc. Natl. Acad. Sci. USA* 93, 8884–8888.

Takeshita, A., Cardona, G. R., Koibuchi, N., Suen, C.-S., and Chin, W. W. (1997). TRAM-1, a novel 160-kDa thyroid hormone receptor activator molecule, exhibits distinct properties from steroid receptor coactivator-1. *J. Biol. Chem.* 272(44), 27629–27634.

Tone, Y., Collingwood, T. N., Adams, M., and Chatterjee, V. K. (1994). Functional analysis of a transactivation domain in the thyroid hormone beta receptor. *J. Biol. Chem.* 269, 31157–31161.

Torchia, J., Rose, D. W., Inostroza, J., Kamei, Y., Westin, S., Glass, C. K., and Rosenfeld, M. G. (1997). The transcriptional co-activator p/CIP binds CBP and mediates nuclear-receptor function. *Nature* 387, 677–684.

Voegel, J. J., Heine, M. J. S., Zechel, C., Chambon, P., and Gronemeyer, H. (1996). TIF2, a 160 kDa transcriptional mediator for the ligand-dependent activation function AF-2 of nuclear receptors. *EMBO J.* 15(14), 3667–3675.

Voegel, J. J., Heine, M. J., Tini, M., Vivat, V., Chambon, P., and Gronemeyer, H. (1998). The coactivator TIF2 contains three nuclear receptor-binding motifs and mediates transactivation through CBP binding-dependent and -independent pathways. *EMBO J.* **17**(2), 507–519.

Wagner, R. L., Apriletti, J. W., McGrath, M. E., West, B. L., Baxter, J. D., and Fletterick, R. J. (1995). A structural role for hormone in the thyroid hormone receptor. *Nature* **378**, 690–697.

Westin, S., Kurokawa, R., Nolte, R. T., Wisely, G. B., McInerney, E. M., Rose, D. W., Milburn, M. V., Rosenfeld, M. G., and Glass, C. K. (1998). Interactions governing nuclear receptor heterodimer–coactivator complexes. *Nature,* in press.

Williams, S. P., and Sigler, P. B. (1998). Atomic structure of progesterone complexed with its receptor. *Nature* **393**, 392–396.

Xu, J., Qiu, Y., DeMayo, F. J., Tsai, S. Y., Tsai, M.-J., and O'Malley, B. W. (1998). Partial hormone resistance in mice with disruption of the steroid receptor coactivator-1 (SRC-1) gene. *Science* **279**, 1922–1925.

Yang, X. J., Ogryzko, V. V., Nishikawa, J., Howard, B. H., and Nakatani, Y. (1996). A p300/CBP-associated factor that competes with the adenoviral oncoprotein E1A. *Nature* **382**(6589), 319–324.

Yao, T.-P., Ku, G., Zhou, N., Scully, R., and Livingston, D. M. (1996). The nuclear hormone receptor coactivator SRC-1 is a specific target of p300. *Proc. Natl. Acad. Sci. USA* **93**, 10626–10631.

Yao, T.-P., Oh, S. P., Fuchs, M., Zhou, N.-D., Ch'ng, L.-E., Newsome, D., Bronson, R. T., Li, E., Livingston, D. M., and Eckner, R. (1998). Gene dosage-dependent embryonic development and proliferation defects in mice lacking the transcriptional integrator p300. *Cell* **93**, 361–372.

Yeh, S., and Chang, C. (1996). Cloning and characterization of a specific coactivator, ARA$_{70}$, for the androgen receptor in human prostate cells. *Proc. Natl. Acad. Sci. USA* **93**, 5517–5521.

Yuan, C.-X., Ito, M., Fondell, J. D., Fu, Z.-Y., and Roeder, R. G. (1998). The TRAP220 component of a thyroid hormone receptor-associated protein (TRAP) coactivator complex interacts directly with nuclear receptors in a ligand-dependent fashion. *Proc. Natl. Acad. Sci. USA* **95**, 7939–7944.

Zhu, Y., Qi, C., Jain, S., Rao, M. S., and Reddy, J. K. (1997). Isolation and characterization of PBP, a protein that interacts with peroxisome proliferator-activated receptor. *J. Biol. Chem.* **272**(41), 25500–25506.

Christian Schindler
Inga Strehlow
Departments of Microbiology and Medicine
College of Physicians and Surgeons
Columbia University
New York, NY 10032

Cytokines and STAT Signaling

I. Introduction

First discovered as the major signal transducer in interferon-mediated gene activation, STATs (signal transducers and activators of transcription) are now known to play a significant role in signal transduction for most cytokines. STATs represent a family of conserved proteins, seven of which have been identified in mammals (i.e., Stat1, 2, 3, 4, 5a, 5b, and 6; Darnell, 1997; Ihle *et al.*, 1994; Schindler and Darnell, 1995). Homologues have also been identified in lower eukaryotes (Hou *et al.*, 1996; Kawata *et al.*, 1997; Yan *et al.*, 1996). JAKs are receptor-associated tyrosine kinases, which mediate the ligand dependent activation of STATs. These two protein families are the defining components of the JAK-STAT pathway.

The JAK-STAT signaling paradigm (see Fig. 1) has been well characterized for many ligands (reviewed in Darnell, 1997; Ihle *et al.*, 1994; Schindler and Darnell, 1995). Briefly, upon binding ligand a receptor will dimerize. This enables receptor-associated JAKs to become activated, which in turn phosphorylate tyrosine motifs in the cytoplasmic tail of the receptor. These receptor tyrosine motifs are recognized by the SH2 domains of STATs, thereby mediating the recruitment of the appropriate STAT to the receptor complex. Once at the receptor, JAKs phosphorylate STATs on a conserved

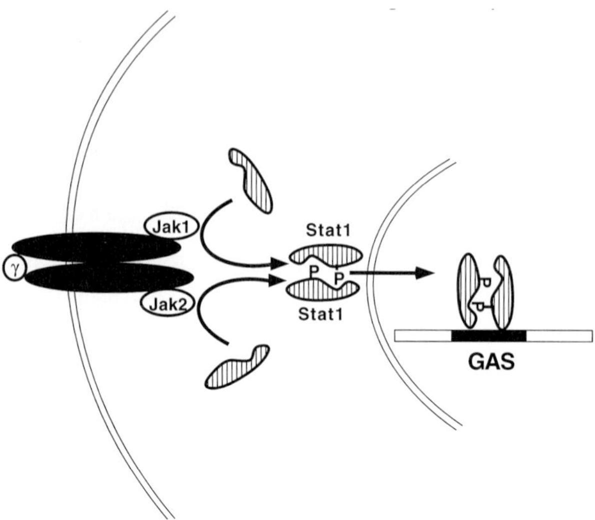

FIGURE I   The IFN-γ stimulated JAK-STAT pathway. When IFN-γ binds its receptor, two-associated tyrosine kinase, Jak1 and Jak2, become activated. These kinases then phosphorylate a receptor tyrosyl residue, which is in turn specifically recognized by the SH2 domain of Stat1. Once recruited to the receptor, Stat1 becomes phosphorylated by the JAKs. Now activated, Stat1 is released from the receptor and forms homodimers, which are competent for nuclear translocation. Once in the nucleus the Stat1 homodimer binds a member of the GAS family of enhancers, culminating in transcription. See text for details.

tyrosine. Activated STATs are released from the receptor and dimerize through the interaction of the SH2 domain of one STAT with the phospho-tyrosine of the other STAT. These dimers translocate to the nucleus, where they bind to members of the GAS (IFN-gamma activation site) family of enhancers, culminating in the transcription of genes. STATs thus transduce high-fidelity signals directly from the cell surface to target genes.

The past 2 years have seen significant progress in the characterization of the JAK-STAT signaling cascade. Important developments have included the establishment of murine "knockout" models and the resolution of the crystal structure of two STATs. Another exciting area of progress has been in the identification of molecules that modify signaling through the STAT pathway. These more recent developments will be the focus of this review.

II. STAT Domains

STATs share a number of functionally conserved domains (see Fig. 2), including an amino terminal domain (NH$_2$), a coiled-coil domain, a DNA binding domain (DBD), an SH2 domain, and a tyrosine activation domain (Y). The transcriptional activation domain (TAD) is carboxy terminal and

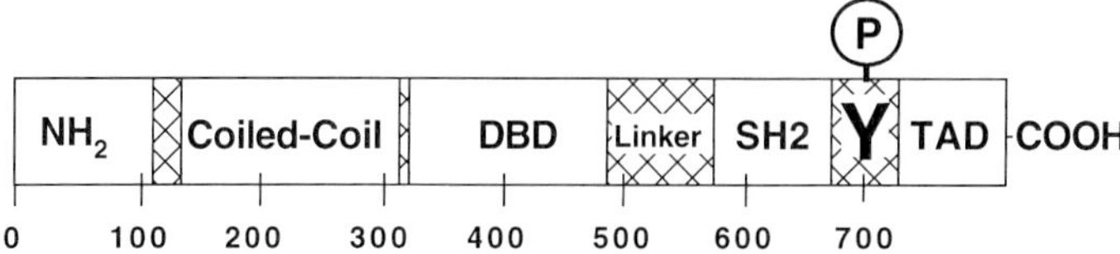

FIGURE 2   STAT structure. STATs share a number of conserved domains, including an amino terminal domain (NH₂), a coiled-coil domain, aDNA binding domain (DBD), a linker domain, an SH2 domain, and a tyrosine activation domain (Y). The sequences carboxy terminal to this tyrosine activation domain are not conserved, but they do encode a transcriptional activation domain (TAD). See text for details.

not well conserved. The recent crystal structures of Stat1 and Stat3 (bound to DNA) have not only confirmed the functional role previously assigned to many of these domains, but provided important insight into how these domains interact with each other (Becker *et al.*, 1998; Chen *et al.*, 1998).

A. Amino Terminus

The amino terminal domain spans ~125 amino acids and is the second most conserved STAT domain. This domain can be removed from the core STAT complex through limited proteolysis, suggesting that it is a physically distinct domain (Vinkemeier *et al.*, 1996). Both functional studies and DNA binding studies have provided compelling evidence that the amino terminus mediates a cooperativity in DNA binding when there are tandem GAS elements (Vinkemeier *et al.*, 1996; Xu *et al.*, 1996). The recent resolution of the structure of the amino terminal fragment has determined that it consists of eight helices that are assembled into a hooklike structure. This domain achieves cooperativity in DNA binding by promoting an interaction between two STAT amino termini (Vinkemeier *et al.*, 1998). Amino termini of STATs have been implicated in other functions. For example, the amino terminus of Stat2 may promote interactions with the IFN-α receptor β-chain (i.e., IFNAR2; Leung *et al.*, 1995, 1996), and the amino terminus of Stat1 may promote association with p300/CBP (Zhang *et al.*, 1996). Additionally, amino terminally chimeric STATs exhibit structural changes and fail to translocate to the nucleus (Strehlow and Schindler, 1998). These chimeric molecules, and an amino terminal point mutant (Shuai *et al.*, 1996), also exhibit a defect in STAT deactivation, suggesting that nuclear translocation and deactivation may be causally linked.

B. Coiled-Coil Domain

The crystal structures of both Stat1 and Stat3 have identified a coiled-coil or four-helix bundle domain that extends from amino acids ~135 to ~315 (Becker *et al.*, 1998; Chen *et al.*, 1998). Consistent with the notion that coiled-coil domains mediate protein–protein interactions, the two four-helix bundles in a STAT dimer protrude laterally (~80 Å) from the core structure and have been shown to associate with other proteins. One of the first and most obvious candidate proteins to bind this region is p48, the DNA binding component of the first characterized STAT transcription factor ISGF3 (interferon stimulated gene factor 3; see Section III,A,1). Although two groups have identified the p48 binding domain, a controversy remains as to whether this interaction is with just Stat2 or both Stat1 and Stat2 (Horvath *et al.*, 1996; Martinez-Moczygemba *et al.*, 1997). More generic searches, with the yeast two hybrid interaction assay, have led to the identification of other interacting proteins, including PIAS3, Nmi and StIP1 (Stat interacting protein-1). Although characterization of PIAS3 suggests that it is a naturally occurring antagonist of Stat3 (Chung *et al.*, 1997; Zhu *et al.*, 1999; see Section VII,C,2), StIP1 appears to play a more general and positive role in STAT signaling (Collum *et al.*, 1999). Studies suggest that StIP1 may serve as a scaffold protein potentiating the interactions between STATs and JAKs.

C. DNA Binding Domain

Consistent with the palindromic structure of GAS elements (see Section VI,A,2), STATs bind DNA as dimers. A symmetric/reciprocal SH2–phosphotyrosine interaction mediates this obligatory dimerization (Gupta *et al.*, 1996; Shuai *et al.*, 1994). The STAT domain that binds DNA was initially identified through the generation of Stat1:Stat3 chimeras (Horvath *et al.*, 1995). This has now been confirmed and extended by the recent solution of the Stat1 and Stat3 crystal structures (Becker *et al.*, 1998; Chen *et al.*, 1998). These structures demonstrate that the DNA binding domain, spanning amino acids ~320–475, includes a β-barrel and has the general architecture of an immunoglobulin fold. As expected, amino acids that are important in the recognition of the GAS element are highly conserved. Both the symmetry and stability of the DNA binding complex are critically dependent on the SH2 dimerization domain and the linker region that connects these two domains. The linker domain provides a critical structural interface between the SH2 and DNA binding domains. Of additional note, Stat2, which is unable to bind GAS elements directly, exhibits a high degree of conservation in each of the regions important for DNA binding.

D. SH2 Domain

SH2 domains, which coevolved with phosphotyrosine-based signal transduction, mediate specific interactions with the tyrosine residues that become phosphorylated during signaling (Hunter, 1995). Consistent with the critical role the STAT SH2 domain plays in signal transduction, this domain (amino acids ~585–685) is among the most highly conserved in the STAT family (Becker *et al.*, 1998; Chen *et al.*, 1998). This SH2 domain mediates several important steps in STAT signal transduction. This includes STAT recruitment to the receptor (Greenlund *et al.*, 1994; Heim *et al.*, 1995; Lin *et al.*, 1995; Stahl *et al.*, 1995), an obligate interaction with the activating JAK (Barahmand-Pour *et al.*, 1998; Gupta *et al.*, 1996), and finally STAT dimerization (Shuai *et al.*, 1994). Dimerization is required for nuclear translocation and DNA binding (Becker *et al.*, 1998; Chen *et al.*, 1998; Strehlow and Schindler, 1998). Remarkably, however, this domain has only limited sequence homology to other SH2 domains, potentially reflecting its relatively early appearance during the evolution of phosphotyrosyl-based signaling. In the case of Stat1, only 16 residues are conserved with the prototypical SH2 domain of v-Src. Yet, the overall structure of the STAT SH2 domain, including the defining arginine at the core of the SH2 domain, is well conserved (Becker *et al.*, 1998; Chen *et al.*, 1998). As previously predicted (Shuai *et al.*, 1994), the STAT structures demonstrate that the phosphotyrosine bound by the SH2 domain is provided by the adjoining partner. This leads to the formation of a pair of crossover connections. Consistent with other SH2 domains, the interaction with the ligand is limited to a phosphotyrosine plus several carboxy terminal residues. For Stat1, where there is better structural detail, this entails an interaction with seven carboxy terminal residues. Residues at $+1$, $+3$, and $+5$ mediate important interactions with the SH2 domain, where residue $+5$ serves an additional important structural role. The considerable variability in the size and chemical properties of these residues among the STATs is likely to be responsible for the high degree of specificity associated with STAT–SH2 interactions.

E. Tyrosine Activation Motif

As outlined in the preceding section, the interaction between the SH2 domain and the activation tyrosine motif is critical for STAT-based signaling. The proximity of this tyrosine to the SH2 domain (i.e., 10–15 amino acids carboxy terminal) effectively prevents it from binding to its own SH2 domain (Becker *et al.*, 1998; Chen *et al.*, 1998). The sequences distal to this residue (i.e., positions $+1$ to $+7$), which contribute to SH2 binding, are highly variable. This is likely to be critical in restricting the SH2 domain's ability to recognize the activation tyrosine, during both receptor recruitment and STAT dimerization (Greenlund *et al.*, 1995). Consistent with this speculation, only a limited number of STAT dimers form, even under conditions where STATs are artificially overexpressed (Schindler and Darnell, 1995). Of note, all STATs have been shown to form homodimers *in vivo*, with the notable exception of Stat2. Stat2 homodimers have only been documented *in vitro* (Bluyssen and Levy, 1997; Gupta *et al.*, 1996). Several STATs also heterodimerize. Stat1:Stat3 heterodimers and Stat5a:Stat5b heterodimers, for example, have been well documented *in vivo* (Darnell, 1997; Schindler and Darnell, 1995). Additional heterodimers have been reported, but their existence is less well documented (Ghislain and Fish, 1996; Li *et al.*, 1996).

F. Carboxy Terminus

The carboxy terminus, which is defined as all sequences carboxy terminal to the tyrosine activation motif, encodes the STAT transcriptional activation domain (TAD). Consistent with other transcription factors, this region varies substantially in both size and sequence (Darnell, 1997; Schindler and Darnell, 1995). The TAD was first mapped to the carboxy terminus during the characterization of two Stat1 isoforms recovered from purified of ISGF3 (see Section III,A; Schindler *et al.*, 1992a). The shorter isoform, Stat1β, which is missing the last 38 carboxy terminal amino acids, was found to arise from an alternative transcript. Curiously, Stat1β was able to fully restore type I IFN, but not type II (i.e., IFN-γ) signaling in a Stat1-deficient cell line. This suggested that a carboxy terminus of full-length Stat1 (Stat1α) is required for IFN-γ signaling, but not for ISGF3 dependent signaling (Müller *et al.*, 1993b). Subsequent studies have confirmed that although the carboxy terminus of Stat1 encodes a TAD, Stat2 provides the TAD for ISGF3 (Qureshi *et al.*, 1996; Wen *et al.*, 1995). TADs have also been carefully mapped to the carboxy termini of Stat3, Stat5, and Stat6 (Azam *et al.*, 1997; Caldenhoven *et al.*, 1996; Lu *et al.*, 1997; Mikita *et al.*, 1996; Moriggl *et al.*, 1996, 1997; Mui *et al.*, 1996). In general, these TADs have been mapped to relatively short regions (i.e., ~50 amino acids) at the end of STATs by deletional studies and "classical" GAL4 fusion assays (Sadowski and Ptashne, 1989). They appear to vary significantly in potency (Park and Schindler, 1999; Moriggl, *et al.*, 1997), but the physiological significance of this observation remains unclear. The localization of STAT TADs to a short carboxy-terminal fragment begets the question as to what function the remaining carboxy terminal sequences serve, especially in Stat2 and Stat6 (Darnell, 1997). Of note, in contrast to all other STATs, the carboxy terminal sequence of Stat2 is not conserved between man and mouse, even though they appear to be functionally conserved (Park and Schindler, 1999). Additional regulation of the TAD is discussed later (see Section VI,B).

Although the potential for Stat1β to serve as a naturally occurring dominant negative isoform remains largely unexplored, a number of studies have examined the potential role of other "naturally occurring" carboxy terminally truncated STATs. Clearly, under conditions of "artificial" overexpression, these isoforms antagonize signaling (Caldenhoven *et al.*, 1996; Minami *et al.*, 1996; Mui *et al.*, 1996; Nakajima *et al.*, 1996; Sasse *et al.*, 1997; Wang *et al.*, 1996). However, the truncated species of Stat3 and Stat5 also appear to be important under more physiological conditions. For example, the naturally occurring truncated isoform of Stat3 (Stat3β), which includes seven unique carboxyl terminal amino acids, exhibits a prolonged pattern of activation (Sasse *et al.*, 1997; Schaefer *et al.*, 1997). Unexpectedly, Stat3β appears to promote the expression of some genes through its unique ability to synergize with c-Jun (Schaefer *et al.*, 1995, 1997; see Section VI). In contrast, the naturally occurring truncated isoforms of Stat5 (p77 and P80) encode no "extra" amino acids, because they are generated by a novel proteolytic processing event (Azam *et al.*, 1995, 1997). These truncated isoforms, and the protease that generates them, have only been identified in immature myeloid tissues (Azam *et al.*, 1997; Bovolenta *et al.*, 1998; Lokuta *et al.*, 1998; Meyer *et al.*, 1998). This has given rise to the speculation that this process may afford ligands an opportunity to transduce a unique set of signals in immature cells.

III. Cytokine Receptors Signal through STATs

The JAK-STAT signaling cascade has been shown to be critical for signaling by many members of the cytokine family of receptors. It may also contribute to signaling by other receptor families, including some receptor tyrosine kinases and G-protein coupled receptors. In each case, these receptors have been shown to activate a distinct subset of JAKs and STATs (see Table I). Cytokine receptors have been placed in functionally related subfamilies that transduce signals through similar sets of JAKs and STATs.

A. Interferon Receptor Family

Interferons are important components of the innate immune system. In addition to providing defense against viral and parasitic infections, they also exhibit antiproliferative and tumoricidal activity (DeMaeyer and DeMaeyer-Giugnard, 1988; Pestka *et al.*, 1987). There are two major classes of interferons (IFNs). The type I IFNs comprise a functionally and structurally related family with more than 20 members, including alpha-IFNs (by far the largest group), IFN-β, IFN-ω, and IFN-τ. They all bind and transduce signals through the type I IFN receptor (Pestka, 1997). The type II family of IFNs consists of a single member, IFN-γ. IFN-γ is a potent immunomodulatory cytokine that binds to a distinct receptor (Farrar and Schreiber, 1993). Both IFN receptors (i.e., type I and type II) consist of two chains, and they constitute the most divergent subfamily of cytokine receptors. This subfamily also includes the receptor for IL-10 (Bazan, 1990).

Genetic and biochemical studies directed at understanding how IFNs stimulate the induction of new genes led to the identification STATs and provided the first insight into JAK function. The first two STATs, Stat1 and Stat2, were identified as components of type I IFN stimulated transcription factor ISGF3 (Fu *et al.*, 1992; Schindler *et al.*, 1992a, 1992b). One of these

TABLE I Summary of JAK and STAT Usage by Extracellular Ligands
(See Text for Details)

Ligands	Jak kinases	STATs
IFN family		
IFNα/β	Tyk2, Jak1	Stat1, Stat2, Stat3, Stat5
IFN-γ	Jak1, Jak2	Stat1, Stat5
IL-10	Tyk2, Jak1	Stat1, Stat3
gp130 family		
IL-6	Tyk2, Jak1, Jak2	Stat3, Stat1
IL-11	Jak1	
OnM	Jak1, Jak2	Stat3, Stat1
LIF	Tyk2, Jak1, Jak2	Stat3, Stat1
CNTF	Tyk2, Jak1, Jak2	Stat3, Stat1
G-CSF	Jak1, Jak2	Stat3, Stat1
IL-12	Tyk2, Jak2	Stat4
Leptin	Jak2	Stat3
γ-C family		
IL-2	Jak1, Jak2, Jak3	Stat3, Stat5
IL-4	Jak1, Jak3	Stat6
IL-7	Jak1, Jak3	Stat3, Stat5
IL-9	Jak1, Jak3	Stat1, Stat3, Stat5
IL-13	Jak1	Stat6
IL-15	Jak1, Jak3	Stat3, Stat5
IL-3 family		
IL-3	Jak2	Stat5
IL-5	Jak2	Stat5
GM-CSF	Jak2	Stat5
Single chain family		
EPO	Jak2	Stat5
GH	Jak2	Stat1, Stat3, Stat5
PRL	Jak2	Stat5
TPO	Jak2, Tyk2	Stat5
Receptor tyrosine kinases		
EGF	Jak1, Jak2	Stat1, Stat3, Stat5
PDGF	Tyk2, Jak1, Jak2	Stat1, Stat3
CSF-1	Tyk2, Jak1	Stat1, Stat3, Stat5
G-protein coupled receptors		
AT1	Tyk2, Jak2	Stat1, Stat2

proteins, Stat1, was subsequently shown to be the major STAT for type II
IFNs (Decker *et al.*, 1991; Schindler *et al.*, 1992b; Shuai *et al.*, 1992). In
contrast, the important role JAKs play in IFN-dependent signaling was
elucidated through the complementation of cells defective in their ability to
respond to IFNs (Müller *et al.*, 1993b; Velazquez *et al.*, 1992; Watling *et
al.*, 1993).

I. Interferon-γ Receptor

IFN-γ stimulates potent antiproliferative, antiviral, and immunomodulatory responses in target cells (Bach *et al.*, 1997; Farrar and Schreiber, 1993; Strehlow and Schindler, 1997). In contrast to IFN-α, IFN-γ function is absolutely required for normal host defenses against several intracellular pathogens (Dalton *et al.*, 1993; Durbin *et al.*, 1996; Huang *et al.*, 1993; Lu *et al.*, 1998; Meraz *et al.*, 1996). IFN-γ is also important for many other aspects of immune response, including T-cell maturation, MHC expression and immunoglobulin (Ig) isotype switching (Bach *et al.*, 1997). Even though IFN-γ is only produced by activated T-lymphocytes and natural killer (NK) cells, most cells express the receptor and can respond to this potent cytokine (Bach *et al.*, 1997; Farrar and Schreiber, 1993; Strehlow and Schindler, 1997). The IFN-γ receptor consists of two subunits, an α-chain and a β-chain (Aguet *et al.*, 1988; Hemmi *et al.*, 1994; Soh *et al.*, 1994). Both chains participate in ligand binding and are required for signal transduction. Mice with targeted disruptions of either of these chains are rendered unresponsive to IFN-γ (Huang *et al.*, 1993; Lu *et al.*, 1998). The α-chain associates with, and mediates the activation of Jak1, whereas the β-chain associates with and mediates activation of Jak2 (Bach *et al.*, 1996; Igarashi *et al.*, 1994; Kaplan *et al.*, 1996a; Kotenko *et al.*, 1995).

Studies with Jak1- and Jak2-deficient cells have confirmed that both kinases are required for signaling (Müller *et al.*, 1993a; Neubauer *et al.*, 1998; Parganas *et al.*, 1998; Rodig *et al.*, 1998; Watling *et al.*, 1993). Once activated, the kinases phosphorylate tyrosine 440 on the α-chain, which in turn mediates the recruitment of Stat1 to the receptor complex (see Fig. 1; Greenlund *et al.*, 1994). Once activated, Stat1 is released from the receptor and forms homodimers (Greenlund *et al.*, 1995; Shuai *et al.*, 1994), which in turn translocate to the nucleus (Shuai *et al.*, 1992). Once in the nucleus, Stat1 homodimers bind members of the IFN gamma activation site (GAS) family of enhancers, culminating in the induction of genes critical to the biological response of type II IFN (Decker *et al.*, 1997). This includes the gene for the transcription factor IRF1 (interferon response factor 1; see later discussion). Mice deficient in Stat1 exhibit the same defect as mice deficient in IFN-γ or the IFN-γ receptor (Dalton *et al.*, 1993; Durbin *et al.*, 1996; Huang *et al.*, 1993; Lu *et al.*, 1998; Meraz *et al.*, 1996), indicating that all the biological response to IFN-γ are dependent on Stat1. More recently, Stat5 homodimers have been implicated in the biological response to IFN-γ, but they appear to function in a much more limited capacity (Meinke *et al.*, 1996).

2. Interferon-α Receptor

IFN-α and IFN-β are the prototypical type I IFNs and play an important role in viral defense (DeMaeyer and DeMaeyer-Giugnard, 1988; Pestka,

1997; Pestka *et al.*, 1987). The IFN-α receptor consists of at least two chains, IFNAR1 and IFNAR2 (Novick *et al.*, 1994; Uzé *et al.*, 1990), which are associated with Tyk2 and Jak1 (Colamonici *et al.*, 1994; Novick *et al.*, 1994). Although the details as to how this receptor recruits STATs remains controversial, it is known to promote the induction of genes through two distinct IFN response elements, the ISRE and GAS (see Section IV,A). This is achieved through the activation of two distinct STAT-dependent signaling pathways.

The ISRE (interferon stimulation response element), which drives the expression of genes that are unique to type I IFNs, binds the atypical STAT based transcription factor ISGF3 (interferon stimulated gene factor 3; Reich *et al.*, 1987). This factor consists of a Stat1–Stat2 heterodimer plus a 48-kDa protein from the IRF (interferon regulatory factor) family of transcription factors (Fu *et al.*, 1992; Schindler *et al.*, 1992a; Veals *et al.*, 1992). p48 is the DNA binding protein in this complex and provides for an overlap in enhancer specificity with other members of the IRF family (e.g., IRF1; Kessler *et al.*, 1988). Stat2 provides the critical transcriptional activation domain (Qureshi *et al.*, 1996).

GAS elements were initially identified as IFN-γ response elements and have been shown to bind Stat1 as well as other STAT dimers. The ability of type I IFNs to signal through Stat1 homodimers in addition to ISGF3 (Pine *et al.*, 1994) is likely to account for some of the functional overlap between type I and type II IFNs. Type I IFNs also promote the formation of Stat3 homodimers and Stat1:Stat3 heterodimers. Recent studies suggest these complexes may also contribute to biological response (Silvennoinen *et al.*, 1993; Yang *et al.*, 1998). Additionally, there is evidence that Stat5 homodimers form in response to stimulation with type I IFNs, but the biological significance of this has not been explored (Meinke *et al.*, 1996). And finally, one study has suggested that a Stat1:Stat2 heterodimer may also bind DNA in response to type I IFNs (Li *et al.*, 1996).

3. Interleukin-10 Receptor

Interleukin (IL)-10 is another important cytokine that modulates immune response. It is one of the major cytokines secreted by the Th2 subpopulation of T-helper cells (Moore *et al.*, 1993; O'Garra and Murphy, 1994). Although this cytokine does stimulate some responses, it is largely recognized for its suppressive effects. For example, IL-10 can block the release of cytokines, especially IFN-γ, from activated Th1 lymphocytes. It will also suppress IL-12 secretion from monocytes and can block T-cell proliferation in response to antigen. Consistent with these observations, mice deficient in IL-10 develop autoimmunity that manifests itself as an ulcerative colitis-like syndrome (Kühn *et al.*, 1993). The receptor for IL-10 consists of two chains that are structurally related to the IFN receptors (Ho *et al.*, 1993; Kotenko *et al.*, 1997). When the gene for one of these receptor chains (i.e.,

CRF2-4) is disrupted, the mice become unresponsive to IL-10 and develop the same type of enterocolitis seen in the IL-10 knockout mice (Spencer *et al.*, 1998). In signaling studies it has been determined that IL-10 signals through the sequential activation of Tyk2/Jak1 and Stat1/Stat3 (Finbloom, 1995; Weber-Nordt *et al.*, 1996; Takeda *et al.*, 1998; Takeda *et al.*, 1999; Riley *et al.*, 1999).

B. Gp130 Receptor Family

The "gp130" family forms a large and pleiotropic subfamily of cytokine (Kishimoto *et al.*, 1995). Each of these receptors transduces signals through a common 130-kDa glycoprotein (gp130), or a gp130-like receptor chain. Members of this family have been shown to play an important role in regulating immune response, inflammatory response, neural growth/development, cardiac growth/development, hepatocyte growth/development, and body fat homeostasis. Gp130 receptors can be divided into two subsets: those that consist of gp130 and a ligand-specific subunit (e.g., receptors for IL-6, IL-11, LIF, CT-1, OSM, and CNTF), and those that consist of a gp130-like homodimer (e.g., receptors for G-CSF, leptin, and IL-12). In receptors from the first group, ligand binding is often mediated by an α-chain. Signaling, which entails the activation of Stat1 and Stat3, is usually mediated by a β chain. In the second group, a single receptor chain is responsible for both ligand binding and JAK-STAT activation. These receptors also signal through the activation of both Stat1 and Stat3. However, despite a compelling body of biochemical evidence, the Stat1 knockout mice indicate that Stat1 is unlikely to significantly contribute to these response. Hence, Stat3 is likely to be the critical signal transducer for this family of receptors. Unfortunately, the embryonic lethal phenotype of the mice has limited critical evaluation of this model (Takeda *et al.*, 1997).

1. Gp130 Receptors

a. IL-6 receptor.   IL-6 and its receptor are the prototypical members of this subfamily and have been shown to mediate important biological responses in many tissues (Hirano *et al.*, 1990; Kishimoto *et al.*, 1995). Many cell types express both the ligand and its receptor. IL-6 has been shown to promote growth arrest and differentiation in myeloid lineages; stimulate B-cell differentiation; serve as a growth factor for plasma cell tumors; and induce the production of acute-phase response proteins in the liver. IL-6 may also serve as an essential growth factor in regenerating livers (Cressman *et al.*, 1996). The receptor consists of an 80-kDa ligand binding chain (IL-6Rα) and a homodimer of gp130. Notably, the carboxy terminus of IL-6Rα appears only to be required to tether the ligand to the membrane, enabling it to interact with gp130. IL-6Rα function can be fully substituted with preparations of soluble receptor chain (Romano *et al.*, 1997). Once bound by the IL-6:IL-6Rα complex, gp130 subunits homodimerize and promote the sequential activation of three JAKs (i.e., Jak1, Jak2 and Tyk2; Lütticken *et al.*, 1994; Stahl *et al.*, 1994) and two STATs (i.e., Stat1 and Stat3). As outlined above, even though both Stat1 and Stat3 are activated by IL-6 (Akira *et al.*, 1994; Bonni *et al.*, 1993; Rothman *et al.*, 1994; Wegenka *et al.*, 1994; Zhong *et al.*, 1994), evaluation of the Stat1 knockout mice indicate that Stat1 is not normally required for this biological response (Meraz *et al.*, 1996). A parallel study, with Stat3 knockout mice, has been uninformative, because they exhibit the same early embryonic lethal phenotype as the gp130 knockout mice (Takeda *et al.*, 1997; Yoshida *et al.*, 1996). However, studies with dominant interfering mutants of Stat3, or gp130 mutants, demonstrated that Stat3 is vital for the biological response to IL-6 (Bonni *et al.*, 1997; Kopf *et al.*, 1994; Minami *et al.*, 1996; Nakajima *et al.*, 1996; Poli *et al.*, 1994; Yamanaka *et al.*, 1996).

b. LIF Receptor.   LIF shares many biological properties with IL-6. Analogous to IL-6, it is produced by many cell types and exerts a wide range of responses in a large number of cell types. This includes hepatocytes, adipocytes, megakaryocytes, neural cells, muscle cells, embryonic stem cells, and osteoclasts (Hinds *et al.*, 1998). However, in contrast to IL-6 knockout mice (Kopf *et al.*, 1994; Poli *et al.*, 1994), those with a targeted disruption of LIF only exhibit a modest defect in hematopoiesis and are defective in blastocyst implantation (Escary *et al.*, 1993; Stewart *et al.*, 1992). The receptor for LIF consists of a heterodimer of gp130 and a unique gp130-like receptor chain referred to as LIF-R (Gearing *et al.*, 1992). LIF-R is also widely expressed. Similar to the gp130 knockout mice, mice with a disruption of LIF-R exhibit a phenotype that is much more severe than that of the LIF knockout, implying a role in other ligands (Ware *et al.*, 1995). LIF-R knockout mice exhibit significant defects in placentation, bone development, neural development, and metabolism, which results in fetal death.

c. CNTF and CT-1 Receptors.   CNTF (ciliary neurotrophic factor) and CT-1 (cardiptropin-1) are two additional ligands that transduce unique signals through the gp130:LIF-R heterodimer. The receptor for CNTF achieves this by employing an additional, unique, GPI-tethered ligand binding chain (CNTF-Rα; Stahl *et al.*, 1994). CNTF and its receptor have been implicated in neural growth and survival. However, CNTF knockout mice are essentially normal (Masu *et al.*, 1993). In contrast, CNTF-Rα knockout mice exhibit profound motor neural defects and die perinatally, suggesting that this receptor may bind another ligand (DeChiara *et al.*, 1995). CT-1 also transduces its signals through a gp130–LIF-R heterodimer. Although there is evidence for an additional ligand-specific receptor component, it has not yet been identified (Robledo *et al.*, 1997). As the name suggests, CT-1 is important in cardiac growth and development (Wollert and Chien,

1997). However, neural, myeloid, and hepatic cells appear to be important targets as well.

d. OSM and IL-11 Receptors.   OSM (oncostatin M) and IL-11 are also pleiotropic members of the IL-6 family. OSM is only secreted by T-cells and macrophages, yet has potent growth regulatory functions on several cell types. This includes fibroblasts, vascular smooth muscle cells, bone cells, and some tumors (Liu *et al.*, 1998b). In human cells, OSM can signal through both a heterodimer consisting of a unique gp130-like OSM receptor α-chain (OSM-Rα) and gp130. It can also signal through a receptor consisting of a gp130-LIF-R heterodimer. The unique OSM-Rα is likely to explain the ability of LIF and OSM to bestow functionally antagonistic signals on the same target cells (Liu *et al.*, 1998). In the murine system, OSM appears to signal exclusively through gp130-OSM-αR heterodimers (Lindberg *et al.*, 1998). IL-11 also employs a unique receptor α-chain (IL-11Rα) that is analogous to IL-6Rα (Kishimoto *et al.*, 1995; Leng and Elias, 1997). The pattern of IL-11Rα expression overlaps significantly with that of IL-6Rα, rendering IL-11 functionally redundant to IL-6 in many cell types. However, IL-11 more effectively stimulates hematopoiesis, especially in the megakaryocytic lineage (Leng and Elias, 1997). Targeted disruption of the IL-11Rα gene yields modest changes, perhaps because other members of this family compensate for the loss of IL-11 function (Nandurkar *et al.*, 1997).

2. Gp130-like Receptors

The second group of gp130 receptors consist of a single gp130-like receptor chain that homodimerizes. This single chain is responsible for both high-affinity ligand binding and activation of the JAK-STAT signaling cascade. Members of this subgroup include the receptors for G-CSF, leptin, and IL-12.

a. G-CSF Receptor.   The receptor for granulocyte-colony stimulating factor (G-CSF) is the prototypical member of this group. G-CSF is important for both basal and stress-induced granulopoiesis (Lieschke *et al.*, 1994). Correspondingly, the expression of the GCSF-R is predominantely limited to those lineages important in granulopoiesis (Ito *et al.*, 1994). Targeted deletion of this receptor leads to a significant, but not an absolute, defect in granulopoiesis (Liu *et al.*, 1996). The G-CSF-R chain mediates both high-affinity ligand binding and signal transduction. Signaling through the JAK-STAT pathway commences with the activation of Jak1 and Jak2. Subsequently Stat1, Stat3, and Stat5 are activated. *In vivo* studies suggest that the activation of Stat3 may be physiologically most important (Nicholson *et al.*, 1995; Tian *et al.*, 1994).

b. Leptin Receptor.   Leptin is a recently discovered hormone that regulates body weight homeostasis, and its receptor is the newest member of this family (Chen *et al.*, 1996; Lee *et al.*, 1996). Leptin was identified as the product of the *obese* gene, which is mutated in *ob* mice (Zhang *et al.*, 1994). The *ob* gene encodes a circulating hormone secreted by fat cells, which binds a receptor in the hypothalamus to regulates appetite and energy expenditure (Friedman, 1997). Consistent with this model, mutation of the leptin receptor, encoded by the *diabetic (db)* gene, gives rise to an obese phenotype (Chen *et al.*, 1996; Lee *et al.*, 1996). Activation of full-length isoforms of this receptor (i.e., in the hypothalamus) has been shown to lead to the activation of Stat1 and Stat3 (Rosenblum *et al.*, 1996). However, many tissues, including the liver and choroid plexus, express alternative receptor splice isoforms that are missing the STAT activation domain (Chen *et al.*, 1996; Lee *et al.*, 1996). The ability of these defective receptors to bind ligand, and their pattern of expression, suggests that they may participate in the regulation of leptin signaling. For example, the truncated receptor expressed in the choroid plexus could promote the translocation of leptin across the blood–brain barrier (Chen *et al.*, 1996).

    c. IL-12 Receptor. IL-12 and its receptor represent a curious divergence in the coevolution between receptor and ligand (Kishimoto *et al.*, 1995). IL-12 is a dimer of a 35- and a 40-kDa subunit. The 35-kDa subunit exhibits a limited pattern of expression and has homology to IL-6. The 40-kDa subunit, expressed by many cell types, has homology to the α-chains of the IL-6 and CNTF receptors. This suggests that it may serve a function analogous to the soluble IL6R-α. IL-12 mediates potent immunomodulatory functions in T-cells, regulating both function and growth. For example, IL-12 is required for the T-cell-dependent production of IFN-γ and plays a critical role in the development of the Th1 subset of T-cells. IL-12 also exhibits potent natural killer (NK) cell stimulating activity. While the complete 70-kDa IL-12 dimer is required for most of these biological responses, p35 and p40 knockout mice suggest that a small subset of responses may be mediated by p40 homodimers (Piccotti *et al.*, 1998). Two highly homologous gp130-like genes, IL-12Rβ1 and IL-12Rβ2, encode the IL-12 receptor. Heterodimers of these two chains appear to be required for the generation of high-affinity IL-12 binding sites (Presky *et al.*, 1996). Targeted disruption of the IL-12Rβ1 genes yields a mouse that is unresponsive to IL-12 (Wu *et al.*, 1997). Consistent with their divergent structure, IL-12 and its receptor stimulate a distinct JAK-STAT signaling pathway. Upon binding ligand, IL-12R associated Tyk2 and Jak2 become activated (Zou *et al.*, 1996). This leads to the subsequent activation of Stat3 and Stat4 (Jacobson *et al.*, 1995). Both biochemical and gene targeting studies suggest that the activation of Stat4 is essential for all of the biological responses to IL-12 (Kaplan *et al.*, 1996; Thierfelder *et al.*, 1996).

C. IL-2 Receptor Family

Another large and important family of cytokine receptors is the IL-2 family. It includes receptors for IL-2, IL-4, IL-7, IL-9, IL-13, and IL-15 (Kishimoto *et al.*, 1994; Leonard and O'Shea, 1998; Zurawski *et al.*, 1993). These ligands all play an important role in lymphocyte function and bind receptors that share a common receptor component, the common gamma chain (γC). In addition, each of these receptors consists of a specific ligand binding chain. Except for the IL-2 and IL-15 receptors, the ligand binding chain also serves an important role in signaling by associating with Jak1 and STATs. The IL-2 and IL-15 receptors employ two chains for ligand binding and signaling. γC is also serves an essential role in signal transduction by mediating the activation of Jak3 (Nakamura *et al.*, 1994; Nelson *et al.*, 1994; Johnston *et al.*, 1994; Witthuhn *et al.*, 1994). Patients that inherit a defect in this chain develop X-linked severe combined immunodeficiency (X-SCID; Leonard, 1996; Noguchi *et al.*, 1993; Russell *et al.*, 1994). The manifestations of this disease, including absence of T-cells and NK cells, and defective B-cells, highlight the important role this receptor chain plays in immune response. Moreover, patients and mice with defects in Jak3 fail to activate downstream targets, therefore exhibiting a similar pattern of immunodeficiency (Leonard, 1996; Nosaka *et al.*, 1995; Park *et al.*, 1995; Russell *et al.*, 1995; Thomis *et al.*, 1995). Based on the STATs they activate, these receptors can be divided into two groups: the receptors for IL-2, IL-7, IL-9, and IL-15, which activate Stat5; and the receptors for IL-4 and IL-13, which activate Stat6 (Hou *et al.*, 1994, 1995; Leonard and O'Shea, 1998; Lin *et al.*, 1995).

1. Receptors for IL-2, IL-7, IL-9, and IL-15

a. IL-2 Receptor. IL-2 is the prototypical member of this family (Leonard, 1996; Taniguchi and Minami, 1993). It plays an important role in both the proliferation and activation of T-cells. It also contributes to B-cell and NK cell responses. The IL-2 receptor consists of three chains. The α-chain (not a member of the cytokine receptor family) and β-chain are required for high-affinity ligand binding. Expression of the α-chain is regulated in T-cells. Only low levels are expressed in naive T-cells. Upon stimulation with antigen, the expression of both IL-2 and the IL-2 receptor α-chain is upregulated in T-cells, setting up an important autocrine loop. Expression of the β and γ chains, which are critical for signal transduction, is not regulated in this manner. In the absence of the IL-2Rβ chain, normal T-cell homeostasis is lost and autoimmunity develops (Suzuki *et al.*, 1995). As outlined previously, loss of the γ-chain leads to X-SCID (Leonard, 1996; Noguchi *et al.*, 1993). Signaling is initiated by the activation of β and γ chain associated Jak1 and Jak3 (Johnston *et al.*, 1994; Miyazaki *et al.*,

1994; Russell *et al.*, 1994; Witthuhn *et al.*, 1994). Whereas the γC appears to be solely required for the activation of Jak3, the β-chain has been shown to promote the recruitment of signaling molecules, including Stat5 (Gaffen *et al.*, 1995; Gilmour *et al.*, 1995; Lin *et al.*, 1995). Consistent with these observations, mice with a targeted disruption of both Stat5 genes (i.e., Stat5a and Stat5b) exhibit a defect in T-cell function that bears a striking resemblance to the IL-2 β-chain knockout mice (Suzuki *et al.*, 1995; Teglund *et al.*, 1998; Moriggl *et al.*, 1999). In some cells, IL-2 has also been shown to activate Stat3, but the functional significance of this activation has not been elucidated.

 b. IL-15 Receptor.   The receptor for IL-15 consists of the β and γ chains of the IL-2 receptor, as well as a unique ligand binding α-chain (Kishimoto *et al.*, 1994; Leonard and O'Shea, 1998). Not surprisingly, IL-2 and IL-15 stimulate the same signaling pathways. However, whereas IL-2 is critical for thymic T-cell development, IL-15 appears to play a more important role in the extrathymic development of both T and NK cells (DiSanto, 1997; Giri *et al.*, 1994; Grabstein *et al.*, 1994).

 c. IL-7 and IL-9 Receptors.   The receptors of IL-7 and IL-9 are heterodimers, each consisting of a unique ligand-binding α-chain and γC. IL-7 and its receptor play a critical role in the early stages of lymphopoiesis (Candeias *et al.*, 1997). Mice with a targeted disruption of the genes for IL-7 or the IL-7 receptor α-chain exhibit profound defects in lymphocyte development (Corcoran *et al.*, 1998; Peschon *et al.*, 1994; vonFreeden-Jeffry *et al.*, 1995). These defects are, however, more severe in the IL-7 receptor knockout mice, because this receptor binds a second ligand, thymic stromal derived lymphopoietin (TSLP; Candeias *et al.*, 1997). IL-7 has been shown to provide several critical early developmental signals. In addition to providing trophic (e.g., antiapoptotic) and proliferative stimuli (Corcoran *et al.*, 1996), it promotes the rearrangement of genes that are critical for lymphocyte development. In T-cells IL-7R signaling is required for the T-cell receptor β-chain (TCRβ) and TCRγ rearrangement (Candeias *et al.*, 1997). In murine B-cells IL-7 receptor signaling is required for immunoglobulin heavy chain (IgH) rearrangement. Intriguingly, human B-cells do not require IL-7. The IL-7R signals through the sequential activation of Jak1/Jak3 and Stat5 (Foxwell *et al.*, 1995). Moreover, mice with a targeted disruption of both Stat5 genes (i.e., Stat5a and Stat5b) exhibit a general defect in lymphopoiesis (Teglund *et al.*, 1998). In contrast to the profound role IL-7 plays in lymphopoiesis, IL-9 appears to function more as an enhancing factor in the development of several hematopoietic lineages. These effects are most important for T-cells (Demoulin and Renauld, 1998). The IL-9 receptor transduces important signals through the sequential activation of Jak1/Jak3 and several STATs (Demoulin *et al.*, 1996). Perhaps consistent with IL-9's more general role, it appears to promote the activation of Stat1 and Stat3 in addition to Stat5 (Bauer *et al.*, 1998).

2. IL-4 and IL-13 Receptors

IL-4 and IL-13 are pleiotropic cytokines that share many functional properties (Abbas *et al.*, 1991). This can be attributed to shared receptor components. Although IL-4 and IL-13 contribute to the regulation of multiple stages in lymphocyte development, they appear to be most important in promoting immune response. IL-4 and its receptor have been characterized in the most detail and are therefore considered the prototypical members of this small family (Leonard and O'Shea, 1998). IL-4 was first identified as a B-cell stimulatory factor, but has subsequently been found to promote immune response that is important for the development of allergies. In resting B-cells this cytokine promotes the expression of a number of cell surface markers that are critical to B-cell function. Yet IL-4's most important function in B-cells may be the induction of immunoglobulin class switching (IgG4 in humans and IgE in both humans and mice). In T-cells, IL-4 plays a critical role in the development of the Th2 subset of T-cells, which in turn promote humoral immunity. In these cells the receptor for IL-4 consists of a unique ligand binding/signal transducing α-chain and γC. This receptor transduces important signals through the sequential activation of Jak1/Jak3 and Stat6 (Hou *et al.*, 1994; Johnston *et al.*, 1994; Witthuhn *et al.*, 1994). Consistent with these observations, targeted disruption of IL-4, IL-4R α-chain (Kopf *et al.*, 1993; Noben-Trauth *et al.*, 1997), or Stat6 (Kaplan *et al.*, 1996; Shimoda *et al.*, 1996; Takeda *et al.*, 1996) lead to profound defects in lymphocyte function. This includes a marked decrease in IL-4 dependent proliferation, a profound decrease Th2 cells, and a decrease in IgG1/IgE production. However, there are many nonhematopoietic cells, which do not express γC or Jak3, yet respond to IL-4. Although the role of these nonhematopoietic responses remains to be elucidated, these and other observations suggest that IL-4 may signal through an additional receptor type (see below; Kammer *et al.*, 1996; Oakes *et al.*, 1996).

IL-13 shares many functional properties with IL-4. However, human T-cells and murine T- and B-cells do not appear to respond to this cytokine (Abbas *et al.*, 1991). The IL-13 receptor shares some components with the IL-4 receptor (Hilton *et al.*, 1996; Orchansky *et al.*, 1997). This receptor consists of a short (i.e., 60 kDa) IL-13 receptor chain (IL-13Rα) and the IL-4 R α-chain. The unique IL-13Rα is required for high-affinity IL-13 binding and signaling (Murata *et al.*, 1998). This heterodimeric receptor transduces signals through the sequential activation of Jak1/Tyk2 and Stat6 (Hilton *et al.*, 1996; Orchansky *et al.*, 1997). Intriguingly, this heterodimer will also bind and transduce signals in response to IL-4 (Murata *et al.*, 1998). This observation not only provides further insight into the functional overlap between these two ligands, but is likely to account for the alternative IL-4 receptor.

D. IL-3 Receptor Family

IL-3, IL-5, and GM-CSF (granulocyte macrophage-colony stimulating factor) constitute another important subfamily of cytokines, which play a crucial role in the proliferation and maturation of myeloid cells (Miyajima *et al.*, 1993). All three cytokines have been implicated in the maturation and activation of several myeloid lineages. IL-5, however, appears to be particularly important in eosinophil and mast-cell biology (Kopf *et al.*, 1996; Sachs, 1993). IL-5 also plays an important role in the B-cell immune response. IL-3 influences the proliferation and maturation of hematopoietic stem cells, as well as progenitors of the granulocyte, macrophage, erythrocyte, eosinophil, megakaryocyte, mast cell, and basophil lineages. In addition, IL-3 serves to activate mature myeloid cells, stimulating phagocytosis and antibody-dependent and -independent cytotoxicity. GM-CSF also functions as a growth factor for progenitors of granulocyte, macrophage, and eosinophil lineages. The receptor for IL-3, IL-5, and GM-CSF each consist of two subunits, a ligand-specific α-subunit and a common signal-transducing β-subunit (Miyajima *et al.*, 1993; Sakamaki *et al.*, 1993). Mice have two β-subunits, both of which exhibit high homology to the human β-chain. The β_{IL3} appears to preferentially associate with the IL-3 receptor α-chain, whereas β_{common} associates with all three α-chains (i.e., IL-3Rα, IL-5Rα, and GM-CSFRα; Hara and Miyajima, 1992). The β-chain, which associates with Jak2, promotes the sequential activation of Jak2 and multiple isoforms of Stat5 (Azam *et al.*, 1995; Liu *et al.*, 1995; Mui *et al.*, 1995; Quelle *et al.*, 1994). Consistent with these observations, fetal myeloid progenitor cells harvested from Jak2 knockout embryos are defective in their ability to respond to IL-3 or GM-CSF (Parganas *et al.*, 1998).

E. Single Chain Receptor Family

The single chain receptors constitute another pleiotropic family that play an important role in both hematopoietic and nonhematopoietic tissues. They include receptors for GH (growth hormone), EPO (erythropoietin), PRL (prolactin), TPO (thrombopoietin), G-CSF, and leptin. However, the receptors for G-CSF and leptin were discussed earlier (Section III,B,2), because of their structural and functional homology to gp130 receptor family. The single-chain receptors discussed in this section signal predominately through the sequential activation of Jak2 and Stat5.

I. GH Receptor

Growth hormone and its receptor have been characterized in most detail and are the prototypical members of this family (Argetsinger and Carter-

Su, 1996; Kishimoto *et al.*, 1994; Wells, 1996). GH plays a critical role in a variety of physiological processes, including bone and muscle growth, as well as overall metabolism. A difference in the pattern of GH secretion in male and female vertebrates contributes to sexual dimorphism (Ram *et al.*, 1996; Udy *et al.*, 1997). Crystallographic studies have demonstrated that the interaction between GH and its receptor promotes receptor dimerization (de Vos *et al.*, 1992). This leads to the activation of receptor-associated Jak2 (Argetsinger *et al.*, 1993), and depending on the cell type, the subsequent activation of Stat1, Stat3, or Stat5 (Gouilleux *et al.*, 1995; Sliva *et al.*, 1994; Yi *et al.*, 1996). Studies with the Stat1 knockout mice provide strong evidence that Stat1 signaling is not important for the biological response to GH (Meraz *et al.*, 1996). Analogous studies on Stat3 remain uninformative (Takeda *et al.*, 1997). However, studies on the Stat5 knockout mice strongly implicate Stat5 in GH signaling (Liu *et al.*, 1997; Teglund *et al.*, 1998; Udy *et al.*, 1997). Not only do Stat5 knockout mice exhibit a reduced adult size (more apparent in males), but they fail to express the appropriate pattern of sexually dimorphic genes. These studies also suggest that Stat5b may play a more important role in GH signaling.

2. EPO Receptor

The essential role erythropoietin (EPO) plays in normal human physiology has prompted a long-standing interest in this cytokine and its receptor (Ihle *et al.*, 1993; Wells, 1996). EPO is essential for definitive erythropoiesis and has become an effective therapeutic for certain anemias. Moreover, a small-molecule mimetic of erythropoietin has been identified. Both EPO and this mimetic bind and promote EPO receptor dimerization, analogous to what had been reported for the GH receptor (Livnah *et al.*, 1996; Wells, 1996b; Wrighton *et al.*, 1996). Targeted disruption of either EPO or the EPO receptor leads to a defect in definitive erythropoiesis and embryonic lethality at day 13 (Wu *et al.*, 1995). These embryos do, however, exhibit primitive erythropoiesis, indicating that EPO is not required for commitment to the erythrocytic lineage. Further supporting a role for the EPO receptor in definitive erythropoiesis is a form of familial erythrocytosis where a single point mutation creates a constitutively active form of this receptor (Chapelle *et al.*, 1993). Like the native receptor, these activated receptors transduce their signals through the sequential activation of Jak2 and Stat5 (Damen *et al.*, 1995; Gouilleux *et al.*, 1995; Witthuhn *et al.*, 1993). Studies on mice with targeted disruptions of the Jak2 and Stat5 indicate that Jak2 must activate important pathways in addition to Stat5. Stat5 knockout mice exhibit a nonlethal anemia, whereas, Jak2 knockout mice die of anemia at embryonic day 13 (Neubauer *et al.*, 1998; Parganas *et al.*, 1998; Teglund *et al.*, 1998; Socolovsky *et al.*, 1999).

3. PRL Receptor

Prolactin is a lactogenic hormone that is synthesized predominately in the pituitary gland. The wide tissue distribution of the PRL receptor, ablation studies, and studies with cultured cells have implicated PRL in many physiological processes (Hennighausen *et al.*, 1997). This includes hematopoiesis, osmoregulation, maternal behavior, fertility (in rodents), and mammary gland function. However, mice with a targeted disruption of either PRL or the PRL receptor only exhibit significant defects in mammary gland development, mammary gland maturation, and fertility (Horseman *et al.*, 1997; Ormandy *et al.*, 1997). Receptor knockout mice have a more severe phenotype, most likely because maternal PRL can cross the placenta.

Like other members of the single-chain receptor family, the PRL receptor transduces signals through the sequential activation of Jak2 and Stat5 (DaSilva *et al.*, 1994; Lebrun *et al.*, 1995; Wakao *et al.*, 1994). Notably, Stat5 was first identified as a PRL dependent transcription factor (Wakao *et al.*, 1994). Consistent with this observation, mice with a targeted disruption of the Stat5 genes exhibit defects that are reminiscent of those in PRL knockout mice, suggesting that this pathway is vital to the biological response promoted by PRL (Horseman *et al.*, 1997; Liu *et al.*, 1997; Teglund *et al.*, 1998). Specifically, these mice fail both to develop the terminal and lateral lobular decorations seen in normal virgin mammary glands and are unable to develop mature mammary glands. Their defect in fertility can be attributed in loss in normal corpus luteal function. These studies also suggest that Stat5a may be more important for PRL dependent signaling (Liu *et al.*, 1997; Teglund *et al.*, 1998).

4. TPO Receptor

Thrombopoietin (TPO) was identified during an intense search for the ligand that bound the cellular homologue of the *mpl* oncogene (Bartley *et al.*, 1994; deSauvage *et al.*, 1994; Lok *et al.*, 1994). Consistent with the characterization of c-*mpl*, TPO was determined to be important in the growth and maturation of the megakaryocyte lineage. However, studies with TPO injections and knockout animals indicate that this ligand has a more pleiotropic role in early hematopoiesis (Cwirla *et al.*, 1997; Solar *et al.*, 1998). In addition to significant thrombocytopenia, c-*mpl* (i.e., TPO-R) and TPO knockout mice also exhibit significant reductions in neutrophil, GM, erythroid, and multi-lineage progenitors (Alexander *et al.*, 1996; Carver-Moore *et al.*, 1996; Gurney *et al.*, 1994). Like other members of this receptor subfamily, the TPO-R transduces signals through the sequential activation of Jak2/Tyk2 and Stat5 (Bacon *et al.*, 1995; Gurney *et al.*, 1995; Pallard *et al.*, 1995). In some cell lines Stat1 and Stat3 are activated by TPO.

F. Noncytokine Receptors

Over the past several years a significant number of publications have implicated STATs in the transduction of signals for receptors that bind ligands from other families. The most compelling evidence comes from tyrosine kinase family of receptors. However, there are several studies implicating other receptor families. Of these, G-protein coupled receptors have been examined most extensively.

1. Receptor Tyrosine Kinases

Receptor tyrosine kinases (RTKs) represent a large, vital, and pleiotropic family of receptors that distinguish themselves from the cytokine receptors by encoding a tyrosine kinase in their cytoplasmic domain. Analogous to the role of JAKs in cytokine signaling, these kinases phosphorylate critical tyrosines in the cytoplasmic receptor tail, which in turn promote the recruitment of signaling molecules. Once at the receptor, many of these signaling molecules themselves become substrates of the receptor tyrosine kinase. Although several STATs have been determined to be activated by RTK members, the potentially redundant role of JAKs in RTK signaling has not been fully elucidated (Vignais *et al.*, 1996). STAT signaling by RTK has been best characterized for the EGF (epidermal growth factor) and PDGF (platelet derived growth factor) receptors.

    a. EGF Receptor.   EGF and its receptor are prototypical members of this family and have been studied extensively for their ability to promote growth in many cell types. Characterization of the ability of EGF (and family members) to stimulate growth has led to the identification of several critical regulatory molecules and signaling pathways. It was somewhat surprising when EGF was determined to also stimulate the activation of Stat1, Stat3, and Stat5 in cultured cells and *in vivo* (Ruff-Jamison *et al.*, 1995; Sadowski *et al.*, 1993; Zhong *et al.*, 1994). Subsequent studies demonstrated that these STATs are activated by phosphorylation of the same critical tyrosine identified during the characterization of cytokine signaling (Shuai *et al.*, 1993; Silvennoinen *et al.*, 1993). Although a compelling body of evidence indicates that STATs are activated by the EGF-R, it has been much more difficult to assign a function to this response. Mice with targeted disruptions in Stat1 and Stat5 do not appear to exhibit any obvious defects in their response to EGF (Durbin *et al.*, 1996; Meraz *et al.*, 1996; Sibilia and Wagner, 1995; Teglund *et al.*, 1998).

    b. PEGF Receptor.   In contrast to their name, PDGF and its receptor mediate important biological responses in many cell types. Depending on the cell type, PDGF appears to stimulate the same pathways as EGF, including the activation of Stat1, Stat3, and Stat5. Moreover, characterization of the ability of PDGF (c-*sis*) to induce the expression of c-*fos* provided the first evidence that STATs transduce signals for the RTK family (Sadowski *et al.*, 1993; Wagner *et al.*, 1990). These studies led to the identification of a novel enhancer in the c-*fos* promoter, the SIE (*sis* inducible element), which binds a novel factor, SIF (*sis* inducible factor). SIF was shown to consist of activated Stat1 and Stat3 (Sadowski *et al.*, 1993; Shuai *et al.*, 1993; Silvennoinen *et al.*, 1993; Wagner *et al.*, 1990; Zhong *et al.*, 1994), making c-*fos* the first bona fide target gene of RTK-dependent STAT signals. Consistent with this observation, the SIE is a member of GAS family of enhancers. However, since the expression of c-*fos* appears to be regulated by several elements, it remained controversial as to whether STATs were important in mediating the biological response to PDGF. This question was addressed by a careful set of studies in which several lines of c-*fos-lacZ* transgenic mice were created, each with a different mutation in the c-*fos* promoter (Robertson *et al.*, 1995). Mice expressing a transgene where the SIE site had been mutated (*mSIE*) failed to express c-*fos* constitutively in skin, bone, and hair (in contrast to the wild-type control), suggesting that SIE is important for constitutive c-*fos* expression (see Section VIII, B; apoptosis). However, since c-*fos* is considered an immediate early response gene, these results are difficult to interpret. In a more physiologic experiment, the induction of c-*fos* expression was evaluated after appropriate stimulation of neural tissues. In these studies the *mSIE* transgene failed to express only in the cortex and CA1 region of the hippocampus. In contrast, mice expressing transgenes with other promoter mutations exhibited more global defects in neural expression after stimulation. These studies provide the best evidence to date that the STATs are important for at least a subset of the biological responses stimulated by RTKs.

2. Angiotensin Receptor

The ligand for the angiotensin (AT_1) receptor is angiotensin II, an 8-amino-acid effector molecule of the renin–angiotensin system, which promotes the elevation of blood pressure. Like many other receptors that bind small ligands, AT_1 is a member of the G-protein coupled family of receptors (Schieffer *et al.*, 1996). These receptors share a number of structural and functional properties including the obligate transduction of signals through a heterotrimeric G-protein. This in turn leads to the activation of a number of signaling pathways, many of which are shared with other receptor types (e.g., RTKs). However, studies have indicated that AT_1 may also transduce signals through the sequential activation of Jak2 and several STATs (i.e., Stat1, Stat2, and Stat3; Bhat *et al.*, 1994; Marrero *et al.*, 1995). Moreover, the Jak2 binding site has been mapped to a 4-amino-acid motif in the cytoplasmic tail of AT_1 (Ali *et al.*, 1997). It will be of interest to determine whether the Jak2 and Stat1 knockout mice exhibit a defect in angiotensin signaling.

IV. Receptor-Dependent STAT Activation ___________

Ligand-dependent dimerization initiates cytokine receptor activation (deVos *et al.*, 1992; Livnah *et al.*, 1996). This entails the activation of receptor-associated tyrosine kinases from the JAK family, through a process of transphosphorylation (Müller *et al.*, 1993a; Watling *et al.*, 1993). The JAKs in turn phosphorylate specific tyrosyl residues in the receptor endodomain, which in turn promotes the recruitment of specific SH2 containing signaling molecules (e.g., STATs) to the receptor complex (see Fig. 1). Once at the receptor, these molecules (especially STATs) become phosphorylated and transduce signals vital to the biological response of the activating cytokine.

A. JAK Activation

Complementation of a type I IFN (i.e., INF-α) unresponsive mutant with the gene encoding Tyk2 provided the first evidence that janus kinases (JAKs) are important for cytokine signaling (Velazquez *et al.*, 1992). To date, four members of this JAK family of "soluble" tyrosine kinases (i.e., Tyk2, Jak1, Jak2 and Jak3) have identified and shown to be critical for cytokine signaling (Ihle *et al.*, 1994a, 1994b; Leonard and O'Shea, 1998). They associate with a poorly defined, often proline-rich domain found in the membrane proximal region of the cytokine receptors (e.g., Ali *et al.*, 1997; Colamonici *et al.*, 1994; Tanner *et al.*, 1995). Homologues that are believed to signal through the JAK-STAT pathway have also been described in other species including *C. elegans* (Adler *et al.*, 1996), zebrafish (Conway *et al.*, 1997), and *Drosophila* (Binari and Perrimon, 1994; Luo *et al.*, 1995). Intriguingly, gain-of-function JAK mutants have been identified in *Drosophila* (Harrison *et al.*, 1995; Luo *et al.*, 1995, 1996). They cause a leukemia-like syndrome, which can be suppressed by a corresponding loss-of-function mutation in the *Drosophila* STAT (D-Stat), indicating that D-Stat is indeed downstream of this JAK (Yan *et al.*, 1996).

I. Receptor Domains Involved in JAK Activation

The endodomains of cytokine receptors encode a functionally conserved region that mediates the interaction with a specific member of the JAK family of tyrosine kinases. The sequence of the JAK binding domain is modestly conserved and for some receptors has been referred to as the box1/box2 region (Ihle *et al.*, 1994a, 1994b; Leonard and O'Shea, 1998). The sequence tends to be a proline-rich and is usually preceded several hydrophobic amino acids (Ali *et al.*, 1997; Colamonici *et al.*, 1994; Quelle *et al.*, 1994; Tanner *et al.*, 1995). This domain is membrane proximal and required for many if not all of the biological signals stimulated by a cytokine receptor.

2. JAK Structure

JAKs have a number of features that distinguish them from other kinases. This includes their large size (molecular weight of 120–135 kDa) and lack of an SH2 domain (Ihle *et al.*, 1994; Leonard and O'Shea, 1998). They also have two domains, JH1 (JAK homology) and JH2, that are conserved with other tyrosine kinase domains. However, only JH1 is catalytically active. The JH2 domain, or pseudokinase domain, lies amino proximal to JH1 and appears to be required for function. The remaining homology among JAKs maps to five more amino proximal JH domains, JH3–JH7. The most amino terminal domains (e.g., JH6 and JH7) are believed to be important in receptor association (Frank *et al.*, 1995; Tanner *et al.*, 1995; Velazquez *et al.*, 1995).

3. JAK Function

Initial studies, evaluating JAK activation by biochemical means, provided significant insight into which JAKs were activated by a given cytokine (Argetsinger *et al.*, 1993; Ihle *et al.*, 1994; Leonard and O'Shea, 1998; Schindler and Darnell, 1995; Witthuhn *et al.*, 1993). Many of these observations were confirmed by more functional studies in Jak1, Jak2, and Tyk2 deficient cell lines, or with kinase-inactive (i.e., dominant interfering) mutants. Although studies in deficient cells lines suggested that many cytokines required the simultaneous activity of two kinases (Müller *et al.*, 1993a; Watling *et al.*, 1993), studies with the kinase-inactive mutants suggested more of a hierarchy to JAK activation (Briscoe *et al.*, 1996; Gauzzi *et al.*, 1996; Zhou *et al.*, 1997). More recently, JAK knockout mice have been evaluated. They have confirmed many of the previous observations, but have also provided additional insight. Importantly, these studies demonstrate that JAKs play a pivotal and often nonredundant role in cytokine signaling.

a. Jak1 Knockout.   Jak1 deficient mice (Jak1−/−) exhibit a perinatal lethal phenotype that has been attributed to a neurological defect that prevents normal suckling (Rodig *et al.*, 1998). Although the defect appears to be due to an inability to transduce signals through gp130-related receptors, it is considerably less severe than the defect found in gp130 knockout mice (Yoshida *et al.*, 1996). This suggests that Jak1 is not required for all gp130-based signals during early development. Evaluation of cytokine signaling in tissues from Jak1−/− mice has confirmed many earlier studies. Jak1−/− mice fail to respond to cytokines from three distinct receptor subfamilies. This includes all class II cytokine receptors (i.e., the IFN receptor subfamily), the gp130 receptor family, and members of the IL-2 receptor family that are γC dependent. The profound defect these mice exhibit in lymphoid development can be largely attributed to a defective IL-7 response.

b. Jak2 Knockout.   Jak2 deficient mice (Jak2−/−) exhibit an embryonic lethal phenotype, dying at embryonic day 12.5 (Neubauer *et al.*, 1998; Parganas *et al.*, 1998). This phenotype can be attributed to a defect in definitive erythropoiesis, analogous to what was observed in the erythropoietin knockout (Wu *et al.*, 1995). As predicted by biochemical studies, fetal myeloid progenitors from Jak2−/− mice fail to respond to thrombopoietin, IL-3, and GM-CSF, but respond normally to G-CSF. Jak2−/− fibroblasts are also defective in their response to IFN-γ, but respond normally to type I IFNs (i.e., IFN-α/β) and IL-6. Consistent with this, Jak2−/− ES cells respond normally to LIF. Additionally, there are no apparent defects in lymphoid development or maturation. Hence, Jak2 knockout mice reveal the pivotal role Jak2-dependent receptors play in the development of important myeloid lineages.

c. Jak3 Knockout. In contrast to the other JAKs, the expression of Jak3 is limited to hematopoietic cells (Johnston *et al.,* 1994; Kawamura *et al.,* 1994; Witthuhn *et al.,* 1994) and appears to exclusively associate with the γC chain receptor (see Section III,C; Leonard, 1996). As anticipated, the Jak3 knockout mice (Jak3−/−) exhibit a profound defect in lymphoid development (Nosaka *et al.,* 1995; Park *et al.,* 1995; Thomis *et al.,* 1995). Specifically, they have a severe block in bone marrow B cell development at the pre-B-cell stage (reminiscent of the IL-7 knockout mice; Peschon *et al.,* 1994; vonFreeden-Jeffry *et al.,* 1995). T-cell maturation is more modestly affected, with a substantial reduction in the number of mature T-cells, especially γδ T-cells. Mature T-cells recovered from Jak3−/− mice fail to respond to IL-2, rendering them functionally ineffective. As anticipated, Jak3−/− lymphocytes are defective in their responses to IL-2, IL-4, and IL-7, leading to a SCID-like phenotype analogous to the γC and IL-7 knockout mice (Leonard, 1996; Peschon *et al.,* 1994; vonFreeden-Jeffry *et al.,* 1995). These studies demonstrate that Jak3 plays a pivotal role in lymphoid development.

B. STAT Activation

Cytokine receptors play a critical role in the activation of STATs. In addition to physically transducing an extracellular signal into the cell, they mediate the specific recruitment and activation of an appropriate set of STATs. Although there appears to be some overlap in the STATs that are activated by different ligands, gene targeting studies indicate a remarkable degree of specificity. They also demonstrate that STATs play a pivotal role in the biological response to many cytokines.

1. Receptor Domains Involved in STAT Activation

Cytokine receptors stimulate the activation of receptor-associated JAKs, which in turn phosphorylate specific tyrosine motifs found in receptor endodomains. Although these tyrosines can lie almost anywhere in the endodomain, they are usually found distal to the JAK binding domain. Once these tyrosines become phosphorylated, they and the 4–5 carboxy proximal amino acids that constitute the SH2 binding/recognition domain, affect the recruitment of the appropriate STAT. The two best-characterized STAT recruitment motifs are found in the α-chain of IFN-γ receptor (Greenlund *et al.,*

1994, 1995) and in gp 130 (Gerhartz *et al.*, 1996; Hemmann *et al.*, 1996; Stahl *et al.*, 1995). A single tyrosine motif is responsible for the recruitment of Stat1 by the IFN-γ receptor. This modular domain has been successfully employed in several chimeric receptors to mediate the specific activation of Stat1 (e.g., Strehlow and Schindler, 1998). In contrast, the gp130 receptor has five motifs, each with a distinct pattern of affinity for Stat1 and/or Stat3. Likewise, these motifs are quite modular and function normally in chimeric receptor constructs. STAT-specific tyrosine motifs have now been identified in most cytokine receptors (see Table II). However, there appear to be exceptions to this model of STAT recruitment/activation. Several groups

TABLE II Cytokine Receptor Tyrosine Motifs Implicated in Mediating the Recruitment of Specific STATs

STAT	Receptor	STAT-binding tyrosine motifs	Reference
Stat1	INF-γ	YDKPH	1
Stat2	INF-α	YVFFP	2
Stat3	IL-6, LIF, IL-10	YXXQ	3–6
Stat1, Stat3	IL-6	YXPQ	3, 5
Stat5	IL-2	YLSLQ	7–9
		YCTFP	7–9
		YFFFH	7–9
	IL-7	YVTMS	7
	IL-9	YLPQE	10, 11
	EPO	YLVLD	8, 12–14
		YTILD	12
	PRL	YLDPT	15
		YVEIH	8
	GH	YVSTD	16, 17
		YFCEA	16, 17
		YITTE	9, 16, 17
		YTSIH	9
	GM-CSF	YLSLP	8
		YLCLP	18
		YVSSA	18
		YVELP	18
		YCFLP	18
Stat 6	IL-4	YKAFS	19, 20
		YKPFQ	19, 20

[a] 1, Greenlund *et al.*, 1994; 2, Yan *et al.*, 1996; 3, Stahl *et al.*, 1995; 4, Hemmann *et al.*, 1996; 5, Gerhartz *et al.*, 1996; 6, Weber-Nordt *et al.*, 1996; 7, Lin *et al.*, 1995; 8, May *et al.*, 1996; 9, Gaffen *et al.*, 1996; 10, Demoulin *et al.*, 1996; 11, Bauer *et al.*, 1998; 12, Gobert *et al.*, 1996; 13, Quelle *et al.*, 1996; 14, Damen *et al.*, 1995; 15, Lebrun *et al.*, 1995; 16, Hansen *et al.*, 1996; 17, Hansen *et al.*, 1997; 18, Itoh *et al.*, 1998; 19, Hou *et al.*, 1994; 20, Ryan *et al.*, 1998.

[b] The tyrosine is not conserved in the murine system.

have reported that mutant receptors, which no longer encode tyrosine motifs, are still able to promote STAT activation (Moriggl *et al.*, 1998; Mui *et al.*, 1995; Okuda *et al.*, 1997; Sotiropoulos *et al.*, 1996; Wang *et al.*, 1995). It has been suggested that in these cases, a tyrosine motif on a JAK may mediate STAT recruitment (Barahmand-Pour *et al.*, 1998; Fujitani *et al.*, 1997; Gupta *et al.*, 1996). Others have suggested that the amino terminal region of STATs may interact with receptor chains in a phosphotyrosine independent manner (Li *et al.*, 1997). Moreover, studies have implicated additional components in the regulation of STAT activation at the receptor. These molecules include the SOCS family (see Section IX; Hilton *et al.*, 1998) and the scaffolding protein StIP1 (Stat interacting protein-1; Collum and Schindler, submitted).

2. Specificity in STAT Activation and Signaling

Stat1 and Stat2 were initially identified as the proteins that mediate the IFN specific induction of a specific set of immediate early genes. Subsequently, additional STATs were identified, either by homology in sequence or by a symmetry in signaling. Although it was possible to demonstrate that these "new" STATs were activated (i.e., tyrosine phosphorylated) in response to stimulation with specific ligands, target genes were often not identified. Based on these types of biochemical studies, each STAT was assigned a role in transducing signals for one or more cytokines (see Table I). However, with the generation of STAT knockout models it has become possible to more carefully assign specificity of function to individual STATs.

a. Stat1 Knockout.   The purification of Stat1 and Stat2 as the components of an IFN-based transcription factor provided compelling evidence for their role in mediating the biological response to IFNs. However, subsequent studies demonstrated that Stat1, along with Stat3, was activated by most members of the IL-6 subfamily as well as by EGF (Bonni *et al.*, 1993; Sadowski *et al.*, 1993; Schindler and Darnell, 1995; Silvennoinen *et al.*, 1993; Zhong *et al.*, 1994). As antibody reagents became more widely available, many additional Stat1 activating ligands were reported. Remarkably, however, the Stat1 knockout mice only exhibit defects in type I and type II IFN signaling (Durbin *et al.*, 1996; Meraz *et al.*, 1996). Furthermore, the defect in IFN-γ signaling bears a striking resemblance to the IFN-γ receptor and IFN-γ knockout mice (Dalton *et al.*, 1993; Huang *et al.*, 1993). These observations suggest that there are few, if any, Stat1 independent signals important for transducing the biological response to IFN-γ. The Stat1 knockout mice do not exhibit a defect in responding to ligands other that signal through both Stat1 and Stat3 (e.g., IL-6 and EGF). Although these studies cannot exclude the possibility that other STATS may compensate for the loss of Stat1, they indicate that Stat1 is not required for the response to non-IFNs ligands.

b. Stat2 Knockout.   Stat2 exhibits a number of features that appear to set it apart from other STATs. Despite containing each of the major STAT functional domains, there is no compelling evidence that Stat2 forms homodimers *in vivo* or binds GAS elements. In addition, there is a remarkable divergence between the human and murine isoforms, with only ~75% sequence identity (vs. ~85% for other STATs) between amino acids 1 and 715. Moreover, the carboxy termini are completely divergent, yet function is conserved (Park and Schindler, 1999).

There is compelling evidence that Stat2 is vital for signal transduction by type I IFNs (Schindler and Darnell, 1995; Strehlow and Schindler, 1997). To date, the best genetic evidence on the biological response stimulated by type I IFNs comes from the type I IFN receptor α-chain knockout mice. These mice exhibit profound defects in antiviral responses, but develop and grow normally under pathogen-free conditions (Broek *et al.*, 1995; Hwang *et al.*, 1995; Müller *et al.*, 1994). Consistent with an important role for Stat2 in this signaling pathway, preliminary studies on Stat2 knockout mice indicate that they fail to respond to type I IFNs (C. Park and C. Schindler, unpublished observation). However, these mice exhibit a few additional unusual features that will require further investigation.

c. Stat3 Knockout.   Biochemical studies have implicated Stat3 in signal transduction for a large number of cytokines, most notably members of the IL-6 family (Akira *et al.*, 1994; Bonni *et al.*, 1993; Raz *et al.*, 1994; Wegenka *et al.*, 1994; Zhong *et al.*, 1994). As outlined previously, Stat1 is usually coactivated with Stat3 in response to IL-6, yet Stat1 knockout mice signal normally in response to these ligands. Consistent with this observation, studies with dominant interfering mutants of Stat3 have determined that Stat3 plays a critical role in mediating the biological response to gp 130-related receptors. For example, dominant interfering mutants of Stat3, but not Stat1, block the ability of IL-6 to induce differentiation in M1 cells (Minami *et al.*, 1996; Nakajima *et al.*, 1996) or to induce IgM production in SKW6.4 cells (Faris *et al.*, 1997). More physiological studies in primary neural cultures have demonstrated that dominant interfering Stat3 mutants block CNTF-induced gliogenesis, but not neurogenesis (Bonni *et al.*, 1997). Likewise, a dominant interfering Stat3 mutant blocks the LIF-dependent self-renewal and promotes differentiation of pluripotent ES cells (Niwa *et al.*, 1998; Raz *et al.*, 1998). Consistent with a potential role for Stat3 in proliferation, several studies have implicated Stat3 in transformation by v-*abl* and v-*src* (Bromberg *et al.*, 1998; Karras *et al.*, 1997; Migone *et al.*, 1995; Turkson *et al.*, 1998; Yu *et al.*, 1995). Additionally, both c-myc and Bcl-X$_L$ have been identified as Stat3 target genes (Kiuchi *et al.*, 1999; Catlett-Falcone *et al.*, 1999). Since heterozygous Stat3 knockout (i.e., −/+) cells appear to grow normally, it has been possible to generate Stat3 null (i.e., −/−) mice (Takeda *et al.*, 1997). Unfortunately, these mice exhibit a lethal phenotype and die at embryonic day 6.5–7.5. The etiology of their demise remains to be elucidated. Consistent with these observations, Stat3 has been shown to be expressed and activated early during embryogenesis (Duncan *et al.*, 1997). More recent studies with tissue specific Stat3 knockout mice indicate that Stat3 is important for IL-6 dependent T-cell proliferation and IL-10 signaling in granulocytes (Takeda *et al.*, 1998; Takeda *et al.*, 1999; Riley *et al.*, 1999).

d. Stat4 Knockout.   Both biochemical and gene targeting studies have determined that Stat4 is predominately involved in signaling for the IL-12 receptor (Jacobson *et al.*, 1995; Kaplan *et al.*, 1996c; Thierfelder *et al.*, 1996). IL-12 has been shown to be required for the critical T-cell independent (i.e., NK cell) production of IFN-γ and the development of T-cell (i.e., Th1) dependent cellular immune response. The Stat4 knockout mice exhibit profound defects in immune response. NK cells fail to produce IFN-γ or kill appropriate target cells in response to IL-12. T-cells are dramatically impaired in their ability to develop into Th1 cells after stimulation with IL-12 or an appropriate infectious agent. Of note, spermatogenesis is normal in Stat4 knockout mice, even though Stat4 is highly expressed in developing spermatozoa (Herrada and Wolgemuth, 1997).

e. Stat5 Knockout.   Stat5a and Stat5b are encoded by two tightly linked and highly homologous genes ($\sim$96% amino acid identity; Azam *et al.*, 1995; Liu *et al.*, 1995; Mui *et al.*, 1995). They diverge only modestly at their carboxy termini and in many assays appear to be functionally redundant. Stat5a and Stat5b were identified during the purification of PRL and IL-3 stimulated transcription factors (Azam *et al.*, 1995; Wakao *et al.*, 1994). Subsequent studies determined that all members of the single-chain IL-2 and IL-3 family of cytokines stimulate the activation of Stat5 (Gouilleux *et al.*, 1995; Hou *et al.*, 1995; Lin *et al.*, 1995; Mui *et al.*, 1995; Pallard *et al.*, 1995; Quelle *et al.*, 1996). In several cases, target genes were identified (Ram *et al.*, 1996; Wakao *et al.*, 1992; Yoshimura *et al.*, 1995, 1996). However, for a long time genetic corroboration of these biochemical observations was based on dominant interfering mutants, or the targeted disruption of a single Stat5 gene (Liu *et al.*, 1997; Moriggl *et al.*, 1996; Mui *et al.*, 1996; Udy *et al.*, 1997; Wang *et al.*, 1996). The dominant interfering Stat5 mutants both confirmed the identity of several putative Stat5 targets genes and indicated that Stat5a and Stat5b are functionally redundant (Moriggl *et al.*, 1996; Wang *et al.*, 1996; Yoshimura *et al.*, 1995, 1996). Despite this functional overlap, the Stat5a and Stat5b single knockout mice exhibited phenotypic defects in their response to PRL and GH, respectively (Liu *et al.*, 1997; Udy *et al.*, 1997). The PRL defect in Stat5a knock mice was most notable for a modest decrease in mammopoiesis and a profound defect in lactation after parturition. The GH defect in the Stat5b knockout mice was most notable for a loss of male sexually dimorphic patterns in weight and hepatic gene expression. The phenotypes of double Stat5a and Stat5b knockout mice were anticipated to be much more severe. Yet these mice exhibit essentially a combined GH and PRL defect (Teglund *et al.*, 1998). The PRL defect is, however, more severe in the double Stat5a knockout. Females are unable to develop a functional corpus luteum and are therefore infertile. They also fail to lactate. Surprisingly, defects in hematopoiesis at first glance seem modest, most notably lymphopenia. However, additional defects are evident in the ability of IL-3, IL-5, GM-CSF, G-CSF, and IL-7 to stimulate bone marrow colony formation *in vitro*. There are also significant defects in IL-2 stimulated T-cell proliferation and cell cycle progression (Moriggl *et al.*, 1999). Finally, due to its ability to mediate Epo dependent induction of Bcl-X_L, Stat5 has been found to be essential for the high erythropoietic rate during fetal development (Socolovsky *et al.*, 1999). Additional evaluation of potential defects in immune response remain to be determined.

f. Stat6 Knockout. As with Stat4, biochemical and genetic studies have only implicated Stat6 in a single major cytokine stimulated signaling pathway, i.e., IL-4 and closely related IL-13 (Hou *et al.*, 1994; Kaplan *et al.*, 1996b; Kontanides and Reich, 1993; Schindler *et al.*, 1994; Shimoda *et al.*, 1996; Takeda *et al.*, 1996). Similarly, IL-4 plays an important role in the development of a subset of T-helper cells, the Th2 population. Hence, a balance between Stat4 and Stat6 signaling pathways must be achieved to obtain an appropriate ratio of Th1 and Th2 dependent responses (i.e., cellular vs. humoral immune responses). IL-4 also plays an important role in the growth and differentiation of B-cells, especially in stimulating the secretion of IgE. Consistent with these observations, B-cells from Stat6 knockout mice fail to proliferate, upregulate MCH II expression, or secrete IgE after stimulation with IL-4. T-cells are also defective, exhibiting a profound inability to differentiate into Th2 cells in response to IL-4 or an appropriate infectious agent. Moreover, this defect is more severe than that seen in the IL-4 knockout mice, suggesting that there are other ligands that regulated Th2 development through Stat6 (e.g., IL-13).

V. STAT Nuclear Translocation

In unstimulated cells, the majority of STATs are found in the cytoplasm. However, after stimulation they rapidly dimerize and translocate to the nucleus. As STATs are too large to passively diffuse into the nucleus, they are transported in an energy-dependent maner (Sekimoto *et al.*, 1996]. Although substantial progress has been made in elucidating how many proteins are transported to the nucleus, STAT nuclear translocation is less well understood.

For many proteins, a known nuclear localization signal or sequence (NLS) promotes interaction with members of a family of nuclear transport proteins (Wozniak *et al.*, 1998). The "classic" NLS consist either of a short stretch of continuous basic amino acids (e.g., the SV40T antigen NLS) or of two short separated stretches of basic amino acids. These sequences are recognized by the α-subunit of a transport protein referred to as α-importin or α-karyopherin. The α-subunit binds the $\beta1$ karyopherin subunit, which associates with Ran GTPase in the nuclear pore complex. Ran plays a critical role in assuring vectoral translocation into the nucleus. A second class of NLSs consist of a conserved 38 amino acid sequence, referred to as the M9, was initially identified in heterogeneous ribonuclear protein A1. This sequence, which regulates both nuclear import and export, directly associates with a $\beta2$-karyopherin. Importation is again energy (probably Ran) dependent, but clearly distinct from the first pathway. Additional karyopherins have been identified that bind a ribosomal NLS and promote Ran dependent nuclear translocation.

Studies on the ability of STATs to translocate to the nucleus have made some progress. First, it was demonstrated that the ligand-stimulated Stat1 nuclear translocation is dependent on Ran GTPase activity (Sekimoto *et al.*, 1996). More recently, it has been determined that activated Stat1 will bind to only one of the two major classes of α-karyopherins and to β-karyopherin (Sekimoto *et al.*, 1997). This contrasts the ability of the SV-40 T-antigen NLS (the "classic" NLS) to bind both major types of α-karyopherin. However, as in the classical pathway, Stat1 association with an α-karyopherin appears to be necessary for its subsequent association with $\beta1$-karyopherin. Although this suggests that STATs may employ a "classical-like" NLS pathway for nuclear translocation, attempts to identify the STAT NLS have been unsuccessful (Sekimoto *et al.*, 1997; I. Strehlow and C. Schindler, unpublished observations). Moreover, it appears that STAT dimerization is important for nuclear translocation (Strehlow and Schindler, 1998). This suggests that a functional STAT NLS may require two active STATs, either because the NLS spans both peptides or because dimerization exposes an otherwise hidden NLS. Consistent with these models, studies have demonstrated that dimerization will rescue the ability of mutant STATs to translocate to the nucleus (Strehlow and Schindler, 1998). Although these observations fail to distinguish between the two possibilities, they do provisionally map the STAT NLS to the amino terminus.

Another potentially important development in our understanding of STAT nuclear translocation is the observation that there is a basal level of nuclear Stat1 and Stat3 in many unstimulated cell types (Schindler *et al.*, 1992b; Zhang *et al.*, 1995). In the past, this basal nuclear staining had been attributed to stimulation by an unknown ligand (i.e., present in culture media). However, more recent studies suggest that a basal level of nuclear STAT may be required to regulate caspase expression (Kumar *et al.*, 1997;

Schindler, 1998). How these inactive STATs get to the nucleus is another intriguing issue. One plausible explanation may be that these STATs are able to translocate to the nucleus and mediate their biological activity by associating with other transcription factors (Schindler, 1998).

VI. STATs as Transcriptional Activators

STATs were initially identified as components of a multiprotein transcription factor ISGF3 (IFN stimulating gene factor 3) that binds to the IFN stimulated response element (ISRE). Subsequent studies have demonstrated that one of these STATs, Stat1, also promotes the induction of genes in response to IFN-γ (Schindler and Darnell, 1995). In contrast to ISGF3, which employs an additional DNA binding protein, IFN-γ stimulated Stat1 homodimers bind DNA directly, the Gamma activation site (GAS). Subsequent studies have determined that the STAT complexes activated by all other cytokines bind to members of the GAS family (Decker *et al.*, 1997; Schindler and Darnell, 1995). These studies have also determined that STAT-based signals may not always work independently, but rather interact, or "network," with other signaling cascades.

A. STAT DNA Binding Specificity

IFNs transduce STAT signals through two distinct transcriptional enhancers, the ISRE and GAS elements (Decker *et al.*, 1997; Schindler and Darnell, 1995). Several members of the IRF-1 family of transcription factors also recognize the ISRE site, providing for an overlap in regulation of these target genes. Likewise, some GAS elements may bind transcription factors that are not members of the STAT family (Dent *et al.*, 1997; Kawamata *et al.*, 1994; Ye *et al.*, 1997).

1. Interferon Stimulated Response Element

Promoter mapping studies led to the identification of a conserved 18-20 bp element (the interferon stimulated response element, ISRE) that was both required for IFN-α stimulated gene induction and bound the IFN-α inducible factor ISGF3 (Levy *et al.*, 1988; Reich *et al.*, 1987). The most effective ISRE was identified in the ISG15 promoter and had the sequence CAGTTTCGGTTTCCC. Comparison with other functionally mapped type I IFN response elements led to the compilation of an ISRE consensus sequence, YAGTTTC(A/T)YTTTYCCC. This was then confirmed by detailed mutagenesis studies (Kessler *et al.*, 1988). These studies also demonstrated that an IRF-1 binding site (i.e., TTTCGGTTTC, actually two tandem binding sites; Escalante *et al.*, 1998) constitutes the core of the ISRE element. Identification of p48, a member of the IRF-1 family, as the DNA binding component of ISGF3 provided an explanation for the overlapping DNA binding specificities of ISGF3 and IRF-1 (Kessler *et al.,* 1990; Veals *et al.,* 1992). Careful DNA binding studies have confirmed that p48 interacts with the "IRF-1 core" in the ISRE and that the Stat1:Stat2 heterodimer promotes contact with the additional flanking nucleotides (Qureshi *et al.,* 1995).

2. Gamma Activation Site

Characterization of the ability of IFN-γ to induce the gene for GBP (guanylate-binding Protein) led to the identification of the first GAS element (Decker *et al.,* 1991; Lew *et al.,* 1991). With the subsequent identification of additional and more robust IFN-γ responsive elements (Khan *et al.,* 1993; Pearse *et al.,* 1993), it became possible to begin to recognize a GAS consensus sequence, TTTCCNGGAAA. Consistent with the palindromic nature of this sequence, subsequent biochemical studies demonstrated that Stat1 binds this element as a homodimer (Gupta *et al.,* 1996; Shuai *et al.,* 1994). Concurrent studies on the ability of other ligands to induce genes (e.g., PDGF, IL-6, and prolactin), independently led to the identification of DNA response elements that were similar in sequence to the gamma activation site (Decker *et al.,* 1997). The demonstration that these ligands also transduced signals through members of the STAT family led to the hypothesis that the GAS elements coevolved with the STAT signaling paradigm (Decker *et al.,* 1997; Sadowski *et al.,* 1993; Wakao *et al.,* 1992; Wegenka *et al.,* 1994). A general consensus for the GAS element of TTCN$_{2-4}$GAA has now emerged. Some STAT dimers exhibit a preference of $N=2$, while for others $N=4$ (Decker *et al.,* 1997; Seidel *et al.,* 1995). Studies directed at characterizing the optimal DNA binding element for Stat1, Stat3, Stat4, and Stat6 homodimers further support this model (Horvath *et al.,* 1995; Schindler *et al.,* 1995; Xu *et al.,* 1996; Yamamoto *et al.,* 1997). Moreover, these studies have also determined that the spacing between palindrome half sites is important in defining STAT binding specificity (Decker *et al.,* 1997). One important exception to these observations is Stat2. Although it contains a conserved DNA binding domain and can form homodimers *in vitro,* (Bluyssen and Levy, 1997; Gupta *et al.,* 1996), it does not appear to bind a known DNA element.

B. STAT Transcriptional Activation

Most transcription factors are modular, consisting of a DNA binding element and a transcriptional activation domain. This latter domain is believed to either directly or indirectly mediate interaction with the basal transcription machinery. Consistent with this model, STATs contain distinct DNA binding and transcriptional activation domains (see Section II). Like many other transcription factors, the STAT transcriptional activation domain (TAD), which has been defined by deletion and domain swapping studies, is poorly conserved (Caldenhoven *et al.,* 1996; Lu *et al.,* 1997;

Mikita *et al.*, 1996; Minami *et al.*, 1996; Moriggl *et al.*, 1997; Mui *et al.*, 1996; Müller *et al.*, 1993b; Qureshi *et al.*, 1996; Wang *et al.*, 1996; C. Park and C. Schindler, manuscript submitted). In addition, studies have shown that the transcriptional activity of STATs can be regulated after activation. This appears to be achieved either through interaction with other transcriptional regulators, or by the direct modification of STATs.

I. Interaction with Transcriptional Regulators

The first evidence that STAT dimers may functionally interact with other transcription factors came from the characterization of ISGF3. In this relatively unusual factor, the Stat1:Stat2 heterodimer fails to bind DNA directly (Kessler *et al.*, 1990). Rather, this dimer associates with and modifies the DNA binding activity of a member of the IRF-1 family of transcription factors. The STAT dimer provides the TAD domain (Müller *et al.*, 1993b; Qureshi *et al.*, 1996). Subsequent studies have determined that STATs interact with several additional transcriptional regulators, often found binding to adjacent elements. For example, even prior to its identification as Stat5, MGF (mammary gland growth factor, a Stat5 homodimer) was shown to antagonize the constitutive negative activity of an adjacent YYI site (in the β-casein promoter; Meier and Groner, 1994). Likewise, earlier studies on the IFN-γ stimulated induction of FcγRI indicated an interaction between Stat1 and other transcription factors (Pearse *et al.*, 1993). Most notable was the role that PU.1, a myeloid specific member of the Ets family of played in restricting INF-γ dependent FcγRI expression to macrophages (Perez *et al.*, 1994). More recently, activated Stat1 has also been shown to synergistically interact with Sp1 in the ICAM-1 (intercellular adhesion molecule 1) promoter (Look *et al.*, 1995), and with NFκB in the IRF-1 promoter (Ohmori *et al.*, 1997; Pine, 1997). Stat1 has also been shown to functionally interact with USF-1, an E-box binding factor, in the promoter of CIITA. CIITA regulates MHC II expression in response to IFN-γ (Muhlethaler-Mottet *et al.*, 1998).

Similar observations have been made for other members of the STAT family. For example, Stat3β, an alternative splice isoform of Stat3 (see below), synergistically interacts with c-*Jun* to promote the Stat3 dependent induction of α2-macroglobulin (Schaefer *et al.*, 1997). Intriguingly, the full-length isoform of Stat3 does not appear to interact with c-*Jun*. In the case of Stat6, its ability to regulate immunoglobulin heavy-chain germ-line ε gene expression, in response to IL-4, has been shown to entail a synergistic interaction with C/EBPβ and NFκB (Mikita *et al.*, 1998; Shen and Stavnezer, 1998). The interaction between Stat6 and NFκB may even occur prior to DNA binding. A similar enhanceosome-like element has been reported in the promoter of the IL-2 receptor α-chain, where Stat5, Elf-1, HMG-I(Y), and a GATA protein synergize to promote IL-2 dependent expression (John *et al.*, 1996; Lécine *et al.*, 1996). Stat5 has also been shown to cooperatively interact with NF-Y and/or the glucocorticoid receptor to promote the PRL dependent induction of genes (Stöcklin *et al.*, 1996; Stoecklin *et al.*, 1997; Subramanian *et al.*, 1998). Notably, the synergy with glucocorticoid receptors appears to be mediated through a single DNA binding element, the GAS (Stoecklin *et al.*, 1997).

STATs have also been shown to interact with transcriptional modifiers from the CBP/p300 family. This interaction was initially identified in a yeast two hybrid screen. A Stat2 prey isolated p300 as an interacting protein (Bhattacharya *et al.*, 1996). These studies suggested that p300 binds to, and is likely to promote, the activity of the Stat2 TAD. A complementary result was obtained during the characterization of the antagonistic of effect IFN-γ on M-CSF induced expression of the scavenger receptor (Horvai *et al.*, 1997). It was determined that Stat1 (activated by IFN-γ) antagonized Ap-1 (activated by M-CSF) through competition for a rate-limiting supply of p300 or CBP. Likewise, the oncogene E1A appears to disrupt Stat1- and Stat2-based signals by competing for p300/CPB (Bhattacharya *et al.*, 1996; Kurokawa *et al.*, 1998). Once stably assembled on the promoter, p300/CBP either directly provides, or as in the case for Stat1, recruits additional histone acetyltransferase activity (e.g., pCIP), believed to be important in transcriptional activation (Korzus *et al.*, 1998). More recent studies have determined that an N-Myc interactor (Nmi) potentiates the association between STATs and p300/CBP (Zhu *et al.*, 1999).

2. STAT Modifications

The initial characterization of Stat1 and Stat2 suggested that STATs undergo three types of modification, tyrosine phosphorylation, serine phosphorylation, and carboxy terminal truncation. Although earlier studies established the important role that tyrosine phosphorylation plays in STAT activation, they failed to appreciate the dynamic nature of STAT serine phosphorylation (Improta *et al.*, 1994; Schindler *et al.*, 1992b; Shuai *et al.*, 1992). However, studies in which the level of STAT tyrosine phosphorylation was found not to correlate directly with biological activity, suggested that an additional modification might be important (Eilers *et al.*, 1993; Lew *et al.*, 1989; Lord *et al.*, 1991; Nakajima and Wall, 1991). Immunoblotting studies with Stat3-specific antibodies lead to the identification of an additional ligand-dependent isoform. This turned out to be a serine phosphorylated isoform of Stat3 (Lütticken *et al.*, 1995; Wen *et al.*, 1995; Zhang *et al.*, 1995). Moreover, phosphorylation of Stat3 as well as Stat1 at serine 727 was shown to be important for a maximal transcriptional response (Eilers *et al.*, 1995; Wen *et al.*, 1995). Since serine 727 lies within a potential MAP kinase motif, it was speculated this site would provide STATs an opportunity to network, or crosstalk, with other signaling pathways. Although several studies suggest that STAT serine phosphorylation is stimulated by "other pathways" (Chung *et al.*, 1998; Kovarik *et al.*, 1998; Ng and Cantrell, 1997), it is likely that this in large part independent of MAP kinases (Chung *et al.*, 1998; Kovarik *et al.*, 1998; Ng and Cantrell, 1997; Zhu *et al.*, 1997). The identity of the STAT serine kinase(s) remains to be elucidated.

Another important STAT modification is the formation of carboxy terminally truncated isoforms. This type of modification was already evident during the characterization of Stat1, where both a full-length (i.e., p91) and carboxy terminally truncated (i.e., p84) isoform were identified in purified preparations of ISGF3 (Schindler *et al.*, 1992a). Subsequent studies mapped the transcriptional activation domain to the STAT carboxy terminus (see Section II,E; Müller *et al.*, 1993b), suggesting that the truncated species might serve as a naturally occurring dominant negative isoform. Although some carboxy terminally truncated species have been shown to behave in a dominant negative fashion (Caldenhoven *et al.*, 1996; Mui *et al.*, 1996; Wang *et al.*, 1996), this is certainly not the case for all truncated species. For example, truncated Stat1 (p84 or Stat1β) is fully functional in the ISGF3 complex (Müller *et al.*, 1993b), where Stat2 provides the transcriptional activation domain (Qureshi *et al.*, 1996). Additionally, truncated Stat3 (Stat3β), which exhibits a prolonged activation profile, appears to synergize with c-Jun to promote transcriptional activation of some genes (Sasse *et al.*, 1997; Schaefer *et al.*, 1997). Likewise, carboxy terminally truncated isoforms of Stat5 have been implicated in transducing a unique set of signals in immature myeloid cells (Azam *et al.*, 1995, 1997; Bovolenta *et al.*, 1998; Lokuta *et al.*, 1998; Meyer *et al.*, 1998). In addition, whereas most carboxy terminally truncated STAT isoforms are generated by an RNA processing event (e.g., intron read-through; Wang *et al.*, 1996), this is not the case for Stat5. In this case the truncated isoform is generated by a protein processing event (Azam *et al.*, 1997; Meyer *et al.*, 1998).

VII. Negative Regulation of STAT Signaling

The important positive role that JAKs and STATs play in the rapid induction of genes has been elucidated in great detail. However, the components of the equally important process of signal decay are less well understood. To date three types of proteins have been implicated in this process: phosphatases, proteases, and "other" antagonists.

A. Phosphatases

Because activation in the JAK-STAT signaling paradigm is mediated by tyrosine phosphorylation, it makes sense to consider phosphatases in the deactivation process. Hence, all three components that are tyrosine phosphorylated during stimulation (i.e., the receptor, JAKs, and STATs) are potential phosphatase substrates. For such a phosphatase to be effective in signal decay, one would imagine that they need to be active at the right time and place. As each of the three potential substrates transits the cytosol, a cytoplasmic phosphatase is an important consideration. Activated STATs, which are uniquely found in the nucleus, could be specifically targeted by a nuclear phosphatase as well. The issue of when phosphatases are active should be considered as well. Either they could be constitutively active (i.e., function as a damper), or their activity could be upregulated in response to the activating stimulus. The best-characerized phosphatases are in fact consistent with this latter possibility.

Two major cytosolic phosphatases that are activated in response to ligands have been described, SHPTP1 (HCP, PTP1C, SHP-1) and SHPTP2 (Syp, PTP1D, SHP-2). As their names suggest, both encode an SH2 domain believed to mediate their receptor recruitment and subsequent activation (Weber-Nordt *et al.*, 1998). SHPTP1 is expressed predominately in hematopoietic cells and has been shown to play a critical role in signal decay in response to several cytokines (Jiao *et al.*, 1997; Klingmüller *et al.*, 1995; Yi *et al.*, 1993). Consistent with these observations, mice with a mutant SHPTP1 locus, i.e., "motheaten," exhibit increased hematopoiesis, leukocytosis, hypersplenism, and immunodeficencies (Shulz *et al.*, 1993). SHPTP1 may also counterregulate type I IFN signaling in some cell types (David *et al.*, 1995). SHPTP2 is more widely expressed and appears to contribute the upregulation of signals in response to many ligands (Ali *et al.*, 1996; Berchtold *et al.*, 1998; Boulton *et al.*, 1994; Stahl *et al.*, 1995). There are additional tyrosine phosphatase studies, but they have yielded conflicting results.

Although the role of phosphatases in counterregulating cytokine signaling near or at the receptor are compelling, other studies support the role of a phosphatases in STAT deactivation as well (David *et al.*, 1993; Haspel *et al.*, 1996; Lee *et al.*, 1997). Moreover, STAT deactivation may not occur until after STATs have translocated to the nucleus (Strehlow and Schindler, 1998). Consistent with this observation, nuclear phosphatases have been reported (David *et al.*, 1993). However, little progress has been made in identifying these phosphatase(s).

B. Proteases

Another important mechanism by which signals are downregulated is by targeting signaling molecules for degradation. Studies have provided support for a model in which activated STATs are targeted for rapid degradation by a ubiquitin/26S proteasome dependent pathway (Kim and Mantiatis, 1996). Although this mechanism serves as a potential alternative to the role of phosphatases in STAT deactivation, both could occur. Studies done under more physiological conditions, with STAT and receptor mutants that fail to be targeted for degradation, are likely to resolve the controversy over the relative role of proteases and phosphatases in STAT signal decay.

C. Antagonists

Several novel proteins that appear to specifically antagonize JAK-STAT signaling have been identified through several approaches (Asao *et al.*, 1997; Chung *et al.*, 1997; Endo *et al.*, 1997; Hilton *et al.*, 1998; Matsumoto *et al.*, 1997; Naka *et al.*, 1997; Starr *et al.*, 1997; Takeshita *et al.*, 1997; Yoshimura *et al.*, 1995). The SOCS family represent the largest subgroup of these antagonists (Hilton *et al.*, 1998). The other two antagonists, STAM and PIAS, are unrelated proteins (Chung *et al.*, 1997; Liu *et al.*, 1998; Takeshita *et al.*, 1997).

I. SOCS

Studies directed at identifying proteins that interact with JAKs, or functionally antagonize IL-6 dependent responses, have converged on the identification of a novel family of suppressors of cytokine signaling (SOCS; Hilton *et al.*, 1998). Although 20 members of a family that encode a common "SOCS box" have been reported, only a subset contain both a SOCS box and SH2 domain and have been implicated in cytokine signaling (Endo *et al.*, 1997; Hilton *et al.*, 1998; Matsumoto *et al.*, 1997; Naka *et al.*, 1997; Starr *et al.*, 1997; Yoshimura *et al.*, 1995). The first member of this family, CIS (cytokine inducible SH2-containing protein), was actually identified in a screen for IL-3 stimulated immediate early genes (Matsumoto *et al.*, 1997; Yoshimura *et al.*, 1995). It was also found to be induced by IL-2, GM-CSF, and EPO, and its expression was found to be STAT dependent (i.e., it is a STAT target gene). Functional studies have determined that CIS stably associates with activated (i.e., tyrosine-phosphorylated) forms of the EPO and IL-3 receptors (i.e., β chain). Moreover, this correlates with an increase in the rate of signal decay. Subsequently, three additional members have been identified. SOCS-1 (JAB, SSI-1) was identified by virtue of its ability to: (1) inhibit IL-6 dependent differentiation of M1 cells (Starr *et al.*, 1997); (2) interact with the JH1 domain of Jak1 (Endo *et al.*, 1997); (3) fortuitously act as a target for an SH2-specific antibody (Naka *et al.*, 1997). SOCS-2,3,4,5,6,7, were identified by their homology to SOCS-1 (Hilton *et al.*, 1998; Starr *et al.*, 1997). SOCS-1 appears to be a STAT target gene and is induced by several cytokines. Although CIS, SOCS-1, SOCS-2, and SOCS3 are all rapidly induced in response to a partially overlapping set of ligands, the expression of SOCS-1 and SOCS-3 is more transient. SOCS-1 and SOCS-3 also share an ability to bind JAKs, leading to a marked reduction in kinase activity. However, kinetic studies indicate that the expression of these SOCS proteins (approximately 1 hour after stimulation) occurs well after JAK activity has decayed (approximately 10–20 minutes after stimulation). Other studies have indicated that SOCS-3 may play an important role in downregulating signals transduced by LIF and leptin in the central nervous system (Auernhammer *et al.*, 1998; Bjorbaek *et al.*, 1998). SOCS knockout and transgenic studies are underway and should provide important insight into this family of counterregulatory molecules.

2. Other Antagonists

Two additional types of cytokine signaling antagonists have been identified. STAM was identified as a substrate of IL-2 dependent tyrosine kinase activity that contained both an SH3 domain and ITAM motif (Asao *et al.*, 1997; Takeshita *et al.*, 1997). Studies with an SH3 deletion mutant of STAM have suggested that the wild-type molecule plays an important role in IL-2 and GM-CSF dependent induction of DNA synthesis and c-*myc* expression.

PIAS proteins are members of a small family of related proteins (Liu *et al.*, 1998a). PIAS1 was identified in a yeast two hybrid screen with a Stat1 bait (Liu *et al.*, 1998a). Surprisingly, however, this protein has affinity for phosphorylated Stat1. PIAS3, a homologue, was found to have affinity for the phosphorylated form of Stat3 (Chung *et al.*, 1997). When overexpressed, these proteins inhibit Stat1 and Stat3 DNA binding, respectively, suggesting that they may directly inhibit STAT signal transduction.

VIII. "Nonclassical" STAT Functions

Although the "classical" role of STATs as receptor-activated transcription factors is well accepted, some studies have ascribed several "nonclassical" functions to members of this family. This includes networking with other signaling cascades and the regulation of apoptosis.

A. Networking with Other Pathways

As outlined in Section VI,B,2, the requirement of serine phosphorylation for full STAT activity indicates that STATs can integrate signals from more than one pathway, i.e., tyrosine phosphorylation from one pathway, and serine phosphorylation from another pathway. Additional examples of how STATs may network with other pathways has come from observations that PKR may sequester Stat1 (Wong *et al.*, 1997) and that Stat3 may serve a scaffold function in IFN-α dependent signaling (Pfeffer *et al.*, 1997).

1. STAT-PKR Interaction

PKR (protein kinase R) is a serine/threonine-specific protein kinase that is induced by IFNs and double stranded RNA (dsRNA; e.g., viral RNA). Once activated, it phosphorylates the α-subunit of the eukaryotic initiation factor 2 (eIF-2α), leading to a block in translation. This is an important defense against viral infection (Proud, 1995). In addition to regulating of translation, studies have implicated PKR in Stat1 mediated transcriptional regulation (Wong *et al.*, 1997). These studies suggest that PKR does not phosphorylate Stat1, but rather binds to the unphosphorylated isoform, yielding a PKR-Stat1 complex. Stimulation of cells with either IFN or dsRNA leads to a release of Stat1, thereby increasing the pool that is available for activation and signaling. The potential role of PKR in regulating other STAT-based signals has not been evaluated.

2. STATs as a Docking Molecule

Like most cytokines, type I IFNs activate more than one signaling cascade. Type I IFNs certainly induce two JAK-STAT pathways (i.e., ISGF3 $\rightarrow$ ISRE and Stat1:1 $\rightarrow$ GAS). Additionally Stat3 has been shown to be activated by type I IFNs in some cell types (Horvath *et al.*, 1996; Silvennoinen *et al.*, 1993; Yang *et al.*, 1996). Consistent with this observation, a potential YXXQ Stat3 recruitment motif (see Table II) can be identified in the α-chain of the type I IFN receptor (IFNAR1; Uzé *et al.*, 1990). Another pathway activated by type I IFNs is the phosphatidylinositol 3-kinase (PI3K) pathway. Surprisingly, studies exploring how the IFN receptor induces this pathway suggest that a phosphotyrosine on Stat3 (i.e., Y^{656}, not Y^{705}!) may serve to recruit the p85 subunit of PI3K to the IFN receptor complex (Pfeffer *et al.*, 1997). These data imply that Stat3 can function as a docking molecule to affect the recruitment and activation of another signaling cascade. However, it remains to be determined whether Y^{656} of Stat3 is actually phosphorylated under these circumstances.

B. STATs and Apoptosis

The ability of cytokines to promote the development of some cell lineages appears to entail both a stimulation of growth and suppression of apoptosis (Fukada *et al.*, 1996; Kinoshita *et al.*, 1995; Minami *et al.*, 1996; Nakajima *et al.*, 1996; Schwarze and Hawley, 1995). Cytokines have also been found to antagonize proliferation in certain cell types. In some cases, this appears to entail the STAT dependent induction of cyclin inhibitors (Chin *et al.*, 1996; Kaplan *et al.*, 1998; Matsumura *et al.*, 1997). It may also entail induction of apoptosis (Deiss *et al.*, 1995). Moreover, one study suggests that STATs may actually regulate programmed cell death in the absence of a "classical" activating stimulus (Kumar *et al.*, 1997). During an investigation of the properties of a Stat1 deficient cell line, it was determined that this line was uniquely resistant to the induction of apoptosis. Not only could this defect in programmed cell death be rescued by the reintroduction of Stat1, but several Stat1 mutants rescued the phenotype as well. This included mutants in the tyrosine activation motif (i.e., Y701F) the SH2 domain (i.e., R602L), and the serine phosphorylation site (i.e., S727A), indicating that "classical" STAT signaling was not involved. Furthermore, a loss in the expression of three caspases (proteases implicated in programmed cell death) was found to correlate with the defect in apoptosis, suggesting that caspase expression does not require the formation of active Stat1 homodimers. How could this be? Perhaps Stat1, which is found at low levels in the nucleus of unstimulated cells (Schindler *et al.*, 1992b; Zhang *et al.*, 1995), associates with other proteins to regulate the expression of caspases (Schindler, 1998). This should be an intriguing area for future investigation.

IX. Concluding Comments

STATs have come a long way from their humble beginnings as IFN dependent transcription factors. Seven members of the family have been identified in mammals as well as additional members in lower eukaryotes. Both biochemical and genetic studies have provided compelling evidence that STATs transduce signals critical to the biological response of most if not all cytokines. In most cases these appear to be responses to an external stress. Particularly gratifying has been the solution of the structures of Stat1 and Stat3 bound to DNA. Fortunately, much remains to be done in identifying the STAT target genes that are responsible for these vital biological effects and in more carefully delineating how STATs network with other signaling cascades.

References

Abbas, A., Lichtman, A. H., and Pober, J. S. (1991). *In* "Cellular and Molecular Immunology," 1st ed. (M. J. Wonsiewicz, ed.). Saunders, Philadelphia.

Adler, K., Gerisch, G., vonHugo, U., Lupas, A., and Schweiter, A. (1996). Classification of tyrosine kinases from *Dictyostelium discoideum* with two distinct, complete or incomplete catalytic domains. *FEBS Lett.* **395**, 286–292.

Aguet, M., Dembic', Z., and Merlin, G. (1988). Molecular cloning and expression of the human interferon-γ receptor. *Cell* **55**, 273–280.

Akira, S., Nishio, Y., Inoue, M., Wang, X.-J., Wei, S., Matsusaka, T., Yoshida, K., Sudo, T., Naruto, M., and Kishimoto, T. (1994). Molecular cloning of APRF, a novel IFN-stimulated gene factor 3 p91-related transcription factor involved in the gp130-mediated signaling pathway. *Cell* **77**, 63–71.

Alexander, W. S., Roberts, A. W., Nicola, N. A., Li, R., and Metcalf, D. (1996). Deficiencies in progenitor cells of multiple hematopoietic lineages and defective megakaryocytopoiesis in mice lacking the thrombopoietic receptor c-mpl. *Blood* **87**, 2162–2170.

Ali, S., Chen, Z., Lebrun, J. J., Vogel, W., Kharitonenkov, A., Kelly, P. A., and Ullrich, A. (1996). PTP1D is a positive regulator of the prolactin signal leading to a β-casein promoter activation. *EMBO J.* **15**, 135–142.

Ali, M. S., Sayeski, P. P., Dirksen, L. B., Hayzer, D. J., Marrero, M. B., and Bernstein, K. E. (1997). Dependence on the motif YIPP for the physical association of Jak2 kinase with the intracellular carboxyl tail of the angiotensin II AT1 receptor. *J. Biol. Chem.* **272**, 23382–23388.

Argetsinger, L. S., and Carter-Su, C. (1996). Mechanism of signaling by growth hormone receptor. *Physiol. Rev.* **76**, 1089–1107.

Argetsinger, L. S., Campbell, G. S., Yang, X., Witthuhn, B. A., Silvennoinen, O., Ihle, J. N., and Carter-Su, C. (1993). Identification of JAK2 as growth hormone receptor–associated tyrosine kinase. *Cell* **74**, 237–244.

Asao, H., Sasaki, Y., Arita, T., Tanaka, N., Endo, K., Kasai, K., Takeshita, T., Endo, Y., Fujita, T., and Sugamura, K. (1997). Hrs is associated with STAM, a signal transducing adaptor molecule. Its suppressive effect on cytokine-induced cell growth. *J. Biol. Chem.* **272**, 32785–32791.

Auernhammer, C. J., Chesnokova, V., Bousquet, C., and Melmed, S. (1998). Pituitary corticotroph SOCS-3: novel intracellular regulation of leukemia-inhibitory factor-mediated proopiomelanocortin gene expression and adrenocorticotropin secretion. *Mol. Endocrinol.* **7**, 954–961.

Azam, M., Erdjument-Bromage, H., Kreider, B. L., Xia, M., Quelle, F., Basu, R., Saris, C., Tempst, P., Ihle, J. N., and Schindler, C. (1995). Interleukin-3 signals through multiple isoforms of Stat5. *EMBO J.* **14**, 1402–1411.

Azam, M., Lee, C., Strehlow, I., and Schindler, C. (1997). Functionally distinct isoforms of Stat5 are generated by protein processing. *Immunity* **6**, 691–701.

Bach, E. A., Tanner, J. W., Marsters, S., Ashkenazi, A., Aguet, M., Shaw, A. S., and Schreiber, R. D. (1996). Ligand induced assembly and activation of the IFNγ receptor in intact cell. *Mol. Cell. Biol.* **16**, 3214–3221.

Bach, E. A., Aguet, M., and Schreiber, R. D. (1997). The IFN gamma receptor: A paradigm for cytokine receptor signaling. *Ann. Rev. Immunol.* **15**, 563–591.

Bacon, C. M., Tortolani, P. J., Shimosaka, A., Rees, R. C., Longo, D. L., and O'Shea, J. J. (1995). Thrombopoietin (Tpo) induces tyrosine phosphorylatin and activation of Stat5 and Stat3. *FEBS Lett.* **370**, 63–68.

Barahmand-Pour, F., Meinke, A., Groner, B., and Decker, T. (1998). Jak2–Stat5 interactions analyzed in yeast. *J. Biol. Chem.* **273**, 12567–12575.

Bartley, T. D., Bogenberger, J., Hunt, P., Li, Y.-S., Lu, H. S., Martin, F., Chang, M.-S., Samal, B., Nichol, J. L., and Swift, S. (1994). Identification and cloning of megakaryocyte growth and development factor that is a ligand for cytokine receptor Mpl. *Cell* **77**, 1117–1124.

Bauer, J., Liu, K., You, Y., Lai, S., and Goldsmith, M. (1998). Heteromerization of the gamma-c chain with the Interleukin-9 receptor alpha subunit leads to STAT activation and prevention of apoptosis. *J. Biol. Chem.* **273**, 9255–9260.

Bazan, J. F. (1990). Shared architecture of hormone binding domains in type I and II interferon receptors. *Cell* **61**, 753–754.

Becker, S., Groner, B., and Müller, C. W. (1998). Three-dimensional structure of the Stat3β homodimer bound to DNA. *Nature* **394**, 145–151.

Berchtold, S., Volarevic, S., Moriggl, R., Mercep, M., and Groner, B. (1998). Dominant negative variants of the SHP-2 tyrosine phosphatase inhibit prolactin activation of JAK2 (janus kinase 2) and induction of Stat5 (signal transducer and activator of transcription 5)-dependent transcription. *J. Mol. Endocrinol.* **12**, 556–567.

Bhat, G. J., Thekkumkara, T. J., Thomas, W. G., Conrad, K. M., and Baker, K. M. (1994). Angiotensin II stimulates sis-inducing factor like DNA binding activity. *J. Biol. Chem.* **269**, 31443–31449.

Bhattacharya, S., Eckner, R., Grossman, S., Oldread, E., Arany, Z., D'Andrea, A., and Livingston, D. (1996). Cooperation of Stat2 and p300/CBP in signaling induced by interferon-α. *Nature* **383**, 344–347.

Binari, R., and Perrimon, N. (1994). Stripe-specific regulation of pair-rule genes by hopscotch, a putative Jak family tyrosine kinase in drosophila. *Genes Devel.* **8**, 300–312.

Bjorbaek, C., Elmquist, J. K., Frantz, J. D., Shoelson, S. E., and Flier, J. S. (1998). Identification of SOCS-3 as a potential mediator of central leptin resistance. *Mol. Cell* **4**, 619–625.

Bluyssen, H. A. R., and Levy, D. E. (1997). Stat2 is a transcriptional activator that requires sequence specific contacts provided by Stat1 and p48 for stable interaction with DNA. *J. Biol. Chem.* **272**, 4600–4605.

Bonni, A., Frank, D. A., Schindler, C., and Greenberg, M. E. (1993). Characterization of a pathway for Ciliary Neurotropic Factor signaling to the nucleus. *Science* 262, 1575–1579.

Bonni, A., SUn, Y., Nadal-Vicens, M., Bhatt, A., Frank, D. A., Rozovosky, I., Stahl, N., Yancopoulos, G. D., and Greenberg, M. E. (1997). Regulation of gliogensis in the central nervous system by the JAK-STAT signaling pathway. *Science* 278, 477–483.

Boulton, T. G., Stahl, N., and Yancopoulos, G. D. (1994). Ciliary neurotrophic factor/leukemia inhibitory factor/interleukin 6/oncostatin M family of cytokines induces tyrosine phosphorylation of a common set of proteins overlapping those induced by other cytokines and growth factors. *J. Biol. Chem.* 269, 11648–11655.

Bovolenta, C., Testolin, L., Benussi, L., Lievens, P. M., and Liboi, E. (1998). Positive selection of apoptosis-resistant cells correlates with activation of dominant-negative Stat5. *J. Biol. Chem.* 273, 20779–20784.

Briscoe, J., Rogers, N. C., Witthuhn, B. A., Watling, D., Harpur, A. G., Wilks, A. F., Stark, G. R., Ihle, J. N., and Kerr, I. M. (1996). Kinase-negative mutants of JAK1 can sustain interferon-gamma-inducible gene expression but not an antiviral state. *EMBO J.* 15, 799–809.

Broek, M. F. V. D., Müller, U., Huang, S., Aguet, M., and Zinkernagel, R. M. (1995). Antiviral defense in mice lacking both alpha/beta and gamma Interferon receptors. *J. Vir.* 69, 4792–4796.

Bromberg, J. F., Horvath, C. M., Besser, D., Lathem, W. W., and Darnell, J. E. (1998). Stat3 activation is required for cellular transformation by v-src. *Mol. Cell. Biol.* 18, 2553–2558.

Caldenhoven, E., vanDijk, T. B., Solari, R., Armstrong, J., Raaijmakers, J. A., Lammers, J.-W., Koenderman, L., and deGroot, R. P. (1996). Stat3β, a splice variant of transcription Stat3, is a dominant negative regulator of transcription. *J. Biol. Chem.* 271, 13221–13227.

Candeias, S., Muegge, K., and Durum, S. K. (1997). IL-7 receptor and VDJ recombination: Trophic versus mechanistic actions. *Immunity* 6, 501–508.

Carver-Moore, K., Broxmeyer, H. E., Luoh, S. M., Cooper, S., Peng, J., Burstein, S. A., Moore, M. W., and Sauvage, F. J. d. (1996). Low levels of erythroid and myeloid progenitors in thrombopietin-and c-mpl-deficient mice. *Blood* 88, 803–808.

Catlett-Falcone, R., Landowski, T. H., Oshiro, M. M., Turkson, J., Levitzki, A., Savino, R., Ciliberto, G., Moscinski, L., Fernandez-Luna, J. L., Nunez, G., Dalton, W. S., and Jove, R. (1999). Constitutive activation of Stat3 signaling confers resistance to apoptosis in human U266 myeloma cells. *Immunity* 10, 105–115.

Chapelle, A. d. l., Traskelin, A. L., and Juvonen, E. (1993). Truncated erythropoietin receptor causes dominantly inherited benign human erythrocytosis. *Proc. Natl. Acad. Sci., USA* 90, 4495–4499.

Chen, H., Charlat, O., Tartaglia, L. A., Woolf, E. A., Weng, X., Ellis, S. J., Lakey, N. D., Culpepper, J., Moore, K. J., Breitbart, R. E., Duyk, G. M., Tepper, R. I., and Morgenstern, J. P. (1996). Evidence that the diabetes gene encodes the Leptin receptor: Identification of a mutation in the Leptin receptor gene in db/db mice. *Cell* 84, 491–495.

Chen, X., Winkemeier, U., Zhao, Y., Jeruzalmi, D., Darnell, J. E., and Kuriyan, J. (1998). Crystal structure of a tyrosine phosphorylated Stat1 dimer bound to DNA. *Cell* 93, 827–839.

Chin, Y. E., Kitagawa, M., Su, W. C., You, Z.-H., Iwamoto, Y., and Fu, X.-Y. (1996). Cell growth arrest and induction of cyclin-dependent kinase inhibitor p21[WAF1/CIP1] mediated by Stat1. *Science* 272, 719–722.

Chung, C. D., Liao, J., Liu, B., Rao, X., Jay, P., Berta, P., and Shuai, K. (1997). Specific inhibition of Stat3 signal transduction by PIAS3. *Science* 278, 1803–1805.

Chung, J., Uchida, E., Grammer, T. C., and Blenis, J. (1998). STAT3 serine phosphorylation by ERK-dependent and -independent pathways negatively modulates its tyrosine phosphorylation. *Mol. Cell. Biol.* 17, 6508–6516.

Colamonici, O. R., Yan, H., Domanski, P., Handa, R., Smalley, D., Mullersman, J., Witte, M., Krishnan, K., and Krolewski, J. (1994). Direct binding and tyrosin phosphorylation of the α-subunit of the type I IFN receptor by the p135-tyk2 tyrosine kinase. *Mol. Cell. Biol.* 14, 8133–8142.

Collum, R., Lee, G., Lee, C., Liu, W., and Schindler, C. (1999). The scaffold protein StIP1 is required for cytokine signal transduction. *Submitted.*

Conway, G., Margoliath, A., Wong-Madden, S., Roberts, R. J., and Gilbert, W. (1997). Jak1 kinase is required for cell migrations and anterior specification in zebrafish. *Proc. Natl. Acad. Sci., USA* **94**, 3082–3087.

Corcoran, A. E., Smart, F. M., Cowling, R. J., Crompton, T., Owen, M. J., and Venkitaraman, A. R. (1996). The interleukin-7 receptor α chain transmits distinct signals for proliferation and differentiation during B lymphopoiesis. *EMBO J.* **15**, 1924–1932.

Corcoran, A. E., Riddell, A., Krooshoop, D., and Venkitaraman, A. R. (1998). Impaired immunoglobulin gene rearrangement in mice lacking the IL-7 receptor. *Nature* **391**, 904–907.

Cressman, D. W., Greenbaum, L. E., DeAngelis, R. A., Ciliberto, G., Furth, E. E., Poli, V., and Taub, R. (1996). Liver failure and defective hepatocyte regeneration in Interleukin-6-deficient mice. *Science* **274**, 1379–1383.

Cwirla, S. E., Balasubramanian, P., Duffin, D. J., Wagstrom, C. R., Gates, C. M., Singer, S. C., Davis, A. M., Tansik, R. L., Mattheakis, L. C., Boytos, C. M., Schatz, P. J., Baccanari, D. P., Wrighton, N. C., Barrett, R. W., and Dower, W. J. (1997). Peptide agonist of the thrombopoietin receptor as potent as the natural cytokine. *Science* **276**, 1696–1699.

Dalton, D. K., Pitts-Meek, S., Keshav, S., Figari, I. S., Bradley, A., and Stewart, T. A. (1993). Multiple defects of immune cell function in mice with disrupted Interferon-γ genes. *Science* **259**, 1739–1742.

Damen, J. E., Wakao, H., Miyajima, A., Krosl, J., Humphries, R. K., Cutler, R. L., and Krystal, G. (1995). Tyrosine 343 in the Erythropoietin receptor postively regulates erythropoietin-induced cell proliferation and Stat5 activation. *EMBO J.* **14**, 5557–5568.

Darnell, J. E. (1997). STATs and gene regulation. *Science* **277**, 1630–1635.

DaSilva, L., Zack-Howard, O. M., Rui, H., Kirken, R. A., and Farrar, W. L. (1994). Growth signaling and Jak2 association mediated by the membrane proximal cytoplasmic regions of prolactin receptors. *J. Biol. Chem.* **269**, 18267–18270.

David, M., Grimely, P. M., Finbloom, D. S., and Larner, A. C. (1993). A nuclear phosphatase downregulates Interferon-induced gene expression. *Mol. Cell. Biol.* **13**, 7515–7521.

David, M., Chen, H. E., Goelz, S., Larner, A. C., and Neel, B. G. (1995). Differential regulation of the alph/bet interferon-stimulated JAK/STAT pathway by SH2 containing tyrosine phosphatase SHPTP1. *Mol. Cell. Biol.* **15**, 7050–7058.

DeChiara, T. M., Vejsada, R., Poueymirou, W. T., Acheson, A., Suri, C., Conover, J. C., Friedman, B., McClain, J., Pan, L., Stahl, N., Ip, N. Y., Kato, A., and Yancopoulos, G. D. (1995). Mice lacking the CNTF receptor, unlike mice lacking CNTF, exhibit profound motor neuron deficits at birth. *Cell* **83**, 313–322.

Decker, T., Lew, D. J., Mirkovitch, J., and Darnell, J. E. (1991). Cytoplasmic activation of GAF, an IFN-γ regulated DNA binding factor. *EMBO J.* **10**, 927–932.

Decker, T., Kovarik, P., and Meinke, A. (1997). GAS elements: A few nucleotides with a major impact on cytokine-induced gene expression. *J. Interferon Cytokine Res.* **17**, 121–134.

Deiss, L. P., Feinstein, E., Berissi, H., Cohen, O., and Kimchi, A. (1995). Identification of a novel serine/threonine kinase and a novel 15 kD protein as potential mediators of the γ interferon-induced cell death. *Genes Devel.* **9**, 15–30.

DeMaeyer, E., and DeMaeyer-Giugnard, J. (1988). "Interferons and Other Regulatory Cytokines." Wiley; New York.

Demoulin, J. B., and Renauld, J. C. (1998). Interleukin 9 and its receptor: An overview of structure and function. *Int. Rev. Immunol.* **16**, 345–364.

Demoulin, J.-B., Uyttenhove, C., VanRoost, E., DeLestre, B., Donckers, D., VanSnick, J., and Renauld, J.-C. (1996). A single tyrosine of the IL-9 receptor is required for STAT activation, antiapoptotic activity and growth regulation by IL-9. *Mol. Cell. Biol.* **16**, 4710–4716.

Dent, A. L., Shaffer, A. L., Yu, X., Allman, D., and Staudt, L. M. (1997). Control of inflammation, cytokine expression, and germinal center formation by BCL-6. *Science* **276**, 589–592.

deSauvage, F. J., Hass, P. E., Spencer, S. D., Malloy, B. E., Gurney, A. L., Spencer, S. A., Darbonne, W. C., Henzel, W. J., Wong, S. C., Kuang, W.-J., Oles, K. J., Hultgren, B.,

Soldberg, L. A., Goeddel, D. V., and Eaton, D. L. (1994). Stimulation of megakaryocytopiesis and thrombopoiesis by the c-mpl ligand. *Nature* **369**, 533–538.

deVos, A. M., Ultsch, M., and Kossiakoff, A. A. (1992). Human growth hormone and extracellular domain of its receptor: Crystal structure of the complex. *Science* **255**, 306–312.

DiSanto, J. P. (1997). Cytokines: Shared receptors, distinct functions. *Curr. Biol.* **7**, R424–R426.

Duncan, S. A., Zhong, Z., Wen, Z., and Darnell, J. E. (1997). STAT signaling is active during early mammalian development. *Develop. Dynamics* **208**, 190–198.

Durbin, J. E., Hackeniller, R., Simon, M. C., and Levy, D. E. (1996). Targeted disruption of the mouse Stat1 gene results in compromised innate immunity to viral disease. *Cell* **84**, 443–450.

Eilers, A., Seegert, D., Schindler, C., Baccarini, M., and Decker, T. (1993). The response of the Gamma IFN Activating Factor (GAF) is under developmental control in cells of the macrophage lineage. *Mol. Cell. Biol.* **13**, 3245–3254.

Eilers, A., Georgellis, D., Klose, B., Schindler, C., Ziemiecki, A., Harpur, A., Wilke, A., and Decker, T. (1995). Differentiation-regulated serine phosphorylation of Stat1 promotes GAF activation in macrophages. *Mol. Cell. Biol.* **15**, 3579–3586.

Endo, T. A., Masuhara, M., Yokouchi, M., Suzuki, R., Sakamoto, H., Mitsui, K., Matsumoto, A., Tanimura, S., Ohtsubo, M., Misawa, H., Miyazaki, T., Leonor, N., Taniguchi, T., Fujita, T., Kanakura, Y., Komiya, S., and Yoshimura, A. (1997). A new protein containing an SH2 domain that inhibits JAK kinases. *Nature* **387**, 921–924.

Escalante, C. R., Yie, J., Thanos, D., and Aggarwal, A. K. (1998). Structure of IRF-1 with bound DNA reveals determinants of interferon regulation. *Nature* **391**, 103–106.

Escary, J.-L., Perreau, J., Dumenil, D., Ezine, S., and Brulet, P. (1993). Leukemia inhibitory factor is necessary for maintenance of haematopoietic stem cells and thymocyte stimulation. *Nature* **363**, 361–364.

Faris, M., Kokot, N., and Nel, A. E. (1997). Involvement of Stat3 in interleukin-6 induced IgM production in a human B-cell line. *Immunology* **90**, 350–357.

Farrar, M. A., and Schreiber, R. D. (1993). The molecular cell biology of Interferon-γ and its receptor. *Annu. Rev. Immunol.* **11**, 571–611.

Finbloom, D. S. (1995). IL-10 induces the tyrosine phosphorylation of tyk2 and Jak1 and the differential assembly of Stat1a and Stat3 complexes in human T cells and monocytes. *J. Immunol.* **155**, 1079–1090.

Foxwell, B. M., Beadling, C., Guschin, D., Kerr, I., and Cantrell, D. (1995). Interleukin-7 can induce the activation of Jak 1, Jak 3 and STAT 5 proteins in murine T cells. *Eur. J. Immunol.* **25**, 3041–3046.

Frank, S. J., Yi, W., Zhao, Y., Goldsmith, J. F., Gilliland, G., Jiang, J., Sakai, I., and Kraft, A. S. (1995). Region of the Jak2 tyrosine kinase required for coupling to the growth hormone receptor. *J. Biol. Chem.* **270**, 14776–14785.

Friedman, J. M. (1997). The alphabet of weight control. *Nature* **385**, 119–120.

Fu, X.-Y., Schindler, C., Improta, T., Aebersold, R., and Darnell, J. E. (1992). The proteins of ISGF3, the IFN-α induced transcription activator, define a new family of signal transducers. *Proc. Natl. Acad. Sci., USA* **89**, 7840–7843.

Fujitani, Y., Hibi, M., Fukada, T., Takahashi-Tezuka, M., Yoshida, H., Yamaguchi, T., Sugiyama, K., Yamanaka, Y., Nakajima, K., and Hirano, T. (1997). An alternative pathway for stat activation that is mediated by the direct interaction between jak and stat. *Oncogene* **14**, 751–761.

Fukada, T., Hibi, M., Yamanaka. Y., Takahashi-Tezuka, M., Fujitani, Y., Yamaguchi, T., Nakajima, K., and Hirano, T. (1996). Two signals are necessary for cell proliferation induced by a cytokine receptor gp130: involvement of STAT3 in anti-apoptosis. *Immunity* **5**, 449–460.

Gaffen, S. L., Lai, S. Y., Xu, W., Gouilleux, F., Groner, B., Goldsmith, M. A., and Greene, W. C. (1995). Signaling through the interleukin 2 receptor β chain activates a STAT-5-like DNA binding activity. *Proc. Natl. Acad. Sci. USA* **92**, 7192–7196.

Gaffen, S. L., Lai, S. Y., Ha, M., Liu, X., Hennighausen, L., Greene, W. C., and Goldsmith, M. A. (1996). Distinct tyrosine residues within the interleukin-2 receptor β chain drive signal transduction specificity, redundancy, and diversity. *J. Biol. Chem.* **271**, 21381–21390.

Gauzzi, M. C., Velazquez, L., McKendry, R., Mogensen, K. E., Fellous, M., and Pellegrini, S. (1996). Interferon-alpha-dependent activation of Tyk2 requires phosphorylation of positive regulatory tyrosines by another kinase. *J. Biol. Chem.* **271**, 20494–20500.

Gearing, D. P., Comeau, M. R., Friend, D. J., Gimpel, S. D., Thut, C. J., McGourty, J., Brasher, K. K., King, J. A., Gillis, S., Mosley, B., Ziegler, S. F., and Cosman, D. (1992). The IL-6 signal transducer, gp130: An oncostatin M receptor and affinity converter for the LIF receptor. *Science* **255**, 1434–1437.

Gerhartz, C., Heesel, B., Sasse, J., Hemmann, U., Landgraf, C., Schneider-Mergener, J., Horn, F., Heinrich, P. C., and Graeve, L. (1996). Differential activation of acute phase response factor/stat3 and stat1 via the cytoplasmic domain of the interleukin 6 signal transducer gp130. *J. Biol. Chem.* **271**, 12991–12998.

Ghislain, J., and Fish, E. N. (1996). Application of genomic DNA affinity chromatography identifies multiple interferon-alpha-regulatted Sta2 complexes. *J. Biol. Chem.* **271**, 12408–12413.

Gilmour, K., Pine, R., and Reich, N. C. (1995). Interleukin 2 activates Stat5 transcription factor (mammary gland factor) and specific gene epression in T lymphocytes. *Proc. Natl. Acad. Sci. USA* **92**, 10772–10776.

Giri, J. G., Ahdieh, M., Eisenman, J., Shanebeck, K., Grabstein, K., Kumaki, S., Namen, A., Park, L. S., Cosman, D., and Anderson, D. (1994). Utilization of the β and γ chains of the IL-2 receptor by the novel cytokine IL-15. *EMBO J.* **13**, 2822–2830.

Gouilleux, F., Pallard, C., Dusanter-Fourt, I., Wakao, H., Haldosen, L.-A., Norstedt, G., Levy, D., and Groner, B. (1995). Prolactin, growth hormone, erythropoeitin and granulocyte-macrophage colony stimulating factor induce MGF-Stat5 DNA binding activity. *EMBO J.* **14**, 2005–2013.

Gobert, S., Chretien, S., Gouilleux, F., Muller, O., Pallard, C., Dusanter-Forte, I., Groner, B., Lacombe, C., Gisselbrecht, S., and Mayeux, P. (1996). Identification of tyrosine residues within the intracellular domain of the erythropoietin receptor crucial for Stat5 activation. *EMBO J.* **15**, 2434–2441.

Grabstein, K. H., Eisenman, J., Shanebeck, K., Rauch, C., Srinivasan, S., Fung, V., Beers, C., Richardson, J., Schoenborn, M. A., Ahdieh, M., Johnson, L., Anderson, M. R., Watson, J. D., Anderson, D. M., and Giri, J. G. (1994). Cloning of a T cell growth factor that interacts with the β chain of the interleukin-2 receptor. *Science* **264**, 965–968.

Greenlund, A. L., Farrar, M. A., Viviano, B. L., and Schreiber, R. D. (1994). Ligand-induced IFN-γ receptor tyrosine phosphorylation couples the receptor to its signal transduction system (p91). *EMBO J.* **13**, 1591–1600.

Greenlund, A., Morales, M. O., Viviano, B. L., Yan, H., Krolewski, J., and Schreiber, R. D. (1995). STAT recruitment by tyrosine phosphorylated receptors: An ordered, reversible affinity driven process. *Immunity* **2**, 677–687.

Gupta, S., Yan, H., Wong, L. H., Ralph, S., Krolewski, J., and Schindler, C. (1996). The SH2 domains of Stat1 and Stat2 mediate multiple interactions in the transduction of IFN-α signals. *EMBO J.* **15**, 1075–1084.

Gurney, A. L., Carver-Moore, K., Sauvage, F. J. d., and Moore, M. W. (1994). Thrombocytopenia in c-mpl-deficient mice. *Science* **265**, 1445–1447.

Gurney, A. L., Wong, S. C., Henzel, W. J., and DeSauvage, F. J. (1995). Distinct regions of c-Mpl cytoplasmic domain are coupled to the JAK-STAT signal transduction pathway and Shc phosphorylation. *Proc. Natl. Acad. Sci. USA* **92**, 5292–5296.

Hara, T., and Miyajima, A. (1992). Two distinct functional high affinity receptors for mouse interleukin-3 (IL-3). *EMBO J.* **11**, 1875–1884.

Harrison, D. A., Binari, R., Nahreini, T. S., Gilman, M., and Perrimon, N. (1995). Activation of *Drosophila* Janus kinase (JAK) causes hematopoietic neoplasia and developmental defects. *EMBO J.* **14**, 2857–2865.

Hansen, J. A., Hansen, L. H., Wang, X., Kopchick, J. J., Gouilleux, F., Groner, B., Nielsen, J. H., Moldrup, A., Galsgaard, E. D., and Billestrup, N. (1997). The role of GH receptor tyrosine phosphorylation in stat5 activation. *J. Mol. Endocrinol.* **3**, 213–221.

Hansen, L. H., Wang, X., Kopchick, J. J., Bouchelouche, P., Nielsen, J. H., Galsgaard, E. D., and Billestrup, N. (1996). Identification of tyrosine residues in the intracellular domain of the growth hormone receptor required for transcriptional signaling and stat5 activation. *J. Biol. Chem.* **271**, 12669–12673.

Haspel, R. L., Salditt-Georgieff, M., and Darnell, J. E. (1996). The rapid inactivation of nuclear tyrosine phosphorylated Stat1 depends upon a protein tyrosine phosphatase. *EMBO J.* **15**, 6262–6268.

Heim, M. H., Kerr, I. M., Stark, G. R., and Darnell, J. E. (1995). Contribution of STAT SH2 groups to specific interferon signaling by the JAK-STAT pathway. *Science* **267**, 1347–1349.

Hemmann, U., Gerhartz, C., Heesel, B., Sasse, J., Kurapkat, G., Grotzinger, J., Wollmer, A., Zhong, Z., Darnell, J. E., Graeve, L., Heinrich, P. C., and Horn, F. (1996). Differential activation of acute phase response factor stat3 and stat1 via the cytoplasmic domain of the interleukin 6 signal transducer gp130. *J. Biol. Chem.* **271**, 12999–13007.

Hemmi, S., Böhni, R., Stark, G., DiMarco, F., and Aguet, M. (1994). A novel member of the interferon receptor family complements functionality of murine interferon γ receptor in human cells. *Cell* **76**, 803–810.

Hennighausen, L., Robinson, G. W., Wagner, K.-U., and Liu, X. (1997). Prolactin signaling in mammary gland development. *J. Biol. Chem.* **272**, 7567–7569.

Herrada, G., and Wolgemuth, D. J. (1997). The mouse transcription factor Stat4 is expressed in haploid male germ cells and is present in the perinuclear theca of spermatozoa. *J. Cell Sci.* **110**, 1543–1553.

Hilton, D. J., Zhang, J.-G., Metcalf, D., Alexander., W. S., Nicola, N. A., and Willson, T. A. (1996). Cloning and characterization of a binding subunit of the interleukin 13 receptor that is also a component of the interleukin 4 receptor. *Proc. Natl. Acad. Sci. USA* **93**, 497–501.

Hilton, D. J., Richardson, R. T., Alexander, W. S., Viney, E. M., Wilson, T. A., Sprigg, N. S., Starr, R., Nicholson, S. E., Metcalf, D., and Nicola, N. A. (1998). Twenty proteins containing a C-terminal SOCS box from five structural classes. *Proc. Natl. Acad. Sci. USA* **95**, 114–119.

Hinds, M. G., Maurer, T., Zhang, J. G., Nicola, N. A., and Norton, R. S. (1998). Solution structure of Leukemia Inhibitory Factor. *J. Biol. Chem.* **273**, 13738–13745.

Hirano, T., Akira, S., Taga, T., and Kishimoto, T. (1990). Biological and clinical aspects of interleukin 6. *Immunol. Today* **11**, 443–449.

Ho, A. S., Liu, Y., Khan, T. A., Hsu, D.-H., and Bazan, J. F. (1993). A receptor for interleukin 10 is related to interferon receptors. *Proc. Natl. Acad. Sci. USA* **90**, 11267–11271.

Horseman, N. D., Zhao, W., Montecino-Rodriguez, E., Tanaka, M., Nakashima, K., Engle, S. J., Smith, F., Markoff, E., and Dorshkind, K. (1997). Defective mammopoiesis, but normal hematopoiesis, in mice with a targeted disruption of the prolactin gene. *EMBO J.* **16**, 6926–6935.

Horvai, A. E., Xu, L., Korzus, E., Brard, G., Kalafus, D., Mullen, T.-M., Rose, D. W., Rosenfield, M. G., and Glass, C. K. (1997). Nuclear integration of JAK/STAT and Ras/AP-1 signaling by CBP and p300. *Proc. Natl. Acad. Sci. USA* **94**, 1074–1079.

Horvath, C. M., Wen, Z., and Darnell, J. E. (1995). A STAT protein domain that determines DNA sequence recognition suggests a novel DNA binding domain. *Genes Devel.* **9**, 984–994.

Horvath, C. M., Stark, G. R., Kerr, I. M., and Darnell, J. E. (1996). Interactions between STAT and non-STAT proteins in the Interferon-Stimulated Gene Factor 3 transcription complex. *Mol. Cell. Biol.* **16**, 6957–6964.

Hou, J., Schindler, U., Henzel, W. J., Ho, T. C., Brasseur, M., and McKnight, S. L. (1994). An Interleukin-4 induced transcription factor: IL-4 Stat. *Science* **265**, 1701–1705.

Hou, J., Schindler, U., Henzel, W. J., Wong, S. C., and McKnight, S. L. (1995). Identification and purification of human Stat proteins activated in response to interleukin-2. *Immunity* **2**, 321–329.

Hou, X. S., Melnick, M. B., and Perrimon, N. (1996). Marelle acts downstream of drosophila HOP/JAK kinase and encodes a protein similar to the mammalian STATs. *Cell* **84**, 411–419.

Huang, S., Hendriks, W., Althage, A., Hemmi, S., Bluethmann, H., Kamijo, R., Vilcek, J., Zinkernagel, R. M., and Aguet, M. (1993). Immune response in mice that lack the Interferon-γ receptor. *Science* **259**, 1742–1745.

Hunter, T. (1995). Protein kinases and phosphatases: The yin and yang of protein phosphorylation and signaling. *Cell* **80**, 225–236.

Hwang, S. Y., Hertzog, P. J. Holland, K. A., Sumarsono, S. H., Tymms, M. J., Hamilton, J. A., Whitty, G., Bertoncello, I., and Kola, I. (1995). A null mutation in the gene encoding a type I interferon receptor component eliminates antiproliferative and antiviral responses to interferons alpha and beta and alters macrophage responses. *Proc. Natl. Acad. Sci. USA* **92**, 11284–11288.

Igarashi, K., Garotta, G., Ozmen, L., Ziemiecki, A., Wilks, A. F., Harpur, A. G., Larner, A. C., and Finbloom, D. S. (1994). Interferon-γ induces tyrosine phosphorylation of Interferon-γ receptor and regulated association of protein tyrosine kinases, Jak1 and Jak2, with its receptor. *J. Biol. Chem.* **269**, 14333–14336.

Ihle, J. N., Quelle, F. W., and Miura, O. (1993). Signal Transduction through the receptor for erythropoietin. *Sem. Immunol.* **5**, 375–389.

Ihle, J., Witthuhn, B., Tang, B., Yi, T., and Quell, F. (1994a). Cytokines receptors and signal transduction. *Bailliere's Clin. Hematol.* **7**, 17–48.

Improta, T., Schindler, C., Horvath, C. M., Kerr, I. M., Stark, G. R., and Darnell, J. E. (1994). Transcription factor ISGF3 formation requires phosphorylated Stat91 protein, but Stat113 protein is phosphorylated independently of Stat91 protein. *Proc. Natl. Acad. Sci. USA* **91**, 4776–4780.

Ito, Y., Seto, Y., Brannan, C. I., Copeland, N. G., Jenkins, N. A., Fukunaga, R., and Nagata, S. (1994). Structural analysis of the functional gene and pseudogene encoding the murine granulocytes colony-stimulating-factor receptor. *Euro. J. Biochem.* **220**, 881–891.

Itoh, T., Liu, R., Yokota, T., Arai, K. I., and Watanabe, S. (1998). Definition of the role of tyrosine residues of the common β subunit regulating multiple signaling pathways of granulocyte-macrophage colony-stimulating factor receptor. *Mol. Cell. Biol.* **18**, 742–752.

Jacobson, N. G., Szabo, S. J., Weber-Nordt, R. M., Zhong, Z., Schreiber, R. D., Darnell, J. E., and Murphy, K. M. (1995). Interleukin 12 signaling in T helper type 1 (Th1) cells involves tyrosine phosphorylation of signal transducer and activator of transcription (Stat)3 and Stat4. *J. Exptl. Med.* **181**, 1755–1762.

Jiao, H., Yang, W., Berrada, K., Tabrizi, M., Shultz, L., and Yi, T. (1997). Macrophages from motheaten and viabole motheaten mutant mice show increased proliferative response to GM-CSF: Detection of potential HCP substrates in GM-CSF signal transduction. *Exp. Hematol.* **25**, 592–600.

John, S., Robbins, C., and Leonard, W. (1996). An IL-2 response element in the human IL-2 receptor α chain promoter is a composite element that binds Stat5, Elf-1, HMG-I(Y) and a GATA family protein. *EMBO J.* **15**, 5627–5635.

Johnston, J. A., Kawamura, M., Kirken, R. A., Chen, Y.-Q., Blake, T. B., Shibuya, K., Ortaldo, J. R., McVicar, D. W., and O'Shea, J. J. (1994). Phosphorylation and activation of the Jak-3 Janus kinase in response to interleukin-2. *Nature* **370**, 151–153.

Kammer, W., Lischke, A., Moriggl, R., Groner, B., Ziemiecki, A., Gurniak, C. B., Berg, L. E., and Friedrich, K. (1996). Homodimerization of Interleukin-4 receptor α chain can induce intracellular signaling. *J. Biol. Chem.* **271**, 23634–23637.

Kaplan, D. H., Greenlund, A. C., Tanner, J. W., Shaw, A. S., and Schreiber, R. D. (1996a). Identification of an Interferon-γ receptor α-chain sequence required for Jak1 binding. *J. Biol. Chem.* **271**, 9–12.

Kaplan, M. H., Schindler, U., Smiley, S. T., and Grusby, M. J. (1996b). Stat6 is required for mediating responses to IL-4 and for the development of Th2 cells. *Immunity* **4**, 313–319.

Kaplan, M. H., Sun, Y.-L., Hoey, T., and Grusby, M. J. (1996c). Impaired IL-12 responses and enhanced development of Th2 cells in Stat4-deficient mice. *Nature* **382**, 174–177.

Kaplan, M. H., Daniel, C., Schindler, U., and Grusby, M. J. (1998). Stat proteins control lymphocyte proliferation by regulating p27Kip1 expression. *Mol. Cell. Biol.* **18**, 1996–2003.

Karras, J. G., Wang, Z., Huo, L., Howard, R. G., Frank, D. A., and Rothstein, T. L. (1997). Signal transducer and activator of transcription-3 (STAT3) is constitutively activated in normal, self-renewing B-1 cells but only inducibly expressed in conventional B lymphocytes. *J. Exptl. Med.* **185**, 1035–1042.

Kawamata, N., Miki, T., Ohashi, K., Suzuki, K., Fukuda, T., Hirosawa, S., and Aoki, N. (1994). Recognition DNA sequence of a novel putative transcription factor, BCL6. *Biochem Biophys. Res. Commun.* **204**, 366–374.

Kawamura, M., McVicar, D. W., Johnston, J. A., Blake, T. B., Chen, Y.-Q., Lal, B. K., Lloyd, A. R., D. J., K., Staples, J. E., Ortaldo, J. R., and O'Shea, J. J. (1994). Molecular cloning of L-Jak, a janus family protein-tyrosine kinase expressed in natural killer cells and activated leukocytes. *Proc. Natl. Acad. Sci., USA* **91**, 6374–6378.

Kawata, T., Shevchenko, A., Fukuzawa, M., Jermyn, K. A., Totty, N. F., Zhukovskaya, N. V., Steriling, A. E., Mann, M., and Williams, J. G. (1997). SH2 signaling in a lower eukaryote: A STAT protein that regulates stalk cell differentiation in *Dictyostelium*. *Cell* **89**, 909–916.

Kessler, D. S., Levy, D. E., and Darnell, J. E. (1988). Two interferon-induced nuclear factors bind a single promoter element in interferon-stimulated genes. *Proc. Natl. Acad. Sci. USA* **85**, 8521–8525.

Kessler, D. S., Veals, S. A., Fu, X.-Y., and Levy, D. E. (1990). Interferon-α regulates nuclear translocation and DNA-binding affinity of ISGF3, a multimeric transcriptional activator. *Genes Devel.* **4**, 1753–1765.

Khan, K. D., Shuai, K., Lindwall, G., Maher, S. E., Darnell, J. E., and Bothwell, A. L. (1993). Induction of the Ly-6A/E gene by Interferon α/β and γ requires a DNA element to which a tyrosine-phosphorylated 91-kDa protein binds. *Proc. Natl. Acad. Sci. USA* **90**, 6806–6810.

Kim, T. K., and Mantiatis, T. (1996). Regulation of Interferon-γ activated Stat1 by the ubiquitin–proteosome pathway. *Science* **273**, 1717–1719.

Kinoshita, T., Yokota, T., Arai, K., and Miyajima, A. (1995). Suppression of apoptotic death in hematopoietic cells by signaling through the IL-3/GM-CSF receptors. *EMBO J.* **14**, 266–275.

Kishimoto, T., Taga, T., and Akira, S. (1994). Cytokine signal transduction. *Cell* **76**, 253–262.

Kishimoto, T., Akira, S., Narazaki, M., and Taga, T. (1995). Interleukin-6 family of cytokines and gp130. *Blood* **86**, 1243–1254.

Kiuchi, N., Nakajima, K., Ichiba, M., Fukada, T., Narimatsu, M., Mizuno, K., Hibi, M., and Hirano, T. (1999). STAT3 is required for the gp130-mediated full activation of the c-myc gene. *J. Exptl. Med.* **189**, 63–73.

Klingmüller, U., Lorenz, U., Cantley, L. C., Neel, B. G., and Lodish, H. F. (1995). Specefic recruitment of SH-PTP1 to the erythropoietin receptor causes inactivation of Jak2 and termination of proliferative signals. *Cell* **80**, 729–738.

Kontanides, H., and Reich, N. C. (1993). Requirement of tyrosine phosphorylation for rapid activation of a DNA binding factor by IL-4. *Science* **262**, 1265–1267.

Kopf, M., LeGros, G., Bachmann, M., Lamers, M. C., Bluethmann, H., and Köhler, G. (1993). Disruption of the murine IL-4 gene blocks TH2 cytokine responses. *Nature* **362**, 245–248.

Kopf, M., Baumann, H., Freers, G., Freudenberg, M., Lamers, M., Kishimoto, T., Zinkernagel, R., Bluethmann, H., and Kohler, G. (1994). Impaired immune and acute-phase responses in interleukin-6-deficient mice. *Nature* **368**, 339–342.

Kopf, M., Brombacher, F., Hodgkin, P. d., Ramsay, A. J., Milbourne, E. A., Dai, W. J., Ovington, K. S., Behm, C. A., Koehler, G., Young, I. G., and Matthaei, K. I. (1996). IL-

5 deficient mice have a developmental defect in CD5+ B-1 cells and lack eosihophilia but have normal antibody and cytotoxic T cell responses. *Immunity* **4**, 15–24.

Korzus, E., Torchia, J., Rose, D. W., Xu, L., Kurokawa, R., McInerney, E. M., Mullen, T. M., Glass, C. K., and Rosenfeld, M. G. (1998). Transcription factor-specific requirements for coactivators and their acetyltransferase functions. *Science* **279**, 703–707.

Kotenko, S., Izotova, L., Pollack, B., Mariano, T., Donnelly, R., Muthukumaran, G., Cook, J., Garotta, G., Silvennoinen, O., Ihle, J., and Pestka, S. (1995). Interaction between the components of the Interferon-γ receptor complex. *J. Biol. Chem.* **270**, 20915–20921.

Kotenko, S. V., Krause, C. D., Izotova, L. S., Pollack, B. P., Wu, W., and Pestka, S. (1997). Identification and functional characterization of a second chain of the interleukin-10 receptor complex. *EMBO J.* **16**, 5894–5903.

Kovarik, P., Stoiber, D., Novy, M., and Decker, T. (1998). Stat1 combines signals derived from IFN-gamma and LPS receptors during macrophage activation. *EMBO J.* **17**, 3660–3668.

Kühn, R., Loehler, J., Rennick, D., Rajewsky, K., and Müller, W. (1993). Interleukin-10 deficient mice develop chronic enterocolitis. *Cell* **75**, 263–274.

Kumar, A., Commane, M., Flickinger, T. W., Horvarth, C. M., and Stark, G. R. (1997. Defective TNF-α induced apoptosis in Stat1 null cells due to low constitutive levels of caspases. *Science* **278**, 1630–1632.

Kurokawa, R., Kalafus, D., Ogliastro, M.-H., Kioussi, C., Xu, L., Torchia, J., Rosenfeld, M. G., and Glass, C. K. (1998). Differential use of CREB binding protein–coactivator complexes. *Science* **279**, 700–703.

Lebrun, J.-J., Ali, S., Goffin, V., Ullrich, A., and Kelly, P. A. (1995). A single phosphotyrosine residue of the prolactin receptor is responsible for activation of gene transcription. *Proc. Natl. Acad. Sci. USA* **92**, 4031–4035.

Lécine, P., Algarté, M., Rameil, P., Beadling, C., Bucher, P., Nabholz, M., and Imbert, J. (1996). Elf-1 and Stat5 Bbind to a critical element in a new enhancer of the human Interleukin-2 receptor α gene. *Mol. Cell. Biol.* **16**, 6829–6840.

Lee, G. H., Proenca, R., Montez, J. M., Carroll, K. M., Darvishzadeh, J. G., Lee, J. I., and Friedman, J. M. (1996). Abnormal splicing of the Leptin receptor in diabetic mice. *Nature* **379**, 632.

Lee, C.-K., Bluyssen, H. A., and Levy, D. E. (1997). Regulation of Interferon-α responsiveness by the duration of janus kinase activity. *J. Biol. Chem.* **272**, 21872–21877.

Leng, S. X., and Elias, J. A. (1997). Interleukin-11. *Int. J. Biochem. Cell Biol.* **29**, 1059–1062.

Leonard, W. J. (1996). The moleuclar basis of X-linked severe combined immunodeficiency: Defective cytokine receptor signaling. *Annu. Rev. Med.* **47**, 229–239.

Leonard, W., and O'Shea, J. J. (1998). JAKS and STATS: Biological implications. *Ann. Rev. Immunol.* **16**, 293–322.

Leung, S., Qureshi, S. A., Kerr, I. M., Darnell, J. E., and Stark, G. R. (1995). Role of STAT2 in the alpha interferon signaling pathway. *Mol. Cell. Biol.* **15**, 1312–1317.

Leung, S., Li, X., and Stark, G. R. (1996). STATs find that hanging together can be stimulating *Science* **273**, 750–751.

Levy, D. E., Kessler, D. S., Pine, R., Reich, N., and Darnell, J. E. (1988). Interferon-induced nuclear factors that bind a shared promoter element correlate with positive and negative transcriptional control. *Genes Devel.* **2**, 383–393.

Lew, D. J., Decker, T., and Darnell, J. E. (1989). Alpha interferon and gamma interferon stimulate transcription of a single gene through different signal transduction pathways. *Mol. Cell. Biol.* **9**, 5404–5411.

Lew, D. J., Decker, T., Strehlow, I., and Darnell, J. E. (1991). Overlapping elements in the guanylate-binding protein gene promoter mediate transcriptional induction by alpha and gamma interferons. *Mol. Cell. Biol.* **11**, 182–191.

Li, X., Leung, S., Qureshi, S., Darnell, J. E., and Stark, G. R. (1996). Formation of STAT1–STAT2 heterodimers and their role in the activation of IRF-1 gene transcription by interferon-α. *J. Biol. Chem.* **271**, 5790–5794.

Li, X., Leung, S., Kerr, I. M., and Stark, G. R. (1997). Functional subdomains of Stat2 required for preassociation with the alpha interferon receptor and for signaling. *Mol. Cell. Biol.* **17**, 2048–2956.

Lieschke, G. J., Grail, D., Hodgson, G., Metcalf, D., Stanley, E., Cheers, C., Fowler, K. J., Basu, S., Zhan, Y. F., and Dunn, A. R. (1994). Mice lacking granulocytes colony-stimulating factor have chronic neutropenia, granulocyte and macrophage progenitor cell deficiency, and impaired neutrophil mobilization. *Blood* **84**, 1737–1746.

Lin, J.-X., Migone, T. S., Tsang, M., Friedmann, M., Weartherbee, J. A., Zhou, L., Yamauchi, A., Bloom, E. T., Mietz, J., John, S., and Leonard, W. J. (1995). The role of shared receptor motifs and common STAT proteins in the generation of cytokine pleitropy and redundancy by IL-2, IL-4, IL-7, IL-13, and IL-15. *Immunity* **2**, 331–339.

Lindberg, R. A., Juan, T. S. C., Welcher, A. A., Sun, Y., Cupples, R., Guthrie, B., and Fletcher, F. A. (1998). Cloning and characterization of a specific receptor for mouse oncostatin M. *Mol. Cell. Biol.* **18**, 3357–3367.

Liu, X., Robinson, G. W., Gouilleux, F., Groner, B., and Hennighausen, L. (1995). Cloning and expression of Stat5 and an additional homolog (Stat5b) involved in prolactin signal transduction in mouse mammary tissue. *Proc. Natl. Acad. Sci. USA* **92**, 8831–8835.

Liu, F., Wu, H. Y., Wesselschmidt, R., Kornaga, T., and Link, D. C. (1996). Impaired production and increased apoptosis of neutrophils in granulocyte colony-stimulating factor receptor-deficient mice. *Immunity* **5**, 491–501.

Liu, X., Robinson, G. W., Wagner, K. U., Garrett, L., Wynshaw-Boris, A., and Hennighausen, L. (1997). Stat5a is mandatory for adult mammary gland development and lactogenesis. *Genes Devel.* **11**, 179–186.

Liu, B., Liao, J., Rao, X., Kushner, S. A., Chung, C. D., Chang, D. D., and Shuai, K. (1998a). Inhibition of Stat1 mediated gene activation by PIASI. *Proc. Natl. Acad. Sci. USA* **95**, 10626–10631.

Liu, J., Hadjokas, N., Mosley, B., Estrov, Z., Spence, M. J., and Vestal, R. E. (1998b). Oncostatin M-specific receptor expression and function in regulating cell proliferation of normal and malignant mammary epithelial cells. *Cytokine* **10**, 295–302.

Livnah, O., Stura, E. A., Johnson, D. L., Middleton, S. A., Mulcahy, L. S., Wrighton, N. C., Dower, W. J., Jolliffe, L. K., and Wilson, I. A. (1996). Functional mimicry of a protein hormone by a peptide agonist: The epo receptor complex at 2.8 A. *Science* **273**, 464.

Lok, S., Kaushansky, K., Holly, R. D., Kuijper, J. L., Lofton-Day, C. E., Oort, P. J., Grant, F. J., Heipel, M. D., Burkhead, S. K., Kramer, J. M., and *et al.* (1994). Cloning and expression of murine thrombopoietin cDNA and stimulation of platlet production *in vivo*. *Nature* **369**, 565–568.

Lokuta, M. A., McDowell, M. A., and Paulnock, D. M. (1998). Cutting edge: Identification of an additional isoform of Stat5 expressed in immature macrophages. *J. Immunol.* **161**, 1594–1597.

Look, D. C., Pelletier, M. R., Tidwell, R. M., Roswit, W. T., and Holtzman, M. J. (1995). Stat1 depends on transcriptonal synergy with Sp1. *J. Biol. Chem.* **270**, 30264–30267.

Lord, K. A., Abdollahi, A., Thomas, S. M., DeMarco, M., Brugge, J. S., Hoffman-Liebermann, B., and Liebermann, D. A. (1991). Leukemia inhibitory factor and interleukin-6 trigger the same immediate early response, inducing tyrosine phosphorylation, upon induction of myeloid leukemia differentiation. *Mol. Cell. Biol.* **11**, 4371–4379.

Lu, B., Reichel, M., Fisher, D. A., Smith, J. F., and Rothman, P. (1997). Identification of the Stat6 domain required for IL-4 induced activation of transcription. *J. Immunol.* **159**, 1255–1264.

Lu, B., Ebensperger, C., Dembic, Z., Wang, Y., Kvatyuk, M., Lu, T., Coffman, R., Pestka, S., and Rothman, P. B. (1998). Targeted disruption of the interferon-γ receptor 2 gene results in severe immune defects in mice. *Proc. Natl. Acad. Sci. USA* **95**, 8233–8238.

Luo, H., Hanratty, W. P., and Dearolf, C. R. (1995). An amino acid substitution in the *Drosophila* hop[tum-l] Jak kinase causes leukemia-like hematopoietic defects. *EMBO J.* **14**, 1412–1420.

Luo, H., Rose, P., Barber, D., Hanratty, W. P., Lee, S., Roberts, T. M., D'Andrea, A. D., and Dearolf, C. R. (1996). Mutation in the Jak kinase JH2 domain hyperactivates drosophila and mammalian JAK-STAT pathways. *Mol. Cell. Biol.* **17**, 1562–1571.

Lütticken, C., Wegenka, U. M., Yuan, J., Buschmann, J., Schindler, C., Ziemiecki, A., Harpur, A. G., Wilks, A. F., Yasukawa, K., Taga, T., Kishimoto, T., Barbieri, G., Pellegrini, S., Sendtner, M., Heinrich, P. C., and Horn, F. (1994). Association of transcription factor APRF and protein kinase Jak 1 with the interleukin-6 signal transducer gp130. *Science* **263**, 89–92.

Lütticken, C., Coffer, P., Yuan, J., Schwartz, C., Caldenhoven, E., Schindler, C., Kruijer, W., Heinrich, P. C., and Horn, F. (1995). Interleukin-6 induced serine phosphorylation of transcription factor APRF: evidence for a role in interleukin-6 target gene induction. *FEBS Lett.* **360**, 137–143.

Marrero, M. B., Schieffer, B., Paxton, W. G., Heerdt, L., Berk, B. C., Delafontaine, P., and Bernstein, K. E. (1995). Direct stimulation of the JAK/STAT pathway by the angiotensin II AT1 receptor. *Nature* **375**, 247–250.

Martinez-Moczygemba, M., Glutch, M. J., French, D. L., and Reich, N. C. (1997). Distinct STAT structure promotes interaction of Stat2 with the p48 subunit of the IFNterferon-α stimulated transcription factor ISGF3. *J. Biol. Chem.* **272**, 20070–20076.

Masu, Y., Wolf, E., Holtmann, B., Sedtner, M., Brem, G., and Thoenen, H. (1993). Disruption of the CNTF gene results in motor neuron degeneration. *Nature* **365**, 27–32.

Matsumoto, A., Masuhara, M., Mitsui, K., Yokouchi, M., Ohtsubo, M., Misawa, H., Miyajima, A., and Yoshimura, A. (1997). CIS, a cytokine inducible sh2 protein, is a target of the jak-stat5 pathway and modulates stat5 activation. *Blood* **89**, 3148–3154.

Matsumura, I., Ishikawa, J., Nakajima, K., Oritani, K., Tomiyama, Y., Miyagawa, J.-I., Kato, T., Miyazaki, H., Matsuzawa, Y., and Kanakura, Y. (1997). Thrombopoietin induced differentiation of a human megakaryoblastic leukemia cell line, CMK, involves transcriptional activation of p21$^{WAF/Cip1}$ by Stat5. *Mol. Cell. Biol.* **17**, 2933–2943.

May, P., Gerhartz, C., Heesel, B., Welte, T., Doppler, W., Graeve, L., Horn, F., and Heinrich, P. C. (1996). Comparative Study on the phosphotyrosine motifs of different cytokine receptors involved in stat5 activation. *Fed. Euro. Biochem. Soc.* **394**, 221–226.

Meier, V., and Groner, B. (1994). The nuclear factor YY1 participated in repression of the β-casein promoter in mammary epithelial cells and is counteracted by mammary gland factor during lactogenic hormone induction. *Mol. Cell. Biol.* **14**, 128–137.

Meinke, A., Barahmand-Pour, F., Woehrl, S., Stoiber, D., and Decker, T. (1996). Activation of different Stat5 isoforms contributes to cell-type-restricted signaling in response to Interferons. *Mol. Cell. Biol.* **16**, 6937–6944.

Meraz, M. A., White, J. M., Sheehan, K. C., Bach, E. A., Rodig, S. J., Dighe, A. S., Kaplan, D. H., Riley, J. K., Greenlund, A. C., Campbell, D., Carver-Moore, K., DuBois, R. N., Clark, R., Aguet, M., and Schreiber, R. D. (1996). Targeted disruption of the Stat1 gene in mice reveals unexpected physiological specificity in the JAK-STAT pathway. *Cell* **84**, 431–442.

Meyer, J., Jücker, M., Ostertag, W., and Stocking, C. (1998). Carboxy-truncated Stat5β is generated by a nucleus-associated serine protease in early hematopoietic progenitors. *Blood* **91**, 1901–1908.

Migone, T.-S., Lin, J.-X., Cereseto, A., Mulloy, J.C., O'Shea, J. J., Franchini, G., and Leonard, W. J. (1995). Constitutively activated JAK-STAT pathway in T cells transformed with HTLV-1. *Science* **269**, 79–81.

Mikita, T., Campbell, D., Wu, P., Williamson, K., and Schindler, U. (1996). Requirement for Interleukin-4 induced gene expression and functional characterization of Stat6. *Mol. Cell. Biol.* **16**, 5811–5820.

Mikita, T., Kurama, M., and Schindler, U. (1998). Synergistic activation of the germline & epsiv; promoter mediated by Stat6 and C/EBPβ. *J. Immunol.* **161**, 1822–1828.

Minami, M., Inoue, M., Wei, S., Takeda, K., Matsumoto, M., Kishimoto, T., and Akira, S. (1996). Stat3 activation is a critical step in gp-130-mediated terminal differentiation and growth arrest of a myeloid cell line. *Proc. Natl. Acad. Sci. USA* **93**, 3963–3966.

Miyajima, A., Mui, A. L., Ogorochi, T., and Sakamaki, K. (1993). Receptors for granulocyte-macrophage colony-stimulat factor, interleukin-3, and interleukin-5. *Blood* **82**, 1960–1974.

Miyazaki, T., Kawahara, A., Fuji, H., Nakagawa, Y., Minami, Y., Liu, Z.-J., Oishi, I., Silvennoinen, O., Witthuhn, B. A., Ihle, J. N., and Taniguchi, T. (1994). Functional activation of Jak1 and Jak3 by selective association with the IL-2 receptor subunits. *Science* **266**, 1045–1047.

Moore, K. W., O'Garra, A., Malefyt, R. d., Vieira, P., and Mosman, T. R. (1993). Interleukin-10. *Ann. Rev. Immunol.* **11**, 165–190.

Moriggl, R., Gouilleux-Gruart, V., Jaehne, R., Berchtold, S., Gatmann, C., Liu, X., Hennighausen, L., Sotiropoulos, A., Groner, B., and Gouilleux, F. (1996). Deletion of the carboxy-terminal transactivation domain of MGF-Stat5 results in sustained DNA binding and a dominant negative phenotype. *Mol. Cell. Biol.* **16**, 5691–5700.

Moriggl, R., Berchtold, S., Friedrich, K., Standke, G. J., Kammer, W., Heim, M., Wissler, M., Stoecklin, E., Gouilleux, F., and Groner, B. (1997). Comparison of the transactivation domains of Stat5 and Stat6 in lymphoid cells and mammary epithelial cells. *Mol. Cell. Biol.* **17**, 3663–3678.

Moriggl, R., Erhardt, I., Kammer, W., Berchtold, S., Schnarr, B., Lischke, A., Groner, B., and Friedrich, K. (1998). Activation of stat6 is not dependent on phosphotyrosine-mediated docking to the interleukin-4 receptor and can be blocked by dominant-negative mutants of both receptor subunits. *Eur. J. Biochem.* **251**, 25–35.

Moriggl, R., Topham, D. J., Teglund, S., Sexl, V., McKay, C., Wang, D., Hoffmeyer, A., Deursen, J. V., Sangster, M. V., Bunting, K. D., Grosveld, G. C., and Ihle, J. N. (1999). Stat5 is required for IL-2-induced cell cycle progression of peripheral T cells. *Immunity* **10**, 249–259.

Muhlethaler-Mottet, A., DiBerardino, W., Otten, L. A., and Mach, B. (1998). Activation of the MHC class II transactivator CIITA by interferon-g requires cooperative interaction between Stat1 and USF-1. *Immunity* **8**, 157–166.

Mui, A., Wakao, H., O'Farrell, A. M., Harada, N., and Miyajima, A. (1995). Interleukin-3, granulocyte-macrophage colony stimulating factor and interleukin-5 transduce signals through two Stat5 homologues. *EMBO J.* **14**, 1166–1175.

Mui, A. L.-F., Wakao, H., Kinoshita, T., Kitamura, T., and Miyajima, A. (1996). Suppression of interleukin-3-induced gene expression by a C-terminal truncated Stat5: Role of Stat5 in proliferation. *EMBO J.* **15**, 2425–2433.

Müller, M., Briscoe, J., Laxton, C., Guschin, D., Ziemiecki, A., Silvennoinen, O., Harpur, A. G., Barbieri, G., Witthuhn, B. A. Schindler, C., Pellegrini, S., Wilks, A. F., Ihle, J. N., Stark, G. R., and Kerr, I. M. (1993a). The protein tyrosine kinase Jak-1 complements defects in the Interferon-α/β and -γ signal transduction pathways. *Nature* **366**, 129–135.

Müller, M., Laxton, C., Briscoe, J., Schindler, C., Improta, T., Darnell, J. E., Stark, G. R., and Kerr, I. M. (1993b). Complementation of a mutant cell lines: Central role of the 91-kDa polypeptide of ISGF3 in the interferon-α and γ signal transduction pathways. *EMBO J.* **12**, 4221–4228.

Müller, U., Steinhoff, U., Reis, L. F., Hemmi, S., Pavlovic, J., Zinkernagel, R. M., and Aguet, M. (1994). Functional role of type I and type II Interferons in antiviral defense. *Science* **264**, 1918–1921.

Murata, T., Taguchi, J., and Puri, R. K. (1998). Interleukin-13 receptor alpha but not alpha chain: A functional component of Interleukin-4 receptors. *Blood* **91**, 3884–3891.

Naka, T., Narazaki, M., Hirata, M., Matsumoto, T., Minamoto, S., Aono, A., Nishimoto, N., Kajita, T., Taga, T., Yoshizaki, K., Akiroa, S., and Kishimoto, T. (1997). Structure and function of a new SAT-induced STAT inhibitor. *Nature* **387**, 924–929.

Nakajima, K., and Wall, R. (1991). Interleukin-6 signals activating junB and TIS11 gene transcription in a B-cell hybridoma. *Mol. Cell. Biol.* **11**, 1409–1418.

Nakajima, K., Yamanaka, Y., Nakae, K., Kojima, H., Ichiba, M., Kiuchi, N., Kitaoka, T., Fukada, T., Hibi, M., and Hirano, T. (1996). A central role for Stat3 in IL-6 induced regulation of growth and differentiation in M1 leukemia cells hybridoma. *EMBO J.* **15**, 3651–3658.

Nakajima, H., Liu, X. W., Wynshaw-Boris, A., Rosenthal, L. A., Imada, K., Finbloom, D. S., Hennighausen, L., and Leonard, W. J. (1997). An indirect effect of stat5a in IL-2-induced proliferation: A critical role for stat5a in IL-2-mediated IL-2 receptor α chain induction. *Immunity* **7**, 691–701.

Nakamura, Y., Russell, S. M., Mess, S. A., Friedmann, M., Erdos, M., Francois, C., Jacques, Y., Adelstein, S., and Leonard, W. J. (1994). Heterodimerization of the IL-2 receptor β- and γ-chain cytoplasmic domains is required for signaling. *Nature* **369**, 330–333.

Nandurkar, H. H., Robb, L., Tarlinton, D., Barnett, L., Kontgen, F., and Begley, C. G. (1997). Adult mice with targeted mutation of the Interleukin-11 receptor (IL-11 Ra) display normal hematopoiesis. *Blood* **90**, 2148–2159.

Nelson, B. H., Lord, J. D., and Greenberg, P. D. (1994). Cytoplasmic domains of the interleukin-2 receptor β and γ chains mediate the signal for T-cell proliferation. *Nature* **369**, 333–336.

Neubauer, H., Cumano, A., Mueller, M., Wu, H., Huffstadt, U., and Pfeffer, K. (1998). Jak2 deficiency defines an essential developmental checkpoint in definitive hematopoiesis. *Cell* **93**, 397–409.

Ng, J., and Cantrell, D. (1997). Stat3 is a serine kinase target in T lymphocytes: Interleukin 2 and T cell antigen receptor signals converge upon serine 727. *J. Biol. Chem.* **272**, 24542–24549.

Nicholson, S. E., Novak, U., Ziegler, S. F., and Layton, J. E. (1995). Distinct regions of the granulocyte colony-stimulating factor receptor are required for tyrosine phosphorylation of the signaling molecules JAK2, Stat3, and p42, p44MAPK. *Blood* **86**, 3698–3704.

Niwa, H., Burdon, T., Chambers, I., and Smith, A. (1998). Self-renewal of pluripotent embryonic stem cells is mediated via activation of Stat3. *Genes Devel.* **12**, 2048–2060.

Noben-Trauth, N., Shultz, L., Brombacher, F., Urban, J. F., Gu, H., and Paul, W. E. (1997). An interleukin 4 (IL-4)-independent pathway for CD4+ T cell IL-4 production is revealed in IL-4 receptor-deficient mice. *Proc. Natl. Acad. Sci. USA* **94**, 10838–10843.

Noguchi, M., Nakamura, Y., Russell, S. M., Ziegler, S. F., Tsang, M., Cao, X., and Leonard, W. J. (1993). Interleukin-2 receptor γ chain: A functional component of the interleukin-7 receptor. *Science* **262**, 1877–1880.

Nosaka, T., vanDeursen, J. M., Tripp, R. A., Thierfelder, W. E., Witthuhn, B. A., McMickle, A. P., Doherty, P. C., Grosveld, G. C., and Ihle, J. N. (1995). Defective lymphoid development in mice lacking Jak3. *Science* **270**, 800–802.

Novick, D., Cohen, B., and Rubinstein, M. (1994). The human Interferon α/β receptor: Characterization and molecular cloning. *Cell* **77**, 391–400.

O'Garra, A., and Murphy, K. (1994). Role of cytokines in determining T-lymphocyte function. *Curr. Opin. Immunol.* **6**, 458–466.

Oakes, S. A., Candotti, F., Johnston, J. A., Chen, Y. Q., Ryan, J. J., Taylor, N., Liu, X., Hennighausen, L., Notarangelo, L. D., Paul, W. E., Blaese, R., and O'Shea, J. (1996). Signaling via IL-2 and IL-4 in Jak3-deficient severe combined immunodeficiency lymphocytes: Jak3-dependent and independent pathways. *Immunity* **5**, 605–615.

Ohmori, Y., Schreiber, R. D., and Hamilton, T. A. (1997). Synergy between intereron-gamma and tumor necrosis factor-alpha in transcriptional activation is mediated by cooperation between signal transducer and activator of transciption 1 and nuclear factor kappaβ. *J. Biol. Chem.* **272**, 14899–14907.

Okuda, K., Smith, L., Griffin, J. D., and Foster, R. (1997). Signaling functions of the tyrosine residues in the beta chain of the granulocyte-macrophage colony-stimulating factor receptor. *Blood* **90**, 4759–4766.

Orchansky, P. L., Ayres, S. D., Hilton, D. J., and Schrader, J. W. (1997). An interleukin (IL)-13 receptor lacking the cytoplasmic domain fails to transduce IL-13-induced signals and inhibits responses to IL-4. *J. Biol. Chem.* **272**, 22940–22947.

Ormandy, C. J., Camus, A., Barra, J., Damotte, D., Lucas, B., Buteau, H., Edery, M., Brousse, N., Babinet, C., Binart, N., and Kelly, P. A. (1997). Null mutation of the prolactin receptor gene produces multiple reproductive defects in the mouse. *Genes Devel.* **11**, 167–178.

Pallard, C., Gouilleux, F., Benit, L., Cocault, L., Souyri, M., Levy, D., Groner, B., Gisselbrecht, S., and Dusanter-Fourt, I. (1995). Thrombopoietin activates a Stat5-like factor in hematopoeitic cells. *EMBO J.* **14**, 2847–2856.

Parganas, E., Wang, D., Stravopidis, D., Topham, D., Marine, J.-C., Teglund, S., Vanin, E. F., Bodner, S., Colamonici, O. R., vanDeursen, J. M., Gorsveld, G., and Ihle, J. N. (1998). Jak2 is essential for signaling through a variety of cytokine receptors. *Cell* **93**, 385–395.

Park, C., and Schindler, C. (1999). Murine Stat2 is uncharacteristically divergent. *Submitted.*

Park, S. Y., Saijo, K., Takahashi, T., Osawa, M., Arase, H., Hirayama, N., Miyake, K., Nakauchi, H., Shirasawa, T., and Saito, T. (1995). Developmental defects of lymphoid cells in Jak3 kinase-deficient mice. *Immunity* **3**, 771–782.

Pearse, R. N., Feinman, R., Shuai, K., Darnell, J. E., and Ravetch, J. V. (1993). Interferon gamma-induced transcription of the high affinity Fc receptor for IgG requires assembly of a complex that includes the 91 kDa subunit of transcription factor ISGF3. *Proc. Natl. Acad. Sci. USA* **90**, 4314–4318.

Perez, C., Coeffier, E., Moreau-gachelin, F., Wietzerbin, J., and Benech, P. D. (1994). Involvement of the transcription factor Pu.1/Spi-1 in myeloid cell restricted expression of an Interferon-inducible gene encoding the human high-affinity Fcγ receptor. *Mol. Cell. Biol.* **14**, 5023–5031.

Peschon, J., Morrissey, P., Grabstein, K., Ramsdell, F., Maraskovsky, E., Gliniak, B., Park, L., Ziegler, S., Williams, D., Ware, C., Meyer, J., and Davison, B. (1994). Early lymphocyte expansion is severely impaired in Interleukin 7 receptor-deficient mice. *J. Exptl. Med.* **180**, 1955–1960.

Pestka, S. (1997). The human Interferon alpha species and hybrid proteins. *Sem. Oncol.* **24**, S9-4 to S9-17.

Pestka, S., Langer, J. A., Zoon, K. C., and Samuel, C. E. (1987). Interferons and their actions. *Ann. Rev. Biochem.* **56**, 727–777.

Pfeffer, L. M., Mullersman, J. E., Pfeffer, S. R., Murti, A., Shi, W., and Yang, C. H. (1997). Stat3 as an adapter to couple phosphatidylinostol 3-kinase to the INFAR1 chain of the type I Interferon receptor. *Science* **276**, 1418–1420.

Piccotti, J. R., Li, K., Chan, S. Y., Ferrante, J., Magram, J., Eichwald, E. J., and Bishop, D. K. (1998). Alloantigen-reactive th1 development in il-12 deficient mice. *J. Immunol.* **160**, 1132–1138.

Pine, R. (1997). Convergence of the TNF-α and IFN-γ signalling pathways through synergistic induction of IRF-1/ISGF-2 is mediated by a composite GAS/κB promoter element. *Nuc. Acids Res.* **25**, 4346–4354.

Pine, R., Canova, A., and Schindler, C. (1994). Tyrosine phosphorylated p91 binds to a single element in the ISGF2/IRF-1 promoter to mediate induction by IFN-α and IFN-γ, and is likely to autoregulate the p91 gene. *EMBO J.* **13**, 158–167.

Poli, V., Balena, R., Fattori, E., Markatos, A., Yamamoto, M., Tanaka, H., Ciliberto, G., Rodan, G. A., and Costantini, F. (1994). Interleukin-6 deficient mice are protected from bone loss caused by estrogen depletion. *EMBO J.* **13**, 1189–1196.

Presky, D. H., Yang, H., Minetti, L. J., Chua, A. O., Nabavi, N., Wu, C. Y., Gately, M. K., and Gubler, U. (1996). A functional interleukin 12 receptor complex is composed of two beta-type cytokine receptor subunits. *Proc. Natl. Acad. Sci. USA* **26**, 14002–14007.

Proud, C. G. (1995). PKR: A new name and new roles. *Trends Biochem. Sci.* **20**, 241–246.

Quelle, F. W., Sato, N., Witthuhn, B. A., Inhorn, R. C., Eder, M., Miyajima, A., Griffin, J. D., and Ihle, J. N. (1994). Jak2 associates with the βc chain of the receptor for granulocyte macrophage colony stimulating factor, and its activation requires the membrane-proximal region. *Mol. Cell. Biol.* **14**, 4335–4341.

Quelle, F. W., Wang, D., Nosaka, T., Thierfelder, W. E., Stravopodis, D., Weinstein, Y., and Ihle, J. N. (1996). Erythropoietin induces activation of Stat5 through association with specific tyrosines on the receptor that are not required for a mitogenic response. *Mol. Cell. Biol.* **16**, 1622–1631.

Qureshi, S. A., Salditt-Georgieff, M., and Darnell, J. E. (1995). Tyrosine phosphorylated Stat1 and Stat2 plus a 48 kDa protein all contact DNA in forming interferon-stimulated-gene factor 3. *Proc. Natl. Acad. Sci. USA* **92**, 3829–3833.

Qureshi, S. A., Leung, S., Kerr, I. M., Stark, G. R., and Darnell, J. E. (1996). Function of Stat2 protein in transcriptional activation by alpha interferon. *Mol. Cell. Biol.* **16**, 288–293.

Ram, P. A., Park, S.-H., Choi, H. K., and Waxman, D. J. (1996). Growth hormone activation of Stat1, Stat3, and Stat5 in rat liver. *J. Biol. Chem.* **271**, 5929–5940.

Raz, R., Durbin, J. E., and Levy, D. E. (1994). Acute phase response factor and additional members of the interferon stimulated gene factor 3 family integrate diverse signal from cytokines, interferons, and growth factors. *J. Biol. Chem.* **269**, 24391–24395.

Raz, R., Lee, C. K., Cannizzaro, L. A., D'Eustachio, P., and Levy, D. E. (1998). Essential role of Stat3 for embryonic stem cell pluripotency. *Proc. Natl. Acad. Sci. USA* **96**, 2846–2851.

Reich, N. C., Evans, B., Levy, D. E., Fahey, D. E., Knight, E., and Darnell, J. E. (1987). Interferon-induced transcription of a gene encoding a 15 kDa protein depends on an upstream enhancer element. *Proc. Natl. Acad. Sci. USA* **84**, 6394–6398.

Remy, I., Wilson, I. A., and Michnick, S. W. (1999). Erythropoietin receptor activation by a ligand-induced conformation change. *Science* **283**, 990–992.

Riley, J. K., Takeda, K., Akira, S., and Schreiber, R. D. (1999). IL-10 receptor signaling through the JAK-STAT pathway: Requirement for two distinct receptor derived signals for anti-inflammatory action. *J. Biol. Chem.* **274**, 16513–16521.

Robertson, L. M., Kerppola, T. K., Vendrell, M., Bocchiaro, C., Morgan, J. I., and Curran, T. (1995). Regulation of c-*fos* expression in transgenic mice requires multiple interdependent transcription control elements. *Neuron* **14**, 241–252.

Robledo, O., Fourcin, M., Chevalier, S., Guillet, C., Auguste, P., Barthelaix, A. P., Pennica, D., and Gascan, H. (1997). Signaling of the cardiotrophin-1 receptor. *J. Biol. Chem.* **272**, 4855–4863.

Rodig, S. J., Meraz, M. A., White, J. M., Lampe, P. A., Riley, J. K., Arthur, C. D., King, K. L., Sheehan, K. C. F., Yin, L., Pennica, D., Johnson, E. M., and Schreiber, R. D. (1998). Disruption of the Jak1 gene demonstrates obligatory and nonredundant roles of the Jaks in cytokine-induced biologic responses. *Cell* **93**, 373–383.

Romano, M., Sironi, M., Toniatti, C., Polentarutti, N., Fruscella, P., Ghezzi, P., Faggionni, R., Luini, W., vanHinsbergh, V., Sozzani, S., Bussolino, F., Poli, V., Ciliberto, G., and Mantovani, A. (1997). Role of IL-6 and its soluble receptor in induction of chemokines and leukocyte recruitment. *Immunity* **6**, 315–325.

Rosenblum, C. I., Tota, M., Cully, D., Smith, T., Collum, R., Qureshi, S., Hess, J. F., Phillips, M. S., Hey, P. J., Vongs, A., Fong, T. M., Xu, L., Chen, H. Y., Smith, R. G., Schindler, C., and Ploeg, L. H. T. V. d. (1996). Functional STAT 1 and 3 signaling by the leptin receptor (OB-R); reduced expression of the rat fatty leptin receptor in transfected cells. *Endocrinology* **137**, 5178–5181.

Rothman, P., Kreider, B., Azam, M., Levy, D., Wegenka, U., Eilers, A., Decker, T., Horn, F., Kashleva, H., Ihle, J., and Schindler, C. (1994). Cytokines and growth factors signal through tyrosine phosphorylation of a family of related transcription factors. *Immunity* **1**, 457–468.

Ruff-Jamison, S., Chen, K., and Cohen, S. (1995). Epidermal growth factor induces the tyrosine phosphorylation and nuclear translocation of Stat5 in mouse liver. *Proc. Natl. Acad. Sci. USA* **92**, 4215–4218.

Russell, S. M., Johnston, J. A., Noguchi, M., Kawamura, M., Bacon, C. M., Friedmann, M., Berg, M., McVicar, D. W., Witthuhn, B. A., Silvennoinen, O., Goldman, A. S., Schmalsteig, F. C., Ihle, J. N., O'Shea, J. J., and Leonard, W. J. (1994). Interaction of the IL-2Rβ and γc chains with Jak1 and Jak3: Implications for XSCID and XCID. *Science* **266**, 1042–1044.

Russell, S. M., Tayebi, N., Nakajima, H., Riedy, M. C., Roberts, J. L., Aman, M. J., Mignone, T.-S., Noguchi, M., Markert, M. L., Buckely, R. H., O'Shea, J. J., and Leonard, W. J. (1995). Mutation of Jak3 in a patient with SCID: Essential role of Jak3 in lymphoid development. *Science* **270**, 797–800.

Ryan, J. J., McReynolds, L. J., Huang, H., Nelms, K., and Paul, W. E. (1998). Characterization of a mobile stat6 activation motif in the human IL-4 receptor. *J. Immunol.* **161**, 1811–1821.

Sachs, L. (1993). The cellular and molecular environment in leukemia. *Blood Cells* **19**, 709–726.

Sadowski, I., and Ptashne, M. (1989). A vector for expressing GAL(1-147) fusions in mammalian cells. *Nuc. Acids Res.* **18**, 7539.

Sadowski, H. B., Shuai, K., Darnell, J. E., and Gilman, M. Z. (1993). A common nuclear signal transduction pathway activated by growth factors and cytokine receptors. *Science* **261**, 1739–1744.

Sakamaki, K., Wang, H.-M., Miyajima, I., Kitamura, T., Todokoro, K., Harada, N., and Miyajima, A. (1993). Ligand-dependent activation of chimeric receptors with the cytoplasmic domain of the Interleukin-3 receptor β subunit (βIL3). *J. Biol. Chem.* **268**, 15833–15839.

Sasse, J., Hemmann, U., Schwartz, C., Schiertshauer, U., Heesel, B., Landgraf, C., Scheider-Mergener, J., Heinrich, P. C., and Horn, F. (1997). Mutational analysis of Acute Response Factor/Stat3 activation and dimerization. *Mol. Cell. Biol.* **17**, 4677–4686.

Schaefer, T. S., Sanders, L. K., and Nathans, D. (1995). Cooperative transcriptional activity of Jun and Stat3β, a short form of Stat3. *Proc. Natl. Acad. Sci. USA* **92**, 9097–9101.

Schaefer, T. S., Sanders, L. K., Park, O. K., and Nathans, D. (1997). Functional differences between Stat3α and Stat3β. *Mol. Cell. Biol.* **17**, 5307–5316.

Schieffer, B., Paxton, W. G., Marrero, M. B., and Bernstein, K. E. (1996). Importance of tyrosine phosphorylation in angiotensin II type 1 receptor signaling. *Hyperten.* **27**, 476–480.

Schindler, C., Fu, X.-Y., Improta, T., Aebersold, R., and Darnell, J. E. (1992a). Proteins of transcription factor ISGF3: one gene encodes the 91- and 84-kDa ISGF3 proteins that are activated by Interferon alpha. *Proc. Natl. Acad. Sci. USA* **89**, 7836–7839.

Schindler, C., Shuai, K., Prezioso, V., and Darnell, J. E. (1992b). Interferon-dependent tyrosine phosphorylation of a latent cytoplasmic transcription factor. *Science* **257**, 809–813.

Schindler, C., Kashleva, H., Pernis, A., Pine, R., and Rothman, P. (1994). STF-IL4: A novel IL-4 induced signal transducing factor. *EMBO J.* **13**, 1350–1356.

Schindler, C., and Darnell, J. E. (1995). Transcriptional responses to peptide ligands: The JAK-STAT pathway. *Ann. Rev. Biochem.* **64**, 621–651.

Schindler, C. (1998). STATs as activators of apoptosis. *Trends Cell Biol.* **8**, 97–98.

Schindler, U., Wu, P., Rothe, M., Brasseur, M., and McKnight, S. L. (1995). Components of a STAT recognizaion code: Evidence for two layers of molecular selectivity. *Immunity* **2**, 689–697.

Schwarze, M. M., and Hawley, R. G. (1995). Prevention of myeloma cell apoptosis by ectopic bcl-2 expression or interleukin 6-mediated up-regulation of bcl-xL. *Cancer Res.* **55**, 2262–2265.

Seidel, H. M., Milocco, L. H., Lamb, P., Darnell, J. E., Stein, R. B., and Rosen, J. B. (1995). Spacing of palindromic half sites as a determinant of selective STAT (signal transducers and activators of transcription) DNA binding and transcriptional activity. *Proc. Natl. Acad. Sci. USA* **92**, 3041–3045.

Sekimoto, T., Nakajima, K., Tachibana, T., Hirano, T., and Yoneda, Y. (1996). Interferon-γ-dependent nuclear import of Stat1 is mediated by the GTPase activity of Ran/TC4. *J. Biol. Chem.* **271**, 31017–31020.

Sekimoto, T., Imamoto, N., Nakajima, K., Tachibana, T., Hirano, T., and Yoneda, Y. (1997). Extracellular signal-dependent nuclear import of Stat1 is mediated by nuclear pore-targeting complex formation with NPI-1, but not Rch1. *EMBO J.* **16**, 7076–7077.

Shen, C. H., and Stavnezer, J. (1998). Interaction of Stat6 and NF-κB: Direct association and synergistic activation of Interleukin-4-induced transcription. *Mol. Cell. Biol.* **18**, 3395–3404.

Shimoda, K., vanDeursen, J., Sangster, M. Y., Sarawar, S. R., Carson, R. T., Tripp, R. A., Chu, C., Quelle, F. W., Nosaka, T., Vignali, D. A., Doherty, P. C., Grosveld, G., Paul, W. E., and Ihle, J. N. (1996). Lack of IL-4 induce Th2 response in IgE class switching mice with disrupted Stat6 gene. *Nature* **380**, 630–633.

Shuai, K., Schindler, C., Prezioso, V., and Darnell, J. E. (1992). Activation of transcription by IFN-γ. Tyrosine phosphorylation of a 91-kDa DNA binding protein. *Science* **258**, 1808–1812.

Shuai, K., Ziemiecki, A., Wilks, A. F., Harpur, A. G., Sadowski, H. B., Gilman, M. Z., and Darnell, J. E. (1993). Polypeptide signaling to the nucleus through tyrosine phosphorylation of Jak and Stat proteins. *Nature* **366**, 580–585.

Shuai, K., Horvarth, C. M., Tsai-Huang, L. H., Quereshi, S. A., Cowburn, D., and Darnell, J. E. (1994). Interferon activation of the transcription factor stat91 involves dimerization through SH2-phosphotyrosyl peptide interactions. *Cell* **76**, 821–828.

Shuai, K., Liao, J., and Song, M. M. (1996). Enhancement of antiproliferative activity of gamma interferon by the specific inhibition of tyrosine dephosphorylation of Stat1. *Mol. Cell. Biol.* **16**, 4932–4941.

Shulz, L. D., Schweitzer, P. A., Rajan, T. V., Yi, T., Ihle, J. N., Mathews, J., Thomas, M. L., and Beier, D. R. (1993). Mutations at the murine motheaten locus are within the hematopoietic cell protein–tyrosine phosphatase (Heph) gene. *Cell* **73**, 1445–1454.

Sibilia, M., and Wagner, F. (1995). Strain-dependent epithelial defects in mice lacking the EGF receptor. *Science* **269**, 234–237.

Silvennoinen, O., Schindler, C., Schlessinger, J., and Levy, D. E. (1993). Ras-independent signal transduction in response to growth factors and cytokines by tyrosine phosphorylation of a common transcription factor. *Science* **261**, 1737–1739.

Sliva, D., Wood, T. J., Schindler, C., Lobie, P. E., and Norstedt, G. (1994). Growth hormone specifically regulates serine protease inhibitor gene transcription via γ-activated sequence-like DNA. *J. Biol. Chem.* **269**, 26208–26214.

Soh, J., Donnelly, R. J., Kotenko, S., Mariano, T. M., Cook, J. R., Wang, N., Emanuel, S., Schwartz, B., Miki, T., and Pestka, S. (1994). Identification and sequence of an accessory factor required for activation of the human Interferon γ receptor. *Cell* **76**, 793–802.

Solar, G. P., Kerr, W., Zeigler, F., Hess, D., Donahue, C., Sauvage, F., and Eaton, D. (1998). Role of c-mpl in early hematopoiesis. *Blood* **92**, 4–10.

Socolovsky, M., Fallon, A. E., Wang, S., Brugnara, C., and Lodish, H. F. (1999). Fetal anemia and apoptosis of red cell progenitors in Stat5a -/- 5b -/- mice: A direct role for Stat5 in Bcl-X1 induction. *Cell* **98**, 1–20.

Sotiropoulos, A., Moutoussamy, S., Renaudie, F., Clauss, M., Kayser, C., Gouilleux, F., Kelly, P. A., and Finidori, J. (1996). Differential activation of stat3 and stat5 by distinct regions of the growth hormone receptor. *J. Mol. Endocrinol.* **10**, 998–1009.

Spencer, S. D., DiMarco, F., Hooley, J., Pitts-Meek, S., Bauer, M., Ryan, A. M., Sordat, B., Gibbs, V. C., and Aguet, M. (1998). The orphan receptor CRF2-4 is an essential subunit of the interleukin 10 receptor. *J. Exptl. Med.* **187**, 571–578.

Stahl, N., Boulton, T. G., Farruggella, T., lp, N. Y., Davis, S., Witthuhn, B. A., Quelle, F. W., Silvennoinen, O., Barbieri, G., Pellegrini, S., Ihle, J. N., and Yancopoulos, G. D. (1994). Association and activation of Jak-Tyk kinases by CNTF-LIF-OSM-IL-6 β receptor components. *Science* **263**, 92–95.

Stahl, N., Farruggella, T. J., Boulton, T. G., Zhong, Z., Darnell, J. E., and Yancopoulos, G. D. (1995). Choice of STATs and other substrates specified by modular tyrosine-based motifs in cytokine receptors. *Science* **267**, 1349–1352.

Starr, R., Willison, T. A., Viney, E. M., Murray, L. J., Rayner, J. R., Jenkins, B. J., Gonda, T. J., Alexander, W. S., Metcalf, D., Nicola, N. A., and Hilton, D. J. (1997). A family of cytokine-inducible inhibitors of signaling. *Nature* **387**, 917–921.

Stewart, C. L., Kaspar, P., Brunet, L. J., Bhatt, H., Gadi, I., Kontgen, F., and Abbondanzo, S. J. (1992). Blastocyst implantation depends on maternal expression of leukemia inhibitory factor. *Nature* **359**, 76–79.

Stöcklin, E., Wissler, M., Gouilleux, F., and Groner, B. (1996). Functional interactions between Stat5 and the glucocorticoid receptor. *Nature* **383**, 726–728.

Stoecklin, E., Wissler, M., Moriggl, R., and Groner, B. (1997). Specific DNA binding of Stat5, but not of glucocorticoid receptor, is required for their functional cooperation in the regulation of gene transcription. *Mol. Cell. Biol.* **17**, 6708–6716.

Strehlow, L., and Schindler, C. (1997). Gamma interferon signaling. *In* "Gamma Interferon in Antiviral Defense" (G. Karupiah, ed.), pp. 61–84. R. G. Landes, Austin, TX.

Strehlow, I., and Schindler, C. (1998). Amino terminal signal transducer and activator of transcription (STAT) domains regulate nuclear traslocation and STAT deactivation. *J. Biol. Chem.* **96,** 5007–5012.

Subramanian, A., Wang, J., and Gil, G. (1998). Stat5 amd nf-y are involved in expression and growth hormone-mediated sexually dimorphic regulation of cytochrome p450 3a10 lithocholic acid 6β-hydroxylase. *Nuc. Acids Res.* **26,** 2173–2178.

Suzuki, H., Kuendig, T. M., Furlonger, C., Wakeham, A., Timms, E., Matsuyama, T., Schmits, R., Simard, J. J. L., Ohashi, P. S., Griesser, H., Tanaguchi, T., Paige, C. J., and Mak, T. W. (1995). Deregulated T cell activation and autoimmunity in mice lacking interleukin-2 receptor β. *Science* **268,** 1472–1476.

Takeda, K., Tanaka, T., Shi, W., Matsumoto, M., Minami, M., Kashiwamura, S., Nakanishi, K., Yoshida, N., Kishimoto, T., and Akira, S. (1996). Essential role of Stat6 in IL-4 signalling. *Nature* **380,** 627–630.

Takeda, K., Noguchi, K., Shi, W., Tanaka, T., Matsumoto, M., Yoshida, N., Kishimoto, T., and Akira, S. (1997). Targeted disruption of the mouse Stat3 gene leads to early embryonic lethality. *Proc. Natl. Acad. Sci. USA* **94,** 3801–3804.

Takeda, K., Kaisho, T., Yoshida, N., Takeda, J., Kishimoto, T., and Akira, S. (1998). Stat3 activation is responsible for IL-6-dependent T cell proliferation through preventing apoptosis: generation and characterization of T cell-specific Stat3-deficient mice. *J. Immunol.* **161,** 4652–4660.

Takeda, K., Clausen, B. E., Kaisho, T., Tsujimura, T., Terada, N., Förster, I., and Akira, S. (1999). Enhanced Th1 activity and development of chronic enteroclolitis in mice devoid of Stat3 in macrophages and neutrophils. *Immunity* **10,** 39–49.

Takeshita, T., Arita, T., Higuchi, M., Asao, H., Endo, K., Kuroda, H., Tanaka, N., Murata, K., and Ishii, N. (1997). STAM, signal transducing adaptor molecule, is associated with janus kinases and involved in signaling for cell growth and c-myc induction. *Immunity* **6,** 449–457.

Taniguchi, T., and Minami, Y. (1993). The IL-2/IL-2 receptor system: A current overview. *Cell* **73,** 5–8.

Tanner, J. W., Chen, W., Young, R. L., Longmore, G., and Shaw, A. S. (1995). The conserved box1 motif of cytokine receptors is required for association with JAK kinases. *J. Biol. Chem.* **270** 6523–6530.

Teglund, S., McKay, C., Schuetz, E., VanDeursen, J. M., Stravopodis, D., Wang, D., Brown, M., Bodner, S., Grosveld, G., and Ihle, J. N. (1998). Stat5a and Stat5b proteins have essential roles and nonessential, or redundant, roles in cytokine responses. *Cell* **93,** 841–850.

Thierfelder, W. E., vanDeursen, J. M., Yamamoto, K., Tripp, R. A., Sarawar, S. R., Carson, R. T., Sangster, M. Y., Vignali, D. A., Doherty, P. C., Grosveld, G., and Ihle, J. N. (1996). Requirement for Stat4 in interleukin-12 mediated response of natural killer and T cells. *Nature* **382,** 171–174.

Thomis, D. C., Gurniak, C. B., Tivol, E., Sharpe, A. H., and Berg, L. J. (1995). Defects in B lymphocyte maturation and T lymphocyte activation in mice lacking Jak3. *Science* **270,** 794–797.

Tian, S.-S., Lamb, P., Seidel, H. M., Stein, R. B., and Rosen, J. (1994). Rapid activation of the Stat3 transcription factor by granulocyte-colony stimulating factor. *Blood* **84,** 1760–1764.

Turkson, J., Bowman, T., Garcia, R., Caldenhoven, E., Groot, R. P. D., and Jove, R. (1998). Stat3 activation by Src induces specific gene regulation and is required for cell transformation. *Mol. Cell. Biol.* **18,** 2545–2552.

Udy, G. B., Towers, R. P., Snell, R. G., Wilkins, R. J., Park, S. H., Ram, P. A., Waxman, D. J., and Davey, H. W. (1997). Requirement of Stat5b for sexual dimorphism of body growth rates and liver gene expression. *Proc. Natl. Acad. Sci. USA* **94,** 7239–7244.

Uzé, G., Lutfalla, G., and Gresser, I. (1990). Genetic transfer of a functional human interferon-α receptor into mouse cells: cloning and expression of its cDNA. *Cell* **60,** 225–234.

Veals, S. A., Schindler, C., Leonard, D., Fu, X.-Y., Aebersold, R., Darnell, J. E., and Levy, D. E. (1992). Subunit of an IFN-α responsive transcription factor is related to interferon regulatory factor and Myb families of DNA binding proteins. *Mol. Cell. Biol.* **12**, 3315–3324.

Velazquez, L., Fellous, M., Stark, G. R., and pellegrini, S. (1992). A protein tyrosine kinase in the Interferon α/β signaling pathway. *Cell* **70**, 313–322.

Velazquez, L., Mogensen, K. E., Barbieri, G., Fellous, M., Uze, G., and Pellegrini, S. (1995). Distinct domains of the protein tyrosine kinase tyk2 required for binding of Interferon-α/β and signal transduction. *J. Biol. Chem.* **270**, 3327–3334.

Vignais, M. L., Sadowski, H. B., Watling, D., Rogers, N. C., and Gilman, M. (1996). Platelet-derived growth factor induces phosphorylation of multiple JAK family kinases and STAT proteins. *Mol. Cell. Biol.* **16**, 1759–1769.

Vinkemeier, U., Cohen, S. L., Moarefi, I., Chait, B. T., Kuriyan, J., and Darnell, J. E. (1996). DNA binding of in vivo activated Stat1α, Stat1β and truncated Stat1: Interaction between NH2-terminal domains stabilizes binding of two dimers to tandem DNA sites. *EMBO J.* **15**, 5616–5626.

Vinkemeier, U., Moarefi, I., Darnell, J. E., and Kuriyan, J. (1998). Structure of the aminoterminal protein interaction domain of Stat-4. *Science* **279**, 1048–1052.

vonFreeden-Jeffry, U., Vieira, P., Lucian, L. A., McNeil, T., Burdach, S. E., and Murray, R. (1995). Lymphopenia in interleukin (IL)-7 gene-deleted mice identifies IL-7 as a nonredundant cytokine. *J. Exptl. Med.* **181**, 1519–1526.

Wagner, B. J., Hayes, T. E., Hoban, C. J., and Cochran, B. H. (1990). The SIF binding element confers sis/PDGF inducibility onto the c-fos promoter. *EMBO J.* **9**, 4477–4484.

Wakao, H., Schmitt-Ney, M., and Groner, B. (1992). Mammary gland-specific nuclear factor is present in lactating rodent and bovine mammary tissue and composed of a single polypeptide of 89 kDa. *J. Biol. Chem.* **267**, 16365–16370.

Wakao, H., Gouilleux, F., and Groner, B. (1994). Mammary gland factor (MGF) is a novel member of the cytokine regulated transcription factor gene family and confers the prolactin response. *EMBO J.* **13**, 2182–2191.

Wang, Y. D., Wong, K., and Wood, W. (1995). Intracellular tyrosine residues of the human growth hormone receptor are not required for the signaling of proliferation or JAK-STAT activation. *J. Biol. Chem.* **270**, 7021–7024.

Wang, D., Stravopodis, D., Teglund, S., Kitazawa, J., and Ihle, J. N. (1996). Naturally occurring dominant negative variants of Stat5. *Mol. Cell. Biol.* **16**, 6141–6148.

Ware, C. B., Horowitz, M. C., Renshaw, B. R., Hunt, J. S., Liggitt, D., Koblar, S. A., Gliniak, B. C., McKenna, H. J., Papayannopoulou, T., Thoma, B., Cheng, L., Donovan, P. J., Peschon, J. J., Bartlett, P. F., Willis, C. R., Wright, B. D., Carpenter, M. K., Davison, B. L., and Gearing, D. P. (1995). Targeted disruption of the low-affinity leukemia inhibitory factor receptor gene causes placental, skeletal, neural and metabolic defects and results in perinatal death. *Develop.* **121**, 1283–1299.

Watling, D., Guschin, D., Müller, M., Silvennoinen, O., Witthuhn, B. A., Quelle, F. W., Rogers, N. C., Schindler, C., Stark, G. R., Ihle, J. N., and Kerr, I. M. (1993). Complementation by the tyrosine kinase Jak2 of a mutant cell line defective in the Interferon-γ signal transduction pathway. *Nature* **366**, 166–170.

Weber-Nordt, R. M., Riley, J. R., Greenlund, A. C., Moore, K. W., Darnell, J. E., and Schreber, R. D. (1996). Stat3 recruitment by two distinct ligand induced tyrosine-phosphorylated docking sites in the interleukin-10 receptor intracellular domain. *J. Biol. Chem.* **271**, 27954–27961.

Weber-Nordt, R. M., Mertelsmann, R., and Finke, J. (1998). The JAK-STAT pathway: Signal transduction involved in proliferation, differentiation, and transformation. *Leukemia Lymphoma* **28**, 459–467.

Wegenka, U. M., Lütticken, C., Buschmann, J., Yuan, J., Lottspeich, F., Mueller-Esterl, W., Schindler, C., Roeb, E., Heinrich, P. C., and Horn, F. (1994). The interleukin-6-regulated acute phase response factor is antigenically and functionally related to members of the signal transducer and activator of transcription (STAT) family. *Mol. Cell. Biol.* **14**, 3186–3196.

Wells, J. A. (1996a). Binding in the growth hormone receptor complex. *Proc. Natl. Acad. Sci. USA* **93**, 1–6.

Wells, J. A. (1996b). Hormone mimicry. *Science* **273**, 449–450.

Wen, Z., Zhong, Z., and Darnell, J. E. (1995). Maximal activation of transcription by Stat1 and Stat3 requires both tyrosine and serine phosphorylation. *Cell* **82**, 241–250.

Witthuhn, B. A., Quelle, F. W., Silvennoinen, O., Yi, T., Tang, B., Miura, O., and Ihle, J. N. (1993). JAK2 associates with the erythropoietin receptor and is tyrosine phosphorylated and activated following stimulation with erythropoietin. *Cell* **74**, 227–236.

Witthuhn, B. A., Silvennoinen, O., Miura, O., Lai, K. S., Cwik, C., Liu, E. T., and Ihle, J. N. (1994). Involvement of the Jak-3 Janus kinase in signalling by interleukins 2 and 4 in lymphoid and myeloid cells. *Nature* **370**, 153–157.

Wollert, K. C., and Chien, K. R. (1997). Cardiotrophin-1 and the role of gp130-dependent signaling pathways in cardiac growth and development. *J. of Mol. Med.* **75**, 492–501.

Wong, A. H.-T., Tam, N. W. N., Yang, Y.-L., Cuddihy, A. R., Li, S., Kirchhoff, S., Hauser, H., Decker, T., and Koromilas, A. E. (1997). Physical association between Stat1 and the interferon-inducible protein kinase PKR and implications for interferon and double-stranded RNA signaling pathways. *EMBO* **16**, 1291–1304.

Wozniak, R. W., Rout, M. P., and Aitchinson, J. D. (1998). Karyopherins and kissing cousins. *Trends Cell Biol.* **8**, 184–188.

Wrighton, N. C., Farrell, F. X., Chang, R., Kashyap, A. K., Barbone, F. P., Mulcahy, L. S., Johnson, D. L., Barrett, R. W., Jolliffe, L. K., and Dower, W. J. (1996). Small peptides as potent mimetics of the protein hormone erythropoietin. *Science* **273**, 458–463.

Wu, H., Liu, X., Jaenisch, R., and Lodish, H. F. (1995). Generation of committed erythroid bfu-e and cfu-e progenitors does not require erythropoietin or the erythropoietin receptor. *Cell* **83**, 59–67.

Wu, C. Y., Ferrante, J., Gately, M. K., and Magram, J. (1997). Characterization of IL-12 receptor β1 chain (IL-12Rβ1)-deficient mice. *J. Immunol.* **159**, 1658–1665.

Xu, X., Sun, Y.-L., and Hoey, T. (1996). Cooperative DNA binding and sequence-selective recognition conferred by the STAT amino-terminal domain. *Science* **273**, 794–797.

Yamamoto, K., Miura, O., Hirosawa, S., and Miyasaka, N. (1997). Binding sequence of STAT4: STAT4 complex recognizes the IFN-gamma activation site (GAS)-like sequence (T/A)TTCC(C/G)GGAA(T/A). *Biochem. Biophys. Res. Commun.* **233**, 126–132.

Yamanaka, Y., Nakajima, K., Fukada, T., Hibi, M., and Hirano, T. (1996). Differentiation and growth arrest signals are generated through the cytoplasmic region of gp130 that is essential for Stat3 activation. *EMBO J.* **15**, 1557–1565.

Yan, R., Luo, H., Darnell, J. E., and Dearolf, C. R. (1996). A JAK-STAT pathway regulates wing vein formation in *Drosophila*. *Proc. Natl. Acad. Sci. USA* **93**, 5842–5847.

Yang, C. H., Shi, W., Basu, L., Murti, A., Constantinescu, S. N., Blatt, L., Croze, E., Mullersman, J. E., and Pfeffer, L. M. (1996). Direct association of STAT3 with the IFNAR-1 chain of the human type I interferon receptor. *J. Biol. Chem.* **271**, 8057–8061.

Yang, C.-H., Murti, A., and Pfeffer, L. M. (1998). Stat3 complements defects in an interferon-resistant cell line: Evidence for an essential role for Stat3 in interferon signaling and biological activities. *Proc. Natl. Acad. Sci. USA* **95**, 5568–5572.

Ye, B. H., Cattoretti, G., Shen, Q., Zhang, J., Hawe, N., Waard, R. D., Leung, C., Nouri-Shirazi, M., Orazi, A., Chaganti, R. S., Rothman, P., Stall, A. M., Pandolfi, P. P., and Dalla-Favera, R. (1997). The BCL-6 proto-oncogene controls germinal-centre formation and Th2-type inflammation. *Nat. Genet.* **16**, 161–170.

Yi, T., Mui, A. L., Krystal, G., and Ihle, J. N. (1993). Hematopoietic cell phosphatase associates with the IL-3 receptor β chain and down-reulates IL-3 induced tyrosine phosphorylation and mitogenesis. *Mol. Cell. Biol.* **13**, 7577–7586.

Yi, W., Kim, S.-O., Jiang, J., Park, S.-H., Kraft, A. S., Waxman, D. J., and Fink, S. J. (1996). Growth hormone receptor cytoplasmic domain differentially promotes tyrosine phosphorylation of signal transducers and activators of transcription 5b and 3 by activated Jak2 kinase. *Molecular Endocrinology* **10**, 1425–1443.

Yoshida, K., Taga, T., Saito, M., Suematsu, S., Kumanogoh, A., Tanaka, T., Fujiwara, H., Hirata, M., Yamagami, T., Nakahata, T., Hirabayashi, T., Yoneda, Y., Tanaka, K., Wang, W. Z., Mori, C., Shiota, K., Yoshida, N., and Kishimoto, T. (1996). Targeted disruption of a gp130, a common signal transducer for the interleukin 6 family of cytokines, leads to myocardial and hematological disorders. *Proc. Natl. Acad. Sci. USA* **93**, 407–711.

Yoshimura, A., Ohkubo, T., Kiguchi, T., Jenkins, N. A., Gilbert, D. J., Copeland, N. G., Hara, T., and Miyajima, A. (1995). A novel cytokine-inducible gene CIS encodes an SH2-containing protein that binds to tyrosine phosphorylated interleukin 3 and erythropoietin receptors. *EMBO J.* **14**, 2816–2826.

Yoshimura, A., Ichihara, M., Kinjyo, I., Moriyama, M., Copeland, N. G., Gilbert, D. J., Jenkins, N. A., Hara, T., and Miyajima, A. (1996). Mouse oncostatin M: An immediate early gene induced by multiple cytokines through the JAK-STAT5 pathway. *EMBO J.* **15**, 1055–1063.

Yu, C.-L., Meyer, D. J., Campbell, G. S., Larner, A. C., Cater-Su, C., Schwartz, J., and Jove, R. (1995). Enhance DNA-binding activity of a Stat3-related protein in cells transformed by the Src oncoprotein. *Science* **269**, 81–83.

Zhang, Y., Proenca, R., Maffel, M., Barone, M., Leopold, L., and Friedman, J. M. (1994). Positional cloning of the mouse obese gene and its human homologue. *Nature* **372**, 425–432.

Zhang, X., Blenis, J., Li, H.-C., Schindler, C., and Chen-Kiang, S. (1995). Requirement of serine phosphorylation for formation of STAT-promoter complexes. *Science* **267**, 1990–1994.

Zhang, J. J., Vinkemeier, U., Gu, W., Chakravarti, D., Horvath, C. M., and Darnell, J. E. (1996). Two contact regions between Stat1 and CBP/p300 in interferon-γ signaling. *Proc. Natl. Acad. Sci. USA* **93**, 15092–15096.

Zhong, Z., Wen, Z., and Darnell, J. E. (1994). Stat3: A STAT family member activated by tyrosine phosphorylation in response to epidermal growth factor and interleukin-6. *Science* **264**, 95–98.

Zhou, Y. J., Hanson, E. P., Chen, Y. Q., Magnuson, K., Chen, M., Swann, P. G., Wange, R. L., Changelian, P. S., and O'Shea, J. J. (1997). Distinct tyrosine phosphorylation sites in JAK3 kinase domain positively and negatively regulate its enzymatic activity. *Proc. Natl. Acad. Sci. USA* **94**, 13850–13855.

Zhu, M.-H., John, S., Berg, M., and Leonard, W. J. (1999). Functional association of Nmi with Stat5 and Stat1 in IL-2 and IFNγ mediated signaling. *Cell* **96**, 121–130.

Zhu, X., Wen, Z., Xu, L. Z., and Darnell, J. E. (1997). Stat1 serine phosphorylation occurs independently of tyrosine phosphorylation and requires an activated Jak2 kinase. *Mol. Cell. Biol.* **17**, 6618–6623.

Zou, J., Presky, D. H., Wu, C. Y., and Gubler, U. (1996). Differential associations between the cytoplasmic regions of the interleukin-12 receptor subunits β1 and β2 and JAK kinases. *J. Biol. Chem.* **272**, 6073–6077.

Zurawski, S., Vega, F., Huyghe, B., and Zurawski, G. (1993). Receptors for interleukin-13 and interleukin-4 are complex and share a novel component that functions in signal transduction. *EMBO J.* **12**, 2663–2670.

John D. Scott*
Mark L. Dell'Acqua*
Iain D. C. Fraser*
Steven J. Tavalin*
Linda B. Lester†

*Howard Hughes Medical Institute
Vollum Institute
Portland, Oregon 97201

†Division of Endocrinology
Oregon Health Sciences University
Portland, Oregon 97201

Coordination of cAMP Signaling Events through PKA Anchoring

I. Introduction

The efficacy of signal transduction events, such as those mediated by the second messenger cyclic 3′,5′-adenosine monophosphate (cAMP), is often taken for granted. Yet the mechanism with which extracellular effectors such as hormones, prostaglandins, or neurotransmitters induce the movement of signals from the inner face of the plasma membrane to specific intracellular targets still remains somewhat enigmatic, particularly when one considers the large number of polypeptides devoted to this process within a eukaryotic cell (Hunter, 1995). This dilemma is compounded by the finding that the cAMP-dependent protein kinase holoenzyme, the principal receptor for cAMP, is able to phosphorylate a wide array of cellular substrates. Thus, mechanisms must clearly exist to organize the correct repertoire of signaling molecules into coordinated units that allow only a subset of the PKA substrates to become phosphorylated (Pawson and Scott, 1997).

One mechanism involves the restriction of PKA to localized sites of action where it can only phosphorylate those substrates in its immediate vicinity. This is achieved in one of two ways: compartmentalized accumulation of the second messenger cAMP, or anchoring of the kinase to subcellular structures and organelles (Faux and Scott, 1996a; Pawson and Scott, 1997). In this chapter we review accumulating data suggesting that the subcellular location of the cAMP-dependent protein kinase is regulated in part through association with a family of A-kinase anchoring proteins called AKAPs.

II. The cAMP-Dependent Protein Kinase

When cAMP was first discovered as a soluble second messenger of hormone stimulated events in the late 1950s (Sutherland and Rall, 1957), it was soon recognized that its primary action in eukaryotic cells was to activate a cAMP-dependent protein kinase (PKA) (Taylor *et al.*, 1990). The PKA holoenzyme is a heterotetramer composed of a regulatory (R) subunit dimer that maintains two catalytic (C) subunits in a dormant state (Corbin and Keely, 1977; Corbin *et al.*, 1973; Potter *et al.*, 1978; Potter and Taylor, 1979). Holoenzyme dissociation ensues upon binding of cAMP to tandem sites in each R subunit (Su *et al.*, 1993, 1995). This alleviates an autoinhibitory contact that releases the active C subunit (Gibbs *et al.*, 1992; Wang *et al.*, 1991). The active kinase is then free to phosphorylate substrates on serine or threonine residues which are presented in a sequence context of Arg-Arg-Xaa-Ser/Thr or Lys-Arg-Xaa-Xaa-Ser/Thr (Kemp *et al.*, 1977; Kemp and Pearson, 1990). Given the frequent occurrence of these sequence motifs in many proteins, it was soon reasoned that unrestricted access of the C subunit to its substrates would lead to indiscriminate phosphorylation. Consequently, several regulatory mechanisms are in place to ensure that cAMP levels and kinase activity are tightly controlled (Adams *et al.*, 1991, 1992; Bacskai *et al.*, 1993; Barsony and Marks, 1990). Second messenger levels are controlled by a balance of adenylyl cyclase (AC) and phosphodiesterase (PDE) activities that generate gradients of cAMP emanating from the plasma membrane (Beavo *et al.*, 1994; Tang and Gilman, 1992). Supplementary signal terminating mechanisms, such as desensitization of AC or compartmentalized activation of PDEs, ensure localized reduction of the second messenger (Cooper *et al.*, 1995; Mons *et al.*, 1995; Shakur *et al.*, 1993; Smith *et al.*, 1996). Although access to cAMP is the primary requirement for PKA activation, additional factors are responsible for driving holoenzyme reformation and returning the kinase to the inactive state. The R subunits are expressed in excess over C subunits favoring rapid reformation of the holoenzyme when cAMP levels return to the basal state (Amieux *et al.*, 1997). In addition, the ubiquitous heat stable inhibitor, PKI, may well serve as a fail-safe device that mops up free C subunit (Krebs and Beavo, 1979).

PKI also facilitates export of the C subunit from the nucleus, which is devoid of R subunits; this suggests a more specific function as a signal terminator for nuclear phosphorylation events (Fantozzi *et al.*, 1992; Wen *et al.*, 1994, 1995).

For some time now it has been thought that the cellular specificity of PKA signaling is related to the existence of multiple C and R subunit isoforms. In mammals, three C subunit isoforms (α, β, and γ) exist, and, although there are subtle differences in the kinetic profiles and cAMP sensitivities for $C\alpha$ and $C\beta$ containing holoenzymes, these two predominant C subunit isoforms are virtually indistinguishable with respect to substrate specificity and interaction with R subunits (Gamm *et al.*, 1996; Scott, 1991; Taylor *et al.*, 1990). In contrast, the dimeric R subunits exhibit both distinct cAMP binding affinities and differential localization within cells (Corbin *et al.*, 1973, 1975). The type I PKA holoenzyme, which contains either $RI\alpha$ or $RI\beta$, is predominantly cytoplasmic and more sensitive to cAMP than type II PKA. In contrast, up to 75% of the type II PKA is targeted to certain intracellular sites through association of the RII subunits, $RII\alpha$ or $RII\beta$, with cellular binding proteins now known as anchoring proteins (for review, see Rubin, 1994; Dell'Acqua and Scott, 1997). Thus, it has been proposed that differences in subcellular targeting of type I and type II PKA are additional factors contributing to specificity in cellular responses.

III. AKAPs

The first RII-binding proteins were identified in the early 1980's by Vallee, De Cammili, Rubin, and Erlichman as contaminating proteins that which copurified with RII after affinity chromatography on cAMP-sepharose (reviewed by Dell'Acqua and Scott, 1997). However, detailed study of these proteins was made possible by the observation of Lohmann and colleagues that many, if not all, of these associated proteins retain their ability to bind RII after they have been immobilized on nitrocellulose filters (Lohmann *et al.*, 1984). As a result, the standard technique for detecting RII-binding proteins is an overlay method that is a slight modification of the Western blot but using radiolabeled RII subunit as a probe (reviewed by Hausken *et al.*, 1997). Using this technique, RII-binding bands ranging in size from 15 to 300 kDa have been detected in a variety of tissues, and it would appear that a typical cell expresses 5–10 distinct binding proteins (Carr *et al.*, 1992a). The RII overlay method has also been refined into an efficient interaction cloning strategy wherein cDNA expression libraries are screened using RII as a probe. This has led to the cloning of numerous RII-binding proteins (Bregman *et al.*, 1989; Carr *et al.*, 1992b; Carr and Scott, 1992; Coghlan *et al.*, 1994; Dransfield *et al.*, 1997b; Lester, 1996; McCartney *et al.*, 1995; Nauert *et al.*, 1997; Lin *et al.*, 1995; Fraser *et al.*, 1998). More recently, these RII-binding proteins were renamed A-kinase anchoring proteins, or AKAPs, to account for their proposed PKA targeting function (Hirsch *et al.*, 1992). A model is presented in Fig. 1 that illustrates the essential domains of AKAPs. Each anchoring protein contains two types of binding site: a common "anchoring motif" that binds the R subunit dimer of PKA, and a unique "targeting domain" that directs the subcellular localization of the PKA–AKAP complex through association with structural proteins, membranes, or cellular organelles.

It was formerly believed that AKAPs exclusively target the type II PKA holoenzyme in which the C subunit is bound to the RII isoform of the regulatory subunit. However, two dual function AKAPs have now been discovered that bind RI or RII (Huang *et al.*, 1997a, 1997b). Although *in vitro* studies indicate that RI binds several AKAPs with a 100-fold lower binding affinity than RII, the submicromolar binding constant interaction lies within the physiological concentration range of RI and AKAPs inside cells. Thus, type I PKA anchoring may be relevant under certain conditions where RII concentrations are limiting, such as in experiments on RIIα knockout mice where Ca^{2+} channel modulation is more sensitive to anchoring inhibitor reagents than in wild-type animals (Burton *et al.*, 1997). Another recently recognized property of AKAPs is that some of the anchoring proteins simultaneously bind PKA and one or more other signaling enzymes. These multivalent AKAPs serve as scaffolds for the assembly of multienzyme signaling complexes consisting of several kinases and phosphatases (Faux and Scott, 1996b). Despite some variations between individual AKAPs, all of these proteins represent a functionally related family of proteins that at least bind PKA and target their complement of kinases and phosphatases to specific subcellular structures. Consequently, the remainder of this chapter will focus on two areas of AKAP research: (1) the role of the anchoring motif as a tool to disrupt PKA anchoring inside cells; and (2) the mapping of AKAP targeting signals that have been used to functionally redirect PKA to specific intracellular locations.

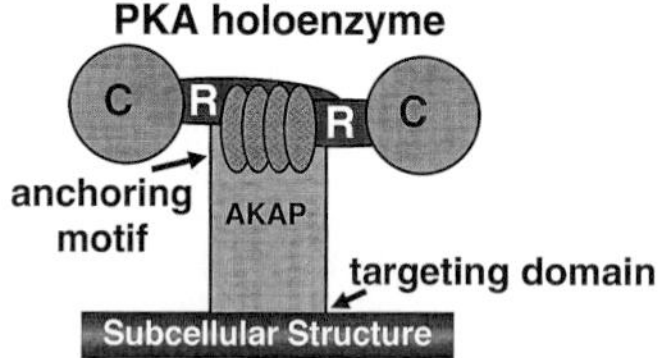

FIGURE 1   Model of the anchored PKA holoenzyme complex. Binding surfaces on the AKAP for association with PKA (anchoring) and for interaction with subcellular organelles or structures (targeting) are indicated.

A. The RII Anchoring Motif

Early work in the field focused on mapping the sites on RII required for interaction with AKAPs. Initially, the minimum region of RII required to bind MAP2 and AKAP75 was defined by screening a family of deletion mutants and chimeric proteins by the overlay assay (Scott *et al.*, 1990; Luo *et al.*, 1990). These studies concluded that RII dimerization was a prerequisite for anchoring and that AKAP-binding required the first 45 residues of RII (Hausken *et al.*, 1994; Luo *et al.*, 1990; Scott *et al.*, 1990). Subsequently, the localization and dimerization determinants were mapped to distinct subsites within this region (Hausken *et al.*, 1994, 1996). Deletion of residues 1 to 5 abolished the anchoring function but had no qualitative effect upon dimerization, and site-directed mutagenesis studies led to the identification of isoleucines at positions 3 and 5 as essential determinants for association with AKAPs (Hausken *et al.*, 1994, 1996). Since leucines and isoleucines are also crucial determinants of the reciprocal binding surface on the AKAP (Glantz *et al.*, 1993) it is possible that RII/AKAP docking may be analogous to the hydrophobic interactions that maintain a leucine zipper in transcription factors such as CEBP and CREB. However, it must be noted that the protein–protein interactions required for RII/AKAP interaction are more elaborate and involve three polypeptide chains, i.e., two RII protomers and a binding surface on the AKAP (Fig. 1). This view is supported by the stoichiometry of the interaction, which suggests that one AKAP binds per RII dimer (Carr *et al.*, 1992a). Furthermore, additional AKAP-binding determinants have been mapped between residues 11 to 25 of each RII molecule (Li and Rubin, 1995).

The rudimentary analysis conducted on RII/AKAP interaction has more recently been confirmed and extended by structural studies that have solved the structure of the AKAP-binding and dimerization surfaces on RIIα. The solution structure of an amino terminal fragment encompassing the first 45 residues of RIIα has been solved by nuclear magnetic resonance (NMR) spectroscopy. Both RII protomers form an antiparallel four-helix bundle with the principal AKAP-binding determinants located within an α-helix between residues 8 and 14, while a second helix between residues 28 and 42 maintains the dimerization contact. In this topology, isoleucines 3 and 5 form hydrophobic regions at either end of the AKAP-binding pocket that would permit their interaction with a reciprocal binding surface on the AKAP (Newlon *et al.*, 1997). Although these residues are principal anchoring determinants, further analysis of an RII/AKAP peptide complex suggests that numerous side chains within the AKAP-binding surface contact the anchoring protein. This may explain the nanomolar binding affinity of RII/AKAP interactions that have been measured by a variety of analytical techniques (Carr *et al.*, 1991, 1992a, 1992b; Hausken *et al.*, 1997). Furthermore, modeling studies suggest that a helical region common to all AKAPs fits snugly into the AKAP-binding surface.

Another level of specificity in PKA signaling may be achieved through the differential localization of RIIα or RIIβ by association with isoform-selective AKAPs. Alternatively, different AKAPs may be selectively expressed in cell types where one PKA isoform predominates. It was reported that RIIα had a sixfold preference for MAP2, whereas RIIβ had a twofold preference for AKAP75 (Leiser *et al.*, 1986). It has also been shown that follicle-stimulating hormone (FSH) treatment of rat granulosa cells induces an 80-kDa RIIα-selective AKAP (Carr *et al.*, 1993). A structural explanation for this differential binding affinity can be reasoned from analysis of the RII 1-45 structure. Sequences of the first 10 amino acids of RIIα and RIIβ are almost identical except for a pair of prolines at positions 6 and 7 in RIIα that is not present in RIIβ. A proline pair at positions 6 and 7 increases the rigidity of the helix–turn–helix motif that forms the AKAP-binding pocket (Newlon *et al.*, 1997). In fact, proline 6 may increase RIIα affinity for certain AKAPs through direct contact with the anchoring proteins. It is interesting to note that another proline pair exists in RIIα at positions 24 and 25 that proceeds the second helix. Interestingly, proline 26 is not conserved in RIIβ and is replaced with an alanine residue. This change is likely to increase the flexibility of the second helix and consequently may alter the dimerization interface in RIIβ. Structure/function studies are currently under way to assess the contribution of this region in the AKAP-binding affinity of RIIβ.

B. The PKA Anchoring Site on AKAPs

Reciprocal studies have identified a common site on AKAPs that binds the R subunit. Deletion analyses located the RII-binding sequences of MAP2 and AKAP150 to short regions of continuous amino acid sequence (Obar *et al.*, 1989; Rubino *et al.*, 1989). However, the nature of the RII-binding motif remained unclear until a human thyroid anchoring protein, called Ht31, was identified (Carr *et al.*, 1991, 1992a). The RII-binding sequence of Ht31 exhibited sequence similarities to both MAP2 and AKAP150 and was predicted to form an α-helix. Helical wheel projections of all three sequences exhibited a striking segregation of hydrophobic and hydrophilic side chains. This led to the proposal that the RII-binding motif of Ht31 and other AKAPs involves an amphipathic helix (Carr *et al.*, 1991). Subsequent studies performed on Ht31 and AKAP79, the human homologue of AKAP150, demonstrated a requirement for this region in RII-binding (Carr *et al.*, 1991, 1992b). The role of helical secondary structure in RII–AKAP interactions was supported by demonstrating that substitution of proline, a residue that perturbs helix formation, at various positions within the RII binding domain abolished RII-binding (Carr *et al.*, 1991, 1992a). These findings were consolidated by the synthesis of peptides encompassing the predicted helical region of Ht31 that were shown to bind either RII or the type II PKA holoenzyme with nanomolar affinity (Carr *et al.*, 1992a; Hausken *et al.*, 1997). Subsequent studies have independently confirmed that similar regions on several other AKAPs are essential determinants for RII-binding (Coghlan *et al.*, 1994; Dransfield *et al.*, 1997b; McCartney *et al.*, 1995; Nauert *et al.*, 1997). The high affinity of these interactions has important consequences for the intracellular localization of PKA. First, the K_D for the RII/AKAP interaction has been calculated from 1 to 11 nM by a variety of analytical methods (Hausken *et al.*, 1997). This affinity constant is well within the intracellular concentration ranges of RII and most AKAPs, suggesting that the RII/AKAP complex will be favored *in situ*. Secondly, the PKA holoenzyme binds Ht31 with the same high affinity as the RII dimer. Thus, the PKA holoenzyme will be anchored in cells when cAMP is at basal levels. Although the involvement of an amphipathic helix has not been definitively proven, analysis of the RII 1-45 structure suggests that an α-helix is the optimal structure to fit into the AKAP-binding pocket (M. Newlon and P. Jennings, personal communication).

C. The Use of AKAP-Derived Peptides inside Cells

Knowledge of the RII-binding domains on several AKAPs has allowed the generation of reagents that alter PKA anchoring within cells. Peptides encompassing the amphipathic helix region of Ht31 (residues 493–515) effectively compete for RII-AKAP interaction *in vitro* and disrupt the subcellular localization of PKA inside cells (Rosenmund *et al.*, 1994). Perfusion of cultured hippocampal neurons with these "anchoring inhibitor peptides" caused a time-dependent decrease in AMPA/kainate-responsive currents, whereas perfusion of control peptides, which were unable to compete for RII-binding, had no effect on channel activity (Rosenmund *et al.*, 1994). Additional controls confirmed that the effects emanated from PKA as perfusion of PKI peptides, which block kinase activity, caused a decrease in channel activity, whereas microinjection of excess C subunit overcame the anchoring inhibitor effect. Collectively, these findings suggested that the Ht31 peptide displaced PKA from anchored sites close to the AMPA/kainate channels, thereby decreasing the probability of channel phosphorylation. Parallel studies by Catterall and colleagues have subsequently shown that Ht31 peptide-mediated disruption of PKA anchoring modulates L-type Ca^{2+} channels in skeletal muscle (Johnson *et al.*, 1994, 1997). Likewise, peptide-mediated disruption of PKA targeting has been implicated in the regulation of smooth muscle calcium activated potassium (K_{CA}) channels (Wang and Kotlikoff, 1996). K_{CA} channels were recorded from tracheal myocytes, showing that introduction of the anchoring inhibitor peptide Ht31 blocked stimulation of the channels that were induced by ATP, whereas the Ht31 control peptide had no adverse effect in channel stimulation. Taken together, each of these studies provide convincing evidence that PKA anchoring may facilitate preferential modulation of physiological PKA substrates. However, there are technical limitations associated with the introduction of bioactive peptides into cells. Although microinjection or microdialysis is suitable for peptide delivery into single cells, the uptake of cell-soluble peptide analogues is necessary to affect many cells. Accordingly, cell-permeant anchoring-inhibitor peptides and Ht31 expression plasmid have been developed to efficiently displace the intracellular location of PKA in a variety of cell types. As will be discussed in detail later, we have used cell soluble versions of the Ht31 peptides to define a role for PKA anchoring in hormone-mediated insulin release from pancreatic beta cells.

D. The Role of PKA Anchoring in GLP-1 Mediated Insulin Secretion

Insulin secretion requires the coordinated action of metabolites, hormones, and neurotransmitters and is regulated by a variety of second-messenger-mediated signaling events that alter the dynamic balance in kinase and phosphatase activity in pancreatic beta cells (Ammala *et al.*, 1994; Sjoholm, 1995). For example, a recently identified hormone, glucagon-like peptide 1 (GLP-1), potentiates glucose-mediated insulin secretion through activation of PKA to favor exocytosis of insulin secretory granules (Drucker *et al.*, 1987; Thorens, 1992; Yaekura *et al.*, 1996). In one study we demonstrate that the subcellular targeting of PKA by a family of targeting proteins called AKAPs is an additional mechanism in the regulation of hormone-mediated insulin secretion (Lester *et al.*, 1997). Using cell-soluble Ht31 anchoring inhibitor peptides and expression of Ht31 plasmid, we demonstrated that the correct targeting of PKA is a determinant in the cAMP response to GLP-1 that induces insulin secretion from pancreatic islets and related cell lines. Lipofectamine was used as a delivery reagent to introduce Ht31, Ht31P control, or a PKA inhibitor peptide, PKI 5-24, into primary rat islets. Concomitant changes in glucose mediated insulin secretion and GLP-1 mediated insulin secretion were monitored by radioimmunoassay. The response of islets treated with any of the bioactive peptides to glucose was significantly different (Fig. 2A). Nontreated islets had similar glucose-mediated insulin secretion of 4.5 ng/well ± 0.57 (Fig. 2A), whereas islets treated with insulinotropic hormone GLP-1 that activates PKA exhibited a (3.5 ± 0.5) fold further increased insulin secretion (Fig. 2B) (Cullinan *et al.*, 1994; Gromada *et al.*, 1995a, 1995b; Yada *et al.*, 1993). Application of the PKA inhibitor peptide (PKI 5-24) blocked the GLP-1 effect (0.7 ± 0.15), confirming a role for the kinase in this process (Fig. 2B). Interestingly, the GLP-1 effect was blocked (0.9 ± 0.1 fold) in islets treated with the Ht31 anchoring inhibitor peptide, whereas the control peptide (Ht31P) had no effect (2.8 ± 0.3 fold increase; Fig. 2B). Similar results were obtained when

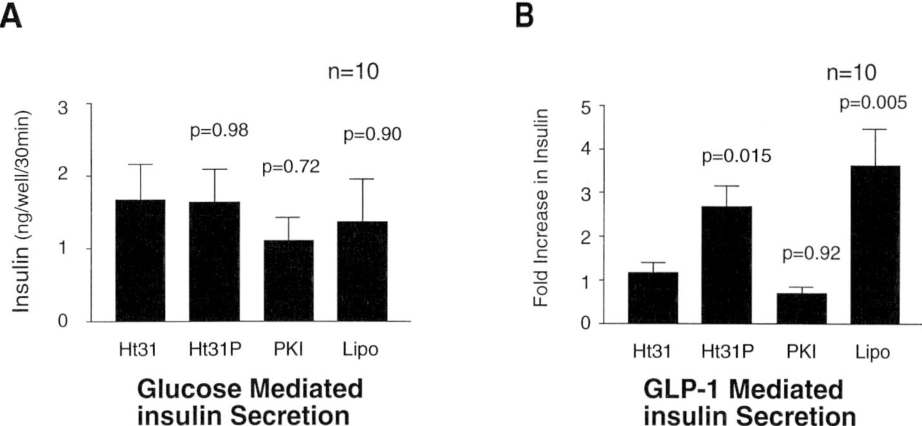

FIGURE 2   Disrupting PKA anchoring inhibits GLP-1 mediated insulin secretion in primary islets. Peptides were introduced into primary cultures of pancreatic islets to assess the role of PKA anchoring on insulin secretion. (A) Glucose-mediated insulin secretion in islets treated with peptides or a lipofectamine control (indicated below each bar) was measured in the culture media by radioimmunoassay. Data from 10 experiments are shown. (B) Changes in GLP-1 mediated insulin secretion were measured by similar methods and are presented as the fold increase over glucose mediated insulin secretion. Data from 10 experiments are indicated.

cell-soluble myristoylated derivatives of these peptides were used in the absence of lipofectamine. Additional controls showed that none of the peptides affected cAMP production in response to GLP-1 or inhibited PKA C subunit activity toward the heptapeptide substrate Kemptide. In sum, these results suggest that disruption of PKA-AKAP targeting attenuates GLP-1 stimulated insulin secretion in pancreatic beta islets.

In order to assess the role of PKA anchoring in regulating insulin secretion by an alternate method, a clonal rat beta cell line, RINm5F, was transfected with plasmids encoding a soluble Ht31 fragment (residues 418–718) or with a mutant form, Ht31P, which was unable to bind RII. Immunochemical analysis demonstrated the diffuse expression of the Ht31 and Ht31P proteins in the appropriate cells while only background staining was seen in the wild-type RINm5F cells (Fig. 3, top panels). The intracellular location of RII was concentrated at perinuclear regions in cells expressing the Ht31P or RINm5F cells (Fig. 3, center column). In contrast, RII was more evenly distributed throughout the cytoplasm in cells expressing the Ht31 fragment (Fig. 3, top-center panel). This redistribution of RII by expression of the anchoring inhibitor protein confirms the *in vitro* disruption of RII–AKAP interactions in islets and RINm5F cells and is consistent with the role of these compounds to alter the subcellular location of the type II PKA holoenzyme (Carr *et al.*, 1991, 1992a). Since RINm5F cells have reduced sensitivities to GLP-1 and glucose (Watanabe *et al.*, 1994), we used dibutyryl cAMP

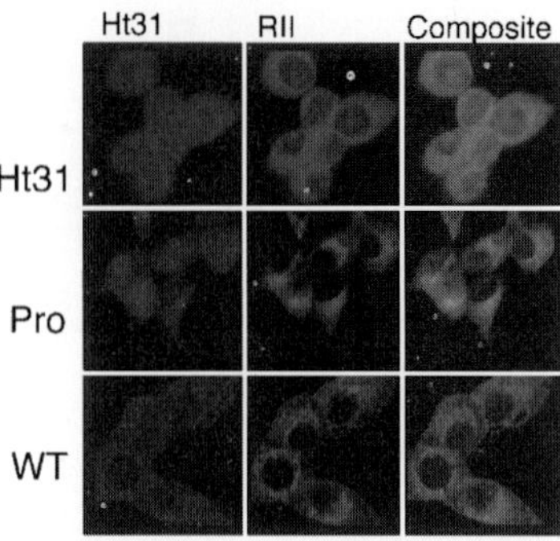

FIGURE 3   Recombinant anchoring inhibitor proteins displace PKA inside cells. The subcellular distribution of Ht31 and RII in RINm5F cells was detected immunochemically using a Leitz Fluovert confocal photomicroscope. The cell lines are indicated next to each row. Ht31 staining (left column), RII staining (center column), and composite images of both signals (right column) are presented.

as an index of GLP-1 action and membrane depolarization with potassium chloride (KCl) as an indicator of PKA independent insulin secretion. The response to 40 mM KCl was similar in all three groups (Fig. 4A), whereas cAMP-mediated insulin secretion was suppressed (1.2 ± 0.03)-fold in cells expressing the active Ht31 fragment (Fig. 4B). However, control cells expressing Ht31P or transfected with vector alone had normal cAMP-mediated insulin responses (Fig. 4B) of approximately (3.0 ± 0.8)-fold above basal.

In an effort to identify how PKA anchoring facilitates cAMP-mediated insulin secretion, we monitored the intracellular calcium response of RINm5F cells expressing Ht31, Ht31P, or vector alone. Increased intracellu-

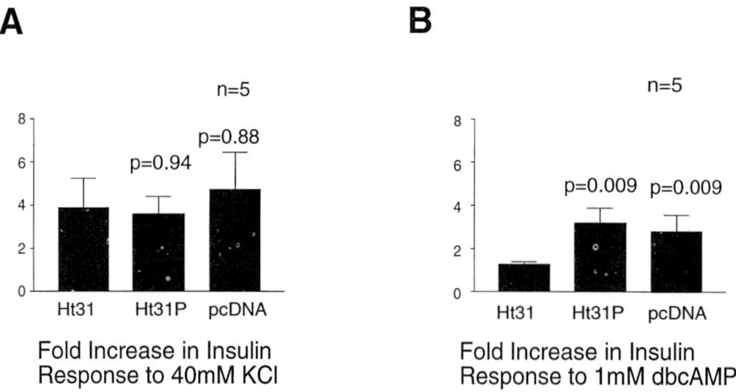

FIGURE 4   Anchoring inhibitor expression blocks cAMP-mediated insulin secretion in beta cells. (A) KCl-mediated insulin secretion and (B) cAMP-mediated insulin secretion by radioimmunoassay. Agonist-induced changes in insulin secretion are indicated as the fold increase over basal levels. Sample sources are indicated below each bar and *P* values for the mean $\pm$ SEM ($n = 5$).

lar calcium levels were detected in response to depolarizing amounts of KCl in all cells (Figs. 5A and 5B). However, the Ht31 expressing cells consistently failed to show an increase in intracellular calcium upon application of dibutyryl cAMP (Figure 5C&D). This suggests that anchored pools of PKA may facilitate cAMP-mediated phosphorylation events that modulate calcium fluxes from intracellular or extracellular sources. Data in pancreatic beta cells have suggested that the L-type Ca^{2+} channel is a site for PKA phosphorylation. This is supported by evidence that PKA anchoring augments Ca^{2+} channel modulation and by data showing that the beta cell L-type Ca^{2+} channel is phosphorylated by the kinase in RINm5F cells (Gray *et al.*, 1997; Johnson *et al.*, 1994, 1997). As will be discussed later in this

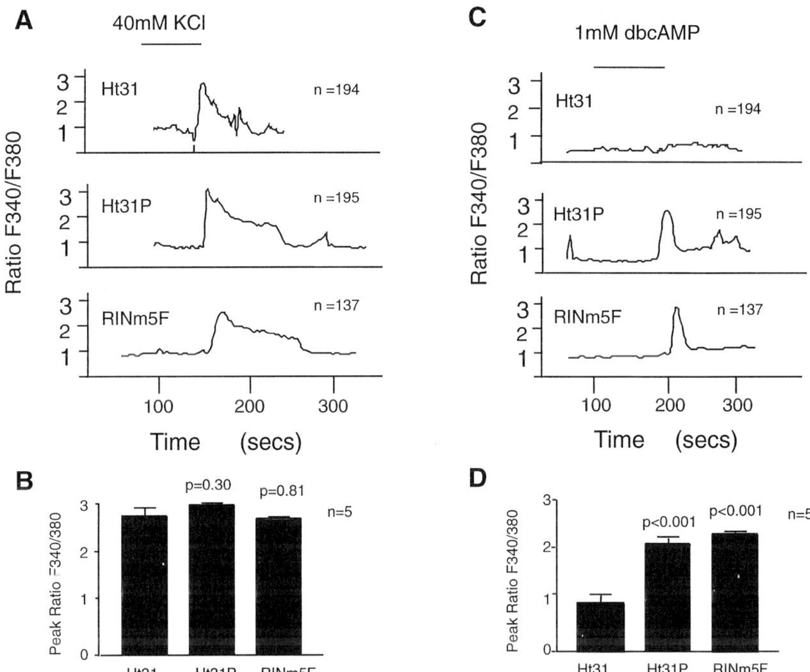

FIGURE 5 Disrupting PKA anchoring blocks rise in intracellular calcium in RINm5F cells. The effect of PKA anchoring on changes in intracellular calcium levels in RINm5F cells and cell lines expressing AKAP fragments was assessed by single-cell microfluorimetry on a Zeiss Axiophot scope. Changes in intracellular calcium (340/380 ratio) were detected by dual wavelength excitation at 340/380 nm with emission at 510 nm in response to (A) 40 mM KCl and (C) 1mM db-cAMP. Recordings were obtained every 4 s in a minimum of 15 cells per field. A representative response is shown for each experimental group and the total number of cells evaluated is indicated in each panel. The cell source is indicated for each sample group. The time of agonist infusion is indicated by the solid bar. The peak F340/380 ratio (from 5 individual experiments) is graphically depicted for the response to (B) 40 mM KCl and (D) 1 mM dbcAMP. Two-tailed *P* values comparing the mean of Ht31 to the other cell lines are indicated above the appropriate bars.

chapter, one site of PKA anchoring may be at or proximal to the L-type calcium channel.

Overall, the conclusions from this study demonstrate the utility of the anchoring inhibitor reagents as cell-based tools to disrupt the intracellular location of PKA by competing for R subunit AKAP interaction. However, despite demonstrating the usefulness of these peptides and providing compelling evidence that the proline derivatives of these compounds are ineffective in the disruption of PKA anchoring, we cannot rule out the possibility that the active anchoring inhibitor reagents are producing some secondary effects *in vivo*. Furthermore, these reagents only serve as vectors to globally disrupt PKA anchoring inside cells. As will become apparent in the latter sections of this chapter, specific subcellular targeting of PKA is achieved through understanding the unique targeting domains on the anchoring proteins and the heterologous expression of compartment-specific AKAPs.

E. AKAP Targeting Interactions

The PKA-anchoring model presented in Fig. 1 proposes that AKAPs should contain a unique targeting site that directs the association of PKA or the signaling complex to subcellular structures. In essence, the targeting domain is a fundamental feature of each AKAP, as it confers specificity by tethering the anchored PKA complex to particular organelles. In most cases, immunochemical and subcellular fractionation techniques have been used to identify AKAPs that are localized to centrosomes (AKAP350) (Keryer *et al.*, 1993), the actin cytoskeleton [Ezrin/AKAP78 (Dransfield *et al.*, 1997a), AKAP250 (Nauert *et al.*, 1997), and AKAP KL (Dong *et al.*, 1998)], the endoplasmic reticulum (AKAP100) (McCartney *et al.*, 1995), the Golgi (AKAP85) Rios *et al.*, 1992), microtubules (MAP2) (Theurkauf and Vallee, 1982), mitochondria (sAKAP84/D-AKAP-1) (Huang *et al.*, 1997; Lin *et al.*, 1995), the nuclear matrix (AKAP95) (Coghlan *et al.*, 1994), the plasma membrane (AKAP15/18 and AKAP79/150) (Bregman *et al.*, 1989; Carr *et al.*, 1992b), and vesicles (AKAP220) (Lester *et al.*, 1996). Less characterized anchoring proteins have been identified in secretory granules, plasma membranes, and the flagella of mammalian sperm (Faux and Scott, 1996a). The subcellular locations of these known AKAPs are indicated in Fig. 6.

More detailed analysis of targeting sequences have been performed on five AKAPs. AKAP220 may be targeted to peroxisomes (Lester *et al.*, 1996), as the last three residues of the protein, Cys-Arg-Leu, conform to a peroxisomal targeting signal 1 (PTS-1) motif that is thought to facilitate the attachment of proteins to the lipid matrix of the peroxisome (Subramani, 1996). AKAP250 is a component of the membrane/cytoskeleton that is enriched in the filopodia of adherent human erythroleukemia cells (Nauert *et al.*, 1997). The amino terminus of AKAP250 contains a consensus myristoylation signal, as well other structural regions that bear some resemblance

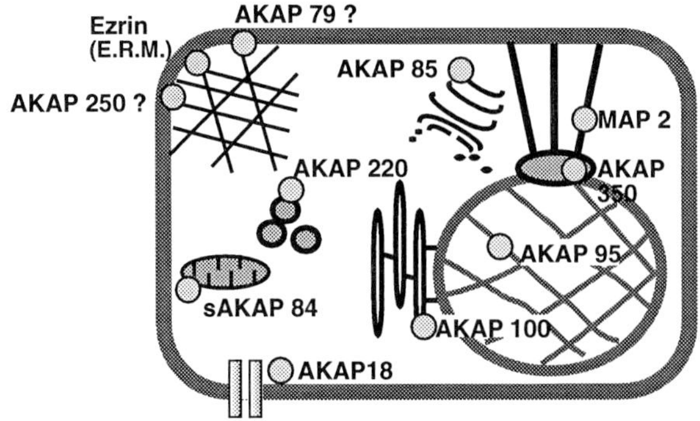

FIGURE 6 Subcellular targeting of AKAPs. A schematic diagram indicating the locations of known AKAPs in a prototypic cell. See text for details.

to actin-binding proteins such as MARCKS and GAP-43 (Aderem, 1992). Likewise, an anchoring protein previously called AKAP78 has now been identified as the cytoskeletal component ezrin (Dransfield *et al.*, 1997a). In fact, ezrin and its two close relatives, radaxin and moesin, bind RII in the overlay assay. All three proteins (E.R.M.) are members of the band 4.1 superfamily of proteins that link the membrane and the cytoskeleton. As will be discussed later, extensive mapping of targeting sequences has been performed on AKAP79 and AKAP18.

I. AKAP79 Targeting

Previous experiments with anchoring inhibitor peptides have demonstrated that AKAP-mediated targeting of PKA is important to regulate excitatory neurotransmitter receptors such as the AMPA and kainate-responsive glutamate receptors (Rosenmund *et al.*, 1994). This function may be fulfilled by AKAP79, a neuronal anchoring protein that has been detected immunochemically in the cell bodies and dendrites of cortical and hippocampal neurons and is enriched in postsynaptic density fractions (Carr *et al.*, 1992b; Glantz *et al.*, 1992; Klauck *et al.*, 1996). AKAP79 also binds the calcium calmodulin-dependent protein phosphatase-2B Calcineurin (CaN) and the calcium and phospholipid-activated protein kinase C (PKC) (Coghlan *et al.*, 1995; Klauck *et al.*, 1996). Hence, it has been proposed that AKAP79 directs the postsynaptic targeting of a multienzyme signaling complex that is involved in coordinating second-messenger-responsive phosphorylation of synaptic proteins (Faux and Scott, 1996b).

The mechanism of AKAP79 targeting to submembrane sites below the plasma membranes and in dendrites has been elucidated (Dell'Acqua *et al.*, 1998). A series of AKAP79 fragments fused to the green fluorescent protein (GFP) were used to define regions in AKAP79 that are sufficient for targeting

(Figs. 7 and 8). Initial experiments confirmed that expression of the full-length AKAP79-GFP fusion in HEK293 cells results in a pattern of membrane localization (Fig. 7C) that is indistinguishable from the localization of the untagged anchoring protein as detected by indirect immunofluorescent staining (Fig. 7B). In contrast, control cells transfected with GFP alone exhibited fluorescence throughout the cytoplasm and the nucleus (Fig. 7A). These results indicate that C-terminal GFP fusion does not adversely affect the subcellular targeting of AKAP79 and confirm the utility of the fluorescent tag approach to map targeting determinants in the anchoring protein.

The N-terminal third of AKAP79 contains three distinct regions of primary structure rich in basic residues: region A, residues 31–52; region B, residues 76–101; and region C, residues 116–145 (Fig. 9A). Previous studies indicated that sequences present in the N-terminal portion of

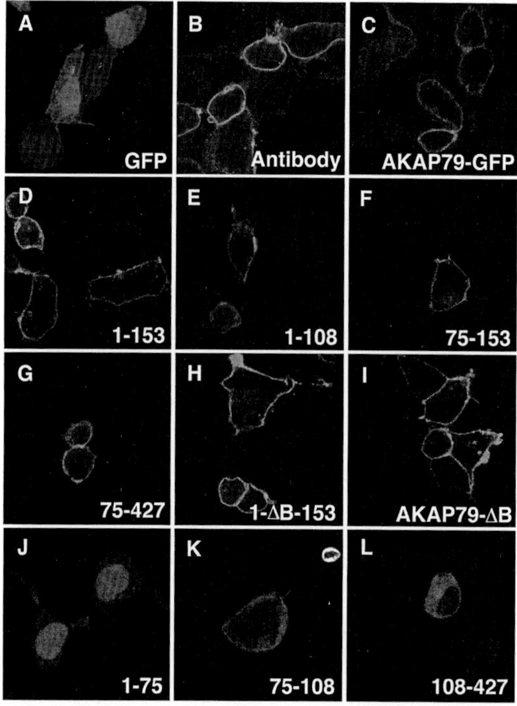

FIGURE 7   Membrane targeting of AKAP79 in HEK-293 cells is mediated by three N-terminal regions. AKAP79-GFP fusion proteins were transiently expressed in HEK-293 cells. Control images for transfections of GFP alone (A) or immunochemical staining (rabbit polyclonal anti-79, FITC) of untagged AKAP79 (B) are also shown. The subcellular localization of each transfected fusion protein (panels C-L) was determined by confocal imaging of GFP fluorescence excited at 490 nm. The images provided correspond to a single confocal plane for a representative field of transfected cells for each fusion construct. The length of each construct is indicated.

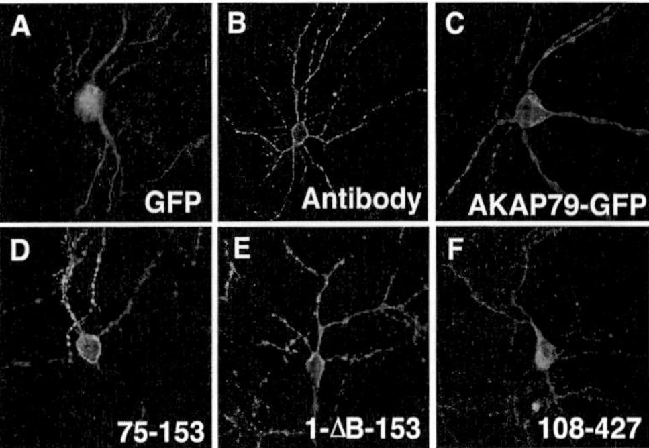

FIGURE 8   Neuronal targeting of AKAP79 is mediated by the N-terminal basic regions. Control neurons expressing GFP alone (A), expressing AKAP-79 GFP (C), or stained immunochemically with rabbit polyclonal anti-150 and FITC-secondary antibodies to visualize endogenous AKAP79/150 (B) are shown. The pattern of cellular GFP fluorescence was imaged in microinjected neurons expressing the 75-153 (D), 1-DB-153 (E), and 108-427 (F) GFP fusion proteins.

AKAP75, termed T1 and T2, which overlap with the A and B basic regions of AKAP79, were involved in targeting in HEK293 cells (Glantz *et al.*, 1993; Li *et al.*, 1996). However, it was previously shown that the A region contains determinants for binding to calmodulin and PKC while other regions interact with CaN (Coghlan *et al.*, 1995; Faux and Scott, 1997; Klauck *et al.*,

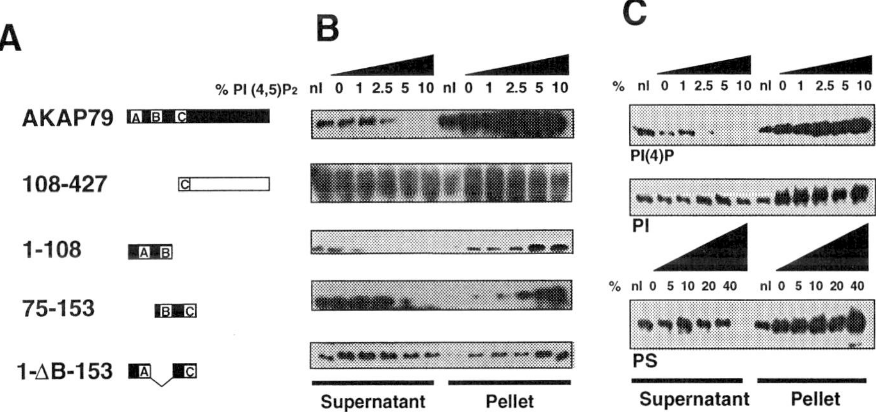

FIGURE 9   AKAP79 basic targeting regions bind to phosphatidylinositol 4,5-bisphosphate. (A) Schematic diagram showing the locations of the A, B, and C membrane targeting regions in AKAP79. (B) A lipid–protein cosedimentation assay was used to detect AKAP79 and fragments that bound to sucrose-loaded phospholipid vesicles. The resulting lipid-bound (pellet) and unbound (supernatant) fractions were detected by immunoblotting with an anti-79 polyclonal antibody. (C) AKAP79 and fragments were analyzed for binding to the vesicles containing the indicated percentages of additional acidic phospholipids: PtdIns(4)P (PI(4)P), PtdIns (PI), and phosphatidylserine (PS).

1996). In order to resolve this issue, AKAP79-GFP fusion proteins containing different combinations of the three basic regions were expressed in HEK293 cells. A fragment encompassing all three basic regions, 1–153, efficiently targeted to the cell periphery, indicating that the N-terminal basic regions alone were sufficient for membrane targeting (Fig. 7D). Likewise, additional constructs containing any combination of two basic regions targeted to the cell periphery (Figs. 7E–7I). At least two of the basic regions were necessary for targeting to the cell periphery, as constructs containing any single basic region failed to effectively target (Figs. 7J–7L). Likewise, microinjection of cDNAs encoding AKAP79-GFP fusion proteins into cortical neurons demonstrated that these basic regions are also responsible for targeting of AKAP79 in neurons (Fig. 8). Collectively, these results suggest that any two of the three AKAP79 basic regions are necessary and sufficient for submembrane targeting in HEK293 cells and cortical neurons.

Although binding determinants for PKC are contained in the A region and those for CaN may overlap with the C region, the results do not support a direct correlation between enzyme binding and membrane targeting (Fig. 9A). Therefore, we concluded that another mechanism must participate in the submembrane targeting of AKAP79. Secondary attachments to the actin cytoskeleton were ruled out, as treatment of HEK293 cells with the actin depolymerizing reagent cytochalasin D did not alter the membrane localization of AKAP79. However, targeting sequences of similar amino acid composition are found in the myristoylated alanine-rich C-kinase substrate protein (MARCKS), GAP43/neuromodulin, and neurogranin. These proteins bind acidic phospholipids such as phosphatidylinositol 4,5-bisphosphate (PtdIns(4,5)P$_2$) and calmodulin to regulate their association with the membranes and cytoskeleton (Aderem, 1992; Blackshear, 1993; Houbre *et al.*, 1991; Lu and Chen, 1997). Hence, the phosphoinositide binding activity of purified recombinant AKAP79 and fragments was assessed using a lipid–protein co-sedimentation assay (Mosior and Newton, 1995) (Fig. 9A). AKAP79 bound to sucrose loaded phospholipid vesicles at concentrations of PtdIns(4,5)P$_2$ as low as 1–2.5% or 1–2.5 μM, suggesting an affinity for PtdIns(4,5)P$_2$ in the low micromolar range (Fig. 9C). AKAP79 also bound PtdIns(4)P, while no binding was seen for PtdIns (PI) (Fig. 9B), thus suggesting some specificity for recognition of the 4-phosphate on the inositol ring. However, AKAP79 also bound the structurally unrelated acidic phospholipid phosphatidylserine (PS), albeit at much higher concentrations (20–40% or 20–40 μM) (Fig. 9C). Again, mapping experiments suggested that combinations of two or more basic regions are able to mediate AKAP79 binding to acidic phospholipids including PtdIns(4,5)P$_2$.

In an attempt to elucidate a mechanism for the regulation of AKAP79 targeting, we focused on a potential role for PKA and PKC phosphorylation. AKAP79 is a substrate for PKA and PKC *in vivo,* and most of the potential sites are contained within the first 153 residues of the protein. Although the

A region does not contain any consensus phosphorylation sites, the B and C regions contain several potential sites for PKA or PKC (Faux and Scott, 1997; Klauck *et al.*, 1996) (Fig. 9A). Interestingly, phosphorylation of recombinant AKAP79 fragments encompassing the B and C regions decreased binding to phospholipid vesicles and promoted the release of a fraction of the anchoring protein from cell membranes. Likewise, the presence of 10 μM calmodulin prevented binding of the native protein to phospholipid vesicles. This suggests that Ca^{2+} calmodulin binding as well as protein phosphorylation may regulate AKAP79 targeting.

One conclusion of these studies is that targeting domains of AKAP79 are functionally distinct from the binding sites for RII, PKC, and CaN (Carr *et al.*, 1992b; Coghlan *et al.*, 1995; Klauck *et al.*, 1996) (Fig. 10). This separation of targeting and enzyme binding functions is especially relevant for one of the proposed functions of AKAP79 in neurons, which is to mediate the postsynaptic localization of a kinase/phosphatase signaling complex (Klauck *et al.*, 1996). Localization of such a multienzyme signaling scaffold would allow coordinated regulation of second-messenger-dependent phosphorylation of synaptic substrates such as calcium channels and neurotransmitter receptors (Gao *et al.*, 1997; Rosenmund *et al.*, 1994). In this model the enzymes are bound to the AKAP79 in the inactive state and are then released in response to second messenger stimulation (Fig. 10). For example, active PKA catalytic subunit is released from the PKA holoenzyme by cAMP binding to the AKAP-anchored inhibitory R subunits, while PKC is most

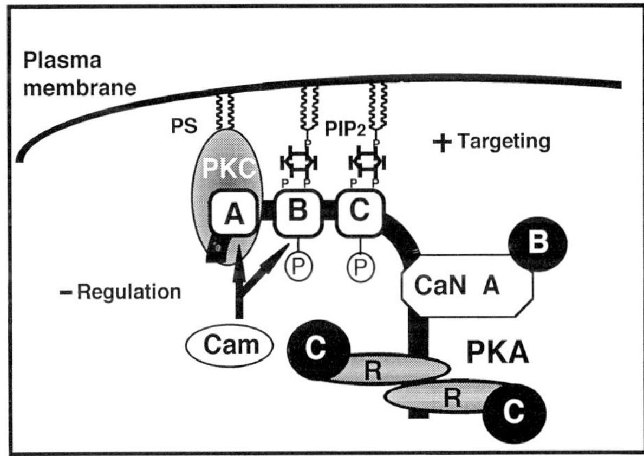

FIGURE 10   A model for regulated membrane targeting of the AKAP79 signaling scaffold. A schematic diagram that indicates the potential domain organization of the membrane targeted AKAP79 signaling complex. Potential positive (+) targeting interactions of the acidic phospholipids, phosphatidylserine (PS) and phosphatidylinositol 4,5-bisphosphate (PIP_2), with PKC and the AKAP79 basic regions (A,B,C) are shown. Possible negative (−) regulation of targeting by protein phosphorylation (PKC and PKA) and calcium signaling (calmodulin) are also depicted.

likely released from inhibition by the anchoring protein in response to intracellular calcium signaling via the competitive binding of calmodulin to AKAP79 (Faux and Scott, 1997).

For the foregoing anchoring model to function properly, it would be important that AKAP79 be targeted at postsynaptic membranes regardless of the docking of individual enzymes to the complex. However, it may be equally important to regulate the AKAP79 targeting interaction. Thus, negative regulatory mechanisms affecting at least two of the three basic domains would be predicted to inhibit membrane binding. This model of regulation (Fig. 10) is consistent with our observations that both protein phosphorylation and Ca^{2+}-calmodulin are able to negatively regulate $PtdIns(4,5)P_2$ binding for full-length AKAP79 and membrane attachment. These findings are reminiscent of the "electrostatic switch" mechanisms proposed for the reversible translocation of MARCKs protein between the membrane and cytosol in response to PKC phosphorylation or Ca^{2+}-calmodulin (McLaughlin and Aderem, 1995). Hence, regulation of AKAP79 location in response to prolonged second-messenger activation of PKA and PKC or elevation of intracellular calcium could also function as an adaptive negative feedback mechanism that would limit the strength or duration of kinase signaling by repositioning the entire AKAP79 scaffold relative to the locations of substrates or second-messenger generation. Alternatively, phosphorylation or Ca^{2+}-calmodulin mediated redistribution of AKAP79 could act as an amplification mechanism that would remove the inhibitory anchoring protein from close proximity to the released active enzymes, thus favoring maintenance of the activated state. It will be of great interest in the future to further characterize the cellular role of phosphoinositides, protein phosphorylation, and calmodulin in regulation of AKAP79 signaling.

2. *AKAP18 Targeting*

A second membrane associated anchoring protein has been identified that is targeted to the plasma membrane through a different mechanism. A small molecular weight anchoring protein called AKAP15/18 was independently cloned by two groups and has been shown to be functionally coupled to the L-type Ca^{2+} channels (Fraser *et al.*, 1998; Gray *et al.*, 1998). Inspection of the AKAP18 sequence identified three putative signals for lipid modification: a myristoylation site at the N-terminal glycine residue and two palmitoylation sites at cysteines 4 and 5 (Fig. 11A). It was postulated therefore that protein–lipid interactions may promote association of AKAP18 with the plasma membrane. Evidence that these residues undergo lipid modification was derived from subcellular fractionation of HEK293 cells transiently transfected with wild-type AKAP18. Cells fractionated in standard hypotonic buffer show that the heterologously expressed AKAP18 protein segregates exclusively with the particulate fraction (Fig. 11B). However, when cells were fractionated in the presence of increasing concentrations of Triton-

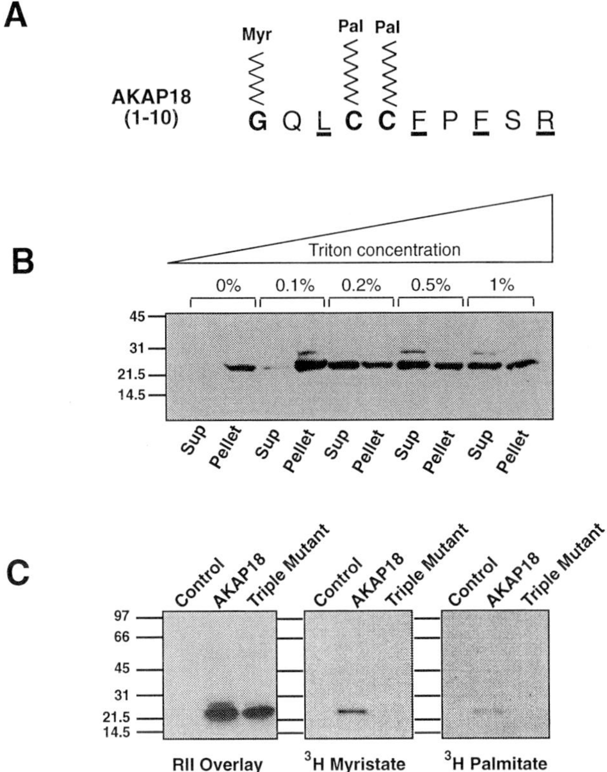

FIGURE 11   AKAP18 is subject to N-terminal lipid modification. (A) The first 10 residues of AKAP18 are shown with the schematic addition of a myristate group on glycine 1 and palmitate groups on cysteines 4 and 5. Hydrophobic residues are underlined. (B) Subcellular fractionation of AKAP18 transfected HEK293 cells in the presence of increasing concentrations of Triton X-100 (0–1%). Soluble (Sup) and particulate (Pellet) fractions at each Triton concentration were separated by SDS–PAGE. AKAP18 protein was detected by immunoblot using a polyclonal antibody raised against recombinant AKAP18. (C) AKAP18 incorporates ³H-myristate and ³H-palmitate in cell culture. HEK293 cells were transfected with wild-type AKAP18 or a mutant with residues Gly[1], Cys[4], and Cys[5] substituted with Ala, Ser, and Ser, respectively (Triple mutant) and labeled with ³H-myristate or ³H-palmitate. The presence of AKAP18 protein in immunoprecipitates was confirmed by RII overlay, and ³H incorporation was detected by fluorography.

X 100, 0.2% detergent was sufficient to relocalize a significant proportion of the AKAP18 from the particulate to the soluble fraction (Fig. 11B). Further studies were carried out to determine directly whether myristate and palmitate were incorporated into AKAP18 transiently expressed in culture. HEK293 cells were transfected with wild-type AKAP18 and a mutant in which residues Gly[1], Cys[4], and Cys[5] were substituted with Ala, Ser, and Ser, respectively (Triple mutant). The cells were incubated in media containing

³H-myristic acid or ³H-palmitic acid, then subjected to immunoprecipitation with anti-AKAP18 antibodies, and the incorporation of ³H in cell extracts was analyzed by SDS–PAGE and fluorography. Figure 11C shows incorporation of both ³H-myristate and ³H-palmitate by a protein migrating at the correct molecular weight for the wild-type AKAP18 expression construct, whereas there was no significant incorporation of label by the AKAP18 triple mutant. An RII overlay of the same protein samples confirmed the presence of an RII binding protein in both extracts, demonstrating effective expression of both AKAP18 constructs (Fig. 11C). Further mutants were generated to test the hypothesis that residues at the N-terminus of AKAP18 are responsible for membrane targeting. The N-terminal glycine residue was substituted by an alanine (G1A), and cysteines 4 and 5 were changed to serine (C4,5S) (Fig. 12A). Together with the triple mutant described earlier, localization of these AKAP18 mutants was first analyzed by transient expression in HEK293 cells followed by subcellular fractionation. As noted previously, the wild-type AKAP18 protein partitioned exclusively to the particulate fraction (Fig. 12B). Removal of the myristoylation signal alone appeared to have little effect on localization, as the G1A mutant remained exclusively in the particulate fraction (Fig. 12B). In contrast, removal of both palmitoylation signals caused a shift of approximately 50% of the C4,5S mutant to the cytosol, while mutation of all three residues (Triple mutant) caused a complete shift of anchoring protein from the particulate to the soluble fraction (Fig. 12B). These findings supported the notion that lipid modification is involved in the localization of AKAP18 to the plasma membrane.

In order to analyze the subcellular localization of wild-type and mutant AKAP18 proteins inside cells, plasmids were constructed to heterologously express the proteins with a C-terminal GFP fluorescent tag. HEK293 cells transfected with GFP alone exhibited fluorescence throughout the cell, whereas expression of the AKAP18/GFP fusion clearly shows a peripheral staining pattern (Fig. 12C). Control experiments demonstrated that the same peripheral localization was observed in cells transfected with an expression construct encoding the wild-type AKAP18 without a GFP tag (Fig. 12C). In the cell shown, AKAP18 was detected by immunochemical staining with a polyclonal antibody raised against the recombinant protein. These results confirm that AKAP18 is targeted to the cell membrane and that C-terminal fusion of the GFP moiety does not affect the membrane association of the full-length anchoring protein. In contrast, gradual delocalization of AKAP18 was apparent as one (G1A), two (C4,5S), and then three (Triple mutant) lipid modification signals were removed (Fig. 12C).

Our studies show that myristoylation of the N-terminal glycine residue and palmitoylation of cysteines 4 and 5 in AKAP18 are involved in attaching the anchoring protein to the cytoplasmic face of the plasma membrane. Posttranslational modification of proteins by acyl groups is now well established as a mechanism for membrane association of signaling proteins

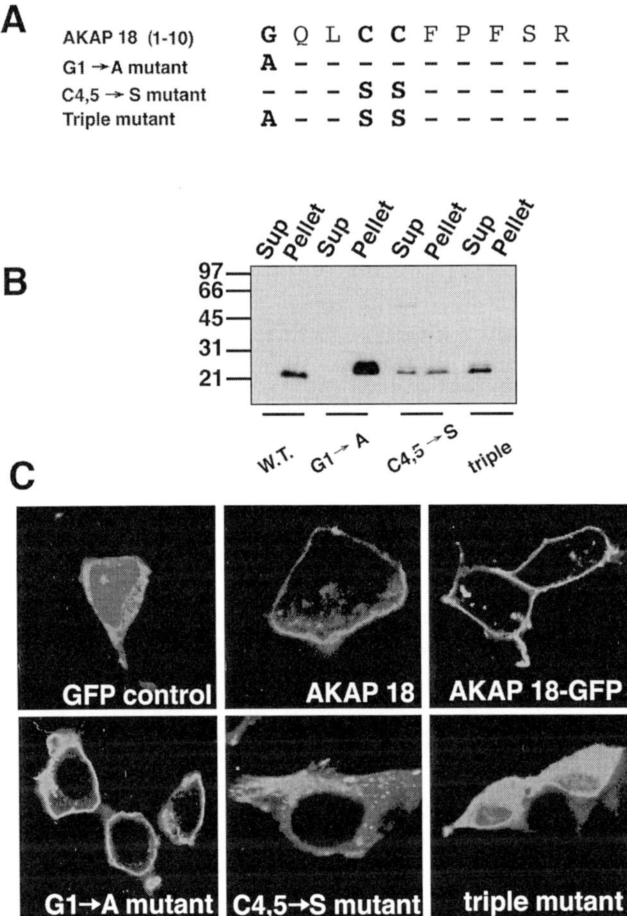

FIGURE 12 Mutations in the first five N-terminal amino acids of AKAP18 disrupt targeting function. (A) Residues 1–10 of AKAP18 showing amino acid changes incorporated for the N-terminal mutants. (B) Subcellular fractionation of HEK293 cells transfected with AKAP18 and the G1A, C4,5S, and triple mutants in the pcDNA3.1/Myc.His expression vector. Soluble (sup) and particulate (pellet) fractions from each construct were separated by SDS–PAGE. AKAP18 proteins were detected by immunoblot using a monoclonal antibody to the c-Myc epitope tag. (C) Confocal microscopy of HEK293 cells transfected with a GFP control plasmid, wild-type AKAP18, and GFP fusions of AKAP18 and the G1A, C4,5S, and triple mutants. The subcellular localization of wild-type AKAP18 was detected by immunochemical staining, whereas the GFP fusions were visualized directly by fluorescent excitation at 490 nm.

(Schlesinger, 1993). One common theme seems to be the presence of multiple sites of lipid modification on the acceptor protein. For example, the Src family of tyrosine kinases contain an N-terminal myristoyl group and one or two palmitoyl groups attached to cysteine residues that contribute to membrane targeting (Kabouridis *et al.*, 1997; Resh, 1994). Likewise, several of the a subunits of heterotrimeric G-proteins are subject to dual acylation by myristate and palmitate (Milligan *et al.*, 1995). Our studies with point mutations of AKAP18 suggest that any two of the three lipid side chains are sufficient to mediate membrane association. The inability of a single lipid moiety to sustain membrane targeting of AKAP18 is in agreement with reports that one acyl group is insufficient to mediate stable attachment of a protein to a lipid bilayer (Resh, 1994). In fact, membrane-associated proteins that are singly acylated are only able to mediate their attachment when another interaction accompanies the lipid modification. For example, the myristoylated alanine-rich C-kinase substrate protein (MARCKS) is membrane targeted through an N-terminal myristate group and a polybasic region that binds acidic phospholipids (Aderem, 1992; Blackshear, 1993), whereas a recently identified Grb2/Sos binding protein, FRS-2, is myristoylated and has been proposed to bind to the FGF receptor through a PTB domain (Kouhara *et al.*, 1997).

3. AKAP18 Is Coupled Functionally to the L-Type Ca^{2+} Channel

It was postulated that membrane targeting of AKAP18 could mediate the localization of PKA in close proximity to transmembrane substrates. Previous studies have suggested that pools of PKA are localized close to skeletal muscle L-type Ca^{2+} channels in order to facilitate rapid and efficient channel phosphorylation (Salvatori *et al.*, 1990). More recent reports have shown that AKAP targeting of the kinase contributes to this process (Burton *et al.*, 1997; Gao *et al.*, 1997; Johnson *et al.*, 1994, 1997). In keeping with this hypothesis, it has been proposed that a low molecular weight AKAP serves to maintain a pool of PKA close to the L-type Ca^{2+} channel (Gray *et al.*, 1997). Given our biochemical evidence that AKAP18 is targeted through protein–lipid interactions to the plasma membrane, we postulated that this anchoring protein may be a physiological partner of the L-type Ca^{2+} channel. A recently established model to test this hypothesis is the reconstitution of PKA modulation of L-type Ca^{2+} channels in HEK293 cells (Gao *et al.*, 1997; Johnson *et al.*, 1997).

Accordingly, whole-cell Ca^{2+} currents were recorded from HEK293 cells transfected with the cardiac α and β Ca^{2+} channel subunits. Okadaic acid (1 μM) was present in the external bath solution and in the pipette solution in order to prevent attenuation of Ca^{2+} current response to cAMP by endogenous phosphatase activity (Gao *et al.*, 1997). Using barium (10 mM) as charge carrier, currents were evoked by depolarization from a holding potential of -80 mV. Whole-cell barium currents activated from -30 mV and peaked at $+10$ to $+20$ mV. Bath application of the cell-permeant cAMP analogue 8-CPTcAMP (1 mM) significantly increased the barium current of cells co-transfected with AKAP18 compared to controls (18.4 $\pm$ 6.5%; $n = 17$ vs 1.1 $\pm$ 2.4%; $n = 12$, $p < 0.05$) (Figs. 13A and 13C). Current augmentation greater than 10% was observed in 9 of 17 cells co-transfected

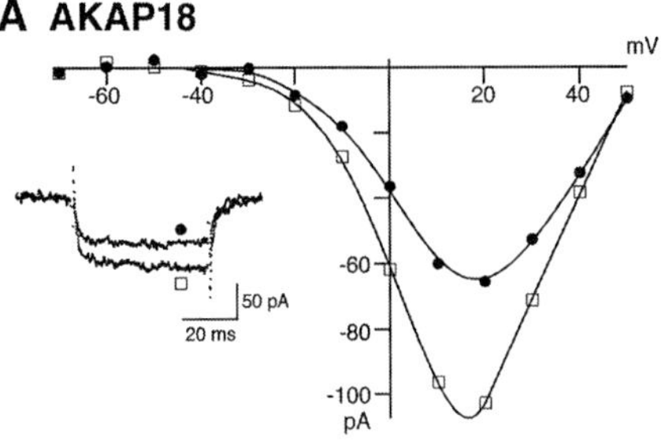

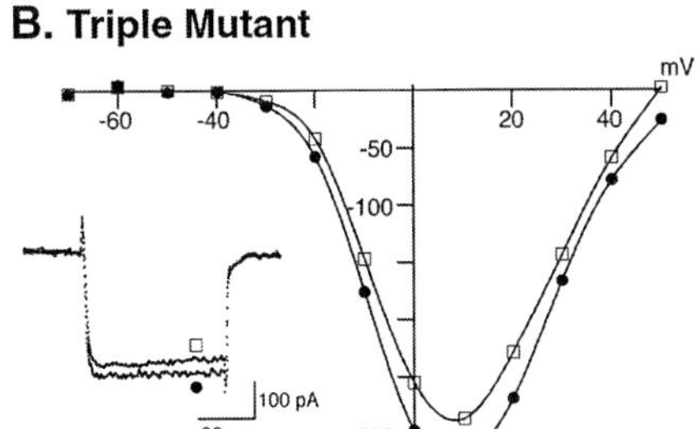

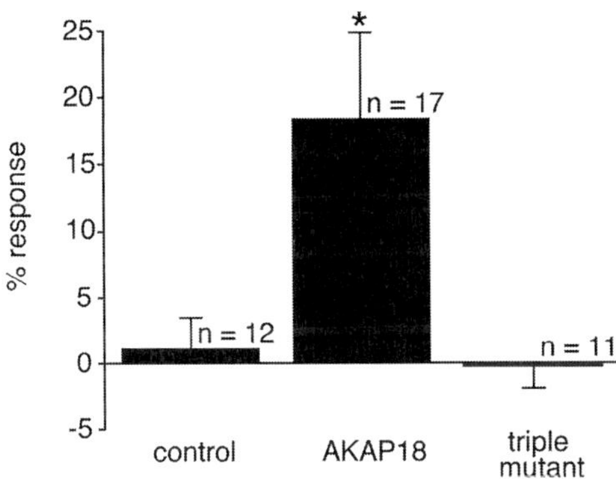

FIGURE 13 Membrane targeting of AKAP18 is required for cAMP-dependent modulation of L-type Ca^{2+} channel currents. (A) Whole-cell current with 10 mM barium as charge carrier was evoked by depolarization from a holding potential of -80 mV. Current–voltage relationship is shown for HEK293 cells transfected with the a_{1c} and b_{2a} cardiac Ca^{2+} channel subunits and wild-type AKAP18. Currents recorded following 2- to 3-min bath application of the cAMP analogue 8-CPTcAMP (1 mM) (open squares) were augmented compared to those before treatment (filled circles). Currents were monitored during drug application by test pulses to 0 mV. Inset shows traces evoked by a voltage step to $+10$ mV. (B) Modulation of the L-type Ca^{2+} channel was not seen following coexpression of a_{1c} and b_{2a} Ca^{2+} channel subunits with the untargeted AKAP18 triple mutant. Current–voltage relation is shown in the absence (filled circles) and presence (open squares) of 8-CPTcAMP (1 mM). Inset shows currents evoked by a voltage step to 0 mV from a holding potential of -80 mV. (C) Percent response to 8-CPTcAMP for each condition tested. Percent response was calculated as [(8-CPTcAMP– control)/control] $\times$ 100%. Response was measured from currents evoked by a voltage step to 0 mV from a holding potential of -80 mV. Barium current through L-type Ca^{2+} channels was significantly ($*p < 0.05$) augmented by 8-CPTcAMP in AKAP18 transfected cells, while cells transfected with the untargeted triple mutant were not significantly different from controls.

with AKAP18 (34.4 $\pm$ 9.3%; $n = 9$), while only 1 of 12 control cells displayed an augmentation. Cells transfected with the triple mutant were not significantly different from controls ($-0.3 \pm 1.6\%$; $n = 11$), with 0 of 11 responding positively to 8-CPTcAMP (Figs. 13B and 13C). These results suggest that AKAP18 contributes to the cAMP-dependent augmentation of L-type Ca^{2+} current and that the membrane targeting domain of AKAP18 appears to be required this for modulation.

4. AKAP18 Alters GLP-1 Mediated Insulin Secretion

As previously discussed, it has been shown that subcellular targeting of PKA by AKAPs is required for efficient hormone-mediated insulin secretion in pancreatic beta cells (Lester *et al.*, 1997). Moreover, it has been shown that a significant site of PKA anchoring could be at or proximal to the L-type Ca^{2+} channel, which has previously been implicated a key mediator of the insulin secretion pathway (Bokvist *et al.*, 1995; Gromada *et al.*, 1997; Safayhi *et al.*, 1997; Suga, 1997). Therefore, the effect of AKAP18 on the process of hormone-mediated insulin secretion was measured. A clonal insulin-secreting rat beta cell line, RINm5F, was transfected with plasmids encoding wild-type AKAP18 and the untargeted triple mutant. As expected, immunocytochemical analysis showed that AKAP18 was concentrated at the periphery of the RINm5F cells, whereas the untargeted triple mutant exhibited a more uniform cytoplasmic staining pattern (Fig. 14A). Importantly, co-staining with RII showed that wild-type AKAP18 was able to mediate a redistribution of PKA to the plasma membrane (Fig. 14A). In untransfected cells lacking exogenous AKAP18 staining, RII exhibits a perinuclear staining pattern (Fig. 14A). As no significant AKAP18 staining is observed in these cells, it would appear that RINm5F cells do not contain endogenous AKAP18. The staining pattern for RII was more diffuse in RINm5F cells expressing the untargeted AKAP18 triple mutant (Fig. 14A), in keeping with analogous experiments where PKA anchoring was disrupted using a cytoplasmic RII binding protein (Lester *et al.*, 1997).

To determine if membrane targeting of PKA by AKAP18 influences hormone-mediated signaling events, insulin secretion was measured upon application of the insulinotropic hormone glucagon-like peptide 1 (GLP-1). Insulin secretion from RINm5F cell lines expressing AKAP18 or the untargeted triple mutant was assessed by radioimmunoassay (Drucker *et al.*, 1987; Yaekura *et al.*, 1996). The increase in insulin secretion over basal levels in response to GLP-1 was significantly higher in cells expressing AKAP18 (34.7 ± 8 pmol/min/10^6 cells, $n = 9$) than in pcDNA transfected controls (16.3 ± 2.6 pmol/min/10^6 cells, $n = 6$) (Fig. 14B). Furthermore, expression of the untargeted triple mutant of AKAP18 resulted in a markedly lower level of GLP-1 stimulated secretion relative to controls (10.5 ± 0.5 pmol/min/10^6 cells, $n = 9$) (Fig. 14B). These results suggest that membrane targeting of PKA through its interactions with AKAP18 can facilitate

A

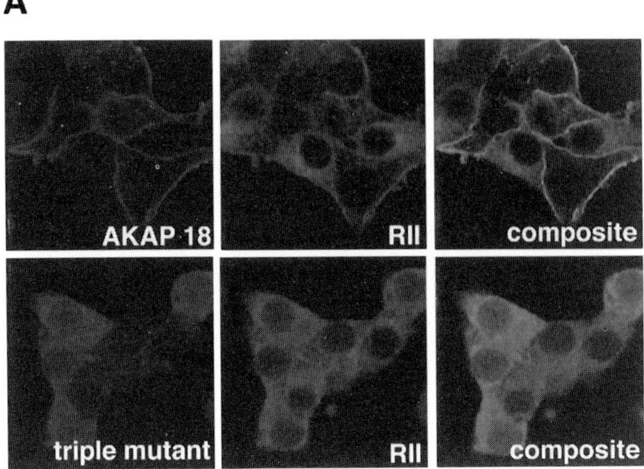

B

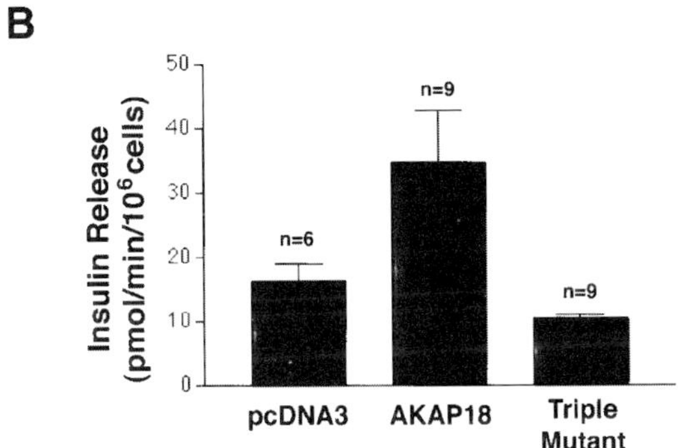

FIGURE 14   AKAP18 mediates a redistribution of RII and influences GLP-1 mediated insulin secretion in RINm5F cells. (A) Confocal microscopy of RINm5F cells transfected with constructs encoding wild-type AKAP18 and the untargeted triple mutant. Immunocytochemical analysis was carried out using polyclonal antibodies raised against recombinant AKAP18 and rat RIIα in rabbit and goat, respectively. Texas Red–conjugated anti-rabbit and FITC-conjugated anti-goat secondary antibodies were used to permit the visualization of both proteins. In the panels showing immunochemical staining of AKAP18 and the triple mutant alone, several untransfected cells are evident because of the nonclonal nature of the AKAP18 transfected RINm5F cell lines. (B) RINm5F cells transfected with wild-type AKAP18, the triple mutant, and a pcDNA3 control plasmid were monitored for GLP-1 mediated insulin secretion. Secretion levels are represented as the increase in pmol insulin released/minute/10⁶ cells relative to basal levels.

GLP-1 mediated insulin secretion. This is consistent with the hypothesis that PKA-mediated membrane events, such as phosphorylation of L-type Ca^{2+} channels, are required for hormone-mediated insulin secretion (Bokvist *et al.*, 1995; Gromada *et al.*, 1997; Safayhi *et al.*, 1997; Suga, 1997).

Collectively, these studies suggest that bringing PKA to submembrane sites via association with AKAP18 permits the modulation of L-type Ca^{2+} channels and augments physiological processes that require Ca^{2+} influx, such as hormone-mediated insulin secretion in pancreatic beta cells. However, it must be stressed that a colocalization of PKA with the channel is likely to represent only one of several anchored PKA pools that participate in this process, as multiple AKAPs have been identified in beta cells (L. B. Lester, unpublished observation). PKA substrates involved in the insulin secretion pathway, such as the Glut-2 glucose transporter and the GLP-1 receptor, may have their own pools of anchored PKA (Gromada *et al.*, 1997; Thorens *et al.*, 1996; Widmann *et al.*, 1995, 1996). Nevertheless, these studies provide compelling evidence to support a targeting hypothesis suggesting that second-messenger-mediated signaling events are controlled not only by the catalytic activities of kinases and phosphatases, but also by where these enzymes are localized within the cell (Faux and Scott, 1996b; Hubbard and Cohen, 1993).

IV. Conclusions and Perspectives

In conclusion, we focus on some emerging themes in AKAP research that are likely to yield valuable new information about this family of signaling proteins. Over the past 2 to 3 years there has been a dramatic increase in the cloning of novel AKAPs. This information should provide a more complete database of sequences and establish if the AKAPs merely represent a convergent group of proteins sharing a common RII-binding motif or whether gene families of anchoring proteins exist. Although most of the current data support the first view, it is worth noting that sAKAP84 is expressed in several forms because of alternative splicing, and a novel family of six related anchoring proteins called the AKAP KL family has been reported (Dong *et al.*, 1998). At this time it appears as if some of the sAKAP84 splice variants are differentially targeted to the endoplasmic reticulum and the mitochondria (Chen *et al.*, 1997). Furthermore, one splice variant, called dAKAP1, was isolated in a two-hybrid screen using an RIα fragment as bait (Huang *et al.*, 1997b). Likewise, independent studies have shown that AKAP79 and Ht31 bind RIα, albeit with a 100-fold lower affinity than RIIα (Burton *et al.*, 1997). This raises the intriguing possibility that certain AKAPs are dual-function anchoring proteins which bind both RI and RII. Not only do these findings provide a molecular mechanism for the compartmentalization of RI (Rubin *et al.*, 1972; Skalhegg *et al.*, 1994), but they significantly expand the original anchoring hypothesis to include all PKA holoenzymes. Undoubtedly, future research will discern the nature of the RI–AKAP interactions and establish the physiological significance of type I PKA anchoring.

Another future perspective in our understanding of the AKAPs is likely to come from structural analysis of AKAP–PKA complexes. The solution structure of an RII fragment that encompasses the AKAP-binding site has been solved (Newlon *et al.*, 1997). Modeling studies with the Ht31 peptide suggest that the RII dimer makes multiple contacts with a hydrophobic face of the AKAP peptide. This type of information may explain why the anchoring inhibitor peptides bind RII with such high affinity and are such potent antagonists of PKA anchoring inside cells. No doubt additional studies will focus on structural elucidation of the entire anchored PKA holoenzyme complex.

Another area of future emphasis will be the characterization of AKAP targeting domains. AKAPs such as ezrin, radaxin, moesin, and gravin appear to be anchoring proteins that are linked to both the plasma membrane and the actin cytoskeleton. This suggests that highly localized PKA phosphorylation events may regulate specific signaling events in very defined microenvironments. For example, both AKAP79 and AKAP18 are able to target PKA to the plasma membrane. But are they capable of maintaining distinct localized pools of the kinase in the same cell? These types of macromolecular organization mediated by anchoring proteins not only would place PKA close to certain substrates, but also would cluster the kinase at sites where it could optimally respond to the generation of specific second messengers. The challenge now facing researchers is to pinpoint which important cellular phosphorylation events are regulated by the anchored kinases and phosphatases.

Acknowledgments

We thank R. Frank for help in all aspects of manuscript preparation. J.D.S. is supported by NIH grant DK44239, I.D.C.F. is supported by Wellcome Trust grant 049076, S.J.T. is supported by NIH grant NS10202, and L.L. is supported by NIH grant DK02353.

References

Adams, S. R., Harootunian, A. T., Buechler, Y. J., Taylor, S. S., and Tsien, R. Y. (1991). Fluorescence ratio imaging of cyclic AMP in single cells. *Nature* **349**, 694–697.

Adams, S. R., Bacskai, B. J., and Taylor, S. S. (1992). Optical probes for cyclic AMP. *In* "Fluorescent Probes for Biological Activity of Living Cells—a Practical Guide" (W. T. Mason and G. Relf, eds.), pp. 36–72. Academic Press, New York.

Aderem, A. (1992). The MARCKS brothers: a family of protein kinase C substrates. *Cell* **71**, 713–716.

Amieux, P. S., Cummings, D. E., Motamed, K., Brandon, E. P., Wailes, L. A., Le, K., Idzerda, R. L., and McKnight, G. S. (1997). Compensatory regulation of RIa protein levels in protein kinase A mutant mice. *J. Biol. Chem.* **272**, 3993–3998.

Ammala, C., Eliasson, L., Bokvist, K., Berggren, P.-O., Honkanen, R. E., Sjoholm, A., and Rorsman, P. (1994). Activation of protein kinases and inhibition of protein phosphatases play a central role in the regulation of exocytosis in mouse pancreatic b cells. *Proc. Natl. Acad. Sci. USA* **91**, 4343–4347.

Bacskai, B. J., Hochner, B., Mahaut-Smith, M., Adams, S. R., Kaang, B.-K., Kandel, E. R., and Tsien, R. Y. (1993). Spatially resolved dynamics of cAMP and protein kinase A subunits in *aplysia* sensory neurons. *Science* **260**, 222–226.

Barsony, J., and Marks, S. J. (1990). Immunocytology on microwave-fixed cells reveals rapid and agonist-specific changes in subcellular accumulation patterns for cAMP or cGMP. *Proc. Natl. Acad. Sci. USA* **87**, 1188–1192.

Beavo, J. A., Conti, M., and Heaslip, R. J. (1994). Multiple cyclic nucleotide phosphodiesterases. *Mol. Pharmacol.* **46**, 399–405.

Blackshear, P. J. (1993). The MARCKS family of cellular protein kinase C substrates. *J. Biol. Chem.* **268**, 1501–1504.

Bokvist, K., Eliasson, L., Ammala, C., Renstrom, E., and Rorsman, P. (1995). Co-localization of L-type Ca^{2+} channels and insulin-containing secretory granules and its significance for the initiation of exocytosis in mouse pancreatic B-cells. *EMBO J.* **14**, 50–57.

Bregman, D. B., Bhattacharya, N., and Rubin, C. S. (1989). High affinity binding protein for the regulatory subunit of cAMP-dependent protein kinase II-b. *J. Biol. Chem.* **264**, 4648–4656.

Burton, K. A., Johnson, B. D., Hausken, Z. E., Westenbroek, R. E., Idzerda, R. L., Scheuer, T., Scott, J. D., Catterall, W. A., and McKnight, G. S. (1997). Type II regulatory subunits are not required for the anchoring-dependent modulation of Ca^{2+} channel activity by cAMP-dependent protein kinase. *Proc. Natl. Acad. Sci. USA* **94**, 11067–11072.

Carr, D. W., and Scott, J. D. (1992). Blotting and band-shifting: techniques for studying protein–protein interactions. *TIBS* **17**, 246–249.

Carr, D. W., Stofko-Hahn, R. E., Fraser, I. D. C., Bishop, S. M., Acott, T. S., Brennan, R. G., and Scott, J. D. (1991). Interaction of the regulatory subunit (RII) of cAMP-dependent protein kinase with RII-anchoring proteins occurs through an amphipathic helix binding motif. *J. Biol. Chem.* **266**, 14188–14192.

Carr, D. W., Hausken, Z. E., Fraser, I. D. C., Stofko-Hahn, R. E., and Scott, J. D. (1992a). Association of the type II cAMP-dependent protein kinase with a human thyroid RII-anchoring protein, cloning and characterization of the RII-binding domain. *J. Biol. Chem.* **267**, 13376–13382.

Carr, D. W., Stofko-Hahn, R. E., Fraser, I. D. C., Cone, R. D., and Scott, J. D. (1992b). Localization of the cAMP-dependent protein kinase to the postsynaptic densities by A-kinase anchoring proteins: characterization of AKAP79. *J. Biol. Chem.* **24**, 16816–16823.

Carr, D. W., DeManno, D. A., Atwood, A., Hunzicker-Dunn, M., and Scott, J. D. (1993). Follicle-stimulating hormone regulation of A-kinase anchoring proteins in granulosa cells. *J. Biol. Chem.* **268**, 20729–20732.

Chen, Q., Lin, R.-Y., and Rubin, C. S. (1997). Organelle-specific targeting of protein kinase AII (PKAII). *J. Biol. Chem.* **272**, 15247–15257.

Coghlan, V. M., Langeberg, L. K., Fernandez, A., Lamb, N. J. C., and Scott, J. D. (1994). Cloning and characterization of AKAP95, a nuclear protein that associates with the regulatory subunit of type II cAMP-dependent protein kinase. *J. Biol. Chem.* **269**, 7658–7665.

Coghlan, V. M., Perrino, B. A., Howard, M., Langeberg, L. K., Hicks, J. B., Gallatin, W. M., and Scott, J. D. (1995). Association of protein kinase A and protein phosphatase 2B with a common anchoring protein. *Science* **267**, 108–112.

Cooper, D. M. F., Mons, N., and Karpen, J. M. (1995). Adenylyl cyclases and the interaction between calcium and cAMP signalling. *Nature* **374**, 421–424.

Corbin, J. D., and Keely, S. L. (1977). Characterization and regulation of heart adenosine 3′:5′-monophosphate–dependent protein kinase isozymes. *J. Biol. Chem.* **252**, 910–918.

Corbin, J. D., Soderling, T. R., and Park, C. R. (1973). Regulation of adenosine 3′,5′-monophosphate–dependent protein kinase. *J. Biol. Chem.* **248**, 1813–1821.

Corbin, J. D., Keely, S. L., and Park, C. R. (1975). The distribution and dissociation of cyclic adenosine 3′:5′-monophosphate–dependent protein kinase in adipose, cardiac, and other tissues. *J. Biol. Chem.* **250**, 218–225.

Cullinan, C. A., Brady, E. J., Saperstein, R., Leibowitz, M. D. (1994). Glucose-dependent alterations of intracellular free calcium by glucagon-like peptide-1 (7-36amide) in individual ob/ob mouse B-cells. *Cell Calcium* **15**, 391–400.

Dell'Acqua, M. L., and Scott, J. D. (1997). Protein kinase A anchoring. *J. Biol. Chem.* **272**, 12881–12884.

Dell'Acqua, M. L., Faux, M. C., Thorburn, J., Thorburn, A., and Scott, J. D. (1998). Membrane targeting sequences on AKAP79 binds phosphatidylinositol 4,5-bisphosphate. *EMBO J.* **17**, 2246–2260.

Dong, F., Feldmesser, M., Casadevall, A., and Rubin, C. S. (1998). Molecular characterization of a cDNA that encodes six isoforms of a novel murine A kinase anchor protein. *J. Biol. Chem.* **273**, 6533–6541.

Dransfield, D. T., Bradford, A. J., Smith, J., Martin, M., Roy, C., Mangeat, P. H., and Goldenring, J. R. (1997a). Ezrin is a cyclic AMP-dependent protein kinase anchoring protein. *EMBO J.* **16**, 101–109.

Dransfield, D. T., Yeh, J. L., Bradford, A. J., and Goldenring, J. R. (1997b). Identification and characterization of a novel A-kinase-anchoring-protein (AKAP120) from rabbit gastric parietal cells. *Biochem. J.* **322**, 801–808.

Drucker, D. J., Philippe, J., Mojsov, S., Chick, W. L., and Habener, J. F. (1987). Glucagon-like peptide I stimulates insulin gene expression and increases cyclic AMP levels in a rat islet cell line. *Proc. Natl. Acad. Sci. USA* **84**, 3434–3438.

Fantozzi, D. A., Taylor, S. S., Howard, P. W., Maurer, R. A., Feramisco, J. R., and Meinkoth, J. L. (1992). Effect of the thermostable protein kinase inhibitory on intracellular localization of the catalytic subunit of cAMP-dependent protein kinase. *J. Biol. Chem.* **267**, 16824–16828.

Faux, M. C., and Scott, J. D. (1996a). More on target with protein phosphorylation: conferring specificity by location. *TIBS* **21**, 312–315.

Faux, M. C., and Scott, J. D. (1996b). Molecular glue: kinase anchoring and scaffold proteins. *Cell* **70**, 8–12.

Faux, M. C., and Scott, J. D. (1997). Regulation of the AKAP79-protein kinase C interaction by Ca^{2+}/calmodulin. *J. Biol. Chem.* **272**, 17038–17044.

Fraser, I. D. C., Tavalin, S. J., Lester, L. B., Langeberg, L. K., Westphal, A. M., Dean, R. A., Marrion, N. V., and Scott, J. D. (1998). A novel lipid-anchored A-kinase anchoring protein facilitates cAMP-responsive membrane events. *EMBO J.* **17**, 2261–2272.

Gamm, D. M., Baude, E. J., and Uhler, M. D. (1996). The major catalytic subunit isoforms of cAMP-dependent protein kinase have distinct biochemical properties *in vitro* and *in vivo*. *J. Biol. Chem.* **271**, 15736–15742.

Gao, T., Yatani, A., Dell'Acqua, M. L., Sako, H., Green, S. A., Dascal, N., Scott, J. D., and Hosey, M. M. (1997). cAMP-dependent regulation of cardiac L-type Ca^{2+} channels requires membrane targeting of PKA and phosphorylation of channel subunits. *Neuron* **19**, 185–196.

Gibbs, C. S., Knighton, D. R., Sowadski, J. M., Taylor, S. S., and Zoller, M. J. (1992). Systematic mutational analysis of cAMP-dependent protein kinase identifies unregulated catalytic subunits and defines regions important for the recognition of the regulatory subunit. *J. Biol. Chem.* **267**, 4806–4814.

Glantz, S. B., Amat, J. A., and Rubin, C. S. (1992). cAMP signaling in neurons: Patterns of neuronal expression and intracellular localization for a novel protein, AKAP 150, that anchors the regulatory subunit of cAMP-dependent protein kinase IIb. *Mol. Biol. Chem.* **3**, 1215–1228.

Glantz, S. B., Li, Y., and Rubin, C. S. (1993). Characterization of distinct tethering and intracellular targeting domains in AKAP75, a protein that links cAMP-dependent protein kinase IIb to the cytoskeleton. *J. Biol. Chem.* **268**, 12796–12804.

Gray, P. C., Tibbs, V. C., Catterall, W. A., and Murphy, B. J. (1997). Identification of a 15-kDa cAMP-dependent protein kinase-anchoring protein associated with skeletal muscle L-type calcium channels. *J. Biol. Chem.* **272**, 6297–6302.

Gray, P. C., Johnson, B. D., Westenbroek, R. E., Hays, L. G., Yates III, J. R., Scheuer, T., Catterall, W. A. and Murphy, B. J. (1998). Primary structure and function of an A kinase anchoring protein associated with calcium channels. *Neuron* **20**, 1017–1026.

Gromada, J., Rorsman, Patrik, Dissing, Steen, Wulff, Birgitte S. (1995a). Stimulation of cloned human glucagon-like peptide 1 receptor expressed in HEK 293 induces cAMP-dependent activation of calcium-induced calcium release. *FEBS Lett.* **373**, 182–186.

Gromada, J., Dissing, S., Bokvist, K., Renstrom, E., Frokjaer-Jensen, J., Wulff, B. S., and Rorsman, P. (1995b). Glucagon-like peptide 1 increases cytoplasmic calcium in insulin secreting betaTC3-cells by enhancement of intracellular calcium mobilization. *Diabetes* **44**, 767–774.

Gromada, J., Ding, W.-G., Barg, S., Renstrom, E., and Rorsman, P. (1997). Multisite regulation of insulin secretion by cAMP-increasing agonists: evidence that glucagon-like peptide 1 and glucagon act via distinct receptors. *Pflugers Arch.* **434**, 515–524.

Hausken, Z. E., Coghlan, V. M., Hasting, C. A. S., Reimann, E. M., and Scott, J. D. (1994). Type II regulatory subunit (RII) of the cAMP dependent protein kinase interaction with A-kinase anchor proteins requires isoleucines 3 and 5. *J. Biol. Chem.* **269**, 24245–24251.

Hausken, Z. E., Dell'Acqua, M. L., Coghlan, V. M., and Scott, J. D. (1996). Mutational analysis of the A-kinase anchoring protein (AKAP)-binding site on RII. *J. Biol. Chem.* **271**, 29016–29022.

Hausken, Z. E., Coghlan, V. M., and Scott, J. D. (1997). Overlay, ligand blotting and band-shift techniques to study kinase anchoring. *In* "Protein Targeting Protocols" (R. A. Clegg, ed.), pp. 47–64. Humana Press, Totowa, NJ.

Hirsch, A. H., Glantz, S. B., Li, Y., You, Y., and Rubin, C. S. (1992). Cloning and expression of an intronless gene for AKAP75, an anchor protein for the regulatory subunit of cAMP-dependent protein kinase IIb. *J. Biol. Chem.* **267**, 2131–2134.

Houbre, D., Duportail, G., Deloulme, J. C., and Baudier, J. (1991). The interactions of the brain-specific calmodulin-binding protein kinase C substrate, neuromodulin (GAP 43), with membrane phospholipids. *J. Biol. Chem.* **266**, 7121–7131.

Huang, L. J., Durick, K., Weiner, J. A., Chun, J., and Taylor, S. S. (1997a). D-AKAP2, a novel protein kinase A anchoring protein with a putative RGS domain. *Proc. Natl. Acad. Sci. USA* **94**, 11184–11189.

Huang, L. J., Durick, K., Weiner, J. A., Chun, J., and Taylor, S. S. (1997b). Identification of a novel dual specificity protein kinase A anchoring protein, D-AKAP1. *J. Biol. Chem.*, **272**, 8057–8064.

Hubbard, M., and Cohen, P. (1993). On target with a mechanism for the regulation of protein phosphorylation. *TIBS* **18**, 172–177.

Hunter, T. (1995). Protein kinases and phosphatases: The yin and yang of protein phosphorylation and signaling. *Cell* **80**, 225–236.

Johnson, B. D., Scheuer, T., and Catterall, W. A. (1994). Voltage-dependent potentiation of L-type Ca^{2+} channels in skeletal muscle cells requires anchored cAMP-dependent protein kinase. *Proc. Natl. Acad. Sci. USA* **91**, 11492–11496.

Johnson, B. D., Brousal, J. P., Peterson, B. Z., Gallombardo, P. A., Hockerman, G. H., Lai, Y., Scheuer, T., and Catterall, W. A. (1997). Modulation of the cloned skeletal muscle L-type Ca^{2+} channel by anchored cAMP-dependent protein kinase. *J. Neurosci.* **17**, 1243–1255.

Kabouridis, P. S., Magee, A. I., and Ley, S. C. (1997). S-acylation of LCK protein tyrosine kinase is essential for its signalling function in T lymphocytes. *EMBO J.* **16**, 4983–4998.

Kemp, B. E., and Pearson, R. B. (1990). Protein kinase recognition sequence motifs. *TIBS* **15**, 342–346.

Kemp, B. E., Graves, D. J., Benjamini, E., and Krebs, E. G. (1977). Role of multiple basic residues in determining the substrate specificity of cyclic AMP–dependent protein kinase. *J. Biol. Chem.* **252**, 4888–4894.

Klauck, T. M., Faux, M. C., Labudda, K., Langeberg, L. K., Jaken, S., and Scott, J. D. (1996). Coordination of three signaling enzymes by AKAP79, a mammalian scaffold protein. *Science* **271**, 1589–1592.

Kouhara, H., Hadari, Y. R., Spivak-Kroizman, T., Schilling, J., Bar-Sagi, D., Lax, I., and Schlessinger, J. (1997). A lipid-anchored Grb2-binding protein that links FGF-receptor activation to the Ras/MAPK signaling pathway. *Cell* **89**, 693–702.

Krebs, E. G. and Beavo, J. A. (1979). Phosphorylation–dephosphorylation of enzymes. *Ann. Rev. Biochem.* **43**, 923–959.

Leiser, M., Rubin, C. S., and Erlichman, J. (1986). Differential binding of the regulatory subunits (RII) of cAMP-dependent protein kinase II from bovine brain and muscle to RII-binding proteins. *J. Biol. Chem.* **261**, 1904–1908.

Lester, L. B., Coghlan, V. M., Nauert B., and Scott J. D. (1996). Cloning and characterization of a novel A-kinase anchoring protein: AKAP220, association with testicular peroxisomes. *J. Biol. Chem.* **272**, 9460–9465.

Lester, L. B., Langeberg, L. K., and Scott, J. D. (1997). PKA anchoring facilitates GLP-1 mediated insulin secretion. *Proc. Natl. Acad. Sci. USA* **94**, 14942–14947.

Li, Y., and Rubin, C. S. (1995). Mutagenesis of the regulatory subunit (RIIb) of cAMP-dependent protein kinase IIb reveals hydrophobic amino acids that are essential for RIIb dimerization and/or anchoring RIIb to the cytoskeleton. *J. Biol. Chem.* **270**, 1935–1944.

Li, Y., Ndubuka, C., and Rubin, C. S. (1996). A kinase anchor protein 75 targets regulatory (RII) subunits of cAMP-dependent protein kinase II to the cortical actin cytoskeleton in non-neuronal cells. *J. Biol. Chem.* **271**, 16862–16869.

Lin, R.-Y., Moss, S. B., and Rubin, C. S. (1995). Characterization of S-AKAP84, a novel developmentally regulated A kinase anchor protein of male germ cells. *J. Biol. Chem.* **270**, 27804–27811.

Lohmann, S. M., DeCamili, P., Enig, I., and Walter, U. (1984). High-affinity binding of the regulatory subunit (RII) of cAMP-dependent protein kinase to microtubule-associated and other cellular proteins. *Proc. Natl. Acad. Sci. USA* **81**, 6723–6727.

Lu, P.-J., and Chen, C.-S. (1997). Selective recognition of phosphatidylinositol 3,4,5-trisphosphate by a synthetic peptide. *J. Biol. Chem.* **272**, 466–472.

Luo, Z., Shafit-Zagardo, B., and Erlichman, J. (1990). Identification of the MAP2- and P75-binding domain in the regulatory subunit (RIIb) of Type II cAMP-dependent protein kinase. *J. Biol. Chem.* **265**, 21804–21810.

McCartney, S., Little, B. M., Langeberg, L. K., and Scott, J. D. (1995). Cloning and characterization of A-kinase anchor protein 100 (AKAP100): A protein that targets A-kinase to the sarcoplasmic reticulum. *J. Biol. Chem.* **270**, 9327–9333.

McLaughlin, S., and Aderem, A. (1995). The myristoyl–electrostatic switch: A modulator of reversible protein–membrane interactions. *TIBS* **20**, 272–276.

Milligan, G., Parenti, M., and Magee, A. I. (1995). The dynamic role of palmitoylation in signal transduction. *TIBS* **20**, 181–186.

Mons, N., Harry, A., Dubourg, P., Premont, R. T., Iyengar, R., and Cooper, D. M. F. (1995). Immunohistochemical localization of adenylyl cyclase in rat brain indicates a highly selective concentration at synapses. *Proc. Natl. Acad. Sci. USA* **92**, 8473–8477.

Mosior, M., and Newton, A. C. (1995). Mechanism of interaction of protein kinase C with phorbol esters. Reversibility and nature of membrane association. *J. Biol. Chem.* **270**, 25526–25533.

Nauert, J. B., Klauck, T. M., Langeberg, L. K., and Scott, J. D. (1997). Gravin, an autoantigen recognized by serum from myasthenia gravis patients, is a kinase scaffold protein. *Curr. Biol.* **7**, 52–62.

Newlon, M. G., Roy, M., Hausken, Z. E., Scott, J. D., and Jennings, P. A. (1997). The A-kinase anchoring domain of Type IIa cAMP-dependent protein kinase is highly helical. *J. Biol. Chem.* **272**, 23637–23644.

Obar, R. A., Dingus, J., Bayley, H., and Vallee, R. B. (1989). The RII subunit of cAMP-dependent protein kinase binds to a common amino-terminal domain in microtubule-associated proteins 2A, 2B, and 2C. *Neuron* **3**, 639–645.

Pawson, T. and Scott, J. D. (1997). Signaling through scaffold, anchoring, and adaptor proteins. *Science* **278**, 2075–2080.

Potter, R. L., and Taylor, S. S. (1979). Relationships between structural domains and function in the regulatory subunit of cAMP-dependent protein kinases I and II from porcine skeletal muscle. *J. Biol. Chem.* **254**, 2413–2418.

Potter, R. L., Stafford, P. H., and Taylor, S. (1978). Regulatory subunit of cyclic AMP-dependent protein kinase I from porcine skeletal muscle: Purification and proteolysis. *Arch. Biochem. Biophys.* **190**, 174–180.

Resh, M. D. (1994). Myristylation and palmitylation of Src family members: The fats of the matter. *Cell* **76**, 411–413.

Rios, R. M., Celati, C., Lohmann, S. M., Bornens, M., and Keryer, G. (1992). Identification of a high affinity binding protein for the regulatory subunit RIIb of cAMP-dependent protein kinase in Golgi enriched membranes of human lymphoblasts. *EMBO J.* **11**, 1723–1731.

Rosenmund, C., Carr, D. W., Bergeson, S. E., Nilaver, G., Scott, J. D., and Westbrook, G. L. (1994). Anchoring of protein kinase A is required for modulation of AMPA/kainate receptors on hippocampal neurons. *Nature* **368**, 853–856.

Rubin, C. S. (1994). A kinase anchor proteins and the intracellular targeting of signals carried by cAMP. *Biochim. Biophys. Acta* **1224**, 467–479.

Rubin, C. S., Erlichman, J., and Rosen, O. M. (1972). Cyclic adenosine 3′,5′-monophosphate–dependent protein kinase of human erythrocyte membranes. *J. Biol. Chem.* **247**, 6135–6139.

Rubino, H. M., Dammerman, M., Shafit-Sagardo, B., and Erlichman, J. (1989). Localization and characterization of the binding site for the regulatory subunit of type II cAMP-dependent protein kinase on MAP2. *Neuron* **3**, 631–638.

Safayhi, H., Haase, H., Kramer, U., Bihlmayer, A., Roenfeldt, M., Ammon, H. P. T., Frosch-mayr, M., Cassidy, T. N., Morano, I., Ahlijanian, M. K., and Striessnig, J. (1997). L-type calcium channels in insulin-secreting cells: Biochemical characterization and phosphorylation in RINm5F cells. *Mol. Endocrinol.* **11**, 619–629.

Salvatori, S., Damiani, E., Barhanin, J., Furlan, S., Giovanni, S., and Margreth, A. (1990). Co-localization of the dihydropyridine receptor and cyclic AMP–binding subunit of an intrinsic protein kinase to the junctional membrane of the transverse tubules of skeletal muscle. *Biochem. J.* **267**, 679–687.

Schlesinger, M. J. (1993). "Lipid Modifications of Proteins." CRC Press, Boca Raton, FL.

Scott, J. D. (1991). Cyclic nucleotide–dependent protein kinases. *Pharmacol. Ther.* **50**, 123–145.

Scott, J. D., Stofko, R. E., McDonald, J. R., Comer, J. D., Vitalis, E. A., and Mangeli, J. (1990). Type II regulatory subunit dimerization determines the subcellular localization of the cAMP-dependent protein kinase. *J. Biol. Chem.* **265**, 21561–21566.

Shakur, Y., Pryde, J. G., and Houslay, M. D. (1993). Engineered deletion of the unique N-terminal domain of the cyclic AMP-specific phosphodiesterase RD1 prevents plasma membrane association and the attainment of enhanced thermostability without altering its sensitivity to inhibition by rolipram. *Biochem. J.* **292**, 677–686.

Sjoholm, A., Honkanen, R. E., and Berggren, P.-O. (1995). Inhibition of serine/threonine protein phosphatases by secretagogues in insulin-secreting cells. *Endocrinology* **136**, 3391–3397.

Skalhegg, B. S., Tasken, K., Hansson, V., Huitfeldt, H. S., Jahnsen, T., and Lea, T. (1994). Location of cAMP-dependent protein kinase type I with the TCR-CD3 complex. *Science* **263**, 84–87.

Smith, J. K., Scotland, G., Beattie, J., Trayer, I. P., and Houslay, M. D. (1996). Determination of the structure of the N-terminal splice region of the cyclic AMP–specific phosphodiesterase RD1 (RNPDE4A1) by ^{1}H NMR and identification of the membrane association domain using chimeric constructs. *J. Biol. Chem.* **271**, 16703–16711.

Su, Y., Taylor, S. S., Dostmann, W. R. G., Xuong, N. H., and Varughese, K. I. (1993). Crystallization of a deletion mutant of the R-subunit of cAMP-dependent protein kinase. *J. Mol. Biol.* **230**, 1091–1093.

Su, Y., Dostmann, W. R. G., Herberg, F. W., Durick, K., Xuong, N.-H., Ten Eyck, L., Taylor, S. S., and Varughese, K. I. (1995). Regulatory subunit of protein kinase A: Structure of deletion mutant with cAMP binding proteins. *Science* **269**, 807–813.

Subramani, S. (1996). Protein translocation into peroxisomes. *J. Biol. Chem.* **271**, 32483–32486.

Suga, S., Kanno, T., Nakano, K., Takeo, T., Dobashi, Y. and Wakui, M. (1997). GLP-1 (7–36) amide augments Ba^{2+} current through L-type Ca^{2+} channel of rat pancreatic B-cell in a cAMP-dependent manner. *Diabetes* **46**, 1755–1760.

Sutherland, E. W., and Rall, T. W. (1957). The properties of an adenine ribonucleotide produced with cellular particles, ATP, Mg^{++}, and epinephrine or glucagon. *J. Amer. Chem. Soc.* **79**, 3608–3610.

Tang, W.-J., and Gilman, A. G. (1992). Adenylyl cyclases. *Cell* **70**, 869–872.

Taylor, S. S., Buechler, J. A., and Yonemoto, W. (1990). cAMP-dependent protein kinase: Framework for a diverse family of regulatory enzymes. *Ann. Rev. Biochem.* **59**, 971–1005.

Theurkauf, W. E., and Vallee, R. B. (1982). Molecular characterization of the cAMP-dependent protein kinase bound to microtubule-associated protein 2. *J. Biol. Chem.* **257**, 3284–3290.

Thorens, B. (1992). Expression cloning of the pancreatic b cell receptor for the gluco-incretin hormone glucagon-like peptide 1. *Proc. Natl. Acad. Sci. USA* **89**, 8641–8645.

Thorens, B., Deriaz, N., Bosco, D., DeVos, A., Pipeleers, D., Schuit, F., Meda, P., and Porret, A. (1996). Protein kinase A-dependent phosphorylation of GLUT2 in pancreatic b cells. *J. Biol. Chem.* **271**, 8075–8081.

Wang, Z. W., and Kotlikoff, M. I. (1996). Activation of K_{Ca} channels in airway smooth muscle cells by endogenous protein kinase A. *Am. J. Physiol.* **15**, L100–L105.

Wang, Y., Scott, J. D., McKnight, G. S., and Krebs, E. G. (1991). A constitutively active holoenzyme from the cAMP-dependent protein kinase. *Proc. Natl. Acad. Sci. USA* **88**, 2446–2450.

Watanabe, Y., Kawai, K., Ohashi, S., Yokota, C., Suzuki, S., and Yamashita, K. (1994). Structure-activity relationships of glucagon-like peptide-1(7–36) amide: Insulinotropic activities in perfused rat pancreases, and receptor binding and cyclic AMP production in RINm5F cells. *J. Endocrinol.* **140**, 45–52.

Wen, W., Harootunian, A. T., Adams, S. R., Feramisco, J., Tsien, R. Y., Meinkoth, J. L., and Taylor, S. S. (1994). Heat-stable inhibitors of cAMP-dependent protein kinase carry a nuclear export signal. *J. Biol. Chem.* **269**, 32214–32220.

Wen, W., Meinkoth, J. L., Tsien, R. Y., and Taylor, S. S. (1995). Identification of a signal for rapid export of proteins from the nucleus. *Cell* **82**, 463–473.

Widmann, C., Dolci, W., and Thorens, B. (1995). Agonist-induced internalization and recycling of the glucagon-like peptide-1 receptor in transfected fibroblasts and in insulinomas. *Biochem. J.* **310**, 203–214.

Widmann, C., Dolci, W., and Thorens, B. (1996). Desensitization and phosphorylation of the glucagon like peptide-1 (GLP-1) receptor by GLP-1 and 4-phorbol 12-myristate 13-acetate. *Mol. Endocrinol.* **10**, 62–75.

Yada, T., Itoh, K., and Nakata, M. (1993). Glucagon-like peptide-1-(7–36) amide and a rise in cyclic adenosine 3′,5′-monophosphate increase cytosolic free Ca^{2+} in rat pancreatic beta-cells by enhancing Ca^{2+} channel activity. *Endocrinology* **133**, 1685–1691.

Yaekura, K., Kakei, M., and Yada, T. (1996). cAMP-signaling pathway acts in selective synergism with glucose or tolbutamide to increase cystolic Ca^{2+} in rat pancreatic b-cells. *Diabetes* **45**, 295–301.

Toru Yamaguchi
Naibedya Chattopadhyay
Edward M. Brown

Endocrine-Hypertension Division
Department of Medicine
Brigham and Women's Hospital and Harvard Medical School
Boston, MA

G Protein-Coupled Extracellular Ca^{2+} (Ca^{2+}_o)-Sensing Receptor (CaR): Roles in Cell Signaling and Control of Diverse Cellular Functions

I. Introduction

Despite having only intermittent access to dietary calcium ions (Ca^{2+}) (Stewart and Broadus, 1987), complex free-living terrestrial organisms, such as mammals, maintain their levels of extracellular ionized calcium concentration (Ca^{2+}_o) within strict limits ($\sim$1.1–1.3 mM). The near constancy of Ca^{2+}_o in humans and other mammals ensures the availability of calcium ions for crucial extracellular roles, such as serving as a cofactor for adhesion molecules, clotting factors and other proteins and regulating neuronal excitability (Stewart and Broadus, 1987; Brown, 1991). Ca^{2+}_o also provides an important source of calcium for this ion's roles as a key second messenger and as a cofactor for various intracellular proteins and enzymes (Pietrobon et al., 1990). The cytosolic free calcium (Ca^{2+}_i) concentration is maintained under basal conditions at a level close to 100 nM. Upon cellular activation, however, Ca^{2+}_i can rise to levels of 1 μM or higher, thereby modulating diverse

Hormones and Signaling

cellular processes, such as neurotransmission, cellular motility, hormonal secretion, muscular contraction and cell division, by interacting with intracellular Ca^{2+}-binding proteins (e.g., calmodulin) (Pietrobon *et al.*, 1990).

Ca^{2+}_o homeostasis is maintained through a complex process involving the parathyroid glands, thyroidal C-cells, kidneys, bone, and intestine (Fig. 1) (Brown, 1991). The precise control of Ca^{2+}_o involves two major types of adjustments of this homeostatic system: (a) regulation of calcium fluxes between extracellular fluid (ECF) and the outside environment via alterations in the renal excretion and/or gastrointestinal absorption of calcium and (b) modulation of the movement of calcium ions between ECF and bone. The parathyroid glands play a central role in this homeostatic system because of their remarkable sensitivity to minute variations in Ca^{2+}_o. There is a steep inverse sigmoidal relationship between circulating levels of PTH and Ca^{2+}_o under normal circumstances (Fig. 2) (Brown, 1983, 1991). The steep slope of this curve ensures that small perturbations in Ca^{2+}_o elicit large changes in PTH secretion. The set-point, or midpoint, of the curve, in turn, contributes importantly to the level at which Ca^{2+}_o is "set" in the ECF. The resultant, Ca^{2+}_o-evoked changes in the circulating levels of PTH directly or indirectly modulate the functions of bone, kidney and intestine as shown in Fig. 1, thereby maintaining Ca^{2+}_o within its normally narrow physiological range. For example, hypocalcemia induces the acute secretion of PTH, which then increases reabsorption of Ca^{2+} from the glomerular filtrate in the distal

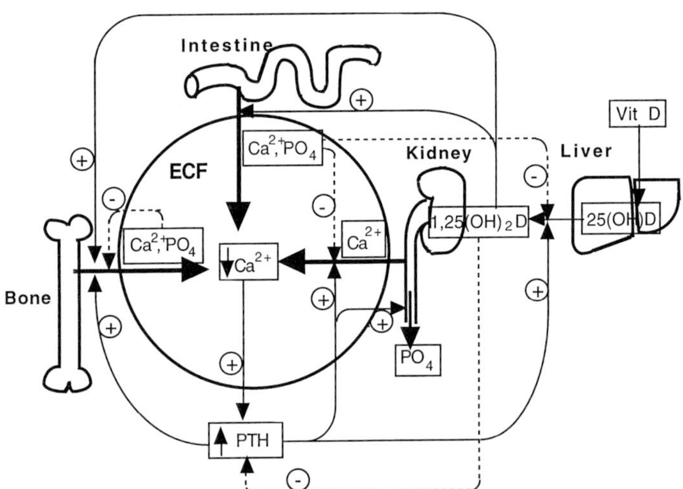

FIGURE I Schematic diagram illustrating the regulation of Ca^{2+}_o homeostasis. The solid lines and arrows indicate the actions of PTH and 1,25-$(OH)_2D_3$; the dotted lines and arrows demonstrate examples of direct effects of Ca^{2+}_o or phosphate ions. Abbreviations are as follows: Ca, calcium; PO_4, phosphate; ECF, extracellular fluid; 1,25-$(OH)_2D$, 1,25-dihydroxyvitamin D; 25(OH)D, 25-hydroxyvitamin D; + signs indicate positive actions while − signs indicate inhibitory effects.

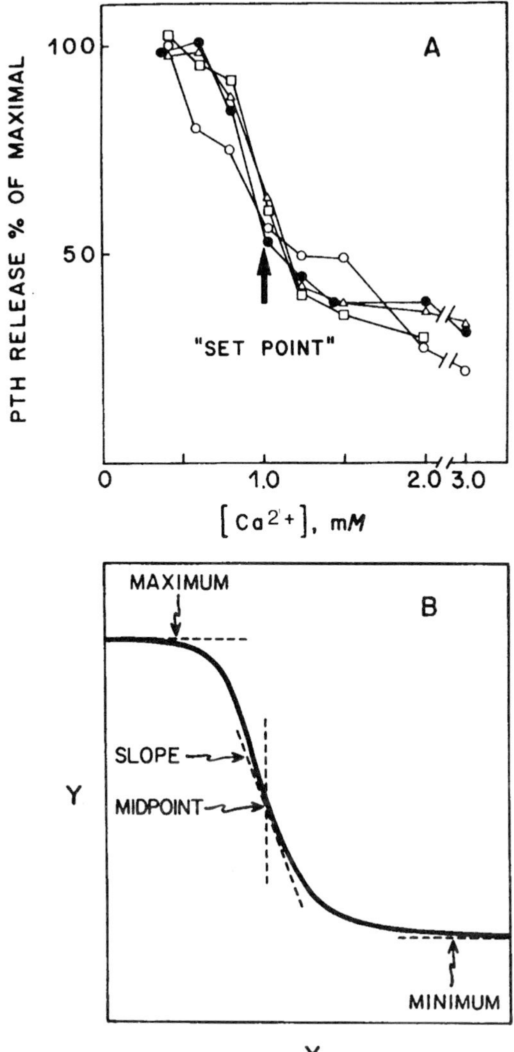

FIGURE 2 Inverse sigmoidal relationship between Ca^{2+}_o and PTH release and four-parameter model describing such curves. The data in the upper panel represent results obtained using dispersed normal human parathyroid cells and are expressed as the percent of the maximal rate of PTH release observed at 0.3 mM Ca^{2+}_o. The set point represents the level of Ca^{2+}_o at which PTH release is half-maximally suppressed. The lower panel illustrates the four parameters that may be used to describe such curves. [Reproduced with permission from Brown, E. M. (1982). PTH secretion *in vivo* and *in vitro*: regulation by calcium and other secretagogues. *Miner. Electrolyte Metab.* **8**, 130–150.]

tubule of the kidney and enhances the resorption of Ca^{2+} from bone. If sufficiently prolonged, hypocalcemia leads to PTH-stimulated 1α-hydroxylation and resultant activation of 25-hydroxyvitamin D_3 in the renal proximal tubule (Weisinger *et al.*, 1989), which, in turn, increases the absorption of Ca^{2+} in the intestine (Fig. 1).

Although there is a substantial body of knowledge available concerning the mechanisms through which calcium ions traverse the plasma membrane via ion channels and other transport proteins, the mechanism(s) enabling cells to recognize and respond to small changes in Ca^{2+}_o was, until recently, poorly understood. A body of indirect evidence had suggested the presence of a Ca^{2+}_o-sensing mechanism in parathyroid cells, C-cells, and kidney cells through which Ca^{2+}_o could act, in effect, as an extracellular first messenger via its own putative G protein-coupled, cell surface receptor. Such evidence, principally accumulated in studies involving dispersed bovine parathyroid cells, included the following actions of elevated Ca^{2+}_o: (a) activation of phospholipase C (PLC) leading to accumulation of inositol 1,4,5-trisphosphate (IP_3) (Brown *et al.*, 1987; Shoback *et al.*, 1988) and consequent release of Ca^{2+} from its intracellular stores (Nemeth and Scarpa, 1987), as well as (b) a pertussis toxin-sensitive inhibition of agonist-stimulated cAMP accumulation (Chen *et al.*, 1989). Further studies also showed that such effects of Ca^{2+}_o could be mimicked by other divalent cations (i.e., Mg^{2+}), trivalent cations (e.g., La^{3+} and Gd^{3+}) (Brown *et al.*, 1990), and even polyvalent cations (neomycin) (Brown *et al.*, 1991; Ridefelt *et al.*, 1992). Because several of these cations had little if any membrane permeability, this result strongly suggested that Ca^{2+}_o could modulate PTH secretion by binding to a cell-surface receptor and not by a mechanism involving transport of calcium or other polyvalent cations across the cell membrane.

II. Cloning of a G Protein-Coupled Ca^{2+}_o-Sensing Receptor (CaR) from Bovine Parathyroid

The body of evidence just outlined suggesting the existence of a Ca^{2+}_o-sensing receptor provided a foundation for the successful use, by Brown *et al.* (1993), of expression cloning in *Xenopus laevis* oocytes to isolate a full-length cDNA encoding the bovine parathyroid Ca^{2+}_o-sensing receptor (abbreviated hereafter as CaR; note that CaSR is an alternative designation for the same receptor). Following the isolation of this novel receptor, nucleic acid hybridization-based techniques permitted the cloning of additional full-length CaRs from human (Garrett *et al.*, 1995a) and chicken (Diaz *et al.*, 1997) parathyroid; rat (Riccardi *et al.*, 1995), human (Aida *et al.*, 1995), and rabbit (Butters *et al.*, 1997) kidney; C-cells (Garrett *et al.*, 1995b); and striatum of rat brain (Ruat *et al.*, 1995). All of these subsequently cloned CaRs were highly homologous to the bovine parathyroid receptor (more than 90% identical in their amino acid sequences), indicating that they represent species and tissue homologues of a common ancestral gene. The predicted amino acid sequences of the human parathyroid CaR (Garrett *et al.*, 1995a) and its various species homologs reveal very similar overall topologies (Fig. 3). The deduced amino acid sequence of the human CaR predicts three major domains: (1) a large amino-terminal hydrophilic extracellular domain (ECD) consisting of 612 amino acids, (2) a hydrophobic transmembrane domain comprising 250 amino acids and containing seven membrane-spanning segments characteristic of the superfamily of G protein-coupled receptors (GPCRs), and (3) a cytosolic carboxy (C)-terminal tail of 217 amino acids. The cDNA for the human CaR also predicted the presence of 11 N-linked glycosylation sites within the extracellular amino-terminal domain (ECD), consistent with the native protein being expressed as a glycoprotein. Within the intracellular domains of the human Ca^{2+}_o-sensing receptor, there are five predicted protein kinase C (PKC) phosphorylation sites. Selected features of CaRs cloned from various species so far are summarized in Table I.

Among the large superfamily of GPCRs, the CaR only shares amino acid sequence homology with the metabotropic glutamate receptors (mGluRs) (Nakanishi, 1992, 1994), the $GABA_B$ receptors (Kaupmann *et al.*, 1997), and the recently identified subfamily of putative pheromone receptors (VRs) isolated from the vomeronasal organ (VNO) of the rat (Matsunami and Buck, 1997; Ryba and Trindell, 1997). All three groups of receptors are related, in turn, to the bacterial periplasmic binding proteins, which serve as extracellular binding proteins involved in chemotaxis and nutrient uptake into bacteria (Adams and Oxender, 1989; Sharff *et al.*, 1992).The mGluRs are a subfamily of GPCRs comprising eight receptor subtypes that are expressed predominantly in the central nervous system and are activated by glutamate, the major excitatory neurotransmitter in the brain (Nakanishi, 1992, 1994). The CaR has only modest identity in its amino acid sequence with the mGluRs (18–24%), but it shares with the latter striking topological similarities. Both classes of receptors possess a large ECD as well as a total of 20 strictly conserved cysteine residues (17 within the extracellular domain and 3 in transmembrane segments or extracellular loops) (Fig. 3). These conserved cysteines may contribute to organizing the respective ECDs into binding pockets appropriate for interacting with these receptors' small charged ligands (e.g., glutamate for the mGluRs and Ca^{2+}_o and other polyvalent cations for the CaR). The CaR's ECD and extracellular loop 2 contain clusters of negatively charged amino acids (i.e., aspartate and glutamate) that might conceivably represent sites contributing to the sensing of Ca^{2+}_o and other polycationic agonists (Fig. 3). Kubo *et al.* (1998) have reported that mGluRs 1α, 3, and 5, but not mGluR2, can sense Ca^{2+}_o over a range similar to that sensed by the CaR, although the concentration response curves for activation of the mGluRs by Ca^{2+}_o are considerably less steep

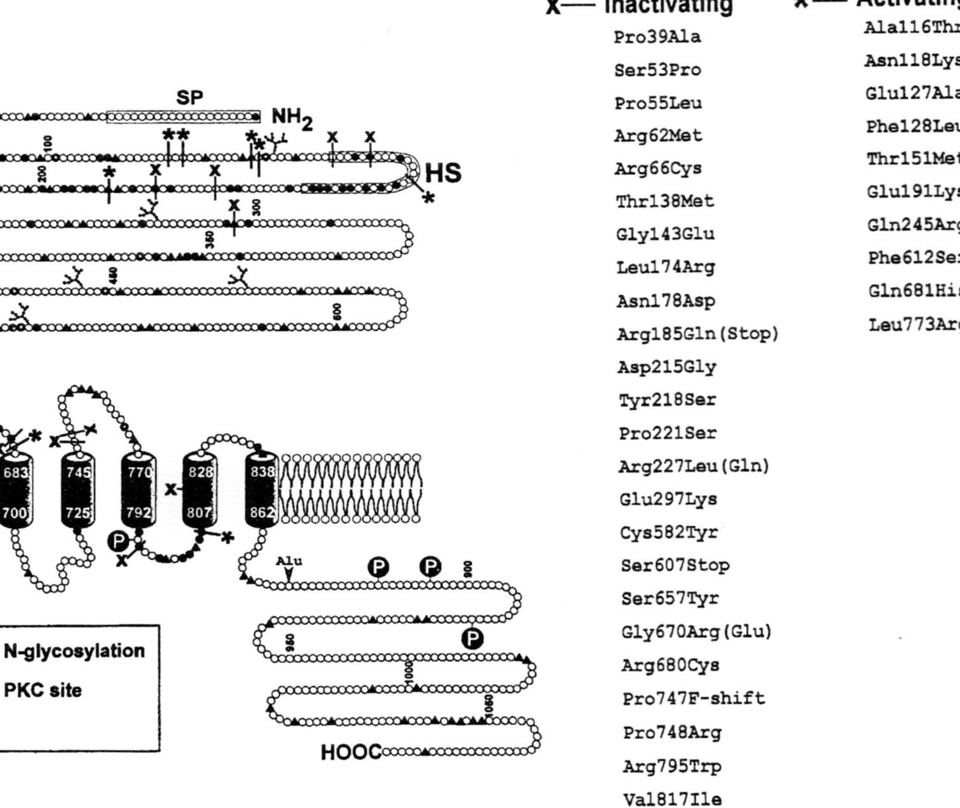

X — Inactivating
Pro39Ala
Ser53Pro
Pro55Leu
Arg62Met
Arg66Cys
Thr138Met
Gly143Glu
Leu174Arg
Asn178Asp
Arg185Gln(Stop)
Asp215Gly
Tyr218Ser
Pro221Ser
Arg227Leu(Gln)
Glu297Lys
Cys582Tyr
Ser607Stop
Ser657Tyr
Gly670Arg(Glu)
Arg680Cys
Pro747F-shift
Pro748Arg
Arg795Trp
Val817Ile
Thr876Alu
* — Activating
Ala116Thr
Asn118Lys
Glu127Ala
Phe128Leu
Thr151Met
Glu191Lys
Gln245Arg
Phe612Ser
Gln681His
Leu773Arg
SP
NH2
HS
Cysteine
Conserved
Acidic
N-glycosylation
PKC site
Alu
HOOC

than that for the CaR. The sensitivities of mGluRs 1α, 3, and 5 to Ca^{2+}_o are determined by a single amino acid residue (Ser166) that is located near the glutamate-binding site. Further studies are required to clarify whether similar Ca^{2+}_o-binding site(s) exist in the corresponding region of the CaR and contribute to its sensing of Ca^{2+}_o.

The $GABA_B$ receptors are G protein-coupled receptors recognizing gamma aminobutyric acid, the principal inhibitory neurotransmitter in the central nervous system, as their physiological ligand (Kaupmann *et al.*, 1997). Finally, the VRs are a multigene subfamily of G protein-coupled receptors that are expressed predominantly, if not exclusively, in the vomeronasal organ. This small sensory organ is thought to be involved in controlling instinctual behavior through input from environmental pheromones (Matsunami and Buck, 1997; Ryba and Trindell, 1997). Although the pheromones activating individual VRs are largely unknown, it is likely that they, like the ligands for the CaR, $GABA_B$ receptor, and mGluRs, are relatively small molecules that interact with the respective large extracellular domains of the individual receptors.

III. Disorders Caused by Inactivating or Activating Mutations of the CaR

Cloning of the CaR was followed almost immediately by the identification of inherited human hyper- or hypocalcemic disorders resulting from loss- or gain-of-function mutations in the CaR, respectively (Pollak *et al.*, 1993, 1994a, 1994b; De Luca *et al.*, 1997). Mutations reducing the activity of the CaR occur in either a heterozygous form that usually causes a mild, generally asymptomatic hypercalcemic disorder [familial hypocalciuric hypercalcemia (FHH)] (Foley *et al.*, 1972; Marx *et al.*, 1981a; Law and Heath, 1985) or in a homozygous form that produces a severe form of hyperparathyroidism in newborn infants [neonatal severe hyperparathyroidism (NSHPT)] (Heath, 1994). In contrast, gain-of-function CaR mutations cause either an autosomal dominant familial form of hypocalcemia or sporadic cases of

---

FIGURE 3 Schematic illustration of the proposed topology of the Ca^{2+}_o-sensing receptor cloned from human parathyroid, depicting locations of activating and inactivating mutations. SP, signal peptide; HS, hydrophobic segment. Also shown are positions of missense and nonsense mutations that cause either familial hypocalciuric hypercalcemia (FHH) or autosomal dominant hypocalcemia; the latter are indicated with the three-letter amino acid code. The normal amino acid is shown before and the mutant amino acid after the number of the relevant codon. Reproduced with permission from Brown, E. M., Bai, M., and Pollak, M. (1997). Familial benign hypocalciuric hypercalcemia and other syndromes of altered responsiveness to extracellular calcium. *In* "Metabolic Bone Diseases," 3rd ed. (S. M. Krane and L. V. Avioli, eds.), San Diego: Academic Press, pp 479–499.

TABLE I Selected Features of CaR Cloned from Various Species

Species	Number of amino acids	Size of protein (kDa)[a]	Putative PKA sites	Putative PKC sites	Putative N-glycosylation sites	References
Human	1079	130–160	2	5	11	Garrett *et al.*, 1995a
Bovine	1085	130–160	0	4	9	Brown *et al.*, 1993
Rat	1079	130–160	2	8	9	Ruat *et al.*, 1995
Rabbit	1070	—	2	6	11	Butters *et al.*, 1997
Chicken	1059	—	2	5	12	Diaz *et al.*, 1997

[a] Most of immunoreactive CaR protein on Western analysis is present as two glycosylated bands with only a small amount of the ~120 kDa nonglycosylated CaR.

hypocalcemia (Chattopadhyay *et al.*, 1996). Both of the latter disorders exhibit a reduction in the set-point of the parathyroid and an excessive degree of hypercalciuria for any given level of Ca^{2+}_o. The locations of inactivating and activating mutations in the human CaR sequence that cause FHH/NSHPT and autosomal dominant or sporadic hypocalcemia, respectively, are shown in Fig. 3.

FHH is a rare, heritable disorder of mineral metabolism, first described in 1972 (Foley *et al.*, 1972). It is characterized by lifelong, mild to moderate but generally asymptomatic hypercalcemia (usually ranging from minimal elevations in Ca^{2+}_o to total serum calcium concentrations of ~12 mg/dl, with occasional families having levels as high as 13–14 mg/dl) (Marx *et al.*, 1981a; Law and Heath, 1985). The degree of elevation of the serum calcium concentration is similar to that seen in primary hyperparathyroidism, but in FHH it is characteristically accompanied by inappropriately low rates of urinary calcium excretion (a calcium to creatinine clearance ratio of <0.01), along with inappropriately normal (i.e., nonsuppressed) circulating levels of PTH given these individuals' hypercalcemia (Marx *et al.*, 1981a; Law and Heath, 1985). The serum phosphate concentration is usually normal, although there may be a tendency toward hypophosphatemia, and mild elevations in serum magnesium level can also be present (Marx *et al.*, 1978, 1981a; Heath, 1989). Hypermagnesemia may be more common in kindreds in which the serum calcium concentration is more elevated, since there is a positive relationship between the serum calcium and magnesium concentrations in FHH; in PHPT, in contrast, there is an inverse relationship between these parameters (Marx *et al.*, 1981a). Serum levels of 25(OH)D and 1,25-(OH)$_2$D are usually normal (Davies *et al.*, 1984; Law *et al.*, 1984; Kristiansen *et al.*, 1987). The disorder is generally considered to be benign, as affected patients have few, if any, of the characteristic symptoms and complications of other disorders causing hypercalcemia. The diagnosis of FHH can thus be made by documenting the combination of a low urinary calcium to creatinine clearance ratio, a normal PTH level, and the autosomal-dominant inheritance of mild, asymptomatic hypercalcemia.

Individuals with FHH show abnormal parathyroid and renal responsiveness to Ca^{2+}_o, which is likely explained both by the presence of mutant CaRs with reduced activity and by a reduced complement of normal CaRs in the plasma membranes of parathyroid and kidney. That is, there does not appear to be any upregulation of the expression of the CaR from the remaining normal allele of the receptor's gene in FHH, at least as judged by the levels of CaR protein expression found in parathyroid and kidney of mice heterozygous for targeted disruption of the CaR gene, which are about half of those in the wild-type mice (Ho *et al.*, 1995). The use of intravenous calcium infusion has shown that patients with FHH have an elevated set point for Ca^{2+}_o-regulated PTH release (Auwerx *et al.*, 1984; Khosla *et al.*, 1993). As a result, for any given level of serum calcium, FHH patients have higher circulating levels of PTH than do normal subjects. To suppress their circulating PTH levels to 50% of their maximal values, these individuals require an increase in serum calcium to a level slightly higher than that necessary to achieve a comparable degree of suppression of PTH in normals. Calcium handling by the kidney is also abnormal in individuals with FHH. These patients show excessively avid renal tubular reabsorption of Ca^{2+} and Mg^{2+} in spite of their hypercalcemia (Marx *et al.*, 1978; Heath, 1989), which causes the low urinary calcium to creatinine clearance ratio that is observed in these patients. This abnormality in renal calcium handling persists even following parathyroidectomy (Attie *et al.*, 1983; Davies *et al.*, 1984), indicating that it is an intrinsic renal defect. Impairment of Ca^{2+}_o-sensing by the CaR in FHH patients was directly proven by studies in which CaRs bearing FHH mutations were expressed in human embryonic kidney (HEK293) cells (Bai *et al.*, 1996, 1997a; Pearce *et al.*, 1996a). The results of these studies revealed an increase in EC_{50} (the level of Ca^{2+}_o required to achieve half-maximal activation of the receptor) and, in many cases, a reduction in maximal activity of the mutant CaRs.

FHH is inherited as an autosomal-dominant trait with a penetrance of well over 90% (Marx *et al.*, 1981a; Law and Heath, 1985; Heath, 1994). Moreover, 90% or more of families exhibit linkage of the disease gene to the locus on chromosome 3 (band q21–24) known to contain the CaR gene (Heath *et al.*, 1993; Pearce *et al.*, 1995). One family, however, showed linkage of a phenotypically indistinguishable disorder to the short arm of chromosome 19, band 19p13.3 (Heath *et al.*, 1993), while another has a disease gene causing an FHH-like disorder that is linked to another locus on chromosome 19 (Lloyd *et al.*, 1999; McMurtry *et al.*, 1992; Trump *et al.*, 1995), illustrating the genetic heterogeneity of the disorder. It is possible, therefore, that either a new form of CaR or some other element playing a key role in Ca^{2+}_o sensing might be located in the additional chromosome loci associated with the FHH phenotype.

The availability of the cloned CaR has also made it possible to show that a number of cases of NSHPT are caused by inactivating mutations of the CaR. The degree of hypercalcemia in NSHPT is usually much more severe than that in FHH, and the disorder has been fatal in some cases when parathyroidectomy was not carried out within the first few weeks of life (Heath, 1994). The marked hypercalcemia is the result of severe primary hyperparathyroidism—with enlargement of all four parathyroid glands—that causes bony demineralization, often accompanied by multiple fractures of long bones and ribs (Eftekhari and Yousefzadeh, 1982; Heath, 1994). Investigations *in vitro* of the function of parathyroid glands resected from infants with NSHPT have revealed markedly abnormal Ca^{2+}_o-regulated PTH release, with substantial increases in set-point and, in some cases, severely impaired inhibition of secretion even at levels of Ca^{2+}_o (e.g., 4 mM) higher than those encountered *in vivo* (Cooper *et al.*, 1986; Marx *et al.*, 1986).

Some infants with NSHPT represent the homozygous form of FHH (Pollak *et al.,* 1994b; Chou *et al.,* 1995; Janicic *et al.,* 1995; Pearce *et al.,* 1995) or, in one case, a compound heterozygote in which a different inactivating CaR mutation was inherited from each of the parents (Kobayashi *et al.,* 1997). Such cases, like the CaR "knockout" mice, illustrate a clear gene dose effect, where loss of one CaR allele produces the mild "resistance" of CaR-expressing tissues to Ca^{2+}_{o} observed in FHH, while loss of both alleles causes much more severe Ca^{2+}_{o} resistance. In several cases, however, NSHPT has been shown to be caused by the presence of heterozygous inactivating mutations of the CaR, either in a familial setting or as a *de novo* mutation in the offspring of normal parents (Marx *et al.,* 1982; Pearce *et al.,* 1995; Bai *et al.,* 1997b). The reasons underlying the more severe clinical manifestations in these infants than those in most infants with heterozygous inactivating CaR mutations and the phenotype of FHH are not well understood, but in some cases the mutant CaR can exert a "dominant" negative action, impairing in some fashion the function of the normal receptor (Bai *et al.,* 1997b).

The recent development of mice heterozygous or homozygous for targeted inactivation of the CaR gene (Ho *et al.* 1995) that share many of the biochemical features of FHH and NSHPT, respectively, has added to the evidence supporting the physiological importance of the CaR in mineral ion metabolism. Mice heterozygous for deletion of the CaR gene via insertion of the neomycin resistance gene into this gene by homologous recombination are phenotypically indistinguishable from their normal littermates and have mild hypercalcemia accompanied by slight increases in their serum levels of PTH—a clinical picture not unlike that of FHH. There is much more severe hypercalcemia, in contrast, accompanied by marked increases in circulating levels of PTH, in homozygous CaR-deficient mice, which usually die within the first few weeks of life (Ho *et al.,* 1995). The availability of these mouse models of FHH and NSHPT will doubtless provide important models for studying the role of the CaR in regulating a variety of cell types and tissues both *in vitro* and *in vivo.*

Activating CaR mutations can cause either a form of autosomal dominant hypocalcemia (Pollak *et al.,* 1994a; Lovlie *et al.,* 1996; Pearce *et al.,* 1996b) or cases of sporadic hypocalcemia (Baron *et al.,* 1996; De Luca *et al.,* 1997; Mancilla *et al.,* 1997), both of which resemble hypoparathyroidism. That is, there is hypocalcemia that ranges in severity from mild to severe, which is accompanied by varying degrees of hyperphosphatemia and may be associated with hypomagnesemia. The presence of hypermagnesemia in some families with FHH (Marx *et al.,* 1981a) and hypomagnesemia in individuals harboring activating mutations of the CaR (Pearce *et al.,* 1996b) indicate that the receptor plays some role in regulating the normal level of Mg^{2+}_{o}. PTH levels in persons with activating mutations of the CaR are inappropriately normal (i.e., they are not increased in response to the hypocalcemia) and are often in the lower half of the normal range or even frankly reduced in some cases (Pollak *et al.*, 1994a; Baron *et al.*, 1996; Pearce *et al.*, 1996b). Therefore, this condition resembles primary hypoparathyroidism in its pathophysiology in that there is an insufficient level of circulating PTH to maintain normocalcemia. The latter is not, however, due to inability of the parathyroid glands to respond to a decrease in Ca^{2+}_o, but rather to a resetting of the parathyroid glands so that they only respond over a lower than normal range of calcium concentrations. In one family with this disorder, a reduction in Ca^{2+}_o from an affected family member's basal serum calcium concentration produced a brisk increase in circulating PTH levels (Estep *et al.*, 1981)—illustrating the reduced set-point for Ca^{2+}_o-regulated PTH secretion that is characteristic of this disorder.

Expression of CaRs harboring activating mutations in HEK293 cells has revealed consistent reductions in the EC_{50}s of the mutant receptors (Bai *et al.*, 1996; De Luca *et al.*, 1997; Mancilla *et al.*, 1997). Therefore, an important factor contributing to the hypocalcemia in patients with activating CaR mutations is the presence of CaRs that are overly sensitive to activation by increases in Ca^{2+}_o. This defect suppresses PTH secretion and increases urinary calcium excretion at inappropriately low levels of Ca^{2+}_o, thereby resetting Ca^{2+}_o downward, with maintenance of stable hypocalcemia *in vivo*. Thus, in contrast to FHH/NSHPT, activating CaR mutations produce "oversensitivity" of CaR-expressing tissues to Ca^{2+}_o. Taken together, available data on the deranged Ca^{2+}_o-regulated PTH secretion and urinary calcium handling in FHH/NSHPT, autosomal dominant hypocalcemia, and CaR-deficient mice strongly support a central role for the receptor in Ca^{2+}_o-regulated PTH secretion and renal tubular reabsorption of calcium, respectively. Further details on the roles of the CaR in regulating the functions of parathyroid, kidney, and other tissues are given later.

IV. Signal Transduction Pathways and Biological Responses Regulated by the CaR

Initial studies of the cloned CaR focused on its roles in modulating signal transduction pathways known for some time to be regulated by other GPCRs, including phospholipases and adenylate cyclase, and associated regulation of rapid cellular responses, such as hormonal secretion (i.e., by parathyroid and C-cells) and ion transport (e.g., in the kidney) (Chattopadhyay *et al.*, 1996). Agonists of the CaR activate phospholipases C, A_2, and D (PLC, PLA_2, and PLD, respectively) in bovine parathyroid cells (Kifor *et al.*, 1997). These actions are likely mediated by the CaR, because high Ca^{2+}_o no longer elicits them in parathyroid cells maintained in culture for 3–4 days, in which the level of CaR expression decreases by 80% or more (Mithal *et al.*, 1995). In addition, CaR agonists activate the same phospholipases in HEK293 cells stably transfected with the CaR, but these agonists are without effect in nontransfected HEK293 cells that do not express an endogenous CaR (Kifor *et al.*, 1997). CaR-mediated stimulation of PLC in bovine parathyroid cells, CaR-transfected HEK cells and most other mammalian cells is thought to reflect coupling of the CaR to PLC via a member of the G_q family of G proteins, probably involving G_{11}, since this effect is not blocked by pertussis toxin. The high Ca^{2+}_o-elicited activations of PLA_2 and PLD, in contrast, are probably indirect, taking place through CaR-mediated, PLC-dependent stimulation of the activity of protein kinase C (PKC), because downregulating or inhibiting PKC substantially diminishes CaR-mediated stimulation of these two phospholipases (Kifor *et al.*, 1997).

The high Ca^{2+}_o-evoked, transient rise in Ca^{2+}_i in bovine parathyroid cells and CaR-transfected HEK293 cells likely results from activation of PLC and the ensuing IP_3-mediated release of Ca^{2+} from its intracellular stores. High Ca^{2+}_o also produces sustained increases in Ca^{2+}_i in CaR-transfected HEK293 cells and parathyroid cells through an incompletely defined influx pathway(s) for Ca^{2+}_o. One such influx pathway may be a high Ca^{2+}_o- and CaR-activated, nonselective cation channel (NCC) in CaR-transfected HEK cells that exhibits substantial permeability to Ca^{2+} (Ye *et al.*, 1996b). A similar NCC in bovine parathyroid cells is activated by high Ca^{2+}_o, presumably through a CaR-dependent pathway, and may contribute to the high Ca^{2+}_o-evoked, sustained elevation in Ca^{2+}_i in this cell type (Chang *et al.*, 1995).

In bovine parathyroid cells, there is a marked, high Ca^{2+}_o-induced inhibition of cAMP accumulation that is pertussis toxin-sensitive (Chen *et al.*, 1989), suggesting that CaR-mediated inhibition of adenylate cyclase involves one or more isoforms of the inhibitory G protein, G_i. Further support for this putative mechanism has been provided by the observation that stable transfection with the CaR confers high Ca^{2+}_o-induced inhibition of cAMP accumulation on HEK293 cells (Chang *et al.*, 1998). Studies in tubules from the medullary thick ascending limb of mouse kidney, however, have shown that high Ca^{2+}_o-evoked inhibition of agonist-stimulated cAMP accumulation (Takaichi and Kurokawa, 1988) can take place through an indirect mechanism involving arachidonic acid (Firsov *et al.*, 1995). Direct addition of arachidonic acid to the tubular suspensions conferred a pertussis toxin-sensitive inhibition of cAMP accumulation (Firsov *et al.*, 1995). More recent studies by these same workers have shown that this mechanism involves a CaR-mediated increase in PLC activity leading to a rise in Ca^{2+}_i owing both to mobilization of intracellular calcium and to influx of Ca^{2+}_o followed by direct inhibition of a calcium-inhibitable isoform of adenylate cyclase present in this nephron segment by the elevated level of Ca^{2+}_i (Ferreira *et al.*, 1998). It is not yet clear whether the high Ca^{2+}_o-evoked inhibition of cAMP accumulation in HEK293 cells stably expressing the cloned CaR (Chang *et al.*, 1998) involves a similar mechanism or whether the CaR directly couples to inhibition of adenylate cyclase via G_i.

More recent studies have emphasized that a wider range of cellular functions are regulated by the CaR, particularly longer term actions, such as the control of cellular proliferation and differentiation. Moreover, investigation of the signal transduction pathways underlying these actions has shown the involvement of not only previously recognized intracellular mediators activated by the CaR but also novel intracellular signaling mechanisms not known to mediate the actions of this receptor, as described later. Available evidence suggests that the CaR can exert either inhibitory or stimulatory actions on cellular proliferation.

The receptor tonically inhibits parathyroid cellular proliferation, because infants with NSHPT due to homozygous inactivating CaR mutations or mice homozygous for targeted disruption of the CaR gene exhibit marked parathyroid cellular hyperplasia (Marx *et al.*, 1982; Fujita *et al.*, 1983; Heath, 1994; Ho *et al.*, 1995). A study showing that the calcimimetic CaR activator, NPS R-568, suppresses parathyroid cellular proliferation in rats with experimentally induced renal insufficiency provides additional support for a direct role for the CaR in suppressing proliferation of parathyroid cells (Wada *et al.*, 1997). High Ca^{2+}_o also inhibits proliferation of the CaR-expressing, human colon cancer-derived cell line, Caco-2 (Kallay *et al.*, 1997), although this action has not yet been proven to be CaR-mediated. Finally, high Ca^{2+}_o inhibits the proliferation and stimulates the differentiation of cultured keratinocytes derived from human or mouse skin—an effect that could also be CaR-mediated, since keratinocytes express the CaR (Bikle *et al.*, 1996).

In contrast, the CaR has been conclusively shown to be involved in the high Ca^{2+}_o-induced proliferation of fibroblasts. CaR agonists stimulate the proliferation of CCL39 hamster fibroblasts transfected with the CaR (Mailland *et al.*, 1997), and transfection of NIH 3T3 cells with a human CaR cDNA harboring an activating mutation induces proliferation and cell transformation (Hoff *et al.*, 1997). In addition, high Ca^{2+}_o stimulates mitogenesis of rat-1 fibroblasts expressing an endogenous CaR through CaR-mediated activation of a mitogen-activated protein (MAP) kinase signaling pathway involving Src kinase and ERK1 (McNeil *et al.*, 1998). Therefore, as recently found for other GPCRs, there can be CaR-mediated cross-talk involving intracellular signaling pathways originally identified as mediating the mitogenic effects of other classes of cell surface receptors (Van Biesen *et al.*, 1996), such as those for insulin-like growth factor-I (IGF-I) or other growth factors.

High Ca^{2+}_o, as well as additional CaR agonists such as Gd^{3+} and neomycin, not only stimulate the proliferation but also induce the chemotaxis of mouse osteoblastic MC3T3-E1 cells (Yamaguchi *et al.*, 1998b), monocyte–macrophage-like J774 cells (Yamaguchi *et al.*, 1998c), and ST-2 stromal cells (Yamaguchi *et al.*, 1998a), and high Ca^{2+}_o is chemotactic for peripheral blood monocytes (Sugimoto *et al.*, 1993). Inhibition of the activation of either G protein or PLC blocks nearly all of the high Ca^{2+}_{o}-induced chemotaxis of MC3T3-E1 cells, whereas inhibition of protein kinase C or phosphoinositide-3-kinase does not (Godwin and Soltoff, 1997). These results suggested that Ca^{2+}_{o}-stimulated chemotaxis of this cell line is linked to the activation of G protein and PLC. We have demonstrated the presence of endogenous CaR expression in MC3T3-E1 murine osteoblastic cells (Yamaguchi *et al.*, 1998b), monocyte–macrophage-like J774 cells (Yamaguchi *et al.*, 1998c), ST-2 murine stromal cells (Yamaguchi *et al.*, 1998a), and human peripheral blood monocytes (Yamaguchi *et al.*, 1998e), raising the possibility that it represents the mechanism through which high Ca^{2+}_{o} stimulates chemotaxis of all three cell types. Others, however, have suggested that a Ca^{2+}_{o}-sensing mechanism distinct from the CaR mediates these actions of CaR agonists on MC3T3-E1 cells (Quarles *et al.*, 1997), as will be described in more detail later (see Section IX). Therefore, additional studies are needed to define unequivocally the role of the CaR in the actions of Ca^{2+}_{o} on these latter three cell types.

V. Tissue Distribution of the CaR

Recent data accumulated since the cloning of the CaR have indicated that it is expressed not only in parathyroid but also in a variety of other tissues, including thyroid C-cells, kidney, intestine, bone, placenta, brain, skin, lens epithelial cells, and breast (for review, see Chattopadhyay and Brown, 1997). A number of these tissues are seemingly uninvolved in systemic Ca^{2+}_{o} homeostasis, suggesting additional roles of the CaR in ion sensing within the microenvironments of each tissue, where Ca^{2+}_{o} could potentially differ from and vary independently of its systemic level. Examples include bone marrow, brain, intestine, and skin, where the receptor could potentially regulate numerous processes. These include, respectively, proliferation and chemotaxis of marrow-derived cells (i.e., CaR-expressing marrow stromal cells or monocytes–macrophages) (Yamaguchi *et al.*, 1998a, 1998b, 1998c, 1998e); activities of ion channels in neurons (Ye *et al.*, 1996a, 1997b), and proliferation of oligodendrocytes (Chattopadhyay *et al.*, 1998b); proliferation of the colonic crypt cells (Kallay *et al.*, 1997); and differentiation of keratinocytes (Bikle *et al.*, 1996) and proliferation of fibroblasts (McNeil *et al.*, 1998), presumably in response to local changes in Ca^{2+}_{o}. The evidence supporting roles for the CaR in regulating these various cell types and processes is given in Sections VI and VII. This widespread tissue distribution of the CaR contrasts with that of the mGluRs, which have been reported to be localized primarily in nervous tissue (Nakanishi, 1992, 1994).

VI. Roles of CaR in Tissues Regulating Mineral Ion Homeostasis

A. CaR in Parathyroid Cells

In spite of its crucial role in controlling parathyroid function, remarkably little is known of the details of how the CaR regulates this cell type. As described earlier, the CaR activates PLC, PLA_2 and PLD and inhibits adenylate cyclase in bovine parathyroid cells (Chen *et al.*, 1989; Bourdeau *et al.*, 1992; Kifor *et al.*, 1997; Chang *et al.*, 1998). After more than two decades of study by several laboratories, however, it remains unclear which, if any, of these pathways represents the key mediator(s) of the inhibitory action of high Ca^{2+}_o on PTH secretion. Furthermore, it is largely unknown how the CaR controls the crucial step(s) in the secretory pathway of the parathyroid cell, from the budding of secretory vesicles at the Golgi apparatus to their eventual exocytosis at the plasma membrane. Therefore, we lack a clear understanding of the intracellular mechanisms distal to the CaR through which high Ca^{2+}_o suppresses PTH secretion, a response that is opposite in direction to that of most other hormone-secreting cells. As noted earlier, the CaR is also thought to mediate the inhibitory action of elevated levels of Ca^{2+}_o on parathyroid cellular proliferation (Wada *et al.*, 1997). It remains to be determined whether it likewise represents a central mediator of other actions of Ca^{2+}_o on parathyroid function, such as inhibition of PTH gene expression.

B. CaR in Thyroid C-Cells

Although PTH and vitamin D are the most important hormones that regulate systemic Ca^{2+}_o homeostasis in humans, calcitonin (CT), which is secreted from the thyroidal C-cells, also exerts a physiologically relevant hypocalcemic action in some species, such as the rat (Brown, 1991). In sharp contrast to the inhibitory action of raising Ca^{2+}_o on PTH secretion, high levels of Ca^{2+}_o stimulate CT secretion (Fried and Tashjian, 1986; Eskert *et al.*, 1989; Fajtova *et al.*, 1991), more akin to the classical, positive relationship between calcium and exocytosis in most other secretory cells. Elevations in Ca^{2+}_o elicit increases in Ca^{2+}_i in both parathyroid cells and C-cells, but the two cell types show clear differences in the mechanisms by which this occurs. Influx of Ca^{2+}_o through voltage-dependent Ca^{2+} channels is the major contributor to high Ca^{2+}_o-evoked increases in Ca^{2+}_i in C-cells (Fried and Tashjian, 1986; Muff *et al.*, 1988; Fajtova *et al.*, 1991), whereas in parathyroid cells there is some controversy regarding the identity of the influx pathways for Ca^{2+}_o (Muff *et al.*, 1988; Pocotte *et al.*, 1995), which more likely involve voltage-insensitive Ca^{2+}-permeable influx pathways, as noted earlier (Chang *et al.*, 1995). Moreover, mobilization of intracellular Ca^{2+} contributes importantly to the initial rise in Ca^{2+}_i in response to elevating Ca^{2+}_o in parathyroid cells (Nemeth and Scarpa, 1987).

The use of RT-PCR, Northern blot analysis, *in situ* hybridization, Western blot analysis, and immunohistochemistry with specific anti-CaR antisera, however, has provided unequivocal evidence that rat C-cells express the same CaR present in parathyroid and kidney (Garrett *et al.*, 1995b; Freichel *et al.*, 1996). Thus, it is likely that activation of this same receptor can either stimulate or inhibit hormonal secretion, depending on the cellular context in which it is expressed. There are, however, some pharmacological differences between the effects of various CaR agonists on parathyroid and C-cells, the basis for which is not well understood. Not all CaR agonists that inhibit PTH secretion evoke CT secretion from C-cells. For instance, while elevating the level of extracellular Mg^{2+} (Mg^{2+}_o) reduces PTH secretion (albeit with a two- to three-fold lower potency than for Ca^{2+}_o) (Brown, 1991), Mg^{2+}_o is without effect on CT secretion from CaR-expressing sheep parafollicular cells (McGehee *et al.*, 1997).

Detailed studies utilizing electrophysiology, measurements of Ca^{2+}_i, and various pharmacological probes have provided evidence for the following sequence of events in the presumably CaR-mediated stimulation of CT secretion by high Ca^{2+}_o in sheep C-cells (McGehee *et al.*, 1997). Initially, CaR-induced activation of phosphatidylcholine-specific PLC yields a source of diacylglycerol for subsequent PKC-mediated stimulation of an NCC. This channel then allows influx of Na^+ and Ca^{2+}, thereby depolarizing the cells, stimulating voltage-dependent, principally L-type Ca^{2+} channels and activating exocytosis of 5-hydroxytryptamine (5-HT) and CT. It is likely that the CaR regulates additional processes in C-cells, including a pertussis toxin-sensitive, protein kinase C-dependent acidification of 5-HT-containing vesicles. Acidification of such vesicles is postulated to play an important role in the mechanism of loading of secretory vesicles with 5-HT, neurotransmitters, and other hormonal products. In contrast, the Ca^{2+}_o-elicited secretion of 5-HT and CT from sheep parafollicular cells is insensitive to pertussis toxin (Tamir *et al.*, 1996).

C. CaR in the Kidney

The kidney plays an essential role in Ca^{2+} and Mg^{2+} homeostasis by regulating the tubular reabsorption of these divalent cations from the glomerular filtrate. PTH and CT, as well as vitamin D, play important roles in regulating divalent mineral handling by the nephron (Brown, 1991). In the absence of these calciotropic factors, however, the steep relationship between plasma and urinary calcium is preserved in both rats (Kurokawa, 1987) and humans (Attie *et al.*, 1983). This suggests that an additional calciotropic factor is involved in regulating renal Ca^{2+} excretion. Recent evidence indicates that Ca^{2+}_o itself, by interacting with the CaR, can function as a calciotropic "hormone" and affords this regulatory function.

Riccardi *et al.* (1996, 1998) found, utilizing RT-PCR with CaR-specific primers as well as immunohistochemistry with specific anti-CaR antisera, that CaR transcripts are present along nearly the entire nephron, including glomerulus, proximal convoluted (PCT) and proximal straight tubule (PST), medullary thick ascending limb (MTAL), cortical thick ascending limb (CTAL), distal convoluted tubule (DCT), cortical collecting duct (CCD), and inner medullary collecting duct (IMCD). In the proximal tubule, the CaR is located principally at the base of the brush border on the apical membrane of the tubular epithelial cells. The CaR also has a predominantly apical distribution in IMCD. In CTAL, in contrast, the receptor is highly expressed on the basolateral membranes of the tubular cells. It is likewise present in a primarily basolateral distribution in MTAL and DCT, although at lower levels than are present in CTAL. In CCD, the CaR is expressed in a substantial fraction of the acid-secreting, type A intercalated cells (Riccardi *et al.*, 1998).

Elucidation of the receptor's distribution along the nephron, taken together with previous studies of the actions of Ca^{2+}_o on various aspects of renal function, have clarified considerably the role of the CaR in the kidney. The "experiments-in-nature" afforded by disorders of Ca^{2+} homeostasis that are caused by inactivating or activating CaR mutations have provided additional clues in this regard (Chattopadhyay *et al.*, 1996). Previous investigations had established that raising peritubular levels of Ca^{2+}_o or Mg^{2+}_o decreased the reabsorption of both of these divalent cations in TAL (Quamme and Dirks, 1980; Quamme, 1982). Reabsorption of Ca^{2+} or Mg^{2+} in CTAL occurs predominantly through a paracellular pathway and is driven by the lumen-positive, transepithelial potential gradient that is generated by the transcellular transport of sodium (Na), potassium (K), and chloride (Cl) by the apical $Na^+/K^+/2Cl^-$ cotransporter, combined with recycling of K^+ ions back into the tubular lumen through an apical K^+ channel (Fig. 4) (De Rouffignac and Quamme, 1994; Hebert *et al.*, 1997). PTH as well as other hormones increasing tubular levels of cAMP in the epithelial cells of the CTAL (e.g., glucagon, calcitonin, and beta-adrenergic catecholamines) enhance Ca^{2+} or Mg^{2+} reabsorption by activating the $Na^+/K^+/2Cl^-$ cotransporter and, in turn, the magnitude of the lumen-positive transepithelial potential gradient (De Rouffignac and Quamme, 1994; Hebert *et al.*, 1997). Studies using patch clamp methodology have shown that Ca^{2+}_o blocks the apical K^+ channel by a mechanism most likely involving CaR-mediated generation of one or more P-450 metabolite(s) of arachidonic acid, probably 20-HETE (Fig. 4) (Wang *et al.*, 1996). The ensuing reduction in apical recycling of potassium depletes luminal levels of K^+, with an attendant reduction in the cotransporter's activity and *pari passu* in paracellular transport of Ca^{2+} and Mg^{2+}. Studies by Ferreira *et al.* (1998) have shown that the CaR also activates PLC in CTAL, leading to an accompanying rise in

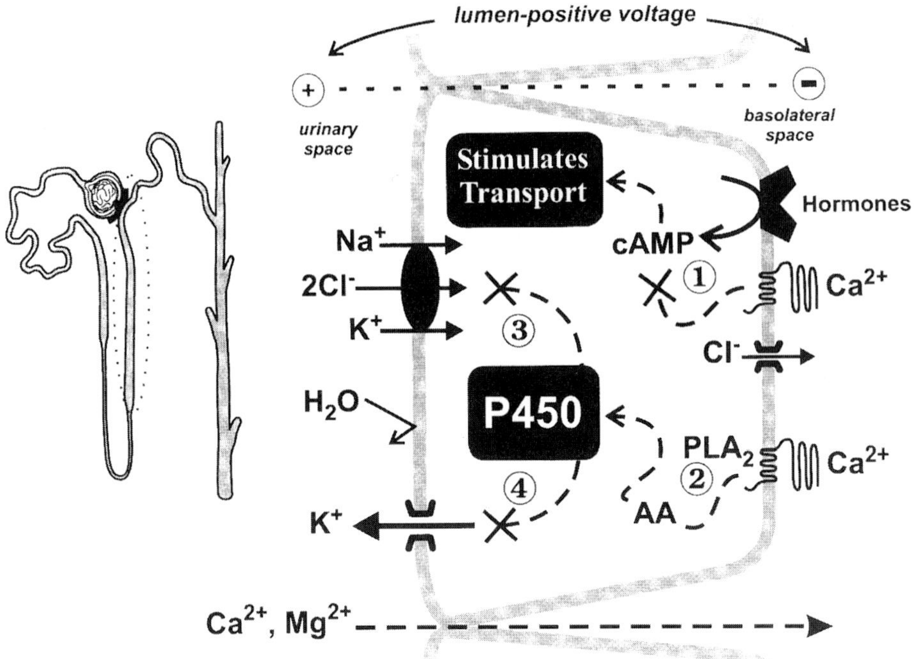

FIGURE 4 Diagram illustrating how the CaR is thought to regulate second messengers and ionic transport in CTAL. Hormones that stimulate cAMP accumulation, such as PTH, increase Ca^{2+} and Mg^{2+} reabsorption via the paracellular pathway by enhancing the lumen-positive transepithelial potential, V_{te}, through stimulation of the $Na^+/K^+/2Cl^-$ cotransporter and the apical K^+ channel. The CaR promotes formation of arachidonic acid (AA) by stimulating PLA_2 (2), which is then metabolized via the P-450 pathway to an active metabolite, likely 20-HETE, that inhibits the K^+ channel (4) and, perhaps, the $Na^+/K^+/2Cl^-$ cotransporter (3). Both actions will decrease overall cotransporter activity, reducing V_{te} and, in turn, the paracellular transport of Ca^{2+} and Mg^{2+}. The CaR also inhibits adenylate cyclase (1), thereby diminishing hormone- and cAMP-activated divalent cation transport. [Reproduced in modified form with permission from Brown, E. M., and Hebert, S. C. (1997). Calcium-receptor regulated parathyroid and renal function. *Bone* 20, 303–309.

$Ca^{2+}{}_i$, which then inhibits adenylate cyclase directly, as noted earlier, rather than via a pertussis-toxin sensitive, G_i-mediated inhibition of adenylate cyclase by the CaR, as is thought to occur in parathyroid. It is probable that, similar to previous results in CaR-transfected HEK cells and bovine parathyroid cells (Kifor *et al.*, 1997), the CaR activates PLA_2 in CTAL indirectly via activation of PLC (Ferreira *et al.*, 1998). Moreover, in addition to reducing divalent cation transport in this nephron segment via the action of metabolites of PLA_2, the CaR also likely inhibits this process through inhibition of hormone (e.g., PTH)-stimulated cAMP accumulation (Ferreira *et al.*, 1998).

Clinical investigation of FHH patients has generated further evidence supporting a central role of the CaR in controlling renal tubular Ca^{2+} reabsorption. These individuals show a reduced capacity to excrete urinary Ca^{2+} in response to their elevated serum Ca^{2+} concentrations, with resultant, inappropriately reduced levels of urinary calcium (Marx *et al.*, 1981a; Law and Heath, 1985). Moreover, this overly avid tubular Ca^{2+} reabsorption in the face of hypercalcemia continues even after parathyroidectomy (Attie *et al.*, 1983; Davies *et al.*, 1984), demonstrating that it is not solely the result of impaired inhibition of PTH secretion by elevated levels of Ca^{2+}_{o}. The only one of the maneuvers normally promoting urinary calcium excretion that does so in FHH is administration of the loop diuretic ethacrynic acid (Attie *et al.*, 1983), further supporting the location in the TAL of the abnormal renal calcium handling in FHH, since the $Na^{+}/K^{+}/2Cl^{-}$ cotransporter inhibited by such loop diuretics is localized solely in MTAL and CTAL. In contrast to persons with FHH, those with activating mutations of the CaR show excessively high urinary levels of Ca^{2+} at any given serum calcium concentration relative those in persons with hypoparathyroidism of other causes, presumably because of "activated" CaRs along the nephron, especially in CTAL (Davies *et al.*, 1995; Baron *et al.*, 1996). Since PTH levels in individuals with CaR activating mutations are often higher than those present in other causes of hypoparathyroidism—which, if anything, should lower urinary calcium excretion—the higher levels of urinary Ca^{2+} in the former condition further stress the important role of renal CaRs in controlling renal calcium handling.

In hypercalcemic persons, CaR-mediated reductions in sodium and calcium transport in the TAL promote increases in the quantities of NaCl and calcium that are excreted in the urine. If maximal arginine vasopressin (AVP)-elicited water reabsorption occurs in the IMCD, luminal calcium concentrations can rise to potentially dangerous levels (>5 mM) that could elevate the risk of precipitation of calcium oxalate and/or calcium phosphate and formation of Ca^{2}-containing renal stones. Hypercalcemia and the resultant hypercalciuria, however, produce a renal concentrating defect characterized by resistance of the epithelial cells of the IMCD to either endogenous or exogenous AVP (nephrogenic diabetes insipidus) (Gill and Bartter, 1961; Suki *et al.*, 1969). The identification of the CaR in the apical surface of the tubular cells in the IMCD has afforded novel insights into how high Ca^{2+} in the urine of the terminal collecting system modulates the urinary concentrating mechanism in this nephron segment (Sands *et al.*, 1997). Perfusion of the lumen of tubules isolated from the rat IMCD with high Ca^{2+}_{o}, probably through activation of the CaR localized on the apical membrane, reversibly diminishes vasopressin-induced, transepithelial water flow by 30–40%. In addition, high Ca^{2+}_{o}-evoked, CaR-mediated inhibition of sodium chloride reabsorption in MTAL (Wang *et al.*, 1996; Hebert *et al.*, 1997) also reduces the degree of hypertonicity that is normally present in the medullary interstitium. The hypertonic medullary interstitium is necessary for passive reabsorption of water consequent to AVP-evoked increases in the availability and/or activity of aquaporin-2 water channels in the apical membrane of the epithelial cells of the IMCD. This action would reduce maximal urinary concentrating ability in hypercalcemic persons. Of interest, patients with FHH are able to concentrate their urine normally despite being hypercalcemic (Marx *et al.*, 1981b). Conversely, persons with activating CaR mutations may develop impaired urinary concentrating ability at normal or even low levels of serum calcium, probably because they are unduly sensitive to the usual inhibitory actions of elevated Ca^{2+}_o on the urinary concentrating mechanism (Pearce *et al.*, 1996b). These latter observations further support the CaR's role as the mediator of the actions of hypercalcemia on urinary concentrating capacity.

D. CaR in the Intestine

The intestine is important for the maintenance of Ca^{2+} homeostasis because of its regulated absorption of dietary Ca^{2+} in response to vitamin D (Brown, 1991). The duodenum is a key locus for 1,25-dihydroxyvitamin D_3 [1,25(OH)$_2$D$_3$]-dependent Ca^{2+} absorption, involving active Ca^{2+} transport that most likely includes the vitamin D-dependent Ca^{2+}-binding protein, calbindin (Pansu *et al.*, 1983; Feher *et al.*, 1989). In contrast, the jejunum and ileum, in addition to absorbing lesser amounts of Ca^{2+}, are known to secrete Ca^{2+}, which may chelate fatty acids and bile salts, thereby forming insoluble "calcium soaps" that could mitigate damage to colonic epithelial cells that might otherwise be caused by soluble, unchelated fatty acids and bile salts (Bronner *et al.*, 1986; Karbach, 1991). Although the major function of colon is to absorb water and Na^+, it also absorbs significant amounts of Ca^{2+} by both vitamin D-dependent and -independent mechanisms (Favus *et al.*, 1981).

We have demonstrated CaR expression in all segments of rat intestine (Chattopadhyay *et al.*, 1998a). It is present at particularly high levels on the basal surface of the absorptive villar cells, small intestinal and colonic crypt cells, and in Auerbach's and myenteric plexuses. Functional CaR expression has also been reported in cell lines derived from various intestinal cells by Gama *et al.* (1997), such as T-84 (human villar cells), HT-29 (human crypt cells), and Caco-2 (human colon carcinoma), where it mobilizes intracellular Ca^{2+} through PLC in response to CaR agonists. Although the CaR's exact role(s) in the intestine remains to be elucidated, several possibilities are that the receptor regulates intestinal motility by modulating neurotransmitter secretion in the enteric nervous system, controls absorptive processes by the epithelial cells of the villi, and/or modulates the secretion of chloride and other ions by crypt cells.

Additional evidence suggests that the CaR may have a role in regulating the proliferation of colonocytes. Raising Ca^{2+}_o inhibits, while reducing Ca^{2+}_o increases, the proliferation of the human colon cancer-derived cell line, Caco-2, as noted earlier, in association with a rise in the expression of the c-*myc* proto-oncogene (Kallay *et al.*, 1997). We have shown that this cell line expresses the CaR (Kallay *et al.*, 1997), which might, therefore, be involved in the low Ca^{2+}_o-induced proliferation of these cells. Application of low levels of Ca^{2+}_o to the apical but not the basolateral cell surface of Caco-2 cells in culture increases c-*myc* expression (Kallay *et al.*, 1997). The lower third of the lumen of the colonic crypts has been postulated to contain reduced levels of Ca^{2+}_o (Whitfield, 1995). Therefore, Ca^{2+}_o-sensing by the CaR might provide a key cellular "switch" that normally inhibits the proliferation and promotes the differentiation of the stem cells at the crypt bases as they migrate up the walls of the crypts.

E. CaR in Bone

Bone, like parathyroid gland and kidney, is involved in systemic mineral ion homeostasis (Brown, 1991), and thus it is possible that the CaR may also play some role within the skeleton by sensing local changes in Ca^{2+}_o caused by bone remodeling. Bone formation during the normal process of skeletal remodeling is initiated by the migration of macrophage-like mononuclear cells to sites of osteoclastic bone resorption, during the "reversal" phase of skeletal turnover that precedes the laying down of new bone, which are then followed by preosteoblasts (Baron, 1996). These preosteoblasts subsequently differentiate into mature osteoblasts and eventually deposit and mineralize osteoid protein. Bone resorption induces local increases in Ca^{2+}_o within the immediate vicinity of osteoclasts that are known to reach levels of 8–40 mM (Silver *et al.*, 1988). The latter could, therefore, provide both macrophage-like mononuclear cells and preosteoblasts with a signal that modulates their subsequent physiological responses, such as migration and proliferation. In fact, high Ca^{2+}_o induces chemotaxis of human peripheral blood monocytes (Sugimoto *et al.*, 1993) and chemotaxis, as well as DNA synthesis of mouse osteoblastic MC3T3-E1 cells (Yamaguchi *et al.*, 1998b). These two cell types have the capacity, respectively, to differentiate into mature osteoclasts under specific culture conditions (Fujikawa *et al.*, 1996) and to differentiate from preosteoblasts to mature osteoblasts in culture (Sudo *et al.*, 1983).

In a previous study, using immunohistochemistry with CaR-specific antisera, we showed expression of the CaR in diverse cell types in human bone marrow, including alkaline phosphatase (ALP)-positive, putative osteoblast precursors, nonspecific esterase (NSE)-positive mononuclear cells, erythroid precursors, and megakaryocytes (House *et al.*, 1997). We subsequently confirmed the presence of both CaR protein and mRNA in the osteoblast-like cell lines, UMR-106 and SAOS-2 (Yamaguchi *et al.*, 1998d), and in human peripheral blood monocytes (Yamaguchi *et al.*, 1998e). We have also found that three different murine cell lines—the monocyte-macrophage-like J774 cell line (Yamaguchi *et al.*, 1998c), the marrow stromal ST2 cell line (Yamaguchi *et al.*, 1998a), and the osteoblastic MC3T3-E1 cell line (Yamaguchi *et al.*, 1998b)—express the CaR. Furthermore, CaR agonists stimulate chemotaxis and proliferation of these three cell lines (Yamaguchi *et al.*, 1998a, 198b, 1998c), suggesting that the CaR could potentially represent the molecular mediator of the actions of high Ca^{2+}_o on these two physiological activities in these marrow-derived cells. Since stromal cells and osteoblasts both belong to the mesenchymal lineage, while monocytes–macrophages belong to the hematopoietic lineage (Caplan and Dennis, 1996), these findings suggest that the CaR is expressed in diverse bone marrow cells, confirming our observation with primary cultures of human and murine bone marrow (House *et al.*, 1997). Thus, our findings are consistent with an important role for the CaR in the key "reversal" phase of bone remodeling through its sensing of calcium ions released by osteoclast-mediated bone resorption, which then ensures the chemotaxis and proliferation of cell populations needed for the orderly transition from bone breakdown to its subsequent replacement by newly formed bone. As described in more detail later, however, other investigators have suggested that the effects of elevated levels of Ca^{2+}_o on MC3T3-E1 cells are mediated by a Ca^{2+}_o-sensing mechanism distinct from the CaR (Quarles *et al.*, 1997). Further studies, therefore, such as the use of bone cells and their precursors that are derived from mice with "knockout" of the CaR or of techniques that downregulate the function of the endogenous CaR in bone cells, are needed to establish definitively the role of this receptor in bone cell function.

F. CaR in Placenta

The placenta plays a key role in fetal mineral ion homeostasis and skeletal growth because all fetal Ca^{2+} must move from mother to fetus via placental transport. Most of the mineralization of the developing fetal skeleton takes place during the third trimester, with deposition of ~30 g of calcium in the newborn's skeleton by the time of birth (Rodda *et al.*, 1992). Ca^{2+}_o-sensing cells are known to be present among the placental cytotrophoblasts and may participate in the control of Ca^{2+} transport between mother and fetus by controlling the secretion of PTHrP by these cells (Hellman *et al.*, 1992). Similar to parathyroid chief cells, raising Ca^{2+}_o increases the level of Ca^{2+}_i in human placental cytotrophoblasts (Juhlin *et al.*, 1990; Bradbury *et al.*, 1996), suggesting that these cells might express a Ca^{2+}_o-sensing mechanism(s) similar to the CaR. Bradbury *et al.* (1998) demonstrated expression of CaR transcripts in cytotrophoblasts from human placenta. In addition to the transcripts present in other CaR-expressing cells that encode the full-length receptor protein, cytotrophoblasts express an additional, alternatively spliced transcript (Bradbury *et al.*, 1998). This RNA species lacks exon 3 and codes for a truncated receptor protein that is probably inactive, because it contains a frame shift, thereby generating a premature stop codon within the CaR's ECD. Human parathyroid cells also express this alternatively spliced CaR transcript in addition to the more abundant transcript that encodes the full length, biologically active receptor. Thus, the CaR represents a potential candidate for the Ca^{2+}_o-sensing mechanism present in placental cytotrophoblasts that modulates Ca^{2+}_i and PTHrP release. These cells also express, however, another putative, Ca^{2+}_o-sensing protein, called gp330 or megalin, that is a member of the LDL receptor superfamily (Hellman *et al.*, 1992; Saito *et al.*, 1994; Hjalm *et al.*, 1996), and the relative importance of the two proteins in mediating the Ca^{2+}_o-sensing capacity of this cell type is not currently known. The fetal parathyroid glands are also known to secrete PTHrP *in utero*. They, too, might participate in the control of placental calcium transport, a possibility that is supported by studies demonstrating alternations in placental calcium transport after parathyroidectomy in fetal sheep (Rodda *et al.*, 1992).

VII. Role of the CaR in Cells Uninvolved in Systemic Mineral Ion Homeostasis

A. CaR in the Brain

In adult rat brain the CaR has been localized to various regions and cell types using immunocytochemistry and/or *in situ* hybridization (Ruat *et al.*, 1995; Chattopadhyay *et al.*, 1997a; Rogers *et al.*, 1997). The receptor is widely expressed in brain, albeit at varying levels. The highest levels are observed in the subfornical organ (SFO) and olfactory bulbs. High expression levels are also present in hippocampus, striatum, cingulate cortex, cerebellum, and the ependymal zones of the cerebral ventricles, as well as in cerebral arteries. The CaR is expressed not only in neurons but also in oligodendroglia (Chattopadhyay *et al.*, 1998b), as well as microglia (Chattopadhyay *et al.*, unpublished observations). As discussed later, although some progress has been made in defining several types of ion channels and cellular responses that may be CaR-regulated in these various cell types, a great deal remains to be learned about the receptor's role in controlling various aspects of brain function, particularly processes involving interactions between brain cells, such as neurotransmission.

Since the SFO is outside of the blood–brain barrier and presumably senses systemic levels of Ca^{2+}_o that are relevant to systemic fluid and electrolyte metabolism, the localization of abundant CaRs in the SFO, which is a key hypothalamic thirst center (Simpson and Routenberg, 1975), could provide a further layer of integration of calcium and water homeostasis beyond that already described in Section VI. High Ca^{2+}_o-elicited, CaR-mediated thirst and the consequent increase in drinking could minimize the dehydration that might otherwise be the consequence of a fixed renal loss of free water owing to CaR-induced renal resistance to the actions of AVP (Sands *et al.*, 1997).

Most CaRs within other areas of the brain are separated from the systemic circulation by the blood–brain barrier, which effectively buffers Ca^{2+}_o in the brain ECF from rapid changes in this parameter that might otherwise occur owing to alterations in systemic levels of Ca^{2+}_o. Rather, it appears most likely that CaRs in the brain as well as in a variety of other tissues uninvolved in systemic ionic homeostasis respond to local changes in Ca^{2+}_o arising from a variety of factors that are described in more detail in Section VIII.

In the brain, it is well documented that substantial alterations in Ca^{2+}_o can take place under conditions of increased neuronal activity and certain pathological states, such as seizures, ischemia, and hypoglycemia. In these circumstances, Ca^{2+}_o within the extracellular fluid of the brain is usually reduced as a result of uptake of Ca^{2+}_o through various types of Ca^{2+}_o-permeable channels in the plasma membrane (Heinemann *et al.*, 1977; Arens *et al.*, 1992; Lucke *et al.*, 1995). Such neuronal activity-dependent changes in Ca^{2+}_o could potentially be sensed by CaRs expressed on neuronal and glial elements in the local microenvironment.

The CaR is highly expressed in synaptic regions of all areas of the hippocampus (Ruat *et al.*, 1995; Chattopadhyay *et al.*, 1997a), although it has not yet been shown with certainty whether it resides principally in a pre- or postsynaptic location. What is the role of the CaR in hippocampal neurons, as well as those in cerebellum and various other regions of the brain? The CaR's overall distribution in hippocampus is similar to those of the mGluRs and the *N*-methyl-D-aspartate (NMDA) type of ionotropic glutamate receptor channels, both of which play important roles in certain forms of long term potentiation (LTP). We have shown that the activity of a Ca^{2+}-permeable NCC is modulated by the CaR expressed in rat hippocampal pyramidal neurons (Ye *et al.*, 1996a). An NCC with similar characteristics is also regulated in a CaR-dependent fashion in HEK293 cells that are stably transfected with the CaR but not in nontransfected HEK293 cells that do not express an endogenous CaR (Ye *et al.*, 1996b). Moreover, spermine and other CaR agonists activate a very similar NCC in neurons cultured from wild-type mice, but not those from mice with "knockout" of the CaR gene (Ye *et al.*, 1997b). The relationship of this NCC to other Ca^{2+}-permeable, nonselective channels, including the NMDA type of ionotropic glutamate receptor channel, is not yet entirely clear. CaR-mediated activation of such NCCs, however, could contribute to changes in neuronal levels of Ca^{2+}_i that are known to be important for the induction of LTP and other forms of synaptic plasticity. Therefore, based on its anatomic distribution in the brain and regulation of calcium-permeable channels, it is conceivable that the CaR contributes to cognitive functions, such as learning and memory. Our observations that large increases in CaR expression take place in the developing rat hippocampus at a time when LTP can first be induced (Chattopadhyay *et al.*, 1997a) could also indicate some involvement of the CaR in such key synaptic functions. Direct evidence for such a role, however, is presently lacking. In view of the recent recognition that several mGluRs are capable of sensing Ca^{2+}_o (Kubo *et al.*, 1998), it will also be of great interest to determine whether the CaR and various mGluRs interact in their sensing of Ca^{2+}_o within regions of the brain involved in cognitive functions that express both types of receptor, such as hippocampus and the role of the CaR in such processes. The use of the CaR "knockout" mouse (Ho *et al.*, 1995), particularly if it could be rescued from the severe hyperparathyroidism that limits the viability of this animal and consequent utility as an experimental model, should be of substantial utility in this regard.

The functions of the CaR in regulating other types of cells within the central nervous system are even less well understood; however, recent data suggest additional roles for the receptor in such cells. We have found that primary cultures of rat oligodendrocytes express a functional CaR (Chattopadhyay *et al.*, 1998b). Moreover, receptor agonists stimulate cell proliferation and the opening of a Ca^{2+}-activated K^+ channel in this cell type (Chattopadhyay *et al.*, 1998b). These putatively CaR-mediated actions are similar to the previously observed, CaR-agonist-evoked increases in the proliferation of bone-marrow-derived cells (Yamaguchi *et al.*, 1998a, 1998b, 1998c), as well as the activation of a Ca^{2+}-activated K^+ channel in parathyroid cells (Kanazirska *et al.*, 1995). Oligodendroglia are known to play a key role in local ionic homeostasis within the brain, particularly in their capacity to buffer large changes in extracellular potassium concentration that might be deleterious to neuronal function. The CaR could potentially play a role in such local ionic regulation by oligodendroglia as follows: Neuronal activity-dependent reductions in Ca^{2+}_o are accompanied by substantial elevation in K^+_o. Since these decreases in Ca^{2+}_o would reduce the activity of CaRs in both oligodendroglia and neurons, they would tend to diminish the activities of Ca^{2+}_o-activated K^+ channels that contributed to the original increase in K^+_o.

B. CaR in Hematopoietic Cells in Bone Marrow

A variety of cell types within bone marrow express the CaR (House *et al.*, 1997), which are not directly involved in forming or breaking down bone (e.g., the function of osteoblasts and osteoclasts, respectively). In the marrow, the CaR is expressed in hematopoietic precursors that likely experience significant variations in the levels of Ca^{2+}_o to which they are exposed as a result in changes in the prevailing rate of bone turnover within the local bone/bone marrow microenvironment. These include red blood cell precursors, megakaryocytes (the precursors of blood platelets), monocytes, and macrophages, all of which express relatively high levels of the CaR, and white blood cells precursors, which express lower levels of the receptor (House *et al.*, 1997). In addition, the murine stromal cell line, ST2, also expresses the CaR (Yamaguchi *et al.*, 1998a). Stromal cells influence both bone cell development and function as well as hematopoietic growth and differentiation by secreting cytokines and growth factors and also through direct cell–cell contact. In stromal cells, high Ca^{2+}_o stimulates cell proliferation and chemotaxis (Yamaguchi *et al.*, 1998a)—actions that could potentially be involved in regulating bone turnover, as noted in Section VI.

The functions of the CaR in hematopoietic cells are not well understood, but there is a limited body of data indicating that changes in Ca^{2+}_o have direct actions on these CaR-expressing hematopoietic elements that are of potential physiological relevance. Elevated levels of Ca^{2+}_o have several effects on macrophages or monocytes. First, high Ca^{2+}_o promotes chemotaxis of peripheral blood monocytes (Sugimoto *et al.*, 1993); therefore, Ca^{2+}_o and the CaR share some of the properties of chemokines and chemokine receptors, respectively (Luster, 1998). Second, Bornefalk *et al.* (1997) have shown that high Ca^{2+}_o stimulates secretion of IL-6 both *in vivo* and *in vitro* from peripheral blood monocytes. Finally, raising Ca^{2+}_o potentiates the fusion of rat alveolar macrophages induced by 1,25-dihydroxyvitamin D_3 (Jin *et al.*, 1990).

Elevating Ca^{2+}_o stimulates the formation of erythroid colonies *in vitro* and raises Ca^{2+}_i in erythroid precursors isolated from uremic patients, an effect that is potentiated by 1,25(OH)$_2$ D_3 (Carozzi *et al.*, 1990), which is known to upregulate CaR expression (Brown *et al.*, 1996). Ca^{2+}_o also modulates several processes in platelets, including stimulating release of arachidonic acid (Matsuoka *et al.*, 1989) and inhibiting the accumulation of cAMP (Siegel and Daly, 1985)—effects that might potentially be CaR-mediated. Interestingly, Ca^{2+}_o has been quantified directly in platelet clumps during platelet activation, and it decreases to a substantial extent (Owen *et al.*, 1995). Thus, although much additional work needs to be carried out defining the possible roles of the CaR in cells of various hematopoietic lineages, available data indicate that (a) several of these cells express the CaR, (b) Ca^{2+}_o modulates their functions, and (c) Ca^{2+}_o can change within their microenvironment—either within the bone marrow as a function of bone turnover or within peripheral blood (e.g., within activated platelet clumps).

C. CaR in Keratinocytes

Human keratinocyte differentiation *in vitro* is triggered by elevations in Ca^{2+}_i in response to increases in Ca^{2+}_o through the involvement of (a) release of Ca^{2+} from intracellular stores (Moscat *et al.*, 1989) and (b) Ca^{2+} influx through NCC (Bikle *et al.*, 1996). Moreover, human keratinocytes express transcripts for the CaR, and the differentiating stimulus of an increase in Ca^{2+}_o leads to an increase in CaR mRNA (Bikle *et al.*, 1996). Therefore, current evidence suggests that the CaR could mediate the actions of Ca^{2+}_o on keratinocyte differentiation through its known capacity to activate PLC and NCC (Ye *et al.*, 1996a, 1996b; Kifor *et al.*, 1997).

D. CaR in Lens Epithelial Cells

High Ca^{2+} has several known effects on lens epithelial cells in culture, including disrupting plasma membrane integrity, reducing the cytoskeletal protein, vimentin, and stimulating Ca-ATPase, actions that are thought to contribute to the formation of cataracts (Delamere and Paterson, 1981). Moreover, clinical observations, such as the development of osmotically induced cataracts in patients with hypocalcemia due to hypoparathyroidism and the high content of calcium in these cataracts, strongly implicate roles for Ca^{2+} in the physiology of lens epithelial cells (Duncan and Bushell, 1975). We have shown that cultured human lens epithelial cells express CaR transcripts and protein and that CaR agonists activate a Ca^{2+}-activated K^+ channel (Chattopadhyay *et al.*, 1997b) similar to that known to be activated by the CaR in parathyroid cells (Kanazirska *et al.*, 1995) and cultured hippocampal pyramidal neurons (Vassilev *et al.*, 1997). Thus, the presence of a similar, CaR-regulated K^+ channel in these three cell types suggests that receptor might be involved in controlling membrane repolarization and a variety of associated, voltage-dependent processes, including cellular metabolism, neurotransmission, and secretion in such CaR-expressing cells.

E. CaR in Ductal Cells of Breast

We have demonstrated that CaR protein is expressed at high levels on ductal epithelial cells of normal breast, fibrocystic breast tissue, and ductal carcinoma of the breast (Cheng *et al.*, 1998). Ca^{2+}_o likely plays important roles in the breast, both in its normal physiologic function(s) and in pathologic states. Ca^{2+} is a key constituent of milk, and lactating mothers elaborate breast milk that contains about 200 mg Ca^{2+}_o daily (Prentice *et al.*, 1995). A diagnostically significant characteristic of ductal carcinoma of the breast that can permit early detection of such cancers by mammography is their propensity to form microcalcifications in the ducts (Galkin *et al.*, 1977). Moreover, *in vitro* studies have demonstrated that elevating Ca^{2+}_o within the physiological range can induce senescence and terminal differentiation of normal human breast epithelial cells in primary culture (McGrath and Soule, 1984).

Although there are limited data available on the factors regulating Ca^{2+} transport into milk, the CaR's presence in the ductal cells of the breast suggests that it could potentially participate in regulating such transport processes. Furthermore, breast cancer cells have a distinct tendency to metastasize to bone. Given our recent documentation that numerous cell types within the bone marrow normally express the CaR (House *et al.*, 1997), the presence of the CaR on malignant breast cells might play some role in the tendency of these and additional CaR-expressing malignant cells to metastasize to the skeleton. Therefore, in view of the importance of Ca^{2+}_o in various aspects of breast function, further investigation of the CaR's putative involvement in such actions will likely provide fruitful avenues of investigation.

VIII. What Are the Signals Recognized by CaRs in Nonhomeostatic Tissues?

A. Spatial Heterogeneity of Ca^{2+}_o and the Concept of Local Ca^{2+}_o Homeostasis

Although the systemic level of Ca^{2+}_o remains within a remarkably narrow range, as will be elaborated upon later, there are a variety of microenvironments where Ca^{2+}_o differs from its systemic level or can vary largely independently of the latter. It is possible that some of the CaR-expressing cells that are seemingly uninvolved in systemic Ca^{2+}_o homeostasis may participate in controlling local Ca^{2+}_o homeostasis by recognizing changes in Ca^{2+}_o in the extracellular fluid within their immediate microenvironments and then adjusting the movements of ions (i.e., divalent cations) or water in order to modulate the local ionic composition in a physiologically relevant manner. Alternatively, some cells may use the information encoded in their extracellular ionic microenvironment to regulate functions unrelated to either systemic or local Ca^{2+}_o homeostasis.

B. Microenvironments in Which Ca^{2+}_o Differs from Its Systemic Level

1. Locations Where Ca^{2+} from the External Environment Produces Variations in Ca^{2+}_o

Since there is only intermittent availability of dietary Ca^{2+}_o to free-living terrestrial organisms, the levels of Ca^{2+}_o as well as other ions in the gastrointestinal tract will change substantially depending on the ionic composition of the diet. Because cells in both stomach (Ray *et al.*, 1997) and small intestine (Chattopadhyay *et al.*, 1998a) express the CaR on their apical cell membranes (e.g., those facing the lumen), they will probably experience significant changes in Ca^{2+}_o that could modulate their functions in a physiologically relevant ways.

2. Changes in Local Levels of Ca^{2+}_o Resulting from Epithelial Transport of Ions and Water

Transport of Ca^{2+} across kidney epithelial cells and elsewhere can occur to varying extents via transcellular [e.g., in DCT (Friedman and Gesek, 1995)] and/or paracellular routes [i.e., in CTAL (De Rouffignac and Quamme, 1994; Hebert *et al.*, 1997)]. In some instances, as in the proximal tubule of the kidney, ion transport takes place in such a way that the ionic composition of the reabsorbed fluid is similar to that present in the tubular fluid (e.g., for Na^+, K^+, Ca^{2+}, and Cl^-). In other cases, in contrast, selective transport of specific ions, occasionally without accompanying water, can modify substantially the concentrations of Ca^{2+}_o and/or other ions in the fluid being reabsorbed and, by extension, that remaining within the tubular lumen. For instance, in the proximal nephron, less Mg^{2+} is reabsorbed than monovalent cations, anions or Ca^{2+} (De Rouffignac and Quamme, 1994). As a result the level of Mg^{2+} in the tubular fluid increases progressively and is about 1.8-fold higher in TAL than that in the initial fluid filtered at the glomerulus (De Rouffignac and Quamme, 1994). Mg^{2+} and Ca^{2+} in the thick ascending limb are reabsorbed both by paracellular and transcellular and routes with little in the way of accompanying water because this nephron segment is remarkably impermeable to water (Kikeri *et al.*, 1989). Therefore, the levels of Ca^{2+}_o and Mg^+_o to which CaRs on the basolateral surface of the tubular epithelial cells of the CTAL are exposed may be significantly higher than those in either tubular fluid or systemic ECF.

In the IMCD, in contrast to TAL, reabsorption of water occurs through water-permeable channels without substantial transport of ions. The IMCD's selective permeability to water allows renal adjustment of the amount of "free" water retained or excreted to levels that are appropriate for the organism's overall state of water balance. As discussed earlier, increased reabsorption of water during antidiuresis could raise Ca^{2+}_o within the tubular fluid of IMCD sufficiently high to promote formation of kidney stones (Sands *et al.*, 1997). CaRs present on the apical surface of IMCD epithelial cells illustrate nicely how this receptor could promote local Ca^{2+}_o homeostasis. Since elevating Ca^{2+}_o in IMCD reduce vasopressin-stimulated water reabsorption, probably via the CaR, Ca^{2+}_o within the luminal fluid feeds back and sets an upper limit for Ca^{2+}_o within the fluid.

3. Alterations in Local Ca^{2+}_o Resulting from Fluxes of Ca^{2+} between Intra- and Extracellular Spaces

Ca^{2+}_o changes substantially in some cases as a result of alterations in calcium fluxes between intra- and extracellular spaces. Significant, neuronal activity-dependent alteration in Ca^{2+}_o occurs within the brain ECF owing to Ca^{2+} influx through calcium-permeable channels (e.g., NMDA channels) (Heinemann *et al.*, 1977; Arens *et al.*, 1992; Lucke *et al.*, 1995). These changes in Ca^{2+}_o are usually accompanied by alterations in Na^+_o and K^+_o due to concomitant influx of sodium ions via voltage-sensitive Na^+ channels and efflux of potassium ions through Ca^{2+}-activated and other K^+ channels. Large changes in the levels of extracellular ions occur, for example, during vigorous electrical stimulation of the cerebellum of an anesthetized rat (Nicholson *et al.*, 1977). In this model, Ca^{2+}_o falls by up to 90% during such stimulation, while K^+_o rises severalfold. These alterations in extracellular ions are rapidly reversible after the stimulation is stopped. Another example where changes in cellular activity lead to alterations in Ca^{2+}_o is the beating heart. When a frog heart is paced electrically *in vitro*, substantial reductions in Ca^{2+}_o take place within the interstitial fluid (Bers, 1983). Even greater alterations in Ca^{2+}_o could conceivably occur in the spatially restricted ECF within the cardiac T-tubular system. The latter are narrow plasma membrane infoldings that invaginate between muscle fibers, thereby providing close contact between the ECF in the T-tubules and intracellular sites where calcium ions participate in stimulus–contraction coupling (Almers *et al.*, 1981).

4. Variations in Local Ca^{2+}_o owing to Ca^{2+} Movements into and out of Extracellular Reservoirs

As noted previously, Ca^{2+}_o underneath actively resorbing osteoclasts can rise as high as 8–40 mM (Silver *et al.*, 1988). Therefore, it is likely that Ca^{2+}_o within the immediate microenvironment of such osteoclasts would change substantially when this calcium is released. Indeed, uncontrolled osteoclastic release of skeletal Ca^{2+}, as in cases where there are extensive skeletal metastases of certain malignancies promoting bone resorption via osteoclast-activating, hormonal factors, such as PTHrP (e.g., breast), even the levels of systemic Ca^{2+}_o can increase well above normal and become life-threatening (Stewart and Broadus, 1987). Ca^{2+}_o in the local skeletal microenvironment is likely to be even higher in this setting. Because breast cancer cells can express abundant CaRs (Cheng *et al.*, 1998), local changes in Ca^{2+}_o occurring within their microenvironment could affect the behavior of metastatic breast cancer cells within the skeleton.

The skeleton also has a substantial ability to take up calcium ions during its normal turnover as well as in some pathological states (Stewart and Broadus, 1987). On a daily basis, several hundred milligrams of Ca^{2+}_o enter the skeleton owing to *de novo* formation of bone by osteoblasts. Local depletion of Ca^{2+}_o will likely take place in the immediate vicinity of osteoblasts actively forming bone. Since osteoblasts can sense Ca^{2+}_o (Quarles *et al.*, 1997; Yamaguchi *et al.*, 1998b, 1998d), their behavior could potentially be modulated by alterations in Ca^{2+}_o occurring in their vicinity. Local rises in Ca^{2+}_o due to bone resorption could initially stimulate proliferation and chemotaxis of osteoblast precursors to sites where bone resorption had recently taken place (Godwin and Soltoff, 1997; Yamaguchi *et al.*, 1998b). Moreover, unlike malignancies causing bone resorption, prostate cancer cells stimulate bone formation (Stewart and Broadus, 1987). Pathological increases in bone formation are in some cases sufficient to lower systemic Ca^{2+}_o, and, therefore, local decreases in Ca^{2+}_o close to sites of abnormal bone formation are likely to be even larger. If metastatic prostate cancer cells can sense Ca^{2+}_o, this could potentially alter their function in ways that could contribute to the pathophysiology of these abnormal states of bone turnover.

C. Are There Endogenous CaR Agonists Other Than Ca^{2+}_o?

As noted earlier, the CaR may also function as a Mg^{2+}_o sensor, perhaps responding to local changes in Mg^{2+}_o within the tubular fluid of the kidney, where the CaR regulates reabsorption of both Ca^{2+} and Mg^{2+} (Hebert, 1996; Hebert *et al.*, 1997). Of the other polycationic agents known to be sensed by the CaR, spermine activates the receptor within a concentration range of 100 μM to 1 mM (Quinn *et al.*, 1997). Since this polyamine is present in some tissues at concentrations within this range, the CaR could potentially act as a spermine receptor within certain local environments. Another exogenous polycation that may activate the CaR under specific circumstances is neomycin (and other antibiotics of the aminoglycoside class). Some of the nephrotoxic actions of these antibiotics, for example, might be the consequence of activation of renal CaRs. Finally, amyloid beta peptides, which are thought to participate in the pathogenesis of Alzheimer's disease, can function as CaR agonists (Ye *et al.*, 1997a), perhaps because the fibrils of amyloid beta fibrils formed *in vitro* have a regular array of positive charges on their surfaces that both contribute to their diagnostically important binding of anionic dyes, such as Congo red, and their activation of the CaR. It is possible, therefore, that the CaR is involved in some way in the pathophysiology of disorders where amyloid beta proteins and other forms of amyloid proteins are deposited in diverse tissues.

IX. Are There Subtypes of the CaR or Additional Forms of Ca^{2+}_o-Sensors/Receptors?

The mGluRs, which as noted earlier share striking topological similarities with the CaR, form a receptor family consisting of at least eight different subtypes, mGluR1–mGluR8 (Nakanishi, 1992). The mGluR subtypes are further subdivided into three subgroups according to their sequence similarities (more than 60%), the signal transduction pathways to which they couple, and their agonist selectivities. Thus, analogous to mGluRs, it is possible that similar receptor subtypes exist for the CaR. Indeed, the genetic heterogeneity of FHH, namely that the disease is linked not only to the CaR locus on chromosome 3 but also to two disease genes on chromosome 19 (Lloyd *et al.,* 1999; McMurtry *et al.,* 1992; Heath *et al.,* 1993; Trump *et al.,* 1995), might indicate the existence of additional calcium sensors encoded by genes in these chromosomal loci.

Quarles *et al.* (1997) have reported that a Ca^{2+}_{o}-sensing mechanism in the mouse osteoblast-like MC3T3-E1 clonal cell line is functionally similar to but molecularly distinct from the CaR. After failing to detect expression of the CaR by Northern analysis and RT-PCR in MC3T3-E1 cells, they identified nucleotide sequences of putative CaR-related receptors (*Casr-rs*) in mouse genomic libraries by PCR (Hinson *et al.,* 1997). The deduced protein sequence of one of these putative receptors (*Casr-rs1*) was 63% similar and 40% identical to the CaR over the available transmembrane region. Although this CaR-related nucleotide sequence was initially identified in MC3T3-E1 cells by RT-PCR and was used as a probe to screen mouse genomic libraries to identify other related sequences, it could only be identified in subsequent analyses of mouse tissues, including MC3T3-E1 cells, by RT-PCR and not by Northern analysis or RNAse protection, suggesting very low levels of expression of uncertain physiological significance. Furthermore, as described earlier, we have recently found that MC3T3-E1 cells, as well as the human and rat osteoblast-like cell lines SAOS-2 and UMR106, respectively, express both CaR transcripts and protein on the basis of RT-PCR, Northern analysis, Western analysis, and immunocytochemistry (Yamaguchi *et al.,* 1998b, 1998d). Clearly, additional studies are necessary to determine whether their CaR-related nucleotide sequences are actually expressed as mature proteins in MC3T3-E1 cells using specific antisera raised to their predicted protein sequences.

Another candidate Ca^{2+}_{o}-sensor that is structurally unrelated to the CaR is a protein called megalin or gp330, which is a member of the LDL receptor superfamily and possesses a higher molecular weight that the CaR (it is >5000 amino acid residues in length) (Juhlin *et al.,* 1990; Lundgren *et al.,* 1994; Saito *et al.,* 1994; Hjalm *et al.,* 1996). Megalin was first identified using monoclonal antibodies that were raised against human parathyroid cells and recognized a large protein present at high levels not only in parathyroid but also in cells of the renal proximal tubular cells as well as in placental cytotrophoblasts. The capacity of some such antibodies to modulate the Ca^{2+}_{o}-sensing properties of these cells provided indirect evidence for its importance as a physiologically relevant Ca^{2+}_{o}-sensor. It is of interest that the CaR is expressed in these same three tissues, and further studies should elucidate whether the two proteins interact in some manner in tissues that express both of them. Moreover, studies involving expression of the full-length megalin molecule will be important to determine whether it indeed senses Ca^{2+}_o.

Evidence suggests that the Ca^{2+}_o-sensing receptor in the osteoclast may be closely related to the ryanodine receptor. Zaidi *et al.* previously showed that agents (e.g., ryanodine or caffeine) known to interact with the ryanodine receptor, which mediates Ca^{2+}_i-induced Ca^{2+} release from intracellular stores (e.g., in skeletal muscle), modulate the process of Ca^{2+}_o-sensing in osteoclasts (Zaidi *et al.*, 1992; Shankar *et al.*, 1995). Moreover, freshly isolated osteoclasts bind [³H]ryanodine, and the latter is displaced by the CaR agonist, Ca^{2+}_o, as well as by the ryanodine receptor antagonist, ruthenium red. Finally, an antibody directed at an epitope that is located within the channel-forming domain of the ryanodine receptor potentiates the actions of Ni^{2+} (which activates the putative osteoclast Ca^{2+}_o-sensing receptor) and labels nonpermeabilized osteoclasts, while an antibody directed at an intracellular epitope produces neither effect (although it does stain permeabilized osteoclasts) (Zaidi *et al.*, 1995). These results are consistent with the presence of a ryanodine-like molecule in the plasma membrane of the osteoclast that potentially acts as a Ca^{2+}_o-sensor or in close association with some other Ca^{2+}_o-sensing molecule. Cloning of this putative Ca^{2+}_o sensor and characterization of its structure and function would clearly be of great interest. It should be noted, however, that studies have demonstrated that freshly isolated rabbit osteoclasts express the same CaR originally cloned from parathyroid (Kameda *et al.*, 1998), suggesting that this receptor could also subserve the function of Ca^{2+}_o-sensing in this cell type.

Taken together, although several lines of evidence suggest the existence of additional forms of Ca^{2+}_o-sensers/receptors structurally unrelated to the CaR, definitive evidence is still lacking for other CaR subtypes similar to those found for the mGluRs. Instead, the CaR shows a widespread tissue distribution compared with that of mGluRs, which are primarily located in neuronal tissues. In this sense, the location and molecular diversity of the CaR at present appear to be distinctly different from those of the mGluRs.

X. Conclusion

Accumulating evidence shows that the CaR provided the molecular basis for a number of the known effects of Ca^{2+}_o on its target tissues involved in mineral ion homeostasis, especially parathyroid and kidney. Although additional studies are needed to prove its mediatory role, the CaR may likewise participate in aspects of the control of intestinal function and bone turnover that are relevant to systemic mineral ion metabolism. This receptor also plays key roles in human disorders such as FHH/NSHPT, autosomal dominant hypocalcemia, and sporadic hypocalcemia/hypoparathyroidism, as well as in the impaired urinary concentrating capacity or, in some cases, frank nephrogenic diabetes insipidus observed in hypercalcemic persons. Findings of the wide distribution of the CaR in tissues uninvolved in systemic Ca^{2+}_o homeostasis, however, indicate that the CaR also participates in modulating neuronal activities in the brain and perhaps intestine as well as a variety of other cellular functions. Therefore, the CaR may subserve previously unrecognized roles in sensing local changes in Ca^{2+}_o in the microenvironments within these tissues that are separated from the systemic circulation. Two of the physiological actions that may be controlled by such local Ca^{2+}_o-sensing are cell proliferation and/or chemotaxis, which are observed in bone marrow-derived cells, fibroblasts, colonocytes, and oligodendrocytes. However, the use of a specific CaR antagonist or the corresponding cells cultured from mice with targeted disruption of the CaR gene will be necessary to establish unequivocally the CaR's role in local Ca^{2+}_o sensing.

References

Adams, M. D., and Oxender, D. L. (1989). Bacterial periplasmic binding protein tertiary structures. *J. Biol. Chem.* **264**, 15739–15742.

Aida, K., Koishi, S., Tawata, M., and Onaya, T. (1995). Molecular cloning of a putative Ca^{2+}-sensing receptor cDNA from human kidney. *Biochem. Biophys. Res. Commun.* **214**, 524–529.

Almers, W., Fink, R., and Palade, P. T. (1981). Calcium depletion in frog muscle tubules: the decline of calcium current under maintained depolarization. *J. Physiol. (Lond.).* **312**, 177–207.

Arens, J., Stabel, J., and Heinemann, U. (1992). Pharmacological properties of excitatory amino acid induced changes in extracellular calcium concentration in rat hippocampal slices. *Can. J. Physiol. Pharmacol.* **70**, S194–S205.

Attie, M. F., Gill Jr, J., Stock, J. L., Spiegel, A. M., Downs Jr, R. W., Levine, M. A., and Marx, S. J. (1983). Urinary calcium excretion in familial hypocalciuric hypercalcemia. Persistence of relative hypocalciuria after induction of hypoparathyroidism. *J. Clin. Invest.* **72**, 667–676.

Auwerx, J., Demedts, M., and Bouillon, R. (1984). Altered parathyroid set point to calcium in familial hypocalciuric hypercalcaemia. *Acta Endocrinologica (Copenh.).* **106**, 215–218.

Bai, M., Quinn, S., Trivedi, S., Kifor, O., Pearce, S. H. S., Pollak, M. R., Krapcho, K., Hebert, S. C., and Brown, E. M. (1996). Expression and characterization of inactivating and activating mutations of the human Ca^{2+}_o-sensing receptor. *J. Biol. Chem.* **271**, 19537–19545.

Bai, M., Janicic, N., Trivedi, S., Quinn, S. J., Cole, D. E. C., Brown, E. M., and Hendy, G. N. (1997a). Markedly reduced activity of mutant calcium-sensing receptor with an inserted Alu element from a kindred with familial hypocalciuric hypercalcemia and neonatal severe hyperparathyroidism. *J. Clin. Invest.* **99**, 1917–1925.

Bai, M., Pearce, S. H. S., Kifor, O., Trivedi, S., Stauffer, U. G., Thakker, R. V., Brown, E. M., and Steinmann, B. (1997b). In vivo and in vitro characterization of neonatal hyperparathyroidism resulting from a de novo, heterozygous mutation in the Ca^{2+}-sensing receptor gene: Normal maternal calcium homeostasis as a cause of secondary hyperparathyroidism in familial benign hypocalciuric hypercalcemia. *J. Clin. Invest.* **99**, 88–96.

Baron, R. (1996). Anatomy and ultrastructure of bone. *In* "Premier on the Metabolic Bone Diseases and Disorders of Mineral Metabolism" (M. J. Favus, ed.), pp. 3–10. Lippincott-Raven, Philadelphia.

Baron, J., Winer, K. K., Yanovski, J. A., Cunningham, A. W., Laue, L., Zimmerman, D., and Cutler, G. B., Jr. (1996). Mutations in the Ca^{2+}-sensing receptor gene cause autosomal dominant and sporadic hypoparathyroidism. *Human Mol. Genet.* **5,** 601–606.

Bers, D. M. (1983). Early transient depletion of extracellular Ca during individual cardiac muscle contractions. *Am. J. Physiol.* **244,** H462–H468.

Bikle, D., Ratnam, A., Mauro, T., Harris, J., and Pillai, S. (1996). Changes in calcium responsiveness and handling during keratinocyte differentiation. Potential role of the calcium receptor. *J. Clin. Invest.* **97,** 1085–1093.

Bornefalk, E., Ljunghall, S., Lindh, E., Bengsten, O., Johansson, A., and Ljunggren, O. (1997). Regulation of interleukin-6 secretion from mononuclear blood cells by extracellular calcium. *J. Bone Miner. Res.* **12,** 228–233.

Bourdeau, A., Souberbielle, J.-C., Bonnet, P., Herviaux, P., Sachs, C., and Lieberherr, M. (1992). Phospholipase-A2 action and arachidonic acid metabolism in calcium-mediated parathyroid hormone secretion. *Endocrinology* **130,** 1339–1344.

Bradbury, R. A., McCall, M. N., Brown, M. J., and Conigrave, A. D. (1996). Functional heterogeneity of human term cytotrophoblasts revealed by differential sensitivity to extracellular Ca^{2+} and nucleotides. *J. Endocrinol.* **149,** 135–144.

Bradbury, R. A., Sunn, K. L., Crossley, M. C., Bai, M., Brown, E. M., Delbridge, L., and Conigrave, A. D. (1998). Expression of the parathyroid Ca^{2+} sensing receptor in cytotrophoblasts from human term placenta. *J. Endocrinol.* **156,** 425–430.

Bronner, F., Pansu, D., and Stein, W. D. (1986). An analysis of intestinal calcium transport across the rat intestine. *Am. J. Physiol.* **250,** G561–G569.

Brown, E. M. (1983). Four parameter model of the sigmoidal relationship between parathyroid hormone release and extracellular calcium concentration in normal and abnormal parathyroid tissue. *J. Clin. Endocrinol. Metab.* **56,** 572–581.

Brown, E. M. (1991). Extracellular Ca^{2+} sensing, regulation of parathyroid cell function, and role of Ca^{2+} and other ions as extracellular (first) messengers. *Physiol. Rev.* **71,** 371–411.

Brown, E. M., Butters, R., Katz, C., and Kifor, O. (1991). Neomycin mimics the effects of high extracellular calcium concentrations on parathyroid function in dispersed bovine parathyroid cells. *Endocrinology* **128,** 3047–3054.

Brown, E. M., Enyedi, P., LeBoff, M., Rothberg, J., Preston, J., and Chen, C. (1987). High extracellular Ca^{2+} and Mg^{2+} stimulate accumulation of inositol phosphates in bovine parathyroid cells. *FEBS Lett.* **218,** 113–118.

Brown, E. M., Fuleihan, G. E.-H., Chen, C. J., and Kifor, O. (1990). A comparison of the effects of divalent and trivalent cations on parathyroid hormone release, 3′,5′-cyclic-adenosine monophosphate accumulation, and the levels of inositol phosphates in bovine parathyroid cells. *Endocrinology* **127,** 1064–1071.

Brown, E. M., Gamba, G., Riccardi, D., Lombardi, D., Butters, R., Kifor, O., Sun, A., Hediger, M. H., Lytton, J., and Hebert, S. C. (1993). Cloning and characterization of an extracellular Ca^{2+}-sensing receptor from bovine parathyroid. *Nature* **366,** 575–580.

Brown, A. J., Zhong, M., Finch, J., Ritter, C., McCracken, R., Morrissey, J., and Slatopolsky, E. (1996). Rat calcium-sensing receptor is regulated by vitamin D but not by calcium. *Am. J. Physiol.* **270,** F454–F460.

Butters, R. R., Jr., Chattopadhyay, N., Nielsen, P., Smith, C. P., Mithal, A., Kifor, O., Bai, M., Quinn, S., Goldsmith, P., Hurwitz, S., Krapcho, K., Busby, J., and Brown, E. M. (1997). Cloning and characterization of a calcium-sensing receptor from the hypercalcemic New Zealand white rabbit reveals unaltered responsiveness to extracellular calcium. *J. Bone Miner. Res.* **12,** 568–579.

Caplan, A. I., and Dennis, J. E. (1996). Mesenchymal stem cells: Progenitors, progeny, and pathways. *J. Bone Miner. Metab.* **14,** 193–201.

Carozzi, S., Ramello, A., Nasini, M. G., Schelotto, C., Caviglia, P. M., Cantaluppi, A., Salit, M., and Lamperi, S. (1990). Bone marrow erythroid precursor Ca^{2+} regulates the response to human recombinant erythropoietin (rHuEPO) in hemodialysis patients. *Int. J. Artific. Organs.* **13**, 747–750.

Chang, W., Chen, T.-H., Gardner, P., and Shoback, D. (1995). Regulation of Ca^{2+}-conducting currents in parathyroid cells by extracellular Ca^{2+} and channel blockers. *Am. J. Physiol.* **269**, E864–E877.

Chang, W., Pratt, S., Chen, T.-S., E, N., Huang, Z., and Shoback, D. (1998). Coupling of calcium receptors to inositol phosphate and cAMP generation in mammalian cells and *Xenopus* oocytes and immunodetection of receptor protein by region-specific antipeptide antisera. *J. Bone Miner. Res.* **13**, 570–580.

Chattopadhyay, N., and Brown, E. M. (1997). Calcium-sensing receptor: Roles in and beyond systemic calcium homeostasis. *Biol. Chem.* **378**, 759–768.

Chattopadhyay, N., Mithal, A., and Brown, E. M. (1996). The calcium-sensing receptor: a window into the physiology and pathophysiology of mineral ion metabolism. *Endocrine Rev.* **17**, 289–307.

Chattopadhyay, N., Légrádi, G., Bai, M., Kifor, O., Ye, C., Vassilev, P. M., Brown, E. M., and Lechan, R. M. (1997a). Calcium-sensing receptor in the rat hippocampus: a developmental study. *Develop. Brain Res.* **100**, 13–21.

Chattopadhyay, N., Ye, C., Singh, D. P., Kifor, O., Vassilev, P. M., Sinohara, T., Chylack Jr, L. T., and Brown, E. M. (1997b). Expression of extracellular calcium-sensing receptor by human lens epithelial cells. *Biochem. Biophys. Res. Commun.* **233**, 801–805.

Chattopadhyay, N., Cheng, I., Rogers, K., Riccardi, D., Hall, A., Diaz, R., Hebert, S. C., Soybel, D. I., and Brown, E. M. (1998a). Identification and localization of extracellular Ca^{2+}-sensing receptor in rat intestine. *Am. J. Physiol.* **274**, G122–G130.

Chattopadhyay, N., Ye, C., Yamaguchi, T., Kifor, O., Vassilev, P. M., Nishimura, R., and Brown, E. M. (1998b). Expression of extracellular calcium-sensing receptor in rat oligodendrocytes: Evidence for its role in proliferation and opening of Ca^{2+}-activated K^+ channels. *Glia*, **24**, 449–458.

Chen, C., Barnett, J., Congo, D., and Brown, E. (1989). Divalent cations suppress $3',5'$-adenosine monophosphate accumulation by stimulating a pertussis toxin-sensitive guanine nucleotide-binding protein in cultured bovine parathyroid cells. *Endocrinology.* **124**, 233–239.

Cheng, I., Klingensmith, M. K., Chattopadhyay, N., Butters, R. R., Soybel, D. I., and Brown, E. M. (1998). Identification and localization of the extracellular calcium-sensing receptor in human breast. *J. Clin. Endocrinol. Metab.* **83**, 703–707.

Chou, Y.-H., Pollak, M., Brandi, M., Toss, G., Arnqvist, H., Atkinson, A., Papapoulos, S., Marx, S., Brown, E., Seidman, J., and Seidman, C. (1995). Mutations in the human Ca^{2+}-sensing receptor gene that cause familial hypocalciuric hypercalcemia. *Am. J. Hum. Genet.* **56**, 1075–1079.

Cooper, L., Wertheimer, J., Levey, R., Brown, E., LeBoff, M., Wilkinson, R., and Anast, C. (1986). Severe primary hyperparathyroidism in a neonate with two hypercalcemic parents: management with parathyroidectomy and heterotopic autotransplantation. *Pediatrics* **78**, 263–268.

Davies, M., Adams, P. H., Lumb, G. A., Berry, J. L., and Loveridge, N. (1984). Familial hypocalciuric hypercalcemia: evidence for continued enhanced renal tubular reabsorption of calcium following total parathyroidectomy. *Acta Endocrinol.* **106**, 499–504.

Davies, M., Mughal, Z., Selby, P., Tymms, D., and Mawer, E. (1995). Familial benign hypocalcemia. *J. Bone Miner. Res.* **10** (**Supplement 1**), S507.

De Luca, F., Ray, K., Mancilla, E. E., Fan, G.-F., Winer, K. K., Gore, P., Spiegel, A. M., and Baron, J. (1997). Sporadic hypoparathyroidism caused by *de novo* gain-of-function mutations in the Ca^{2+}-sensing receptor. *J. Clin. Endocrinol. Metab.* **82**, 2710–2715.

De Rouffignac, C., and Quamme, G. A. (1994). Renal magnesium handling and its hormonal control. *Physiol. Rev.* **74**, 305–322.

Delamere, N. A., and Paterson, C. A. (1981). Hypocalcaemic cataract. *In* "Mechanisms of Cataract Formation in the Human Lens" (G. Duncan, ed.), pp. 219–236. Academic Press, New York.

Diaz, R., Hurwitz, S., Chattopadhyay, N., Pines, M., Yang, Y., Kifor, O., Einat, M., Butters, R., Hebert, S., and Brown, E. (1997). Cloning, expression and tissue localization of the calcium-sensing receptor in the chicken (*Gallus domesticus*). *Am. J. Physiol.* **273**, R1008–R1016.

Duncan, G., and Bushell, A. R. (1975). Ion analyses of human cataractous lenses. *Exp. Eye Res.* **20**, 223–230.

Eftekhari, F., and Yousefzadeh, D. (1982). Primary infantile hyperparathyroidism: clinical, laboratory, and radiographic features in 21 cases. *Skeletal Radiol.* **8**, 201–208.

Eskert, R., Scherubl, H., Petzelt, C., Friedhelm, R., and Ziegler, R. (1989). Rhythmic oscillations of cytosolic calcium in rat C-cells. *Mol. Cell. Endocrinol.* **64**, 267–270.

Estep, H., Mistry, Z., and Burke, P. (1981). Familial idiopathic hypocalcemia. *In* "Proceedings and Abstracts of the 63rd Annual Meeting of the Endocrine Society," p. 275 (abstract). Endocrine Society, Cincinnati.

Fajtova, V., Quinn, S., and Brown, E. (1991). Cytosolic calcium responses of single rMTC 44-2 cells to stimulation with external calcium and potassium. *Am. J. Physiol.* **261**, E151–158.

Favus, M. J., Kathpalia, S. C., and Coe, F. L. (1981). Kinetic characteristics of calcium absorption and secretion by rat colon. *Am. J. Physiol.* **240**, G350–G354.

Feher, J. J., Fullmer, C. S., and Fritzch, G. K. (1989). Comparison of the enhanced steady-state diffusion of calcium by calbindin D9K and calmodulin: Possible importance in intestinal absorption. *Cell Calcium.* **10**, 189–203.

Ferreira, M. D. d. J., Helies-Toussaint, C., Imbert-Teboul, M., Bailly, C., Verbavatz, J.-M., Bellanger, A.-C., and Chabardes, D. (1998). Co-expression of a Ca^{2+}-inhibitable adenylyl cyclase and of a Ca^{2+}-sensing receptor in the cortical thick ascending limb of the rat kidney. *J. Biol. Chem.* **273**, 15192–15202.

Firsov, D., Aarab, L., Mandon, B., Siaume-Perez, S., Rouffignac, S. D., and Chabardes, D. (1995). Arachidonic acid inhibits hormone-stimulated cAMP accumulation in the medullary thick ascending limb of the rat kidney by a mechanism sensitive to pertussis toxin. *Pfleugers Arch.* **429**, 636–646.

Foley T., Jr., Harrison, H., Arnaud, C., and Harrison, H. (1972). Familial benign hypercalcemia. *J. Pediatr.* **81**, 1060–1067.

Freichel, M., Zinc-Lorenz, A., Hollishi, A., Hafner, M., Flockerzi, V., and Raue, F. (1996). Expression of a calcium-sensing receptor in a human medullary thyroid carcinoma cell line and its contribution to calcitonin secretion. *Endocrinology* **137**, 3842–3848.

Fried, R., and Tashjian, A. J. (1986). Unusual sensitivity of cytosolic free Ca^{2+} to changes in extracellular Ca^{2+} in rat C-cells. *J. Biol. Chem.* **261**, 7669–7674.

Friedman, P., and Gesek, F. (1995). Cellular calcium transport in renal epithelia: measurement, mechanisms, and regulation. *Physiol. Rev.* **75**, 429–471.

Fujikawa, Y., Quinn, J. M. W., Sabokbar, A., McGee, J. O. D., and Athanasou, N. A. (1996). The human osteoclast precursor circulates in the monocyte fraction. *Endocrinology* **137**, 4058–4060.

Fujita, T., Watanabe, N., Fukase, M., Tsutsumi, M., Fukami, T., Imai, Y., Sakaguchi, K., Okada, S., Matsuo, M., and Takemine, H. (1983). Familial hypocalciuric hypercalcemia involving four members of a kindred including a girl with severe neonatal primary hyperparathyroidism. *Miner. Electr. Metab.* **9**, 51–54.

Galkin, B. M., Feig, S. A., Patchefsky, A. S., Rue, J. W., Gamblin Jr, W. J., Gomez, D. G., and Marchant, L. M. (1977). Ultrastructure and microanalysis of "benign" and "malignant" breast calcifications. *Radiology* **124**, 245–249.

Gama, L., Baxenlale-Cox, L. M., and Breitwieser, G. E. (1997). Ca^{2+}-sensing receptor in intestinal epithelium. *Am. J. Physiol.* **273**, C1168–C1175.

Garrett, J. E., Capuano, I. V., Hammerland, L. G., Hung, B. C. P., Brown, E. M., Hebert, S. C., Nemeth, E. F., and Fuller, F. (1995a). Molecular cloning and functional expression of human parathyroid calcium receptor cDNAs. *J. Biol. Chem.* **270,** 12919–12925.

Garrett, J. E., Tamir, H., Kifor, O., Simin, R. T., Rogers, K. V., Mithal, A., Gagel, R. F., and Brown, E. M. (1995b). Calcitonin-secreting cells of the thyroid express an extracellular calcium receptor gene. *Endocrinology* **136,** 5202–5211.

Gill, J. J., and Bartter, F. (1961). On the impairment of renal concentrating ability in prolonged hypercalcemia and hypercalciuria in man. *J. Clin. Invest.* **40,** 716–722.

Godwin, S. L., and Soltoff, S. P. (1997). Extracellular calcium and platelet-derived growth factor promote receptor-mediated chemotaxis in osteoblasts through different signaling pathways. *J. Biol. Chem.* **272,** 11307–11312.

Heath, H. I. (1989). Familial benign (hypocalciuric) hypercalcemia. A troublesome mimic of primary hyperparathyroidism. *Endocrinol. Metab. Clin. North Am.* **18,** 723–740.

Heath, D. A. (1994). Familial hypocalciuric hypercalcemia. *In* "The Parathyroids" (J. P. Bilezikian, R. Marcus, and M. A. Levine, eds.), pp. 699–710. Raven Press, New York.

Heath, H., Jackson, C., Otterud, B., and Leppert, M. (1993). Genetic linkage analysis of familial benign (hypocalciuric) hypercalcemia: Evidence for locus heterogeneity. *Am. J. Hum. Genet.* **53,** 193–200.

Hebert, S. C. (1996). Extracellular calcium-sensing receptor: Implications for calcium and magnesium handling in the kidney. *Kidney Int.* **50,** 2129–2139.

Hebert, S. C., Brown, E. M., and Harris, H. W. (1997). Role of the Ca^{2+}-sensing receptor in divalent mineral ion homeostasis. *J. Exp. Biol.* **200,** 295–302.

Heinemann, U., Lux, H. D., and Gutnick, M. J. (1977). Extracellular free calcium and potassium during paroxysmal activity in cerebral cortex of the rat. *Expt. Brain Res.* **27,** 237–243.

Hellman, P., P, R., Juhlin, C., Akerstrom, G., Rastad, J., and Gylfe, E. (1992). Parathyroid-like regulation of parathyroid-hormone-related protein release and cytoplasmic calcium in cytotrophoblast cells of human placenta. *Arch. Biochem. Biophys.* **293,** 174–180.

Hinson, T. K., Damodaran, T. V., Chen, J., Zhang, X., Qumsiyeh, M. B., Seldin, M. F., and Quarles, L. D. (1997). Identification of putative transmembrane receptor sequences homologous to the calcium-sensing G-protein-coupled receptor. *Genomics* **45,** 279–289.

Hjalm, G., Murray, E., Crumley, G., Harazim, W., Lundgren, S., Onyango, I., Ek, B., Larsson, M., Juhlin, C., Hellamn, P., Davis, H., Akerstron, G., Rask, L., and Morse, B. (1996). Cloning and sequencing of human gp330, a Ca^{2+}-binding receptor with potential intracellular signaling properties. *Eur. J. Biochem.* **239,** 132–137.

Ho, C., Conner, D. A., Pollak, M., Ladd, D. J., Kifor, O., Warren, H., Brown, E. M., Seidman, C. E., and Seidman, J. G. (1995). A mouse model for familial hypocalciuric hypercalcemia and neonatal severe hyperparathyroidism. *Nature Genet.* **11,** 389–394.

Hoff, A. O., Cote, G. J., Fritsche Jr, H. A., Schultz, P., Khorana, S., Parthasarathy, R., and Gagel, R. F. (1997). An activating mutation of the calcium-sensing receptor creates a new oncogene. *J. Bone Miner. Res.* **12 (suppl.1),** S143.

House, M. G., Kohlmeier, L., Chattopadhyay, N., Kifor, O., Yamaguchi, T., LeBoff, M. S., Glowacki, J., and Brown, E. M. (1997). Expression of an extracellular calcium-sensing receptor in human and mouse bone marrow cells. *J. Bone Miner. Res.* **12,** 1959–1970.

Janicic, N., Pausova, Z., Cole, D. E. C., and Hendy, G. N. (1995). Insertion of an Alu sequence in the Ca^{2+}-sensing receptor gene in familial hypocalciuric hypercalcemia and neonatal severe hyperparathyroidism. *Am. J. Hum. Genet.* **56,** 880–886.

Jin, H. J., Miyaura, C., Tanaka, H., Takito, J., Abe, E., and Suda, T. (1990). Fusion of mouse alveloar macrophages induced by 1-alpha, 25-dihydroxyvitamin D3 involves extracellular, but not intracellular, calcium. *J. Cell. Physiol.* **142,** 434–439.

Juhlin, C., Lundgren, S., Johansson, H., Rastad, J., Akerstrom, G., and Klareskog, L. (1990). 500 kilodalton calcium sensor regulating cytoplasmic Ca^{2+} in cytotrophoblast cells of human placenta. *J. Biol. Chem.* **265,** 8275–8280.

Kallay, E., Kifor, O., Chattopadhyay, N., Brown, E. M., Bischoff, M. G., Peterlik, M., and Cross, H. S. (1997). Calcium dependent c-*myc* proto-oncogene expression and proliferation of Caco-2 cells: A role for a luminal extracellular calcium-sensing receptor. *Biochem. Biophys. Res. Commun.* **232,** 80–83.

Kameda, T., Mano, H., Yamada, Y., Takai, H., Amizuka, N., Kobori, M., Izumi, N., Kawashima, H., Ozawa, H., Ikeda, K., Kameda, A., Hakeda, Y., and Kumegawa, M. (1998). Calcium-sensing receptor in mature osteoclasts, which are bone-resorbing cells. *Biochem. Biophys. Res. Commun.* **245,** 419–422.

Kanazirska, M. P. V., Vassilev, P. M., Ye, C. P., Francis, J. E., and Brown, E. M. (1995). Extracellular Ca^{2+}-activated K^+ channels modulated by variations in extracellular Ca^{2+} in dispersed bovine parathyroid cells. *Endocrinology* **136,** 2238–2243.

Karbach, U. (1991). Segmental heterogeneity of cellular and paracellular calcium transport across rat duodenum and jejunum. *Gastroenterology* **100,** 47–58.

Kaupmann, K., Huggel, K., Heid, J., Flor, P. J., Bischoff, S., Kickel, S. J., McMaster, C., Angst, C., Bittiger, H., Froestl, W., and Bettler, B. (1997). Expression cloning of GABA$_B$ receptors uncovers similarity to metabotropic glutamate receptor. *Nature* **386,** 239–246.

Khosla, S., Ebeling, P. R., Firek, A. F., Burritt, M. M., Kao, P. C., and Heath III, H. (1993). Calcium infusion suggests a "set-point" abnormality of parathyroid gland function in familial benign hypercalcemia and more complex disturbances in primary hyperparathyroidism. *J. Clin. Endocrinol. Metab.* **76,** 715–720.

Kifor, O., Diaz, R., Butters, R., and Brown, E. M. (1997). The Ca^{2+}-sensing receptor (CaR) activates phospholipases C, A$_2$, and D in bovine parathyroid and CaR-transfected, human embryonic kidney (HEK293) cells. *J. Bone Miner. Res.* **12,** 715–725.

Kikeri, D., Sun, A., Zeidel, M. L., and Hebert, S. C. (1989). Cell membranes impermeable to NH_3. *Nature* **339,** 478–480.

Kobayashi, M., Tanaka, H., Tsuzuki, K., Tsuyuki, M., Igaki, H., Ichinose, Y., Aya, K., Nishioka, N., and Seino, Y. (1997). Two novel missense mutations in calcium-sensing receptor gene associated with neonatal severe hyperparathyroidism. *J. Clin. Endocrinol. Metab.* **82,** 2716–2719.

Kristiansen, J. H., Rodbro, P., Christiansen, C., Johansen, J., and Jensen, J. T. (1987). Familial hypocalciuric hypercalcemia. III: Bone mineral metabolism. *Clin. Endocrinol. (Oxf.).* **26,** 713–716.

Kubo, Y., Miyashita, T., and Murata, Y. (1998). Structural basis for a Ca^{2+}-sensing function of the metabotropic glutamate receptors. *Science* **279,** 1722–1725.

Kurokawa, K. (1987). Nephrology forum: Calcium-regulating hormones and the kidney. *Kidney Int.* **32,** 760–771.

Law W. M., Jr., and Heath III, H. (1985). Familial benign hypercalcemia (hypocalciuric hypercalcemia). Clinical and pathogenetic studies in 21 families. *Ann. Int. Med.* **105,** 511–519.

Law W. M., Jr., Bollman, S., Kumar, R., and Heath III, H. (1984). Vitamin D metabolism in familial benign hypercalcemia (hypocalciuric hypercalcemia) differs from that in primary hyperparathyroidism. *J. Clin. Endocrinol. Metab.* **58,** 744–747.

Lloyd, S. E., Pannett, A. A., Dixon, P. H., Whyte, M. P., Thakker, R. V. (1999). Localization of familial benign hypercalcemia, Oklahoma variant (FBHOk), to chromosome 19q13. *Am. J. Hum. Genet.* **64,** 189–195.

Lovlie, R., Eiken, H. G., and Sorheim, H. (1996). The Ca^{2+}-sensing receptor gene (PCAR1) mutation T151M in isolated autosomal dominant hypoparathyroidism. *Hum. Genet.* **98,** 129–133.

Lucke, A., Kohling, R., Straub, H., Moskopp, D., Wassman, H., and Speckmann, E.-J. (1995). Changes of extracellular calcium concentration induced by application of excitatory amino acids in the human neocortex in vitro. *Brain Res.* **671,** 222–226.

Lundgren, S., Hjalm, G., Hellman, P., Ek, B., Juhlin, C., Rastad, J., Klareskog, L., Akerstrom, G., and Rask, L. (1994). A protein involved in calcium sensing of the human parathyroid and placental cytotrophoblast cells belongs to the LDL-receptor protein superfamily. *Exp. Cell Res.* **212**, 344–350.

Luster, A. D. (1998). Chemokines: Chemotactic cytokines that mediate inflammation. *New Engl. J. Med.* **338**, 436–445.

Mailland, M., Waelchli, R., Ruat, M., Boddeke, H. G. W. M., and Seuwen, K. (1997). Stimulation of cell proliferation by calcium and a calcimimetic compound. *Endocrinology* **138**, 3601–3605.

Mancilla, E. E., De Luca, F., Ray, K., Winer, K. K., Fan, G.-F., and Baron, J. (1997). A Ca^{2+}-sensing receptor mutation causes hypoparathyroidism by increasing receptor sensitivity to Ca^{2+} and maximal signal transduction. *Pediatr. Res.* **42**, 443–447.

Marx, S., Spiegel, A., Brown, E., Koehler, J., Gardner, D., Brennan, M., and Aurbach, G. (1978). Divalent cation metabolism. Familial hypocalciuric hypercalcemia versus typical primary hyperparathyroidism. *Am. J. Med.* **65**, 235–242.

Marx, S. J., Attie, M. F., Levine, M. A., Spiegel, A. M., Downs, R. W., Jr., and Lasker, R. D. (1981a). The hypocalciuric or benign variant of familial hypercalcemia: Clinical and biochemical features in fifteen kindreds. *Medicine (Baltimore).* **60**, 397–412.

Marx, S. J., Attie, M. F., Stock, J. L., Spiegel, A. M., and Levine, M. A. (1981b). Maximal urine-concentrating ability: Familial hypocalciuric hypercalcemia versus typical primary hyperparathyroidism. *J. Clin. Endocrinol. Metab.* **52**, 736–740.

Marx, S., Attie, M., Spiegel, A., Levine, M., Lasker, R., and Fox, M. (1982). An association between neonatal severe primary hyperparathyroidism and familial hypocalciuric hypercalcemia in three kindreds. *N. Engl. J. Med.* **306**, 257–284.

Marx, S., Lasker, R., Brown, E., Fitzpatrick, L., Sweezey, N., Goldbloom, R., Gillis, D., and Cole, D. (1986). Secretory dysfunction in parathyroid cells from a neonate with severe primary hyperparathyroidism. *J. Clin. Endocrinol. Metab.* **62**, 445–449.

Matsunami, H., and Buck, L. B. (1997). A multigene family encoding a diverse array of putative pheromone receptors in mammals. *Cell* **90**, 775–784.

Matsuoka, I., Nakahata, N., and Nakanishi, H. (1989). Inhibitory effect of 8-bromo cyclic AMP on an extracellular Ca^{2+}-dependent arachidonic acid liberation in collagen-stimulated rabbit platelets. *Biochem. Pharmacol.* **38**, 1841–1847.

McGehee, D. S., Aldersberg, M., Liu, K.-P., Hsuing, S.-C., Heath, M. J. S., and Tamir, H. (1997). Mechanism of extracellular Ca^{2+} receptor-stimulated hormone release from sheep thyroid parafollicular cells. *J. Physiol.* **502**, 31–44.

McGrath, C. M., and Soule, H. D. (1984). Calcium regulation of normal human mammary epithelial cells in culture. *In Vitro Dev. Dev. Biol.* **20**, 652–662.

McMurtry, C., Schranck, F., Walkenhorst, D., Murphy, W., Kocher, D., Teitelbaum, S., Rupich, R., and Whyte, M. (1992). Significant developmental elevation in serum parathyroid hormone levels in a large kindred with familial benign (hypocalciuric) hypercalcemia. *Am. J. Med.* **93**, 247–258.

McNeil, S. E., Hobson, S. A., Nipper, K. D., and Rodland, K. D. (1998). Functional calcium-sensing receptors in rat fibroblasts are required for activation of SRC kinase and mitogen-activated protein kinase in response to extracellular calcium. *J. Biol. Chem.* **273**, 1114–1120.

Mithal, A., Kifor, O., Kifor, I., Vassilev, P., Butters, R., Krapcho, K., Simin, R., Fuller, F., Hebert, S. C., and Brown, E. M. (1995). The reduced responsiveness of cultured bovine parathyroid cells to extracellular Ca^{2+} is associated with marked reduction in the expression of extracellular Ca^{2+}-sensing receptor messenger ribonucleic acid and protein. *Endocrinology* **136**, 3087–3092.

Moscat, J., Fleming, T. P., Molloy, C. J., Lopez-Barahona, M., and Aaronson, S. A. (1989). The calcium signal for Balb/MK keratinocyte terminal differentiation induces sustained alterations in phosphoinositide metabolism without detectable protein kinase-C activation. *J. Biol. Chem.* **264**, 11228–11235.

Muff, R., Nemeth, E. F., Haller-Brem, S., and Fischer, J. A. (1988). Regulation of hormone secretion and cytosolic Ca^{2+} by extracellular Ca^{2+} in parathyroid cells and C-cells: Role of voltage-sensitive Ca^{2+} channels. *Arch. Biochem. Biophys.* **265**, 128–135.

Nakanishi, S. (1992). Molecular diversity of glutamate receptors and implications for brain function. *Science* **258**, 597–603.

Nakanishi, S. (1994). Metabotropic glutamate receptors: synaptic transmission, modulation and plasticity. *Neuron* **13**, 1031–1037.

Nemeth, E., and Scarpa, A. (1987). Rapid mobilization of cellular Ca^{2+} in bovine parathyroid cells by external divalent cations. *J. Biol. Chem.* **202**, 5188–5196.

Nicholson, C., Ten Bruggencate, G., Steinberg, R., and Strokle, H. (1977). Calcium modulation in brain extracellular microenvironment demonstrated with ion-selective microelectrode. *Proc. Natl. Acad. Sci. USA* **74**, 1287–1290.

Owen, W. G., Bichler, J., Ericson, D., and Wysokinski, W. (1995). Gating of thrombin in platelet aggregates by pO_2-linked lowering of extracellular Ca^{2+} concentration. *Biochemistry* **34**, 9277–9281.

Pansu, D., Ballaton, C., and Bronner, F. (1983). Duodenal and ileal calcium absorption in rat and effects of vitamin D. *Am. J. Physiol.* **244**, G32–G37.

Pearce, S., Trump, D., Wooding, C., Besser, G., Chew, S., Heath, D., Hughes, I., and Thakker, R. (1995). Calcium-sensing receptor mutations in familial benign hypercalcaemia and neonatal hyperparathyroidism. *J. Clin. Invest.* **96**, 2683–2692.

Pearce, S. H. S., Bai, M., Quinn, S. J., Kifor, O., Brown, E. M., and Thakker, R. V. (1996a). Functional characterization of calcium-sensing receptor mutations expressed in human embryonic kidney cells. *J. Clin. Invest.* **98**, 1860–1866.

Pearce, S. H. S., Williamson, C., Kifor, O., Bai, M., Coulthard, M. G., Davies, M., Lewis-Barned, N., McCredie, D., Powell, H., Kendall-Taylor, P., Brown, E. M., and Thakker, R. V. (1996b). A familial syndrome of hypocalcemia with hypercalciuria due to mutations in the calcium-sensing receptor. *N. Engl. J. Med.* **335**, 1115–1122.

Pietrobon, D., Di Virgilio, F., and Pozzan, T. (1990). Structural and functional aspects of calcium homeostasis in eukaryotic cells. *Eur. J. Biochem.* **120**, 599–622.

Pocotte, S., Ehrenstein, G., and Fitzpatrick, L. (1995). Role of calcium channels in parathyroid hormone secretion. *Bone* **16**, S365–S372.

Pollak, M., Brown, E. M., Chou, Y. H., Hebert, S. C., Marx, S. J., Steinmann, B., Levi, T., Seidman, C. E., and Seidman, J. G. (1993). Mutations in the human Ca^{2+}-sensing receptor gene cause familial hypocalciuric hypercalcemia and neonatal severe hyperparathyroidism. *Cell* **75**, 1297–1303.

Pollak, M., Brown, E., Estep, H., McLaine, P., Kifor, O., Park, J., Hebert, S., Seidman, C., and Seidman, J. (1994a). Autosomal dominant hypocalcaemia caused by Ca^{2+}-sensing receptor gene mutation. *Nat. Genet.* **8**, 303–307.

Pollak, M., Chou, Y. H., Marx, S. J., Steinmann, B., Cole, D. E. C., Brandi, M. L., Papapoulos, S. E., Menko, F., Hendy, G. N., Brown, E. M., Seidman, C. E., and Seidman, J. G. (1994b). Familial hypocalciuric hypercalcemia and neonatal severe hyperparathyroidism. effects of mutant gene dosage on phenotype. *J. Clin. Invest.* **93**, 1108–1112.

Prentice, A., Jarjou, L. M. A., Cole, T. J., Stirling, D. M., Dibba, B., and Fairweather-Tait, S. (1995). Calcium requirements of lactating Gambian mothers, effects of a calcium supplement on breast-milk calcium concentration, maternal bone mineral content and urinary calcium excretion. *Am. J. Clin. Nutr.* **62**, 58–67.

Quamme, G. A. (1982). Effect of hypercalcemia on renal tubular handling of calcium and magnesium. *Can. J. Physiol. Pharmacol.* **60**, 1275–1280.

Quamme, G. A., and Dirks, J. H. (1980). Intraluminal and contraluminal magnesium on magnesium and calcium transfer in the rat nephron. *Am. J. Physiol.* **238**, F187–F198.

Quarles, D. L., Hartle II, J. E., Siddhanti, S. R., Guo, R., and Hinson, T. K. (1997). A distinct cation-sensing mechanism in MC3T3-E1 osteoblasts functionally related to the calcium receptor. *J. Bone Miner. Res.* **12**, 393–402.

Quinn, S. J., Ye, C. P., Diaz, R., Kifor, O., Bai, M., Vassilev, P., and Brown, E. M. (1997). The calcium-sensing receptor: A target for polyamines. *Am. J. Physiol.* **273**, C1315–C1323.

Ray, J. M., Squires, P. E., Curtis, S. B., Meloche, M. R., and Buchan, A. M. J. (1997). Expression of the calcium-sensing receptor on human antral gastrin cells. *J. Clin. Invest.* **99**, 2328–2333.

Riccardi, D., Park, J., Lee, W.-S., Gamba, G., Brown, E. M., and Hebert, S. C. (1995). Cloning and functional expression of a rat kidney extracellular calcium/polyvalent cation-sensing receptor. *Proc. Natl. Acad. Sci. USA* **92**, 131–135.

Riccardi, D., Lee, W.-S., Lee, K., Segre, G. V., Brown, E. M., and Hebert, S. C. (1996). Localization of the extracellular Ca^{2+}-sensing receptor and PTH/PTHrP receptor in rat Kidney. *Am. J. Physiol.* **271**, F951–F956.

Riccardi, D., Hall, A. E., Chattopadhyay, N., Xu, J., Brown, E. M., and Hebert, S. C. (1998). Localization of the extracellular Ca^{2+}/(polyvalent cation)-sensing receptor protein in rat kidney. *Am. J. Physiol.* **274**, F611–F622.

Ridefelt, T., Hellman, P., Wallfelt, C., Akerstrom, G., Rastad, J., and Gylfe, E. (1992). Neomycin interacts with Ca^{2+}-sensing of normal and adenomatous parathyroid cells. *Mol. Cell. Endocrinol.* **83**, 211–218.

Rodda, C. P., Caple, I. W., and Martin, T. J. (1992). Role of PTHrP in fetal and neonatal physiology. *In* "Parathyroid Hormone Related Protein: Normal Physiology and Its Role in Cancer" (B. P. Halloran and R. A. Nissenson, eds.), pp. 169–196. CRC Press, Boca Raton, FL.

Rogers, K. V., Dunn, C. E., Brown, E. M., and Hebert, S. C. (1997). Localization of calcium receptor mRNA in the adult rat central nervous system by in situ hybridization. *Brain Res.* **744**, 47–56.

Ruat, M., Molliver, M. E., Snowman, A. M., and Snyder, S. H. (1995). Calcium sensing receptor: molecular cloning in rat and localization to nerve terminals. *Proc. Natl. Acad. Sci. USA* **92**, 3161–3165.

Ryba, N. J. P., and Trindell, R. (1997). A new multigene family of putative pheromone receptors. *Neuron* **19**, 371–379.

Saito, A., Pietromonaco, S., Loo, A. K., and Farquhar, M. G. (1994). Complete cloning and sequencing of rat gp330/"megalin," a distinctive member of the low density lipoprotein receptor gene family. *Proc. Natl. Acad. Sci. USA* **91**, 9725–9729.

Sands, J. M., Naruse, M., Baum, M., Jo, I., Hebert, S. C., Brown, E. M., and Harris, W. H. (1997). Apical extracellular calcium/polyvalent cation-sensing receptor regulates vasopressin-elicited water permeability in rat kidney inner medullary collecting duct. *J. Clin. Invest.* **99**, 1399–1405.

Shankar, V. S., Pazianis, M., Huang, C. L., Simon, B., Adebanjo, C. A., and Zaidi, M. (1995). Caffeine modulates Ca^{2+} receptor activation in isolated rat osteclasts and induces intracellular Ca^{2+} release. *Am. J. Physiol.* **268**, F447–454.

Sharff, A. J., Rodseth, L. E., Spurlino, J. C., and Quiocho, F. A. (1992). Crystallographic evidence for a large ligand-induced hinge-twist motion between the two domains of the maltodextrin binding protein involved in active transport and chemotaxis. *Biochemistry* **31**, 10657–10663.

Shoback, D., Membreno, L. A., and McGhee, J. (1988). High calcium and other divalent cations in increase inositol trisphosphate in bovine parathyroid cells. *Endocrinology* **123**, 382–389.

Siegel, A. M., and Daly, J. W. (1985). Receptor (norepinephrine), P-site (2′,5′-dideoxyadenosine), and calcium-mediated inhibition of prostaglandin and forskolin-activated cAMP generating systems in human platelets. *J. Cyclic Nucleotide Protein Phosphorylation Res.* **10**, 229–246.

Silver, I. A., Murrils, R. J., and Etherington, D. J. (1988). Microlectrode studies on the acid microenvironment beneath adherent macrophages and osteoclasts. *Exp. Cell. Res.* **175**, 266–276.

Simpson, J. B., and Routenberg, A. (1975). Subfornical organ lesions reduce intravenous angiotensin-induced drinking. *Brain Res.* **88**, 154–161.

Stewart, A. F., and Broadus, A. E. (1987). Mineral metabolism. *In* "Endocrinology and Metabolism" (P. Felig, J. D. Baxter, A. E. Broadus, and L. A. Frohman, eds.), pp. 1317–1453. McGraw-Hill, New York.

Sudo, H., Kodama, H., Amagai, Y., Yamamoto, S., and Kasai, S. (1983). In vitro differentiation and calcification in a new clonal osteogenic cell line derived from newborn mouse calvaria. *J. Cell. Biol.* **96**, 191–198.

Sugimoto, T., Kanatani, M., Kano, J., Kaji, H., Tsukamato, T., Yamaguchi, T., Fukase, M., and Chihara, K. (1993). Effects of high calcium concentration on the functions and interactions of osteoblastic cells and monocytes and on the formation of osteoclast-like cells. *J. Bone Miner. Res.* **8**, 1445–1452.

Suki, W. M., Eknoyan, G., Rector Jr, F. C., and Seldin, D. W. (1969). The renal diluting and concentrating mechanism in hypercalcemia. *Nephron* **6**, 50–61.

Takaichi, K., and Kurokawa, K. (1988). Inhibitory guanosine triphosphate-binding protein-mediated regulation of vasopressin action in isolated single medullary tubules of mouse kidney. *J. Clin. Invest.* **82**, 1437–1444.

Tamir, H., Liu, K. P., Aldersberg, M., Hsuiung, S. C., and Gershon, M. D. (1996). Acidification of serotonin-containing secretory vesicles induced by a plasma membrane calcium receptor. *J. Biol. Chem.* **271**, 6441–6450.

Trump, D., Whyte, M. P., Wooding, C., Pang, J. T., Pearce, S. H. S., Kocher, D. B., and Thakker, R. V. (1995). Linkage studies in a kindred from Oklahoma, with familial benign (hypocalciuric) hypercalcaemia (FBH) and developmental elevations in serum parathyroid hormone levels, indicate a third locus for FBH. *Hum. Genet.* **96**, 183–187.

Van Biesen, T., Luttrell, L. M., Hawes, B. E., and Lefkowitz, R. J. (1996). Mitogenic signaling via G protein-coupled receptors. *Endocrine Rev.* **17**, 698–714.

Vassilev, P. M., Ho-Pao, C. L., Kanazirska, M. P., Ye, C., Hong, K, Seidman, C. E., Seidman, J. G., and Brown, E. M. (1997). Cao-sensing receptor (CaR)-mediated activation of K^+ channels is blunted in CaR gene-deficient mouse neurons. *Neuroreport* **8**, 1411–1416.

Wada, M., Furuya, Y., Sakiyama, J.-I., Kobayashi, N., Miyata, S., Ishii, H., and Hagano, N. (1997). The calcimimetic compound NPS R-568 suppresses parathyroid cell proliferation in rats with renal insufficiency. *J. Clin. Invest.* **100**, 2977–2983.

Wang, W.-H., Lu, M., and Hebert, S. C. (1996). Cytochrome P-450 metabolites mediate extracellular Ca^{2+}-induced inhibition of apical K^+ channels in the TAL. *Am. J. Physiol.* **271**, C103–C111.

Weisinger, J. R., Favus, M. J., Langman, C. B., and Bushinsky, D. (1989). Regulation of 1,25-dihydroxyvitamin D_3 by calcium in the parathyroidectomized, parathyroid hormone-replete rat. *J. Bone. Miner. Res.* **4**, 929–935.

Whitfield, J. (1995). Calcium as differentiator and killer—colon cells. *In* "Calcium in Cell Cycles and Cancer" (J. F. Whitfield, ed.), pp. 153–177. CRC Press, New York.

Yamaguchi, T., Chattopadhyay, N., Kifor, O., and Brown, E. M. (1998a). Extracellular calcium (Ca^{2+}_o)-sensing receptor in a murine bone marrow-derived stromal cell line (ST2): Potential mediator of the actions of Ca^{2+}_o on the function of ST2 cells. *Endocrinology,* **139**, 3561–3568.

Yamaguchi, T., Chattopadhyay, N., Kifor, O., Butters, R. R., Sugimoto, T., and Brown, E. M. (1998b). Mouse osteoblastic cell line (MC3T3-E1) expresses extracellular calcium (Ca^{2+}_o)-sensing receptor and its agonists stimulate chemotaxis and proliferation of MC3T3-E1 cells. *J. Bone Miner. Res.,* **13**, 1530–1538.

Yamaguchi, T., Kifor, O., Chattopadhyay, N., Bai, M., and Brown, E. M. (1998c). Extracellular calcium (Ca^{2+}_o)-sensing receptor in a mouse monocyte-macrophage cell line (J774): Potential mediator of the actions of Ca^{2+}_o on the Function of J774 cells. *J. Bone Miner. Res.,* 1390–1397.

Yamaguchi, T., Kifor, O., Chattopadhyay, N., and Brown, E. M. (1998d). Expression of extracellular calcium (Ca^{2+}_o)-sensing receptor in the clonal osteoblast-like cell lines, UMR-106 and SAOS-2. *Biochem. Biophys. Res. Commun.* **243**, 753–757.

Yamaguchi, T., Olozak, I., Chattopadhyay, N., Butters, R. R., Kifor, O., Scadden, D. T., and Brown, E. M. (1998e). Expression of extracellular calcium (Ca^{2+}_o)-sensing receptor in human peripheral blood monocytes. *Biochem. Biophys. Res. Commun.* **246**, 501–506.

Ye, C., Kanazirska, M., Quinn, S., Brown, E. M., and Vassilev, P. M. (1996a). Modulation by polycationic Ca^{2+}-sensing receptor agonists of nonselective cation channels in rat hippocampal neurons. *Biochem. Biophys. Res. Commun.* **224**, 271–280.

Ye, C., Rogers, K., Bai, M., Quinn, S. J., Brown, E. M., and Vassilev, P. M. (1996b). Agonists of the Ca^{2+}-sensing receptor (CaR) activate nonselective cation channels in HEK293 cells stably transfected with the human CaR. *Biochem. Biophys. Res. Commun.* **226**, 572–579.

Ye, C., Ho, C., Kanazirska, M., Quinn, S., Rogers, K., Seidman, C. E., Seidman, J. G., Brown, E. M., and Vassilev, P. M. (1997a). Amyloid β proteins activate Ca^{2+}-permeable channels through calcium-sensing receptors. *J. Neurosci. Res.* **47**, 547–554.

Ye, C., Ho-Pao, C. L., Kanazirska, M., Quinn, S., Seidman, C. E., Seidman, J. G., Brown, E. M., and Vassilev, P. M. (1997b). Deficient cation channel regulation by extracellular Ca^{2+} in neurons from mice with targeted disruption of the Ca^{2+}_o-sensing receptor gene. *Brain Res. Bull.* **44**, 75–84.

Zaidi, M., Shankar, V. S., Alam, A. S. T., Moonga, B. S., Pazianis, M., and Huang, C. L. (1992). Evidence that a ryanodine receptor triggers signal transduction in the osteoclast. *Biochem. Biophys. Res. Commun.* **188**, 1332–1336.

Zaidi, M., Shankar, V. S., Tunwell, R., Adebanjo, O. A., Mackrill, J., Pazianis, M., O'Connell, D., Simon, B. J., Rifkin, B. R., and Venkitaraman, A. R. (1995). A ryanodine receptor-like molecule expressed in the osteoclast plasma membrane functions in extracellular Ca^{2+} sensing. *J. Clin. Invest.* **96**, 1582–1590.

Debra E. Bramblett
Hsiang-Po Huang
Ming-Jer Tsai

Department of Cell Biology
Baylor College of Medicine
Houston, Texas 77030

Pancreatic Islet Development

I. Introduction

The World Health Organization estimates that, in 1997, 143 million people were afflicted with diabetes worldwide, 16 million of them in the United States, and that these numbers will likely double by the year 2025. Diabetes is one of the most common afflictions of the aged, but it is not restricted entirely to the elderly, as it is often associated with obesity. Therefore, as the average adult lifespan increases and as the population as a whole becomes more sedentary, we are faced with the daunting task of prevention, treatment, and (one hopes) finding a cure for one of the most debilitating diseases of our time. Appropriately, research with emphasis on the endocrine pancreas has grown; in fact, a virtual explosion of discoveries regarding pancreas development has occurred over the past few years. The goal for all of us in this field is to identify the molecular events that govern pancreas functionality and the genetic components that delineate these events. Ulti-

Hormones and Signaling

mately, these findings will lead to the better treatment and prevention of pancreatic disease. This review summarizes many of the recent findings related to the molecular and developmental biology of the pancreas. As will become apparent, the emphasis of the most recent studies, and consequently this survey, is the specification and differentiation of pancreatic endocrine cells during ontogeny.

A. The Pancreas

The pancreas is composed of two different glandular components, the exocrine pancreas and the endocrine pancreas. The exocrine portion comprises about 90% of the gland and secretes digestive juice into the duodenum. The remainder of the pancreas is composed of endocrine cells that secrete hormones essential for carbohydrate homeostasis into the blood stream. There are four different endocrine cell types in the pancreas: α, β, δ, and PP-cells. Aggregations containing each of the four cell types form the islets of Langerhans that are embedded in the exocrine tissue. Primarily, the β-cells secrete insulin in response to high levels of glucose in the blood. Although insulin plays an important role in many aspects of cell physiology, its most prominent role is to stimulate glucose uptake in peripheral tissues after a meal. In contrast, the α-cells secrete glucagon in response to low blood-glucose levels. This stimulates the release of glucose from stores, thereby restoring blood-glucose levels to normal. The δ-cells secrete the hormone somatostatin, whose role in the pancreas is to inhibit the secretion of insulin, glucagon, and pancreatic polypeptide. Pancreatic polypeptide (pp) is secreted by PP-cells. The role of PP is not absolutely clear, but it appears to be involved in the regulation of other islet hormones and possibly in food intake (Clark *et al.*, 1984). As glucose is the major energy source for all cells, the role of pancreatic hormones in the maintenance of blood glucose homeostasis is essential to normal cell function and thus to life.

B. Insulin Action

Insulin has a multitude of functions in the cell that affect metabolism. Insulin reduces the levels of circulating fatty acids, amino acids, and blood glucose by promoting their conversion to storage forms of each and by inhibiting gluconeogenesis and lipolysis. Insulin also has positive and negative effects on gene expression and has been shown to suppress apoptosis and to induce DNA synthesis. Most cells of the body rely on insulin for the uptake of glucose because it cannot penetrate the cell membrane. Thus, by inducing the deposition of glucose transporter molecules (GLUT4) on the cell surfaces in peripheral tissues, insulin directs cellular uptake of glucose, lowering blood-glucose levels.

Diabetes mellitus results from insufficient insulin action and is characterized by high blood glucose and altered cellular metabolism. There are two forms of the disease. Type I diabetes results from an autoimmune reaction that causes the destruction of β-cells and subsequently insulin insufficiency. Type II diabetes is characterized by a resistance to insulin action in peripheral tissues, impaired insulin action in the liver to inhibit glucose production (gluconeogenesis), and deregulated insulin secretion. Though these conditions are, in part, genetically determined, the etiology of diabetes is complex and the genetic components, for the most common forms of the disease, are still unknown.

The molecular events of insulin signaling have been well studied and have been reviewed (Holman and Kasuga, 1997; Myers and White, 1996); therefore, this topic will not be covered in detail here. Briefly, insulin signaling is initiated with the secretion of insulin by β-cells; this is followed by diffusion to other tissues through the bloodstream, and finally the physical interaction with the insulin receptor. After insulin binds to its receptor, the receptor is autophosphorylated and its intrinsic tyrosine kinase activity is induced. The activation of the tyrosine kinase activity of the insulin receptor results in the activation of several different insulin receptor substrate (IRS) proteins, which are associated with the receptor at the cell membrane. In turn, the activated IRS molecule activates a variety of downstream effector molecules such as phosphoinositol 3-kinase (PI 3-kinase). This leads to a cascade of intracellular signaling events, many of which remain unknown, that culminate in the physiological effects of insulin action.

One example of insulin action is the deposition of the glucose transporter (GLUT) isoforms on the cell surface. This requires IRS activation via the insulin receptor tyrosine kinase. Because they are components in the insulin-signaling cascade, IRS molecules have been considered excellent candidate players in diabetes etiology, but no direct correlation has been made between mutations in these molecules and the natural occurrence of the disease. However, a mouse strain with a disruption in the IRS-2 gene has been generated, and it has a diabetic phenotype reminiscent of the type II diabetes found in humans (Withers *et al.*, 1998).

The IRS-2 deficient pups have high blood glucose and transiently high insulin levels as early as 3 days after birth, and they succumb to diabetes by 10–16 weeks (Withers *et al.*, 1998). Observed defects in insulin-stimulated PI 3-kinase activation in the IRS-2 null mice are believed to underlie the abnormalities in glucose homeostasis in these animals.

Interestingly, inactivation of the IRS-2 gene in the mouse causes defects not only in insulin action, but also in insulin secretion. The defect in insulin secretion is what sets the IRS-2 mutant phenotype apart from the IRS-1 mutant phenotype. The IRS-1 mutant does display insulin resistance in peripheral tissues, but never displays a true diabetic state.

In humans, compensatory increases in insulin secretion and β-cell hypertrophy normally ensue in the face of chronic high blood glucose. Indeed, insulin resistance is invariably associated with obesity; however, these individuals do not all become diabetic. Type II diabetes does result when β-cells do not secrete enough insulin to compensate for the resistance to insulin action in the peripheral tissues (Kahn, 1998). β-Cell failure may stem from the effects of chronic high blood glucose (glucose toxicity) that eventually causes the β-cells to lose sensitivity to glucose stimulation. Thus, the type II diabetic may be able to compensate for the insulin resistance initially by hypersecreting insulin, but eventually the β-cells become desensitized and fail to function. Similarly, the IRS-2 mutant mouse displays transiently high insulin levels combined with hyperglycemia, reflecting insulin resistance in the peripheral tissues. In addition, IRS-2 mutant neonates have reduced β-cell mass and do not elicit a proper compensatory β-cell hyperplasia in response to high blood glucose, reflecting a type II diabetes–like syndrome.

The reduced β-cell mass that is found in the IRS-2 null neonate is suggestive of a developmental disorder. In the developing pancreas, IRS-2 protein is detectable in insulin-expressing cells and in pancreatic ducts, the site of endocrine cell neogenesis. Thus, there is a potential role for IRS-2 in endocrine cell maintenance in addition to its well-established role in insulin signaling. Eventually, the IRS-2 mutant mice show progressive deterioration of glucose homeostasis due to insulin resistance in the liver and skeletal muscle as well as β-cell function. The concept that mutations in the IRS-2 gene manifest in a type II diabetes-like syndrome is indeed provocative, but no correlation between alterations (polymorphisms) in the IRS-2 gene and the manifestation of the diabetes in human populations has been reported.

Alterations in several genetic loci, other than IRS-2, have been associated with the manifestation of diabetes in humans. Some examples are the HNF-4α gene in MODYI (maturity onset diabetes of the young), the Glucokinase gene in MODYII, the HNF1α gene in MODYIII, and the PDX-1 gene in pancreatic agenesis (Stoffers *et al.*, 1997; Vionnet *et al.*, 1992; Yamagata *et al.*, 1996a, 1996b). However, these forms of diabetes are relatively rare. Thus, further research on pancreatic gene regulation and pancreas development is deemed necessary to determine the more common cause(s) of the diabetic condition.

C. Regulation of Gene Expression in the Endocrine Pancreas

The pancreas is an outstanding system for studying cell-specific gene expression, and in the following sections the regulation of the expression of three pancreas-specific genes (insulin, somatostatin, and glucagon) will be discussed (Fig. 1). Analysis of pancreatic gene promoters has revealed that members of several transcription-factor families that are integral partici-

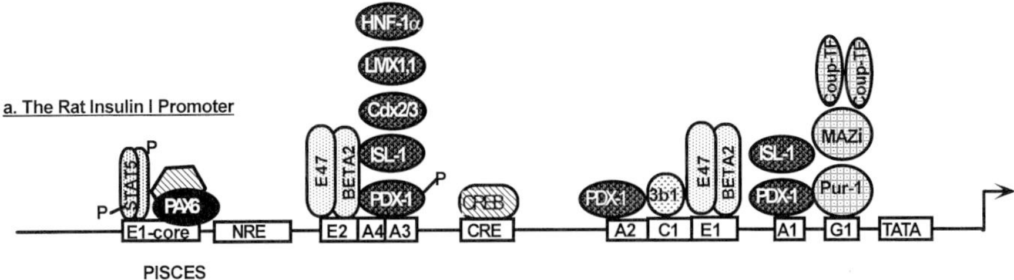

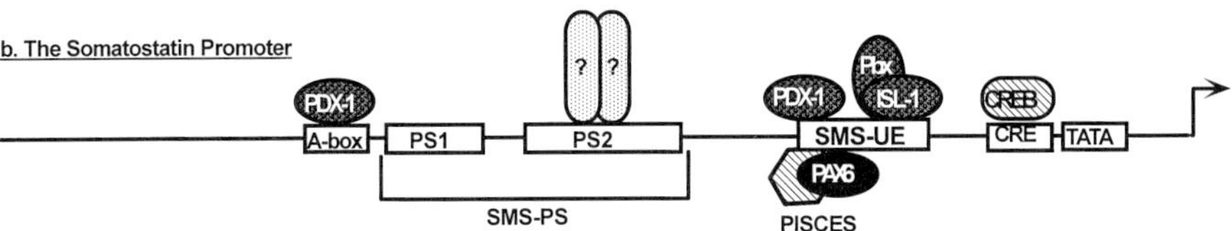

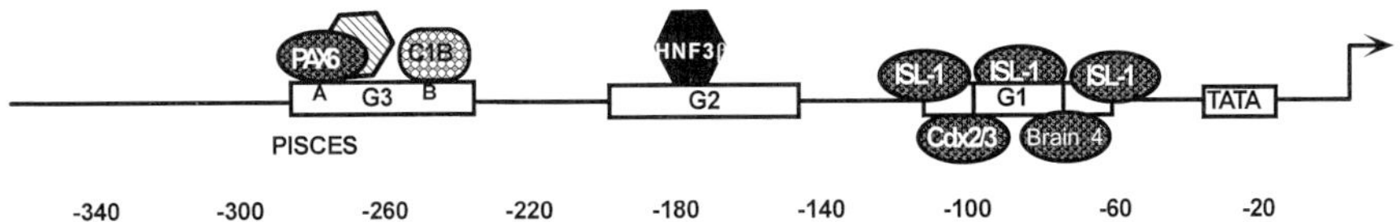

FIGURE I   Islet hormone gene promoters. The insulin promoter (a), somatostatin promoter (b), and glucagon promoter (c) share some, but not all, tissue-specific enhancer elements. In all three panels, DNA binding factors of the same family are indicated by ovals or circles with the same fill pattern. Boxes represent the regulatory elements that reside in each promoter region. See text for details.

pants in pancreas-specific gene regulation in the adult pancreas. These include the homeodomain family of transcriptional regulators, the basic helix–loop–helix family, and the zinc-finger family (both type I and type II zinc finger transcription factors) (Fig. 1). Interestingly, many of the transactivators that contribute to pancreas-specific gene expression play an equally important role in pancreatic organogenesis. Thus, the following mini-review of pancreas-specific gene expression can also serve as a handbook for factors crucial to pancreas development.

I. The Insulin Promoter

Expression of the insulin gene is controlled by its promoter, which may extend as far as 4 kilobases (kb) upstream of its coding region. Both rats and mice have two insulin genes because of a duplication that likely occurred in an ancestral rodent from which both mice and rats evolved (Soares *et al.*, 1985). The insulin I gene shares approximately 520 bp of the upstream ancestral insulin II promoter and lacks one of two introns that are present in the ancestral insulin II gene (Soares *et al.*, 1985). Though some potential regulatory elements may exist in one or the other of the two insulin gene promoters, the expression pattern of the two genes is virtually identical. Generally, the transcriptional regulatory elements, which are essential for appropriate insulin promoter regulation, have been evolutionarily conserved between the rat insulin I and insulin II promoters and across species. Thus, for simplicity, only the rat insulin I gene promoter is depicted in Fig. 1a, utilizing the nomenclature for insulin promoter elements that has been described by German *et al.* (1995).

The insulin promoter possesses a TATA box motif, TATAA, around 20 to 30 nucleotides upstream of the transcriptional start site. The TATA box, found in most genes, facilitates the assembly of the RNA polymerase initiation complex. Tissue-specific expression, mimicking that of the insulin gene, can be conferred to any reporter gene by 360 base pairs (bp) most proximal to the transcriptional start site of the insulin gene (Edlund *et al.*, 1985; Walker *et al.*, 1983). This region provides binding sites for several tissue-specific and ubiquitously expressed transcriptional regulators that work in concert to determine appropriate insulin expression.

Two multipartite enhancers in the first 300 bp of the rat insulin I promoter can confer low levels of tissue specificity on heterologous promoters (German *et al.*, 1992; Karlsson *et al.*, 1987). These enhancers are each composed of an E-box element and more than one A-box element that function synergistically to provide full activity of the insulin enhancer. Interestingly, both the rat insulin promoters, I and II, retain the most proximal enhancer, but rat insulin promoter II does not have a second E-box-containing element. Previously, our laboratory and others showed that the β-cell-specific enhancer (referred to as RIPE3) in the rat insulin II promoter is composed of at least four binding sites, A2/C1, E1, and A1 (Fig. 1a). RIPE3 can confer β-cell specific gene expression onto a heterologous promoter in transient transfection experiments and in transgenic mice (Crowe and Tsai, 1989; Hwung *et al.*, 1990; Shieh and Tsai, 1991; Stellrecht *et al.*, 1997).

The two E-box motifs, CANNTG, are located at -105 (E1) and -231 (E2) in the rat insulin I promoter (German *et al.*, 1992, 1995; Karlsson *et al.*, 1987). The E-box elements are crucial to insulin gene expression, as mutation of either E-box in the rat insulin I gene, or the single E-box element in the rat insulin II gene, leads to a dramatic reduction of promoter activity in transfection experiments (Crowe and Tsai, 1989; Karlsson *et al.*, 1987).

E-box motifs are found in many gene promoters and are bound by members of the basic helix–loop–helix (bHLH) family of transcription factors. Basic HLH factors usually bind tissue-specific E-boxes as heterodimers, composed of one ubiquitous (class A) and one tissue-specific (class B) bHLH protein that cannot bind to DNA strongly as a homodimer. E-box binding complexes in the insulin promoter are no exception in that they are each composed of one class-A bHLH factor (E47 or the related BETA1) and one class-B bHLH factor (Cordle *et al.*, 1990; German *et al.*, 1991; Park and Walker, 1992; Peyton *et al.*, 1994; Shieh and Tsai, 1991). The class B component of the E1-binding complex, BETA2 (beta-cell E-box transcriptional activator), has been cloned and characterized by our laboratory (Naya *et al.*, 1995). BETA2 binds to the E1 element in a complex with the ubiquitously expressed factors E47 or BETA1, resulting in insulin gene activation specifically in insulin secreting β-cells (Naya *et al.*, 1995).

Interestingly, BETA2 was also isolated by another group and identified as NeuroD based on its ability to induce neuronal cell differentiation (Lee *et al.*, 1995). However, for the remainder of this chapter, this will be referred to as BETA2. Thus, BETA2 is similar to other members of the tissue-specific subgroup of the bHLH family of proteins that are frequently involved in cell fate determination. Several additional bHLH factors are involved in directing neurogenesis (Bray, 1997; Caudy *et al.*, 1988; Gradwohl *et al.*, 1996; Lee, 1997; Schwab *et al.*, 1998). There are several additional differentiation processes that are known to involve the action of bHLH factors, including hematopoiesis (Mellentin *et al.*, 1989), myogenic conversion of primitive mesodermal cells (Weintraub, 1993), sex determination (Caudy *et al.*, 1988), and osteogenic commitment (Kazhdan *et al.*, 1997). The best-known example is the myogenic regulatory factor (MRF) family: MyoD, myogenin, MRF4/6, and Myf-5, which are a set of cell-type-specific bHLH proteins whose expression is limited to skeletal muscle (Abmayr and Keller, 1998). Individually, these proteins are capable of converting fibroblasts into myoblast (muscle progenitor cells) by initiating the program for muscle-specific gene expression. Because BETA2 has the ability to induce neural differentiation and because BETA2 appears to play a role in pancreas development, it is tempting to speculate that a similar family of tissue-specific bHLH factors exists in the pancreas, which takes part in aspects of pancreatic development.

BETA2 clearly contributes to the tissue-specific expression of the insulin gene (Naya *et al.*, 1995). However, for more than one reason, the tissue specificity of the insulin gene must be defined by additional factors. First, BETA2/NeuroD is expressed in all four pancreatic endocrine cells, suggesting that either the insulin promoter contains a negative regulatory element(s) that restricts expression in non-β cells, or β-cells contain additional tissue-specific factor(s) that contribute to β-cell specific insulin expression. In addition, though it is significantly reduced, insulin expression is not eliminated in mice that lack a functional copy of the BETA2 gene. Thus, BETA2 is crucial for maximal islet cell-specific expression, but other factors are capable of activating the insulin gene and these are required to achieve appropriate tissue-specific expression of the insulin promoter.

There is a bHLH family subgroup whose members act as negative regulators through the inactivation of positive acting factors. These factors suppress gene expression by the sequestration of transcriptional activator proteins into inactive complexes by non-DNA-binding proteins. One example is the well-known transcriptional repressor Id (Benezra *et al.*, 1990). Id is a specialized member of the bHLH family that has no basic DNA-binding domain that has been shown to be expressed in pancreatic cell lines and can affect insulin expression by sequestering E47 (Cordle *et al.*, 1990; Madsen *et al.*, 1996). Also, we have reported the cloning of a class B bHLH factor, BETA3, that does not appear to bind the insulin E-box, although it was cloned from insulin-expressing HIT cells (Peyton *et al.*, 1996). In a similar fashion to Id, BETA3 appears to squelch the activity of BETA2 by competing for dimerization with E47 and preventing it from binding to the insulin promoter (Peyton *et al.*, 1996).

The E-box elements appear to function synergistically with A-box elements (named for their AT-rich sequence) to activate the insulin promoter. Homeodomain proteins typically bind such elements with a core TAAT sequence. Furthermore, homeodomain proteins specify the architectural layout of developing embryos across a broad spectrum of species (Botas, 1993; Gehring *et al.*, 1994). The rat insulin I promoter contains four A-boxes, A1–A4. The rat insulin II promoter shares three of these (A1 to A3). The A2 element overlaps with a C-rich element C1. The A1, A3, and A4 and the combined A2/C1 element function synergistically with E1 to generate maximal β-cell-specific gene expression of the insulin II promoter (German *et al.*, 1995; Hwung *et al.*, 1990; Shieh and Tsai, 1991). The homeodomain protein PDX-1 is expressed highly in the β-cells of the adult animal, has been shown to bind to the A1 and A2 elements in the β-cell-specific enhancer, and is a potent activator of insulin gene transcription (Ohlsson *et al.*, 1993b; Peers *et al.*, 1994; Petersen *et al.*, 1994). Similar to the requirement for BETA2, high-level insulin expression requires PDX-1. But, as also found in mice lacking BETA2, the insulin gene is expressed in the pancreas, though at low levels, in the absence of PDX-1 (Ahlgren *et al.*, 1996).

Several additional homeodomain-containing proteins (Cdx-2/3, HNF1α, ISL-1, LMX1.1, Nkx2.2, Nkx6.1, Prox-1, PAX6) are expressed in β-cells and are capable of binding to the A elements of the insulin gene (Emens *et al.*, 1992; German *et al.*, 1992; Oliver *et al.*, 1993; Jensen *et al.*, 1996; Karlsson *et al.*, 1990; Sander *et al.*, 1998). Among these additional homeodomain-containing factors, however, only PAX6 has been demonstrated to directly enhance the tissue specific activation of the insulin promoter by binding to a insulin promoter element (now referred to PISCES) and to be required for normal insulin expression *in vivo*. (Sander *et al.*, 1997). Alterations in the HNF1α gene have been linked to the manifestation of MODYII. However, the diabetic phenotype observed in mice lacking the HNF1α gene has been shown to be due to defects in insulin secretion and not reduced insulin gene transcription (Pontoglio *et al.*, 1998). The pancreatic roles of several of the homeodomain-containing factors just listed will be emphasized later, but any role they might play in insulin gene activation *in vivo* is inconclusive.

The topic of nutrient regulation of the insulin gene is broad and out of the scope of this review; however, it is interesting to note that the A3 element of the human insulin promoter is reported to be responsible for glucose-stimulated transcriptional activation (Petersen *et al.*, 1994). Furthermore, it has been suggested that glucose stimulation of the insulin gene is mediated via phosphorylation of PDX-1, which binds to insulin A box elements (MacFarlane *et al.*, 1994). Conversely, chronically high blood glucose can lead to the loss of β-cell function or glucose toxicity. The repression of insulin expression in response to supraphysiological glucose concentrations is reportedly due to reduced protein binding to insulin promoter A-boxes and C1 (Poitout *et al.*, 1996). Studies showing that PDX-1 expression is depressed under similar conditions further supports a role for PDX-1 in glucose toxicity (Olson *et al.*, 1995; Sharma *et al.*, 1995). In addition, C/EBPβ has also been reported to contribute to the repression of the insulin gene under chronic high glucose conditions via an interaction between E47 and C/EBPβ, which prevents the bHLH factor from dimerization and/or binding to DNA (Lu *et al.*, 1997).

There are several additional regulatory elements in the insulin promoter that likely mediate its induction in response to external stimuli. The G1 element of the insulin promoter, between -57 and -40, is bound by zinc-finger containing factors. For example, the orphan steroid hormone receptor, COUP-TF, can bind to this element (Hwung *et al.*, 1988). The zinc-finger protein Pur-1 is another example that is particularly interesting because it has been genetically linked to IDDM (insulin dependent diabetes mellitus) (Kennedy *et al.*, 1995; Kennedy and Rutter, 1992). Other members of the Pur-1 family are expressed in β-cells, such as MAZi (myc-associated zinc finger protein of human islets), which enhances the transcriptional activity of the c-*myc* gene in human islet cells (Tsutsui *et al.*, 1996). However, Pur-1 is capable of inducing insulin expression in both pancreatic and non-pancreatic cell types. Thus, the role of the zinc-finger protein(s) that bind to the promoter in the transcriptional regulation of insulin needs further investigation.

A negative regulatory element (NRE) has been identified in the human and rat insulin I and II promoters (Boam *et al.*, 1990; Clark *et al.*, 1995; Goodman *et al.*, 1996; Laimins *et al.*, 1986; Nir *et al.*, 1986; Sander and German, 1997; Shiran *et al.*, 1993; Whelan *et al.*, 1989). Sequence analysis of the human NRE localized between -258 and -279 revealed a 73% homology with the negative glucocorticoid response element (nGRE) consensus sequence. In support of the proposal that the insulin NRE is a nGRE, dexamethasone treatment of cells containing an insulin NRE-CAT reporter construct decreased CAT gene expression by 48% (Goodman *et al.*, 1996). The insulin promoter NRE may serve to restrict expression of the insulin gene to the islets of Langerhans. Some investigators have found that the NREs in the rat and mouse insulin promoters are active only in β-cell lines and not in non-β-cell lines (Clark *et al.*, 1995; Leshkowitz *et al.*, 1992). However, others have identified an NRE that is active in cells expressing insulin as well as non-insulin-producing cells, but is overridden by dominant-positive transacting factors present in insulin-producing cells (Nir *et al.*, 1986). The controversy is strengthened by studies in which deletion of the NRE had either no effect or actually reduced promoter activity in both β-cell tumor lines and primary islet-cell cultures (Walker *et al.*, 1983). Interestingly, a cassette containing five copies of the NRE linked to a heterologous promoter does have a negative effect on transcription in non-β-cell tumor lines and in primary non-β-cells (Sander and German, 1997). However, the NRE has also been demonstrated to function as an activator of transcription in primary cultures of rat β-cells (Sander and German, 1997). Thus, the conflict as to the tissue specificity of the insulin NRE remains unresolved and may reflect the disparity often observed between different cell lines and primary cultures.

Finally, human, mouse, and both rat insulin promoters contain a cyclic AMP (cAMP) response element (CRE) that responds to hormone stimulation (Docherty and Clark, 1994; Philippe and Missotten, 1990). The CRE is located between -177 and -184 in the rat insulin I gene. CREB, the CRE-binding protein, is thought to activate insulin transcription in response to hormone stimulation as follows. Hormonal stimulation causes adenylate-cyclase activation, causing an increase in cAMP levels. The elevated cAMP level leads to protein kinase A activation, which in turn phosphorylates CREB, thereby stimulating its binding to the insulin CRE (Philippe and Missotten, 1990). Another mechanism has been proposed for the activation of the rat insulin I gene by growth hormone. In this case, insulin stimulation is mediated through the JAK/STAT pathway. The phosphorylation of STAT5 monomers leads to their dimerization and binding to a distal element in the rat insulin I promoter previously referred to as the E1-core element (Galsgaard *et al.*, 1996; Madsen *et al.*, 1996). There are likely many more regulatory elements in the insulin promoter that modulate insulin induction to a variety of external stimuli that have not been mentioned here or have not yet been identified.

2. The Somatostatin Promoter

A 500-bp region upstream of the transcriptional start site confers δ-cell specific activity to the rat somatostatin promoter as illustrated in Fig. 1b (Leonard *et al.*, 1993). PDX-1 was originally isolated and characterized based on its ability to bind and thereby activate the somatostatin promoter through binding to two tissue-specific promoter element sites, between -86 and -104 and between -286 and -303 (Leonard *et al.*, 1993). A third tissue-specific promoter element, more distal from the start of transcription, has been reported, but the two sites most proximal to the start of transcription are reported to be the most sensitive to PDX-1 action (Leonard *et al.*, 1993). Synergistic activities of PDX-1 bound to the site between -86 and -104 (referred to here as SMS-UE), and a cyclic AMP-response element (CRE), located at -32 to -56, are sufficient to confer tissue-specific expression to the somatostatin promoter (Andrisani *et al.*, 1987; Montminy *et al.*, 1996; Powers *et al.*, 1989; Vallejo *et al.*, 1992a, 1992b). It appears that the SMS-UE is a tripartite transcriptional element with three binding sites for homeodomain proteins (Vallejo *et al.*, 1992a, 1992b). In addition to PDX-1, the homeodomain factor, ISL-1, binds to the SMS-UE enhancer at an overlapping position (at -85 to -99). ISL-1 is reported to function synergistically with CREB to stimulate high-level somatostatin expression (Leonard *et al.*, 1992). Thus, both these factors have been proven capable of activating the somatostatin gene in transfection assays (Leonard *et al.*, 1992, 1993). However, others have demonstrated that PDX-1, in association with another homeodomain factor Pbx, a human proto-oncogene with extensive sequence homology to the *Drosophila* protein extradenticle (Rauskolb *et al.*, 1993), comprises the predominant fraction of the DNA binding activity at this site (Peers *et al.*, 1995). Finally, the SMS-UE also overlaps with the PISCES element that serves as a binding site for the homeodomain factor PAX6 together with a ubiquitous winged-helix factor (Diedrich *et al.*, 1997). The PISCES element appears to be common to the glucagon and insulin promoters and will be discussed further (Sander *et al.*, 1997).

The restriction of somatostatin expression to δ-cells may be maintained, in part, by a compound negative regulatory element (SMS-PS) bound by at least two complexes, PS1 (-237 to -220) and PS2 (-208 to -189) (Powers *et al.*, 1989; Vallejo *et al.*, 1995). A distal silencer located between -425 and -345 (Vallejo *et al.*, 1995) may also contribute to the restriction of somatostatin expression (not included in Fig. 1b). Significantly, PS1 and PS2 can individually repress the enhancer effects of SMS-UE in both somatostatin producing and nonproducing cells (Vallejo *et al.*, 1995). Thus, it appears that the cell-specific expression of somatostatin is dependent, at least in part, on the compensatory effect exerted by the cell-specific SMS-UE over the SMS-PS. The protein(s) that comprises the SMS-PS-binding complex has not been identified. Though the PS2 element contains an E-box motif, the bHLH family member(s) that can bind to this site and exert a negative effect on the promoter activity has not been found. Neither the bHLH activator Pan1 nor the bHLH repressor Id has an effect on the somatostatin promoter, and likewise complexes that do bind PS1 and PS2 do not contain other known bHLH factors such as Pan1, E12/E47, or *myc* (Vallejo *et al.*, 1995).

It is possible that the function of PS1 is developmentally regulated. PS1 does not function as somatostatin gene repressor in certain insulin producing β-cell lines, whereas it does function in δ-cell lines. In contrast, the PS2 element is functional in both β and δ-cell lines (Vallejo *et al.*, 1995). During development, insulin and somatostatin are coexpressed in a subset of pre-δ-cells that will eventually differentiate into either PP-cells or somatostatin-producing δ-cells. It has been proposed that, in order for somatostatin to be active in insulin-producing pre-δ-cells, the somatostatin promoter must first be partially released from repressor activity before δ-cell-specific transcriptional activators are fully active (Vallejo *et al.*, 1995).

3. *The Glucagon Promoter*

In mammals, the glucagon gene produces a single mRNA transcript that is translated and processed differently in brain, pancreatic islets, and the intestine (Drucker, 1998). Proglucagon is cleaved into two glucagon-like peptides, GLP-1 and GLP-2, that are secreted by enteroendocrine cells of the small and large intestine, while glucagon is specifically secreted from the pancreatic α-cells. GLP-1 and GLP-2 appear to be specifically secreted in the intestine because of the tissue-specific expression of prohormone convertases (PCs) in the enteroendocrine cells (Drucker, 1998). Therefore, the tissue specificity of the glucagon gene products is, in part, regulated by posttranslational mechanisms.

Alpha (α)-cell-specific activity of the glucagon promoter requires at least three composite elements, referred to as G1 (-100 to -65 in the rat promoter), G2 (-192 to -174 in the rat promoter), and G3 (-248 to -241 in the rat promoter) (Philippe *et al.*, 1988), as illustrated in Fig. 1b. G1 is required for low-level transcriptional activity of the glucagon promoter and is referred to as the glucagon promoter. G2 and G3 are referred to as enhancers because their activity is dependent on a functional G1 element. G1 contains three AT-rich sequences important for promoter function and has been implicated in α-cell-specific expression of the glucagon gene (Morel *et al.*, 1995). The AT-rich sequences are indicative of homeodomain transcription factor binding sites. In fact, the caudal-related homeodomain protein Cdx2/3 binds to the G1 element with high affinity (Laser *et al.*, 1996). Cdx2/3 binds to the G1 element, as one subunit of a complex initially referred to as B3, to activate glucagon gene transcription in both the intestine and the pancreas (Laser *et al.*, 1996; Wang and Drucker, 1996). Also, the homeodomain factor ISL-1 can bind to the tripartite G1 element at all three A/T-rich sites, but it can activate glucagon transcription only through binding to the two most upstream sites (Wang and Drucker, 1995). Finally, the POU domain transcription factor brain 4 has been shown to be a major constituent of the G1-binding proteins in α-cells and contributes to glucagon promoter activation.

Although their activity is dependent on an intact G1 element, the G2 and G3 enhancers induce maximal glucagon expression. G2 and G3 can act independently of each other and can function in all islet cell types. G3 contains two domains, A and B. More detailed analysis of the G3 element revealed that the A domain is bound by two complexes referred to as C1A and C1B, with the C1A complex being islet cell specific. Interestingly, the inhibitory effect of insulin on the secretion and biosynthesis of glucagon appears to be mediated through the G3A element (Philippe *et al.*, 1995).

The A domain of the composite G3 element (-265 to -245) is similar in sequence to elements found in the insulin I (-329 to -307) and somatostatin (SMS-UE, -103 to -75) promoters (Diedrich *et al.*, 1997; Knepel *et al.*, 1991; Sander *et al.*, 1997). This pancreatic islet cell-specific enhancer sequence, now referred to as PISCES, is considered an important regulatory element for the activity of all three islet hormone gene promoters that may coordinate islet hormone expression (Diedrich *et al.*, 1997; Knepel *et al.*, 1991; Wrege *et al.*, 1995b). Initially, the PISCES element was described as being modular in structure, bound by a factor highly expressed in pancreatic islets together with a ubiquitously expressed winged-helix protein (Diedrich *et al.*, 1997; Knepel *et al.*, 1991; Wrege *et al.*, 1995a).

The homeodomain factor that binds to the PISCES element in all three promoters is now believed to be the factor PAX6 (Ohlsson *et al.*, 1993b; Sander *et al.*, 1997). The specific binding of PAX6 to the A site of the G3 element in the glucagon promoter can be competed by the insulin I element (E1-core element) as well as the somatostatin upstream element (UE). Thus, all three of these elements contain the PISCES element (Sander *et al.*, 1997; Wrege *et al.*, 1995a). In each of these promoters, the PISCES site is part of a modular element in which PISCES contributes to islet-specific activity (Sander *et al.*, 1997). The ubiquitously expressed factor that binds to the glucagon G3 element has only been characterized as a winged-helix factor; it has not been conclusively identified.

Interestingly, the winged-helix family member HNF3β is known to bind to the upstream glucagon G2 element to repress glucagon transcription (Philippe *et al.*, 1994). HNF3β is known for its early developmental expression in the gut endoderm and is believed to play a role in the development of several organs in the gut. Three isoforms of HNF3β (HNF3β1–3) are expressed in α-cells. Furthermore, the relative levels of the HNF3β-1 isoform, as opposed to the HNF3β-2 and -3 isoforms in α-cells, which appear to have opposing effects on glucagon expression, may be important in regulating glucagon expression (Diedrich *et al.*, 1997; Philippe, 1995; Philippe *et al.*, 1994). In addition, a protein-kinase C response element has been mapped to the glucagon G2 element, and a functional interaction between HNF3β and an unknown factor bound to a closely associated site has been determined to mediate the protein kinase C responsiveness.

In summary, several of the same transcriptional activators bind to the promoter regions of insulin, glucagon, and somatostatin genes. The means by which the islet specific expression of these promoters is further restricted to a particular cell type likely involves unique cooperative interactions between these factors. Indeed, there are several examples of cooperation between islet-specific transcriptional activators. For example, high-level tissue-specific expression of both the insulin and the somatostatin genes appears to involve a cooperative interaction between PDX-1 and factors that bind nearby. In the somatostatin promoter, PDX-1 cooperatively binds to the somatostatin promoter with Pbx (Peers *et al.*, 1995), whereas in insulin promoter, PDX-1 was demonstrated to function cooperatively with E47 to activate transcription (Odagiri *et al.*, 1996; Peers *et al.*, 1995; Serup *et al.*, 1995). Also, the LIM homeodomain protein, LMX1.1, and the bHLH factor, E47, can act synergistically to activate the insulin promoter. This is mediated through a direct interaction between the Lim2 domain of LMX1.1 and the bHLH domain of E47, while they are bound to the A3/A4 and E2 elements, respectively (German *et al.*, 1992). Additionally, the BETA2/E47 hetero-dimer complex, bound to the E1 box, requires a synergistic interaction with an unknown islet factor that binds the C1 element, referred to as islet-specific-factor 3b1, to direct β-cell specific the insulin expression (Naya *et al.*, 1995; Peers *et al.*, 1994). Thus, factors that bind to the pancreatic hormone promoters, particularly the members of the homeodomain and bHLH families, have a functional interdependency to direct cell specific gene activation. Indeed, such cooperation between bHLH and homeodomain proteins to generate cell-specific gene expression has been documented on other promoters and in other tissues. For example, BETA2 and Ptx1, a Bicoid-related homeodomain protein, function synergistically to ensure corticotroph-cell-specific expression of the proopiomelanocortin (POMC) gene in the brain (Peers *et al.*, 1995; Poulin *et al.*, 1997).

II. Development of the Endocrine Pancreas

The study of islet hormone gene expression in the adult pancreas has provided insight into the factors involved in pancreatic development. Extensive use of targeted gene mutagenesis (knockout) technology has revealed key developmental roles for each of these factors. This section examines several of the most recent discoveries in pancreas development obtained through targeted gene disruption, following a brief morphological description of mouse pancreas development.

A. Pancreatic Morphogenesis

1. Endodermal Origin of the Pancreas

It is often necessary to break things down to the most rudimentary parts to understand a system of interest. Thus, it is appropriate to begin this section with a description of pancreatic morphogenesis. The pancreas is formed from gut endoderm. Initially, the mouse embryo is U-shaped with the germ layers in an inverted orientation. At this stage, the ectoderm (neural tube and surface ectoderm) lies on the inside and the endoderm (gut) lies on the outside (Hogan *et al.*, 1994). As a result of a process called turning, the germ layers assume the typical conformation of the vertebrate embryo, with the gut endoderm on the inside forming the gut tube (Hogan *et al.*, 1994). Prior to the formation of the gut tube, the lateral margins of the definitive endoderm are continuous with the visceral yolk sac endoderm. Medially, the ventral surface of the notochord is intercalated into the roof of the gut endoderm (Hogan *et al.*, 1994). Eventually the notochord detaches from the endoderm and is separated from the endoderm by the paired dorsal aortae, and it forms a midline rod. The pharynx forms at the rostral end of the gut tube. Posterior to the pharynx, the digestive tube forms a series of constrictions, giving rise to the esophagus, followed by the stomach, the small intestine, and the large intestine. Additional out-pocketings from the pharyngeal region give rise to several organs (the thyroid, lung buds, and liver).

Later, after approximately 9 days of gestation, a series of endodermal evaginations, or tubes of epithelium, project out into the surrounding mesenchyme caudal to the stomach. This region of the duodenum, sometimes referred to as the hepatopancreatic ring, has extraordinary potential for forming glandular tissue. The first projection extending out ventrally into the mesenchyme between embryonic day (E) 8.5 and 9 (Zaret, 1996) is called the hepatic duct. It is induced to proliferate and branch, thereby forming the glandular epithelium of the liver. On the dorsal side of the duodenum, in opposition to the hepatic duct, is the dorsal pancreatic bud. This column of epithelium that emerges at E9.5 in the mouse is referred to as the duct of Santorini (Pictet and Rutter, 1972). The dorsal pancreatic bud emerges prior to the condensation mesenchyme (Ahlgren *et al.*, 1996). Soon thereafter, the epithelium is induced to branch and differentiate into exocrine structures by signals emanating from the surrounding mesenchyme. The appearance of the dorsal bud slightly precedes that of the duct of Wirsung that projects out from the ventral side of the gut tube near the junction of the hepatic duct and will eventually become the ventral pancreas (Pictet and Rutter, 1972).

The adult pancreas is formed from the fusion of the dorsal and ventral pancreatic diverticula. During embryogenesis, the stomach rotates relative to the notochord. This apparent rotation does not involve actual movement, but rather is due to the increased growth of the duodenum relative to the body wall (Drews, 1995). As the stomach rotates, the ventral pancreas is repositioned, juxtaposing it with the dorsal pancreas in a new dorsal position. As a result of this repositioning of the dorsal and ventral pancreas, which occurs at around E16 in the mouse (Wessells, 1967), the duct of Santorini and the duct of Wirsung are fused to form one main pancreatic duct and the ventral and dorsal pancreas become integrated. In summary, the body of the pancreas is formed from the fusion of dorsal and ventral pancreatic primordia derived from two distinct diverticula extending from the definitive embryonic gut endoderm into the surrounding mesenchyme.

The acinar structures that compose the exocrine pancreas can be detected at the termini of the elongating ducts at E15 (Bock *et al.*, 1997). The adult acini are composed of linear arrays of polarized cells, pyramidal in shape, that have basal nuclei filled with apical zymogen (secretory) granules (Motta, 1997). The apical surfaces form the lumen of the acini. Acinar cells secrete a variety of digestive proenzymes into pancreatic ducts that empty into the duodenum. There are many digestive enzymes produced from these cells, including nucleases, proteases, amylases, and lipases that usually become activated in the duodenum by proteolytic cleavage.

Around E16.5 to E17, clusters of endocrine cells appear, embedded in the exocrine tissue. These clusters compose the endocrine portion of the adult pancreas and are referred to as the islets of Langerhans. Each islet is composed of four hormone-secreting cell types: the insulin-producing β-cell, the somatostatin-producing δ-cell, the glucagon-producing α-cell, and the pancreatic peptide–producing PP-cell. Islet cells are arranged such that the β-cells form a tight spherical cluster surrounded by α-cells, δ-cells, and PP-cells. The islets are composed of 80% β-cells, 15% α-cells, and 5% δ-cells (Pictet and Rutter, 1972). Overall, the pancreas is composed mostly of exocrine tissue with the endocrine portion comprising only 1–2%.

2. Morphogenesis of the Pancreatic Epithelium

The dorsal pancreas and the ventral pancreas both develop three epithelial compartments, the ductal system, the exocrine system, and endocrine portion. The endodermal projections of pancreatic diverticula are surrounded by a cap of condensed mesenchyme and are composed of folded strands of epithelial cells that represent the primitive pancreatic ducts branching from the main duct. It has long been thought that the differentiation of pancreatic epithelium is dependent on signals secreted by the surrounding mesenchyme (Golosow and Grobstien, 1962). Indeed, the pancreatic epithelium does not grow and differentiate into exocrine cells and acini in the absence of mesenchyme (Golosow and Grobstien, 1962; Miralles *et al.*, 1998; Wessells and Cohen, 1967). However, prior to the morphogenesis of the pancreatic bud (roughly E8.5 to E9), glucagon, insulin, and somatostatin transcripts can be detected in the precise area of the dorsal duodenum that will give rise to the pancreas. This is approximately 10–12 hours prior to mesenchymal condensation (Alpert *et al.*, 1988; Gittes and Rutter, 1992).

More recently, elegant culture experiments performed by Gittes *et al.* (1996) defined the role of mesenchyme in pancreas development more clearly. In these experiments, pancreatic epithelium from E11 mouse embryos was cultured *in vitro* under various conditions. When pancreatic epithelium was cultured with its corresponding mesenchyme, acinar, ductal, and islet structures developed. In contrast, the pancreatic epithelium cultured alone did not develop these structures. Furthermore, when pancreatic epithelium was grown in Matrigel, which is rich in basement membranes, cystic structures staining positive for pancreatic duct cell markers developed, but endocrine or exocrine cell markers were not detected. Thus, basement membranes, or components of them, were shown to be important in the determination of ductal morphogenesis and duct cell differentiation. Interestingly, when pancreatic epithelium was placed under the renal capsule of a syngeneic adult mouse along with mesenchyme, mature pancreatic tissue with acini, ducts, and mature islets showing central insulin expression developed. In contrast, when the pancreatic epithelium was grown under the same conditions without mesenchyme, no ductile or acinar structures could be detected, but mature islet cells were formed. Thus, the program necessary for the differentiation of pancreatic epithelium into adult islets is already established by E11 in the mouse embryo, and the default setting for pancreatic epithelium at this stage may be to form islets in the absence of mesenchymal signals. Therefore, the pancreatic mesenchyme is required for the differentiation of acini and ducts, but not for pancreatic endocrine cell development at this stage. Since, in the absence of embryonic mesenchymal signals, endocrine cell differentiation occurred only *in vivo*, some external factors, present in the subcapular environment of the kidney may allow for this differentiation process. The early specification of the mature islet cell fate and the apparent requirement of mesenchyme signals for duct and acinar development and not endocrine cell development at this stage has been confirmed by other investigators (Ahlgren *et al.*, 1997; Gittes *et al.*, 1996; Kim *et al.*, 1997).

Endocrine cell development appears to occur in two waves. This was initially proposed because the developmental accumulation of insulin is biphasic, with the production of insulin being considerably higher in the second phase than the first (Pictet and Rutter, 1972). The first wave of endocrine cell development occurs between E9.5 and E12.5. During this time, cells expressing insulin and/or glucagon appear in small clusters of like cells associated with the ductile epithelium (Alpert *et al.*, 1988). Insulin-producing cells that appear between E9.5 and E12.5 do not express certain mature β-cell markers. Most if not all coexpress glucagon, and these cells appear to be incapable of aggregating into true islets (Alpert *et al.*, 1988; Jackerott *et al.*, 1996; Pang *et al.*, 1994; Pictet and Rutter, 1972). Between the first and second wave of β-cell development, little endocrine cell proliferation occurs, but the second wave of endocrine cell development at E15 is characterized by a huge increase in β-cell numbers. These cells express mature markers, do not coexpress glucagon, and are capable of aggregating into islets of Langerhans, which become detectable at E16.5 (Miralles *et al.*, 1998).

As mentioned, the β-cells that arise early, between E9.5 and E12, do not express some mature β-cell-specific markers such as the glucose transporter isoform GLUT2. GLUT2 and Rad3A (a marker for secretory granules) are expressed in mature β-cells and have been used to monitor the appearance of mature β-cells during development. Comparing the expression patterns of GLUT2 and insulin during ontogeny led Pang *et al.* (1994) to propose that two sets of β-cell precursors develop from the pancreatic epithelium during the two waves of development. GLUT2 is expressed in the endoderm prior to pancreatic morphogenesis. Until E12.5, GLUT2 and insulin-expressing cells are mutually exclusive, with most of the GLUT2-expressing cells in the ductile epithelium and the insulin-expressing cells in small interstitial clusters (Miralles *et al.*, 1998; Pang *et al.*, 1994). Three days later (E15.5) when the second wave of β-cell neogenesis occurs, there is a dramatic increase in the number of GLUT2 positive cells that express insulin, such that they comprise a large part of the ductile epithelium. In contrast to the first wave of β-cell differentiation, cells in the second wave resemble more mature β-cells, in that they do not coexpress glucagon and do express Rad3A and GLUT2. Moreover, cells bearing these markers are found in islet-like structures just before birth and in the adult (Pang *et al.*, 1994). The early insulin-expressing cells that also express GLUT2 are never detected, and proliferation of the small number of insulin-expressing cells that arise early is believed to be insufficient to account for the huge accumulation of β-cells during the second wave (Miralles *et al.*, 1998). Therefore, the distinctive expression pattern of this mature β-cell marker (GLUT2) reinforces the theory the second wave of β-cell development involves a secondary round of β-cell differentiation, rather than enhanced proliferation of preexisting β-cells.

Mesenchyme may actually repress mature pancreatic endocrine cell development. In a series of organoculture experiments in which the mesenchymal cap from E12.5 pancreatic buds was removed, the number of endocrine cells that developed increased fourfold in comparison to pancreatic buds that retained their mesenchymal cap (Miralles *et al.*, 1998). At the same time, exocrine tissue development was repressed by mesenchyme depletion, as has been observed by other groups. In addition, immunohistological experiments revealed that the β-cell population enhanced by mesenchyme depletion expresses GLUT2 and Rad3A. Also, these cells were capable of forming genuine islets. In contrast, E12.5 pancreatic buds cultured with their mesenchymal caps intact developed a smaller number of insulin-producing cells that did not express GLUT2 and were not capable of forming islets, and 50% of these cells expressed glucagon.

Thus, the β-cell population whose development was enhanced by mesenchyme depletion at E12.5 had a more mature phenotype, resembling cells that normally develop during the second wave of β-cell proliferation (Miralles *et al.*, 1998). The observed increase in β-cell mass in response to mesenchyme depletion was deemed due to an increase in differentiation and not proliferation, reinforcing the idea that β-cells develop in two waves (Miralles *et al.*, 1998). Together, these observations suggest that mesenchyme-derived factors may repress the second wave of β-cell development while enhancing exocrine development. Follistatin could represent such a factor because it is expressed in pancreatic mesenchyme during development. Furthermore, it is known to inhibit cell-differentiation factors of the TGFβ family, BMP7 and activin, which are expressed by pancreatic epithelial cells during early development (Furukawa *et al.*, 1995; Lyons *et al.*, 1995). Indeed, follistatin is capable of mimicking both the repressive and the inductive effects of mesenchyme on the endocrine and exocrine tissues, when tested on pancreatic epithelium grown in culture (Miralles *et al.*, 1998). It is certainly possible that mesenchymal factors like follistatin modulate the repressive effects of activin on Sonic hedgehog (Shh), and it would be interesting to study whether mesenchymal–epithelial interactions participate in Shh repression after notochord separation (see section III for further discussion).

B. Developmentally Regulated Pancreas Gene Expression: The Endocrine Cell Lineage Model

All pancreatic exocrine and endocrine cell types are believed to derive from a common endodermal precursor cell in the ductile epithelium (Gittes *et al.*, 1996; Pictet and Rutter, 1972). Several lines of evidence support this hypothesis. First, endocrine cells that are monospecific for the expression of one islet hormone in the adult tend to coexpress islet hormones during ontogeny (Alpert *et al.*, 1988; Herrera *et al.*, 1991; Teitelman *et al.*, 1993; Upchurch *et al.*, 1994). As pancreatic development proceeds cells that coexpress more than one hormone disappear while the number of monospecific cells increases. Also, the existence of mixed ductile/acinar/islet cell phenotypes in culture supports a lineage relationship between the three pancreatic systems (Beck and Madsen, 1989; Drucker *et al.*, 1987; Gu *et al.*, 1994; Gu and Sarvetnick, 1993; Jensen *et al.*, 1996; Madsen *et al.*, 1986). The nature of the stem cells, which give rise to exocrine and endocrine cells in the developing embryo, has not been determined. However, adult duct tissue does maintain an extraordinary regenerative capacity, showing both endocrine cell and exocrine cell renewal. This indicates the presence of pancreatic stem cells in the adult pancreas. However, these cells may actually be of a different phenotype than the stem cells in the embryonic pancreas, leaving the question as to the nature of the primordial pancreatic stem cell unanswered (Bonner-Weir *et al.*, 1993; Dudek *et al.*, 1991; Rosenberg *et al.*, 1996; Teitelman, 1996; Wang *et al.*, 1995).

The Teitelman laboratory (Fig. 2) established a model predicting a lineage relationship between all endocrine cell types. This model was initially based on the timing and pattern of islet hormone expression during ontogeny, as determined by immunohistochemical experiments (Alpert *et al.*, 1988). These experiments showed that the four primary islet cell types appear in a staggered

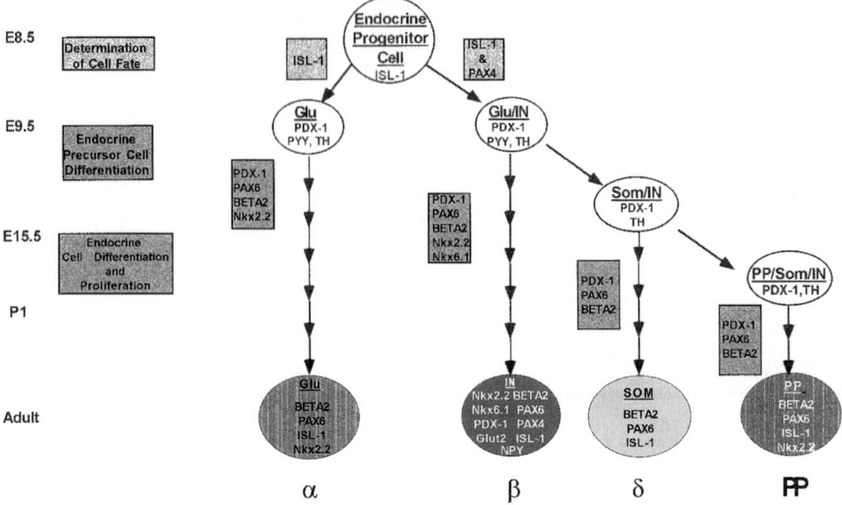

FIGURE 2   The endocrine lineage model. Open circles represent precursor endocrine cells. Filled circles represent mature pancreatic endocrine cells. Boxes display the transcriptional regulator factor(s) involved in endocrine cell development. Arrows indicate steps or branch points in the endocrine cell lineage. Adapted from Teitelman (1996).

fashion during pancreas development and often coexpress more than one hormone until later in development. Considering the time at which each hormone made its appearance and its subsequent coexpression pattern in the endocrine cell precursors, Teitelman projected that mature single positive cells derive from precursor cells that often coexpress insulin.

The first endocrine cells that appear at E9.5 coexpress insulin and glucagon (Alpert *et al.*, 1988; Guz *et al.*, 1995b; Pictet and Rutter, 1972; Teitelman *et al.*, 1993). Insulin-expressing cells that do not express glucagon gradually appear around E14. This is followed by the appearance of pre-δ-cells that express somatostatin at E14.5, and 36% of these cells were found to express insulin also. Finally, at P1, cells coexpressing pancreatic polypeptide (PP) appear, and 25% of these cells were found to coexpress insulin. Cells expressing PP and somatostatin and insulin also appear transiently. Thus, insulin is expressed by all the primary endocrine cell types during development. These results indicate that all pancreatic endocrine cell types arise from a common progenitor cell that can activate the insulin promoter and that differentiation to the mature islet-cell phenotype correlates with the loss of insulin coexpression. Furthermore, upon their appearance, each of the endocrine cell types transiently express the marker TH (tyrosine hydroxylase), further substantiating their lineage relationship (Alpert *et al.*, 1988; Guz *et al.*, 1995b).

Expression of the homeodomain protein PDX-1 can be detected in the pancreatic primordia as early as E8.5 and is coexpressed with each of the pancreatic hormones and the exocrine markers during development (Guz *et al.*, 1995b). Interestingly, monospecific pre-α-cells that express only glucagon and not insulin also have been detected at the same time as the earliest insulin-coexpressing cells appear (E9.5) (Guz *et al.*, 1995a; Roberts *et al.*, 1995). Furthermore, these early pre-α-cells also express the mature β-cell specific transcription factor PDX-1, transiently. PDX-1 expression is gradually restricted to the β-cells in the developing endocrine pancreas, though its expression is maintained throughout the surrounding duodenal epithelium and in the exocrine cells through adulthood (Serup *et al.*, 1995). This indicates that PDX-1 plays a role in the initial stages of endocrine cell differentiation and is important for the maintenance of the fully differentiated β-cells as well as exocrine cells and cells of the duodenal mucosa. Thus, the endocrine lineage model projects that a common pancreatic stem cell that expresses PDX-1 gives rise to both exocrine and endocrine progenitor cells that continue to express PDX-1. Precursor cells that express either glucagon alone or both glucagon and insulin are derived from the endocrine progenitor cell, and these coexpressing cells subsequently give rise to the mature monospecific β, δ, and PP cells (Guz *et al.*, 1995b).

Teitelman's model was initially based solely on the observed immunohistochemical colocalization of islet hormones during fixed times of pancreatic development; it was not a formal proof for the cell-lineage relationship. Thus, the possibility that the hormone-coexpressing cells are independently derived from a progenitor population and bear no relationship to the adult, monospecific islet cell types, could not be ruled out. However, the theory behind the cell lineage model is that the sequential activation and/or inactivation of key regulatory factors, which operate on endocrine genes, allows the establishment of pancreatic endocrine cells from a common precursor cell. Subsequently, as maturation proceeds, these factors direct selective inactivation of inappropriate transcriptional activators and consequently certain hormone gene products. Overtime, this would result in the cell-specific expression pattern of each islet hormone that characterizes each of the mature islet-cell phenotypes. Thus, the expression pattern of PDX-1 and other transcriptional regulators, which will be discussed in further detail, provides further support for Teitelman's endocrine lineage model.

Study of the differentiation of pluripotent pancreatic cells in culture also supports the endocrine cell lineage model. PDX-1 is expressed in heterogeneous rat-islet-tumor cell cultures derived from a liver metastasis referred to as MSL cells (Madsen *et al.*, 1993, 1996, 1997). Clonal populations of glucagon-producing or insulin-producing cell lines can be derived from the same clonal cell line derived from the MSL cell population. Interestingly, the PDX-1 gene is not expressed in the resulting glucagonoma cells, whereas PDX-1 expression was maintained in the insulinoma cells. Therefore, glucagon-expressing or insulin-expressing cell lines can be derived from the same clonal parental cell line in culture, again supporting the theory that such a common precursor for α- and β-cells exists *in vivo* (Blume *et al.*, 1995; Madsen *et al.*, 1988, 1991, 1993, 1996). In addition, the inactivation of PDX-1 in the glucagonoma cell line agrees with the *in vivo* observation that the loss of PDX-1 expression coincides with the differentiation of non-β-cell types (Guz *et al.*, 1995b).

Still, there is a difference of opinion over the type of hormone-expressing cell that appears first in the developing pancreas and which of these cells gives rise to the rest of the endocrine lineage. Herrera *et al.* (1991) claimed that pancreatic polypeptide (PP) is expressed very early in the mouse pancreas. PP belongs to the neuropeptide Y family of regulatory peptides, which includes pancreatic polypeptide (PP), NPY, and PYY. Others have demonstrated that the antisera raised against members of this family often demonstrate a high level of cross reactivity (Jackerott and Larsson, 1997; Teitelman *et al.*, 1993, Teitelman 1993). Rigorous characterization of antibodies, specific to each neuropeptide family member, has established that PYY is the first member of this family of peptide hormones to be expressed at E12 in rats (about E10.5 in mice) (Myrsén-Axcrona *et al.*, 1997; Upchurch *et al.*, 1994). Also, these experiments confirmed that PP is not expressed at a detectable level until after birth and that NPY is not expressed until E17 in rats (about E16 in mice) (Myrsén-Axcrona *et al.*, 1997; Teitelman, 1993).

The expression patterns of NPY and PYY by islet cells during development also gives further credence to the current endocrine-cell-lineage model. There is a near-total coexistence of PYY with glucagon during development that coincides with the gradual separation of glucagon and insulin immunoreactivity into separate cell types (Myrsén-Axcrona *et al.*, 1997). That is, PYY, insulin, and glucagon are said to coexist within the same islet cells early in development, but as the monospecific hormone-producing cells appear, PYY expression is detected only in cells expressing glucagon and not in cells expressing only insulin. This suggest that insulin-containing cells differentiate from cells coexpressing glucagon and PYY. NPY is almost exclusively expressed in non-glucagon-expressing insulin-positive cells as they appear until just before birth, making it a good marker for the mature β-cells. However, the role these peptides play during islet cell differentiation is not clear. Though NPY is expressed in nearly all β-cells just before birth, just after birth NPY immunoreactivity rapidly declines. However, NPY is known to inhibit glucose-stimulated insulin secretion and is mitogenic for vascular cells (Erlinge *et al.*, 1994; Moltz and McDonald, 1985; Petterson *et al.*, 1987; Skoglund *et al.*, 1991). Thus, it is possible that NPY serves as a growth factor or a modulator of insulin secretion during islet development (Myrsén-Axcrona *et al.*, 1997).

Given that the nature of the endocrine progenitor cell is still unclear, Herrera *et al.* (1994, 1998) proceeded to use two unique transgenic animal approaches to investigate the endocrine cell lineage. In the first approach, either glucagon, insulin, or PP cells were ablated through promoter-targeted expression of the diphtheria toxin A chain (DT)-coding region in transgenic mice (Herrera *et al.*, 1994, 1998). Interestingly, the loss of glucagon- or insulin-expressing cells due to the expression of DT under the control of a cell-specific promoter during development did not significantly affect the development of the remaining endocrine cell types in these transgenic animals. Strikingly, however, transgenic animals in which PP cells were ablated lacked insulin- and somatostatin-containing cells. These results suggest that PP-expressing cells are necessary for the β- and δ-cell development and that glucagon- and insulin-expressing cells are not required for the development of the other endocrine cell types. The presence of α-cells in normal numbers in mice carrying the PP-DT transgene demonstrated the cell specificity of the DT-transgene product and the lack of cell ablation due to bystander effects or a leaky promoter. These surprising results do not agree with the Teitelman endocrine cell lineage model. However, it can not be ruled out that the PP-DT transgene is ectopically expressed in cells that do not express the PP gene. Herrera *et al.* (1994) proposed two possible interpretations of these results. First, PP-expressing cells may produce a growth factor necessary for the development and/or maintenance of β-cells. Secondarily, a lineage relationship may relate β- and δ-cells to PP-expressing cells such that ablation of PP-cells precludes the development of β- and δ-cells.

More recently, Herrera *et al.* (1998) developed a transgenic animal approach to assay for the coexpression of islet-cell markers. This secondary approach involves a Cre/loxP bigenic system in which one strain of mouse, carrying an insulin-promoter-driven reporter gene whose expression depends upon the presence of Cre recombinase, is crossed with a transgenic mouse strain carrying either a glucagon or PP-promoter-driven Cre transgene. In resultant bigenic animals, the reporter gene can be expressed only in cells in which both the transgene promoters are active. Interestingly, no reporter gene activity is detected in bigenics carrying the glucagon-promoter-driven CRE transgene. Thus, both this approach and the previous transgenic approach suggest that neither glucagon nor insulin gene-expressing cells are the precursors to the remainder of the endocrine cell lineage and that PP gene-expressing cells are indispensable for β- and δ-cell differentiation (Herrera *et al.*, 1998). Even though these transgenic approaches do address the question as to the nature of the primordial endocrine stem cell in an interesting and unique fashion, it is certainly possible that the islet hormone promoter fragments regulating these transgenes do not absolutely recapitulate the expression pattern of the endogenous gene. Integration site effects may effect transgene expression or the loss of regulatory element(s) important for gene expression during development present in the endogenous gene.

Further attention must be given to the identification of the endocrine precursor cell and to resolve the discrepancies between the data derived from transgenic animal and immunohistochemical approaches. Several investigators have sought to determine roles of transcriptional regulators that regulate pancreas-specific genes in pancreas development. Analysis of loss-of-function mutants generated by gene targeting has generated several mouse models for defects in pancreatic development. Interestingly, much of the data gleaned from these experiments seems to support aspects of Teitelman's endocrine lineage model.

C. Dual and Overlapping Roles of Transcriptional Regulators in the Definition of Islet Cell Phenotypes, Revealed through Knockout Technology

It is well established that cells that compose particular layers, regions, or segments of the early embryo can acquire the competency to become a certain tissue or organ. Furthermore, extracellular signaling between different embryonic germ layers is essential for determining cell fate. In other words, in response to environmental cues, cells become programmed to follow a particular differentiation pathway. In this way, the developing embryo is patterned. Historically, most of the research on embryonic patterning has focused on the neural tube and limb, and little is known about the extracellular signals that are responsible for patterning definitive gut endoderm, from which the pancreas is derived. As mentioned earlier, the field of pancreas development has acquired several new tools that will be useful in determining how pancreas cell fate is determined. A variety of transcriptional regulators, essential to pancreas-specific gene expression, have been isolated through the use of advanced molecular biology techniques (Table I). Further analysis has shown that the expression pattern of several of these genes appears to corroborate the endocrine lineage model. Even more significantly, the absolute requirement of these same genes for pancreas development has been revealed in studies conducted with transgenic animals and by direct gene targeting. Thus, it seems that we are on the brink of a myriad of discoveries relating to definitive gut endoderm patterning and pancreas development, through the functional analysis of several transcriptional regulators *in vivo*.

1. The bHLH Family

a. BETA2   BETA2, a basic helix–loop–helix (bHLH) protein isolated from hamster insulinoma cells, has been shown to be an important activator of insulin gene transcription as mentioned earlier (Naya *et al.*, 1995). Interestingly, BETA2 has also been found to important for the expression of the hormone secretin in insulin-producing β-cells during development (Mutoh *et al.*, 1997). BETA2 was also cloned from embryonic stem cells and named NeuroD. Subsequently, it was identified in postmitotic neurons during neuronal development and was shown to be capable of determining neuronal cell fates when overexpressed in *Xenopus* (Lee *et al.*, 1995).

The expression pattern of BETA2 in the developing mouse pancreas has been determined by X-gal histochemistry. The BETA2 expression pattern was easily visualized in heterozygous mice carrying a knockin mutation in one BETA2 allele, in which the β-galactosidase gene replaced the majority of the BETA2 coding region. BETA2 was first detected in a subset of cells in the pancreatic primordium at E9.5. The majority of the β-galactosidase-positive cells also coexpress glucagon, suggesting that BETA2 is present in the earliest endocrine cell precursors. By E12.5, BETA2-positive cells were found in both the dorsal and ventral pancreatic lobes. Later on, BETA2-positive cells began to detach from the pancreatic epithelium and form distinct clusters, consistent with previous observations that islet cells originate from the pancreatic duct epithelium (Pictet and Rutter, 1972). At E17.5, these clusters were organized into spherelike structures. Further examination of BETA2 expression using double immunofluorescence microscopy revealed that these spheres were indeed islets of Langerhans, with insulin-positive cells in the center and glucagon-positive cells in the periphery. More importantly, we found that BETA2 expression was exclusively confined in endocrine cells of the pancreas, suggesting that BETA2 might play a role in maintaining islet cell differentiation. In the adult pancreas, BETA2 is expressed in all four pancreatic islet cell types. In addition to the pancreas, BETA2 is also expressed in intestinal endocrine cells, in the pituitary, in the retina, and in the central and peripheral nervous system. The importance of BETA2 in these organs is now under investigation in many groups, including our own.

Mice homozygous for the deletion of the BETA2 gene appear malnourished and dehydrated within 2 days after birth and die 3–5 days postpartum. Examination of the blood-glucose levels of mice at 2 days of age showed that homozygous mutant mice exhibited marked hyperglycemia compared to age-matched heterozygotes or wild-type mice. Also, ketonuria was found in some of the mutant mice, suggesting that these mice suffered from severe diabetes. These results support our hypothesis that, *in vivo*, BETA2 is important for both insulin gene transcription and normal islet function. However, the fact that insulin is still expressed in cells lacking BETA2 may cast doubt on whether it is really important in this process. Yet the results are consistent with at least an important role in islet function and/or development. X-gal histochemistry analysis of the pancreas of postnatal day 2(P2) mutant mice revealed marked reduction of β-galactosidase-positive cells. More importantly, no mature islets of Langerhans were found in the BETA2 mutant pancreas using double-labeled immunofluorescence microscopy. Indeed, all four types of islet cells were still present, although the numbers of β-, α-, and δ-cells decreased considerably (a 74% reduction in β-cells, a 39% reduction in α-cells, and an 18% reduction in δ-cells; Naya *et al.*, 1997). Thus, BETA2 is essential for both the formation of mature islets and the maintenance of proper islet cell numbers; however, it may not be absolutely required for the differentiation of the islet cell lineage.

TABLE I Pancreas Specific Transcription Factors Analyzed by Targeted Gene Disruption

Gene	Spatial expression pattern	Temporal expression pattern	Knockout pancreatic phenotype
BETA2	All pancreatic and gut endocrine cells. (Naya *et al.,* 1997)	E9.5 to adult (Naya *et al.,* 1997)	Disorganized islets of Langerhans and reduced islet cell numbers. Failure of secretin- and cholecystokinin-producing enteroendocrine cells to develop. Abnormal polarity of exocrine cells and failure of these cells to secrete zymogen granules. Postnatal lethality at 3–5 days due to insufficient insulin resulting in hyperglycemia and ketoacidosis and possibly the lack of secretion of zymogen granules (Naya *et al.,* 1997).
PDX-1	First in dorsal gut endoderm. In exocrine and all endocrine cell types during development. Gradually restricted to β-cells by birth. Maintained in the duodenum (Miller *et al.,* 1994; Guz *et al.,* 1995; Offield *et al.,* 1996).	E8.5 to adult (Offield *et al.,* 1996)	Pancreatic agenesis (Jonsson *et al.,* 1994). Primordial pancreatic buds emerge, but pancreatic epithelium fails to grow or differentiate (Offield *et al.,* 1996; Ahlgren *et al.,* 1996). Glucagon- and insulin-positive cells present early. Postnatal lethality due to insufficient insulin (Jonsson *et al.,* 1994) or malnutrition due to a lack of digestion in the absence of a pancreas and a functional rostral duodenum (Offield *et al.,* 1996).
Nkx2.2	Expressed in α-, β-, PP-cells, but not δ-cells.		No insulin-producing β-cells and reduced α- and PP-cells. Large accumulation of cells displaying β-cell markers but not insulin. Postnatal lethality likely due to insufficient insulin.

Nkx6.1	Majority of cells in the pancreatic bud. Restricted to the β-cells in the adult pancreas (Sander *et al.*, 1997, 1998)		Drastic reduction of insulin-producing cells at birth. Normal numbers of α-, δ-, and PP-cells.
PAX6	The gut endoderm prior to pancreatic morphogenesis. Coexpressed with each of the islet hormones during development until in adult hood (Turque *et al.*, 1994; St.-Onge *et al.*, 1997; Sander *et al.*, 1997).	E8-E9 to adult (Sander *et al.*, 1997).	No glucagon-producing α-cells were found in the PAX6 $-/-$ embryo (St. Onge *et al.*, 1997). However, PAX6 Sey[Neu] mutants have reduced numbers of all four hormone-producing cells and disorganized islets (Sander *et al.*, 1997). PAX6 Sey[Neu] mutation and PAX6 knockout results in perinatal lethality, likely due to a neuronal phenotype (Hogan *et al.*, 1988, Schmahl *et al.*, 1993, Stoykova *et al.*, 1996).
PAX4	Restricted to insulin-producing cells (Sosa-Pineda *et al.*, 1997).	E9.5 to adult (Sosa-Pineda *et al.*, 1997).	Lack mature β- and δ-cells but have increased numbers of α-cells. Normal at birth followed by growth retardation and dehydration by 48 hours due to hyperglycemia. Death by 3–4 days (Sosa-Pineda *et al.*, 1997).
ISL-1	All classes of pancreatic cells and mesenchymal cells surrounding the dorsal pancreatic bud. Expressed in all four islet cells in the adult (Karlsson, 1990).	E9 to adult (Ahlgren, 1997)	No dorsal pancreatic mesenchyme. Pancreatic exocrine tissue fails to differentiate specifically in dorsal pancreas due to the lack of mesenchyme. Complete loss of endocrine cell differentiation (Ahlgren, 1997). Developmentally arrested after E9.5 (Pfaff, 1996).

To determine exactly when and possibly how these defects occur during pancreatic development, the ontogeny of the BETA2 mutant pancreas from E9.5 to E17.5 was analyzed by visualizing the insulin- and glucagon-expressing cells using double immunofluorescence microscopy. Interestingly, before E14.5, no obvious differences in morphology or in the number of endocrine cells could be detected between the mutant, heterozygous, and wild-type pancreata. However, at E17.5, instead of forming islet-like spherical structures found in heterozygous and wildtype pancreata, the endocrine cells in the mutant pancreas aggregated randomly as small clusters of cells. In addition, the total number of endocrine cells in the BETA2 knockout pancreas decreased by 57%. This reduction may have been caused by excessive programmed cell death. TUNEL assays on the mutant pancreas at this stage showed a significant increase in the number of apoptotic cells compared to the heterozygous and wild-type pancreata, with a fivefold difference in the E17.5 embryo and a 20-fold difference in the neonate. In contrast, there were no obvious differences in proliferation in the mutant and wild-type endocrine pancreas as judged by staining with the proliferative cell nuclear antigen, PCNA (Naya *et al.*, 1997).

The multiple pancreatic defects in BETA2 knockout mice may be casually related to each other. For instance, the lack of islet formation and proper cell–cell interactions may be the cause of the enhanced apoptosis that was observed in the mutant pancreas at E17.5. We cannot rule out the possibility that the lack of mature islet formation can be attributed to insufficient numbers of functional endocrine cells. However, enhanced apoptosis is not the likely cause of poor islet formation, since at E16.5 the number of apoptotic endocrine cells was similar in wild-type and knockout embryos. In addition, in the normal pancreas the process of islet formation is nearly complete by E17.5. Thus, the observed increase in apoptosis at E17.5 seemed to occur just after an unsuccessful attempt to form mature islets.

It is not clear why the endocrine cells in BETA2 mutant pancreas fail to form proper islets. Islet formation involves many events, including cell–cell sorting, cell migration, and cell reorganization (Slack, 1995). The initial clustering of β-cells occurs between E13.5 and E14.5, whereas the first islet-like structures become evident in the pancreas at E17.5. Cell adhesion molecules (CAMs) expressed in the islet cells, such as E-, N-, and R-cadherins, and neural cell adhesion molecule (NCAM), have been proposed to play important roles in the organization of endocrine cells into islets (Cirulli *et al.*, 1994; Hutton *et al.*, 1993; Moller *et al.*, 1992). In aggregation studies, using cultured rat islet cells, it has been demonstrated that E-cadherin, a calcium-dependent CAM that is weakly expressed in all islet cell types, is important for the aggregation of both β- and non-β-cells. In contrast, NCAM, a calcium-independent CAM that is enriched in non-β-cells in the rats is more important for the sorting of non-β-cells (Rouiller *et al.*, 1991). However, there has been a controversy over the role of NCAM, since no differential expression of NCAM was found in mouse islets (Dahl *et al.*, 1996), arguing against NCAM alone being responsible for sorting different types of islet cells in the mouse. More recent studies have shown that the initial clustering of β-cells is perturbed in transgenic mice expressing a dominant-negative form of E-cadherin in β-cells, and later in development islet formation is inhibited (Dahl *et al.*, 1996). Furthermore, the normal concentration of N-cadherin and endogenous E-cadherin at cell–cell contacts is disrupted, suggesting that deregulation of more than one cadherin or CAM molecule may contribute to the aggregation defect in these transgenic mice.

Furthermore, the requirement of N-CAM and N-cadherin for different aspects of pancreatic islet architecture is being studied currently through analysis of N-CAM and N-cadherin knockout mice (Semb *et al.*, 1998). The normal peripheral localization of α-cells is reported to be randomized in the N-CAM knockout as compared to controls, while N-cadherin embryos die around E9.5 and may have defects in early pancreatic organogenesis. Further analysis of these N-CAM and N-cadherin knockout mice should help determine the relationship between the observed phenotype and the function of these proteins. However, the current evidence suggests that CAMs and cadherins are important for islet morphogenesis.

Little is known about the initial events triggering apoptosis of pancreatic endocrine cells. In other systems, it has been recognized that lack of cell adhesions or cell–cell interactions is sufficient to cause cell death (Peluso, 1997). Another possibility is that BETA2, like MyoD, may induce the activity of important cell cycle regulators, such as p21 (Halevy *et al.*, 1995), that result in egress from the cell cycle. Therefore, the loss of functional BETA2 may prepare cells to reenter the cell cycle. However, if other differentiation signals are present and antagonize the reentry status, cells may not manage to finish a new cell cycle round and instead undergo apoptosis. Other explanations for the increased level of apoptosis observed in the BETA2 knockout mice may exist, and resolution of this question requires further studies to determine whether it is associated with the failure to form mature islets.

b. Other bHLH Transcription Factors   There are other bHLH factors expressed in the pancreas, but little is known about their roles in pancreas development. For example, BETA2 activation of the insulin promoter can be repressed by BETA3, a bHLH factor that shares some homology with the BETA2 family and is also expressed in a restricted manner in the mouse brain and pancreas during development (Bramblett and Tsai, unpublished results). Neurogenin 3 is another bHLH family member that is specifically expressed in the pancreas and hypothalamus (Sommer and Ma, 1998). Interestingly, a related gene, neurogenin 1, is an upstream regulator of BETA2 in neuronal cells. Since neurogenin 3 is expressed earlier than BETA2 in the pancreatic islets, it also may be upstream of BETA2 in the pancreatic signaling pathway (Fode *et al.*, 1998; Ma *et al.*, 1998; Sommer and Ma,

1998). Experiments are currently underway in our laboratory to determine
if neurogenin transcriptionally activates the BETA2 promoter. However, it
appears that the BETA2 promoter is stimulated by neurogenin 3 in cotransfection experiments (Huang and Tsai, unpublished results), and promoter-
mapping experiments are currently underway to define the element(s) bound
by neurogenin3.

The bHLH factor Mist-1 is the first bHLH factor that is principally
expressed in the gastrointestinal tract. Mist-1 mRNA is first observed at
E10.5 in the primitive gut and in the developing pancreatic bud until E16.5
(Lemercier *et al.*, 1997). The Mist-1 protein is unique in that it can bind to
an E-box element as a heterodimer with the class A bHLH family members
E12 and E47, but it lacks a functional transcriptional activation domain,
suggesting it may function as a repressor of transcription. Mist-1 is restricted
to the exocrine epithelium, making it more likely to play a role in exocrine
cell than endocrine cell development.

Finally, Islet-Brain-1 (IB1) is a transcriptional activator that contains
both a basic helix–loop–helix (bHLH) domain and a phosphotyrosine interacting domain (PID). This protein was cloned from a rat cDNA library in
an attempt to identify DNA-binding proteins necessary for β-cell-specific
gene activation (Bonny *et al.*, 1998). IB1 is related to JIP-1, a murine inhibitor
of the c-Jun amino terminal kinase (JNK) activated pathway; is highly
expressed in pancreatic β-cells; and is a transactivator of the GLUT2 gene
(Bonny *et al.*, 1998). Though a role for IB1 in pancreas development has not
been established, its possession of bHLH and PID domains is characteristic of
a factor that is involved in cell determination pathways. Thus, it is likely
to play an important role in pancreas development.

Thus, members of the bHLH family are highly expressed in the pancreas.
Furthermore, the contribution of BETA2 to the regulation of insulin and
to the morphogenesis of the islets of Langerhans was suggested by *in vitro*
experiments and confirmed by gene-targeting experiments. Because of the
important role bHLH factors frequently play in the establishment of differentiated cell lineages and because BETA2 serves an important role in neuronal
differentiation, we suspect that BETA2 and other bHLH factors act as key
determinants of the endocrine cell fate. The presence of each of the pancreatic
endocrine cell types in the BETA2 mutant mouse may actually reflect the
existence of redundant bHLH factors that compensate for the loss of BETA2.
Indeed, additional bHLH factors expressed in the endocrine pancreas have
already been identified (BETA3, BETA1, E47, Neurogenin3, IB1). Therefore,
further investigation of bHLH family members expressed in the pancreas
will certainly reveal key aspects of pancreatic gene regulation and ontogeny.

2. The Homeobox Family

Several members of the homeobox family of transcription factors that
regulate pancreas-specific genes are also essential components of pancreas

development. Homeobox factors recognize a canonical TAAT motif in the upstream, noncoding regions of many genes by way of a DNA-binding motif referred to as the homeodomain. The homeodomain DNA-binding domain, which confers the specificity for the canonical TAAT motif, was first identified in homeotic selective genes in *Drosophila melanogaster* (Gehring, 1987). The homeotic factors play a role in committing cells within a certain region of the fly embryo to a particular identity such as a wing or eye. In general, members of this family of transcription factors display critical spatial–temporal expression patterns during development and regulate the expression of key genes that ensure the specification of cells to a particular cell fate or phenotype. Homeodomain factor family members have now been demonstrated to play a similar roles in many aspects of vertebrate development (Boncinelli and Mallamaci, 1995; Joyner, 1996; Tickle and Eichele, 1996).

 a. PDX-1 The homeodomain factor XIHbox8, isolated from *Xenopus laevis,* was found to be highly expressed within endodermal cells in a narrow band of the duodenum early in development and later restricted to the developing pancreas (Wright and Schnegelsberg, 1988). The mouse, rat, and human homologs of XIHbox8, which have since been identified by several groups, are expressed in the pancreas and the duodenum (Jonsson *et al.,* 1994; Leonard *et al.,* 1993; Miller *et al.,* 1994; Offield *et al.,* 1996; Ohlsson *et al.,* 1993a; Peers *et al.,* 1994). Though this factor has been identified by many names (IPF-1, PDX-1, IDX-1, or STF-1) it will be referred to herein as pancreatic duodenal homeobox protein-1, or PDX-1.

 PDX-1 is expressed in the pancreatic anlage just prior to its evagination from the definitive gut at approximately E8.5 (Ahlgren *et al.,* 1996). PDX-1 was found to activate several pancreas-specific, genes including the insulin, somatostatin, glucokinase, IAPP, and GLUT2 genes (Leonard *et al.,* 1993; Petersen *et al.,* 1994; Waeber *et al.,* 1996; Watada *et al.,* 1996a, 1996b) by binding to TAAT containing A-boxes in their promoters. Teitelman's endocrine cell lineage model, discussed previously, proposes that islet cell maturation is characterized by the selective inactivation of inappropriate hormone gene products (Alpert *et al.,* 1988; Guz *et al.,* 1995b; Teitelman, 1993). Accordingly, PDX-1 is transiently coexpressed with each of the four pancreatic hormones early in islet cell differentiation, but as differentiation proceeds, PDX-1 is progressively restricted to the β-cells (Ohlsson *et al.,* 1993b) and some δ-cells. Thus, it was hypothesized that the endocrine cell lineage was defined in part by the sequential downregulation of PDX-1 in all non-β-cells. Since PDX-1 is continuously expressed in all epithelial cells from a very early point in pancreas development, it has been proposed to be the master regulator of pancreas development. Indeed, the disruption of the PDX-1 gene through homologous recombination in the mouse resulted in pancreatic agenesis (Jonsson *et al.,* 1994). Furthermore, a single nucleotide deletion in PDX-1, which prevents the formation of a functional activator, appears to cause pancreatic agenesis in humans (Stoffers *et al.*, 1997).

Upon further inspection of the PDX-1-deficient mice, between E10 and E12, a rudimentary dorsal evagination from the gut tube appears at the level of the presumptive pancreas anlage. Even though the initial budding of the pancreatic anlage occurs in the PDX-1 mutants, the subsequent growth and branching is arrested. However, expression of both insulin and glucagon in specific cells could be detected in the PDX-1 null mouse (Ahlgren *et al.*, 1996; Offield *et al.*, 1996). Thus, PDX-1 is not required for the specification of the insulin- or glucagon-expressing cells that can normally be detected at E9.5. However, by E13 very few glucagon- and insulin-positive cells were found when compared to age-matched wild-type embryos, and at no time could the exocrine marker, amylase, be detected in the PDX-1 mutant pancreases. Interestingly, the β-cell marker GLUT2 was highly expressed in the epithelial cells of the dorsal pancreas of the PDX-1 mutant (Offield *et al.*, 1996). This may indicate that islet cell differentiation from duct cell precursors has been arrested in PDX-1 mutants. Alternatively, this may reflect a lack of β-cell expansion, migration, and aggregation into islets. Thus, PDX-1 is not essential for the primary specification of the gut endoderm to the pancreatic fate, but it is required for the subsequent morphogenesis of the pancreatic anlage and endocrine cell expansion.

Interestingly, the growth and development of pancreatic mesenchyme from PDX-1 mutant mice is not affected by the loss of PDX-1, reflecting that pancreatic mesenchyme does not require pancreatic PDX-1-dependent epithelial signals for development. Tissue recombination experiments confirmed that PDX-1 mutant mesenchyme is perfectly capable of inducing the growth and differentiation of wild-type epithelium, whereas presumptive pancreatic epithelium from PDX-1 mutant embryos is not rescued by wild-type mesenchyme (Ahlgren *et al.*, 1996). Thus, PDX-1 is not required for the development of pancreatic mesenchyme, and the lack of epithelial growth is not due to defective mesenchyme, but rather to the lack of PDX-1 expression in the epithelial cells.

In summary, PDX-1 is required for the differentiation of pancreatic epithelium into exocrine tissue and ducts. On the other hand, factors other than PDX-1 are required for the initial induction of gut endoderm cells to acquire the pancreatic fate and for the induction of pancreatic budding. Also, even though PDX-1 is important for appropriate activation of the insulin gene, it is not required for its expression. Furthermore, functional PDX-1 is not required for the early differentiation of hormone-producing cells in the pancreas, as insulin- and glucagon-expressing cells can be detected in PDX-1 mutants. However, PDX-1 does appear to be necessary for the progression of islet-cell differentiation to the mature phenotype.

b. Members of the Nk Family: Nkx2.2 and Nkx6.1   The role of two members of the Nkx subfamily of homeobox factors, Nkx2.2 and Nkx6.1, in pancreas development is currently under investigation. Nkx factors are vertebrate factors that have homeodomains homologous to the *Drosophila* NK-2 gene (Kim and Nirenberg, 1989). These factors regulate critical steps of organogenesis during vertebrate development. The majority of the NK family have been found to play roles either in neuronal cell specification in restricted regions of the CNS or in the differentiation of the heart and pharyngeal endoderm (Bober *et al.*, 1994; Buchberger *et al.*, 1996; Durocher *et al.*, 1997; Pabst *et al.*, 1997; Price *et al.*, 1992; Reecy *et al.*, 1997; Rinkwitz-Brandt *et al.*, 1995). Additional roles for NK family members, in male urogenital system, have also been reported (Sciavolino *et al.*, 1997). Nkx.2.2 is known for its essential role in motor neuron differentiation (Pfaff *et al.*, 1996), but Nkx2.2, as well as the family member Nkx6.1, has proven to be crucial to pancreatic islet-cell development (Sussel *et al.*, 1998b; Jensen *et al.*, 1996; Sander *et al.*, 1998).

Nkx6.1 is initially expressed throughout the pancreatic bud, but later it becomes restricted to β-cells (Madsen *et al.*, 1997; Sander *et al.*, 1998). Mice lacking the Nkx6.1 gene have a drastic reduction in insulin-producing cells at birth but have normal numbers of glucagon, somatostatin, and PP-producing cells (Sander *et al.*, 1998). The small number of insulin-expressing cells that do develop are thought to be differentiated as indicated by the expression of mature β-cell markers IAPP (islet amyloid polypeptide), PC1 (prohormone convertase-1), and PDX-1. It is believed that β-cells form normally in the Nkx6.1 mutant pancreas up to E13, but after E13 β-cell development appears to be retarded because β-cells fail to increase in number (personal communication from M. Sander). The fact that Nkx6.1-positive cells are also detected in the pancreatic ducts suggests that this factor is involved in the differentiation of endocrine cells from ductile precursors. Also, because β-cell development in Nkx6.1 mutant mice appears normal until E13, it is possible that Nkx6.1 is required for the second wave of β-cell neogencsis from ductile precursor cells (Miralles *et al.*, 1998; Pang *et al.*, 1994; Sander *et al.*, 1998; personal communication, M. Sander). The Nkx6.1 null mutant phenotype provides further evidence supporting the theory that there are at least two rounds of β-cell differentiation. However, further experiments must be conducted to confirm the identity and developmental state of the pancreatic endocrine cells in the Nkx6.1 mutant mouse.

Nkx2.2 expression in the pancreas is different from Nkx6.1, in that it is expressed in α-cells, β-cells, and PP-cells, but not in δ-cells of the adult pancreas. Nkx2.2 null mutants lack insulin-producing β-cells, and there is a reduction of α-cells and PP-cells (Sussel *et al.*, 1998a, 1998b). The fact that insulin is not expressed at any stage in the development of the Nkx2.2 animals suggests that this factor may serve to activate insulin expression directly; however, there is no evidence supporting such a role. Nkx2.2 mutants develop severe hyperglycemia and die shortly after birth. Interestingly, there are a large number of cells in the Nkx2.2 mutant pancreas that do not produce any of the four pancreatic hormones but do express early endocrine cell markers such as PDX-1 and PAX6. Many of these cells also express the β-cell markers IAPP and PC1, suggesting that the lack of hormone production does not reflect the absence of the cell types that would normally produce insulin. However, these cells do lack the expression of GLUT2 and Nkx6.1, indicating that there is a block in β-cell differentiation in the Nkx2.2 null mice resulting in an accumulation of pre-β-cells that are not capable of expressing insulin. Thus, given the large number of underdifferentiated cells, Nkx2.2 may be required for β-cell differentiation, but not the specification and maintenance of this cell type. However, the reduction in α-, δ-, and PP-cell numbers may indicate a different role of Nkx2.2 in these cell types, such that the Nkx2.2 null mutation disrupts either the specification or the maintenance of these cell types.

Nkx2.2 appears to be upstream of Nkx6.1 in the β-cell lineage, since Nkx6.1 is no longer expressed in the presumptive pre-β-cells that appear to be arrested in the Nkx2.2 mutant. Also, Nkx2.2 and Nkx6.1 differ with regard to the range of their effects on the endocrine cell lineage, with Nkx2.2 affecting the development of α, β, and PP islet cell types and Nkx6.1 affecting specifically β-cells. Little has been reported about the ability of either Nkx2.2 or Nkx6.1 to directly regulate pancreas-specific gene expression. However, it is likely that Nkx2.2 regulates genes involved in the development of each of the islet cells while Nkx6.1 regulates β-cell specific genes.

c. PAX6   The homeobox factor PAX6 contains a conserved sequence motif, the paired box, which encodes a DNA-binding domain similar to the PAX (paired-box) family of vertebrate genes (Mansouri *et al.*, 1996). Members of the PAX family have been shown to dictate tissue-specific gene expression as well as development in such tissues at a very primordial level (Mansouri *et al.*, 1996). Analogously, PAX6 is expressed in the nervous system, the eye, and the endocrine pancreas during development and in the adult (Madsen *et al.*, 1997; Turque *et al.*, 1994), and it influences tissue-specific gene expression and development of these tissues (Cvekl *et al.*, 1995; Ericson *et al.*, 1997; Sander and German, 1997).

The PAX6 homeodomain contributes to the lens-specific expression of the crystallin genes (Cvekl *et al.*, 1995). Correspondingly, mutations in the PAX6 gene, such as Small eye (Sey[neu]), have been shown to cause disruptions in eye development in mice, humans, and *Drosophila* (Glaser *et al.*, 1994; Hill *et al.*, 1991; Jordan *et al.*, 1992; Matsuo *et al.*, 1993; Quiring *et al.*, 1994). Furthermore, this PAX6 mutation causes neonatal death likely due to brain abnormalities and disruption of eye and nasal development.

PAX6 is expressed in all four of the hormone-producing cells of the pancreas from the earliest stages. Sander *et al.* (1997) investigated the effect of the Sey[neu] on pancreas development and found a three- to fourfold reduction in all four hormone-producing cell types. They also found that the islets had abnormal morphology, similar to the BETA2 knockout mice. Although the number of endocrine cell clusters is relatively normal in the pancreas of homozygous Sey^{neu} mice, the hormone-producing cells fail to organize into normal islet structures with β-cells surrounded by α-, δ-, and PP-cells. When hormone levels in age-matched littermates were compared, they found that hormone protein and mRNA levels were reduced in mutant pancreata to a greater degree than could be accounted for by the reduction of islet-cell number. The reduction in islet hormone levels correlates with the contribution of PAX6 to the activation of the insulin, glucagon, and somatostatin promoters through the PISCES element. It was concluded that the lack of a functional PAX6 protein results in reduced hormone mRNA transcription, a reduced number of hormone-producing cells as early as E10.5, and a disruption in islet morphology (Sander *et al.*, 1997).

It is interesting that both the PAX6 mutant and the BETA2 knockout have reduced hormone expression and disorganized islets. However, the islet morphology defect in PAX6 and BETA2 mutant mice is not likely due to insufficient hormone secretion. SPC2 and SPC3 (PC3/PC1) are the primary pancreatic prohormone endoproteases. Without them, the processing of proinsulin, proglucagon, and prosomatostatin is severely impaired (Furuta *et al.*, 1997). Though adult mice lacking the prohormone convertase SPC2 (PC2) also have disrupted pancreatic islet morphology, the SPC2 mutant mice display no disturbed islet morphology at birth (Furuta *et al.*, 1997). This phenotype is attributed to the chronic hypoglycemia in SPC2 mutant mice due to the lack of glucagon that leads to α-cell hyperplasia and β-cell depletion. Therefore, the islet morphology defects seen in the SPC2 mutants are due to a response to chronic severe glucagon deficiency and not to a disruption in pancreatic development. Furthermore, mice completely lacking both insulin genes are reported to have basically normal islet development, though slightly hyperplastic, with each of the four pancreatic cell types present (Duvillic *et al.*, 1997). Therefore, the defect in PAX6 and BETA2 knockout mice is clearly the result of disrupted developmental mechanisms and not due to a pancreatic hormone deficiency.

The disrupted islet phenotype may be attributed to improper cell–cell interactions. PAX6 mutant mice were found to have altered adhesive properties of cortical neuronal cells and have considerably reduced expression of R-cadherin in the developing brain (Stoykova *et al.*, 1997). Thus, deregulation of cadherins or other CAMs may be responsible for the lack of proper islet formation in both the PAX6 and BETA2 mutant mice. Neither N-CAM nor E-cadherin downregulation has been observed in the pancreas of BETA2 mutant mice (unpublished observations, Huang and Tsai). Furthermore, no significant reduction in the level of N-CAM or N-cadherin was observed in the PAX6 mutant pancreas (Sander *et al.*, 1997). However, PAX6 reportedly binds the promoter region of the NCAM-L1 gene, a member of the immuno-globulin superfamily with a broad distribution of expression (Chalepakis *et al.*, 1994). Thus, because of the likelihood that NCAM- or cadherin-like molecules are influenced by PAX6 and BETA2, we are currently conducting studies to determine which of these molecules are deregulated in the BETA2 mutant mice, and the extent to which BETA2 and PAX6 can cross-regulate each other.

A PAX6 knockout mouse has also been generated using targeted gene disruption technology; however, the reported phenotype is somewhat different from the Sey[neu] mutant mouse strain. No glucagon-producing α-cells can be detected in the PAX6 knockout mouse, and the authors propose that PAX6 is required for the specification of pancreatic α-cells (St-Onge *et al.*, 1997). The concept that PAX6 directs the specification of the branch point within the endocrine cell lineage that lead to the α-cell population is an appealing proposal because it supports the current model for the endocrine cell lineage. There are several possible causes for the discrepancy between the Sey[neu] and the PAX6 homozygous null mouse. First, the Sey[neu] mutant mouse may be a hypomorph in which the mutant allele codes for a protein with partial biological activity. This is a possibility since the Sey[neu] mutation contains a point mutation in a splice donor site upstream of the homeodomain. Incorrect splicing results in the inclusion of an intronic sequence in the mRNA that introduces a stop codon downstream of the activation domain (Glaser *et al.*, 1994). Sander *et al.* (1997) discounts this possibility, with evidence that the mutant protein does not effect the transcriptional activity of the insulin, somatostatin, or glucagon genes in transfection experiments. Because no functional activity has been attributed to the Sey[neu], it is more likely that the disparity between the two mutant phenotypes is due to genetic background differences, as this has contributed to the phenotypic variation observed in other mutants (Dunn *et al.*, 1997; Johnson *et al.*, 1997; Kash *et al.*, 1997; Kent *et al.*, 1997; Ludwig *et al.*, 1996; Silva *et al.*, 1997; Xu *et al.*, 1997).

d. PAX4   Homozygous null mutants of the paired-box family member PAX4 have been described, and these mice have a pancreatic defect also (Sosa-Pineda *et al.*, 1997). Mutant PAX4 neonates are indistinguishable from their littermates at birth, but after 48 hours they appear to be growth retarded and dehydrated and die within 3 days after birth. Analysis of PAX4 mutants was facilitated by fusing the β-galactosidase gene in frame with the amino terminus of the PAX4 gene. By staining for β-galactosidase, PAX4 expression could be detected in the PAX4 heterozygotes beginning at around E10.5 in the dorsal pancreas and at E11 in the ventral pancreas. Interestingly, insulin-, glucagon-, and PDX-1 producing cells are present in the pancreas of PAX4 mutant embryos at E10.5. However, at birth, virtually no somatostatin-, PDX-1-, or insulin-producing cells are detectable the pancreas of the PAX4 null mutant (Sosa-Pineda *et al.*, 1997). Thus, neonatal knockout mice are completely deficient in mature β-cells even though β-cell markers are detectable earlier in embryogenesis. In striking contrast, α-cells are present in abnormally high numbers in the PAX4 mutant mice (Sosa-Pineda *et al.*, 1997). These data support the hypothesis that in the absence of PAX4, maintenance of the β-cell lineage is disrupted. As a consequence, δ-cell differentiation is also affected, because δ-cells arise from a branch point in the β-cell lineage (Fig. 2). Sosa-Pineda *et al.* (1997) propose that early on in the development of the PAX4 mutant, endocrine cells lose their commitment to the β-cell phenotype and acquire the α-cell phenotype instead (Sosa-Pineda *et al.*, 1997). These data suggest that differentiation to the α-cell phenotype may be the default pathway in the absence of PAX4.

e. *ISL-1*   The LIM homeodomain protein ISL-1 was originally identified by screening for proteins capable of binding to a DNA probe from the rat insulin gene enhancer (Karlsson *et al.*, 1990). As mentioned, ISL-1 is a transcriptional activator of several pancreas-specific genes. It is expressed in the dorsal pancreatic epithelium and in cells of the lateral mesenchyme beginning at embryonic day E9, and its expression can be detected in the mesenchyme surrounding the dorsal pancreatic bud when it begins to evaginate at E9.5. ISL-1 is expressed in cells expressing glucagon when they appear at E9.5 and in each of the other islet cell types when they appear. This corresponds well with the fact that ISL-1 positively regulates glucagon, somatostatin, and insulin gene transcription (German *et al.*, 1992; Leonard *et al.*, 1992; Wang and Drucker, 1995). Furthermore, ISL-1 expression in the ventral pancreatic epithelium correlates with the appearance of islet cells at E11 (Ahlgren *et al.*, 1997).

Disruption of ISL-1 has a global effect on embryonic development as a whole. ISL-1 mutants display defects in vascular endothelium, particularly a disruption in the formation of the dorsal aorta, and also perturbations in the development of motor neurons, sensory neurons of the dorsal root and cranial sensory ganglia, and the splanchnic mesenchyme, as well as the endocrine pancreas (Pfaff *et al.*, 1996). Furthermore, analysis of ISL-1 mutant pancreata, at E9.5, revealed the requirement of functional ISL-1 for the development of dorsal pancreatic mesenchyme and for the generation of glucagon-expressing cells from the dorsal pancreatic epithelium (Ahlgren *et al.*, 1997). Because of multiple vascular and neurological defects, the growth of the mutant ISL-1 embryo arrests at E9.5; thus, development of the pancreas in the ISL-1 mutant pancreas could not be analyzed *in vivo* after the initial budding stage.

To determine how far the ISL-1 mutant pancreas could have developed given the full gestational time frame, ISL-1 nulls were studied in embryonic tissue explants cultured *in vitro*. Gut explants containing the pancreatic primordium from E9.5 ISL-1 mutant embryos, grown for 7 days in culture, gave rise to cells that were negative for glucagon, insulin, and somatostatin, yet positive for the exocrine marker amylase. Similar cultures from E9.5

heterozygous animals produced glucagon-, insulin-, somatostatin-, and amylase-secreting cell types. After 7 days, supplementation of ISL-1 mutant epithelium with wild-type mesenchyme did not restore the differentiation of islet cells in the mutant epithelium, confirming that the lack of endocrine cells in ISL-1 mice was not due to a lack of pancreatic mesenchyme. Interestingly, half-gut explants from ISL-1 mutants at E9.5, which excluded the ventral pancreatic anlage, did not produce amylase exocrine cells when grown in culture. Thus, the exocrine cells present in the whole gut explants actually arose from the ventral pancreatic anlage. When cultured with mesenchyme from a heterologous region, dorsal pancreatic epithelium from both wild-type and ISL-1 mutant embryos could be induced to generate amylase-positive exocrine cells. Thus, it was concluded that ISL-1 expression in the epithelium is required for the specification of the entire endocrine cell lineage, but not for exocrine cell differentiation. In addition, ISL-1 expression is required for the development of dorsal pancreatic mesenchyme that subsequently induces cells of the dorsal pancreatic epithelium to differentiate and express exocrine-cell markers (Ahlgren *et al.*, 1997).

Interestingly, it seems that the ventral pancreas does not require the same set of inductive signals as the dorsal pancreas. In contrast to the dorsal pancreas, ISL-1 is not expressed in the ventral pancreatic mesenchyme, which remains virtually unaffected by the loss of ISL-1. This uneven expression pattern reflects the existence of a separate mechanism for dorsal/ventral patterning in the pancreas. Furthermore, the expression of marker genes in the ventral pancreas, such as PDX-1, are unaffected by the loss of functional ISL-1, whereas in the dorsal pancreatic epithelium, PDX-1 was greatly reduced. Furthermore, the elegant microdissection experiments performed by Kim *et al.* (1997), on chicks, showed that removal of the notochord early in chick embryo development, at a stage prior to the separation of the notochord from the endoderm, prevents both exocrine- and endocrine-specific gene expression in the dorsal pancreatic bud. However, gene expression in the ventral pancreas is not affected. Thus, the environmental signals that induce dorsal and ventral pancreas development derive from unique sources but have surprisingly similar results.

In summary, factors from the bHLH and homeobox transcriptional regulatory families (BETA2, ISL-1, PDX-1, PAX6, PAX4, Nkx2.2, and Nkx6.1) have proven to be essential for the development of the pancreas as demonstrated by targeted gene disruption. Analyses of these mutant strains revealed that each of these regulators affects the pancreatic endocrine cell lineage at different points. ISL-1 appears to be necessary very early in the endocrine lineage, at the point when glucagon- and insulin-producing cells differentiate from ductile endocrine progenitor cells. PDX-1 is the first marker gene to be expressed in the pancreatic anlage, and it appears to play a broad range of roles in the pancreatic development. It is necessary for the growth and branching of the pancreatic epithelium as well as for the differentiation of both exocrine and endocrine precursor cell types. However, its role in the endocrine lineage appears to follow that of ISL-1. PDX-1 takes part in the differentiation of α and β endocrine precursor cells, only after they have differentiated from a common endocrine progenitor cell, as indicated by the presence of glucagon- and insulin-expressing cells in the PDX-1 mutant. In contrast, the ISL-1 mutation blocks the development of all the hormone-producing cells.

The endocrine lineage appears to occur gradually and to require many factors, which are important for the different aspects of development. Besides ISL-1, all the mutants described (PAX6, PAX4, PDX-1, BETA2, Nkx2.2, Nkx6.1) seem to have hormone-expressing cells early in development, but deficits in the normal population of endocrine cells are apparent at birth. Thus, it is possible that factors such as PAX6, BETA2, Nkx2.2, and Nkx6.1 not required for the primary specification of the endocrine progenitor cell or the endocrine precursor cells, but are required for the second round of endocrine cell differentiation from ductal precursors that occurs during late in gestation. The Nkx2.2 mutant, which never expresses insulin, might appear to be the exception. However, the NKx2.2 mutants develop large numbers of pancreatic cells that display many of the mature β-cell phenotypic markers other than insulin, simply demonstrating that the lack of insulin expression does not reflect the absence of the cell type that would normally express insulin.

III. Signals Influencing Pancreas Development ———————

A. Patterning of the Gut Endoderm

Little is known about the events that lead to endodermal gut patterning. Discussed briefly earlier, the vertebrate embryo has organizing centers such as the zone of polarizing activity (ZPA), the notochord, and the floor plate that induce tissues to take on a particular polarity and identity (Placzek *et al.*, 1990, 1991). Signals that derive from the ZPA induce limb development. Signals from the notochord, a rod of mesoderm along the center of the embryo, induce the floor plate in the adjacent neural tube to differentiate. In turn, signals emanating from the floor plate induce motor neurons and orient commissural axon outgrowth (Placzek and Furley, 1996). Also, signals between endoderm and mesoderm govern the specification and patterning of the respiratory and digestive organs, including the trachea, lungs, stomach, intestines, and pancreas. Interestingly, the dorsal side of the hepatopancreatic ring is in contact with the notochord prior to pancreatic morphogenesis, until the cells of the pancreatic anlage have acquired their cell fate. The significance of this positioning has not gone unnoticed. New evidence supports the notion that the notochord influences endoderm differentiation, pancreas-specific gene activation, and budding of the pancreatic epithelium.

B. The Notochord Influences Pancreas Development

Thus far, neither the specification of endodermal cells to the pancreas cell fate nor the capacity of the pancreatic anlage to bud has been completely eliminated by targeted disruption. This suggests that the primary event(s) responsible for specification of gut endoderm to the pancreas cell fate have not been altered in these mutants. *In vitro* tissue culture experiments showed that commitment to the pancreatic cell fate occurs very early. It precedes the detachment of the ventral side of the notochord from the dorsal gut endoderm, suggesting that communication with the notochord may be necessary for specification (Gittes and Rutter, 1992; Kim *et al.*, 1997). The notochord is positioned contiguous to the ventral side of the neural tube and the dorsal side of the gut endoderm until E8.5 in mice (13 somites or stage 14 in chickens), at which time it separates from the endoderm and the dorsal aortae obstruct contact between the two germ layers. The proximal location of the notochord provides this organizing tissue ample opportunity to signal the induction of the pancreatic anlage. But, the requirement of notochord signaling for pancreatic development was not demonstrated until recently.

Kim *et al.* (1997) described strong evidence that signaling from the notochord is required for the development of competent gut endoderm into a dorsal pancreatic bud. Elegant explanation experiments in chicks demonstrated that removal of the notochord before the pancreas is specified prevents pancreas-specific gene expression. Stage 11 chicks have no morphological sign of pancreatic bud formation and the ventral side of the notochord and dorsal endoderm are still in contact. Though PAX6, ISL-1, and HNF3β are already expressed in the stage11 pancreatic anlage, glucagon, PDX-1, insulin, and the exocrine cell marker carboxypeptidase A are expressed later, just before the formation of the pancreatic bud, at stage 15. Notably, there are some variations in the temporal expression pattern of some genes between the mouse and chicken embryos. PAX6 and ISL-1 expression precedes PDX-1 expression in chicks, whereas the converse is true in the mouse, but the significance of this variation has not been addressed. After growing stage 11 chick embryos, with their notochord removed, for two days (from stage 11 to stage 20) *in vitro,* ISL-1, PAX-6, glucagon, PDX-1, insulin, and carboxypeptidase-A could not be detected by RT-PCR even though all of these markers were detected in similarly grown control embryos with intact notochords.

Taken alone, one might question whether removal of the notochord exposes pancreatic endoderm to repressive factors that prevent pancreatic gene expression. However, additional germ layer culture experiments showed that the notochord does indeed induce pancreas genes. In these experiments, dorsal endoderm, which included the pancreatic anlage, was grown in culture either alone or with notochord. Endoderm from stage 12 chick embryos already expresses PAX6, ISL-1, and HNF3β, but not glucagon

PDX-1 or insulin. After culturing stage 12 endoderm without a notochord in collagen matrix for 3 days, PAX6, ISL-1, HNF3β, and glucagon were detected, but PDX-1 and insulin were not. However, coculturing with notochord tissue resulted in additional expression of PDX-1 and insulin. These data suggest that the notochord can initiate expression of pancreatic genes like PDX-1 and insulin in competent endoderm. Furthermore, stage 13 (19 somites) endoderm grown *in vitro* without notochord expressed PAX6, ISL-1, HNF-3β, and glucagon and PDX-1, whereas similarly grown stage 14 (22 somites) endoderm expressed insulin in addition to all other pancreas marker genes. These data confirm that acquisition of pancreatic cell fate is a gradual progression and that its completion is coincident with notochord detachment from the endoderm at stage 14 (22 somites) in chick embryos (Kim *et al.*, 1997).

C. Candidate Factors Important for Gut Tube Regionalization and Pancreas Development

I. HNF3β

The fork head/winged helix transcription factor, HNF3β, is one of a group of genes (HNF3α, β, and γ) known to be required for hepatocyte-specific gene expression, and it is suspected to be important for endoderm regionalization. Although HNF3β is expressed from the earliest point in endoderm generation, embryos that have lost a functional copy of this gene retain a small population of endodermal cells. Therefore, HNF3β is not required for the initial specification of the endodermal germ layer, but it is required for its further development (Zaret, 1996). Secretion of Shh by the notochord can induce HNF3β expression in the adjacent neural tube floor plate (Echelard *et al.*, 1993). Once HNF3β gene is activated, the gene product helps maintain its own synthesis and is believed to initiate Shh expression in the floorplate. Comparison of HNF3β and Shh expression patterns has revealed a striking overlap, indicating that these factors are likely expressed in the same cells. However, HNF3β is expressed shortly before Shh, and it has been speculated that Shh and HNF3β actually maintain each other's expression in the notochord and neural tube. It has been demonstrated that Shh produced by the endoderm, as opposed to the notochord, is an inductive signal acting on the visceral mesoderm to induce BMP-4 (a TGFβ family member) and Hox genes during induction and regionalization of the chick hindgut (Roberts *et al.*, 1995). Also, it has been suggested that HNF3β may lead to transcription of Shh within the gut endoderm, contributing to gut regionalization.

Comparison of the developmental expression pattern of HNF3 family members (HNF3 α, β, and γ) during embryogenesis revealed that these genes are sequentially activated in an overlapping pattern that progresses during endodermal development (Zaret, 1996). This observation led to the proposition that HNF3 family members define the regionalization of the gut endoderm. Furthermore, HNF3β is likely to be involved in the initial specification of pancreatic endoderm differentiation and maintenance of the differentiated phenotype, because it is expressed in the pancreas anlage even before the expression of any known pancreatic markers and in the dorsal pancreas throughout development. There are reports that HNF3β contributes to the islet-cell-specific transcription of PDX-1, α-amylase, and glucagon through direct binding to their promoters (Cockell *et al.*, 1995; Diedrich *et al.*, 1997; Philippe *et al.*, 1994, 1995; Sharma *et al.*, 1997; Wu *et al.*, 1997). However, our knowledge of the extent to which HNF3β plays a role in pancreatic endoderm specification is limited and requires further investigation.

2. TGFβ

Transforming growth factor β (TGFβ) family of cytokines regulates cell proliferation, differentiation, and recognition, as well as development, tissue recycling, and repair. These effects are mediated by the serine/threonine kinase activity of the TGFβ membrane receptors (Massague, 1996). In this fashion, TGF-β has been implicated in the patterning of the anterior/posterior axis of the gut endoderm in the frog, with a specific signaling function in the establishment of the anterior endoderm (Henry *et al.*, 1996). Interestingly, three TGF-β isoforms, TGFβ-1, TGFβ-2, and TGFβ-3, have been detected in pancreatic islets, acinar cells, and ductal cells by immunohistochemistry and *in situ* hybridization techniques (Yamanaka *et al.*, 1993). Also, the TGFβ family members activin and BMP7 have been found to be expressed by pancreatic epithelial cells during early development (Furukawa *et al.*, 1995; Lyons *et al.*, 1995). Furthermore, activin has been reported to activate the transcription of the *Xenopus* homologue of PDX-1, X1hbox8 (Gamer and Wright, 1995).

TGFβ acts as a mitogen in mesenchymal cells while it is antiproliferative in epithelial cells (Yingling *et al.*, 1995). In the pancreas, TGF-β serves to negatively control the growth of pancreatic acinar cells and is essential for the maintenance of the undifferentiated acinar phenotype in the exocrine pancreas (Bottinger *et al.*, 1997). It has been proposed that the negative effects of TGFβ on pancreatic cell growth may be mediated by a zinc-finger transcription factor TIEG (TGFβ-inducible early gene). TIEG is regulated by TGFβ1 and is expressed in pancreatic acinar and in duct cells. Overexpression of TIEG results in apoptosis in exocrine pancreas cells, which has been shown to play a role in pancreatic remodeling during normal pancreatic development in the rat (Scaglia *et al.*, 1997). Thus, there is evidence that TGFβ negatively regulates pancreatic cell growth mediated through TIEG.

In another study , TGFβ activity in the pancreas was eliminated in transgenic mice carrying a dominant-negative mutant TGFβII receptor (DNR mice) (Bottinger *et al.*, 1997). The loss of responsiveness to TGFβ

in DNR mice had no effect on endocrine cell differentiation, but did have an effect on exocrine cells. DNR mice showed increased acinar cell proliferation and apoptosis as well as an overaccumulation of ductlike structures and the loss of acini. The coincident reduction in acini with the abnormal accumulation of ductlike structures suggested that the maintenance of the differentiated acinar cell type has been disturbed. The authors proposed that acinar cells were dedifferentiating into duct cells in the DNR mice and that acinar dedifferentiation is the consequence of increased proliferation of normally quiescent cells. The perturbed proliferation and differentiation of pancreatic acinar cells, due to the loss of TGFβ responsiveness, has many implications for pancreatic cancer.

3. Shh

Shh is expressed in the anterior and posterior ends of the endoderm at the earliest point in gut formation and then becomes restricted during development. As mentioned earlier, endodermal-Shh has been implicated in the determination of the fate of adjacent mesoderm at different regions of the gut tube (Apelquist *et al.*, 1997). However, from E10.5 of mouse development onward, Shh is excluded from the gut endoderm in the region of the pancreatic anlage, suggesting that Shh must be eliminated for the proper development of the pancreas. To test this hypothesis, Apelquist *et al.* (1997) conducted transgenic mouse experiments in which the Shh coding region was placed under the control of the PDX-1 promoter (*PDX-1-Shh*). Significantly, they found that the ectopic Shh expression in the pancreatic endoderm converted pancreatic mesoderm into intestinal mesenchyme.

Histological analysis of the *PDX-1-Shh* mice revealed that cells from the region of pancreatic endoderm expressed smooth muscle cell markers and exhibited other characteristics of smooth muscle. However, ectopic Shh expression did not totally prevent pancreatic endoderm differentiation. The transgenic *PDX-1-Shh* pancreas epithelium had a mixed pancreatic–duodenal phenotype, with cells expressing glucagon, insulin, and amylase as well as muscle cell markers, although the endocrine cells were not organized into islets and the exocrine tissue did not form distinct acinar structures.

In summary, conversion of pancreatic mesenchyme into duodenal mesenchyme, as occurs in the *PDX-1-Shh* mice, is not sufficient to prevent later events in pancreatic development and endocrine and exocrine cell differentiation, but does disrupt proper morphogenesis of the pancreatic epithelium. Thus, the data from the *PDX-1-Shh* mice suggest that the inductive signal(s) for commitment to the pancreatic cell lineage may still be in place even when Shh is ectopically expressed. Thus, it was concluded that spatial restriction of Shh expression observed at different anterior/posterior levels of the gut generates distinct mesodermal derivatives and that the exclusion of Shh from pancreatic endoderm allows for proper pancreas morphogenesis (Apelquist *et al.*, 1997).

Finally, although signals emanating from the notochord appear to be required for specification of endodermal cells to the pancreatic cell fate, the signal(s) that confers the pancreatic identity on the gut endoderm does not appear to be the predominant signaling molecule from the notochord, Shh. Doug Melton's group has reported that Shh was unable to induce pancreatic budding when added to isolated stage 11.5 dorsal endoderm in culture (Kim *et al.*, 1997). Moreover, signals emanating from the notochord have been shown to repress Shh expression in the pancreatic endoderm. Accordingly, separation of the notochord from pancreatic endoderm by microdissection allows endodermal Shh expression, which in turn correlates with the lack of PDX-1 and insulin expression (Hebrok *et al.*, 1998). Others have shown that the caudal notochord adjacent to the pancreatic anlage expresses the TGF-β family member activin during the period when the prepancreatic endoderm seems to require signals from the notochord. Hebrok *et al.* (1998) have now identified two TGFβ family members, activin-βB and FGF2, as notochord factors that can repress endodermal Shh, thereby permitting pancreas development. Therefore, establishment of a specific pattern of Shh expression along the anterior/posterior axis of the gut tube prior to E10.5 (in the chick) is important for aspects of future organogenesis within the embryonic gut tube and the exclusion of Shh expression required for the initiation of pancreatic organogenesis. However, repression of Shh expression alone is not sufficient for pancreatic development outside of the pancreatic anlage. Thus, further investigation of endoderm and mesoderm communication is necessary to determine what combination of factors induce the competence of the pancreatic anlage to differentiate (Hebrok *et al.*, 1998).

4. NGF

Nerve growth factor (NGF) is well known for its important role in the differentiation and survival of neurons and neural crest-derived cells; however, it has also been implicated in pancreatic development. The high-affinity NGF receptor Trk-A has been shown to be expressed in the developing rat pancreas, β-cell lines, and primary islet cultures (Kanaka-Gantenbein *et al.*, 1995a). Also, the expression pattern of Trk-A in the pancreas is developmentally regulated, shifting from the duct epithelium in early fetal life to the β-cells postnatally (Kanaka-Gantenbein *et al.*, 1995b). This pattern of expression is suggestive of a role for Trk-A in islet cell differentiation as they migrate from the ducts. However, the role of Trk-A in the pancreas is not clear because mice lacking a functional Trk-A gene exhibit normal pancreas morphogenesis even though the null mutation leads to perinatal lethality (Smeyne *et al.*, 1994).

An interesting *in vitro* model of islet morphogenesis has been characterized, however (Kanaka-Gantenbein *et al.*, 1995a), and used to examine the influence of NGF and the NGF receptors on islet morphogenesis. In this system, E21 rat fetal pancreases were mildly collagenase-digested and grown in culture. During this primary culture, exocrine pancreatic tissue gradually degenerated, fibroblast-like cells proliferated in a monolayer, and spherical islet-like structures emerged from this monolayer. The monolayer was concluded to be non-endocrine; however, the spherical structures that emerged all expressed RNAs characteristic of islet cells. Thus, islet formation could be induced in culture, allowing for the study of gene expression during islet morphogenesis.

The authors found that the NGF receptor was expressed specifically in the cells of the developing islet while the corresponding ligand was expressed only in the surrounding, non-endocrine cells. Therefore, the authors proposed a model in which NGF or other neurotrophins secreted by non-endocrine pancreatic cells act in a paracrine manner on islet-cell precursors to induce morphogenesis. Their model was supported by the negative effect of K252, a NGF inhibitor, which inhibited *in vitro* islet morphogenesis (Kanaka-Gantenbein *et al.*, 1995a). Though these experiments suggest a role for NGF in islet morphogenesis *in vitro,* additional experiments are necessary to demonstrate that it has a role *in vivo.*

Finally, there is indirect evidence that NGF signaling may influence the transcriptional regulation of BETA2. Neurite outgrowth by NGF is a transcription-dependent process that is mediated by the bHLH transcriptional repressor HES-1. It has been shown that HES-1 DNA-binding activity is posttranslationally inhibited during NGF signaling and that inhibition of HES-1 mediates in the induction of neurite outgrowth by NGF signaling (Strom *et al.*, 1997). Interestingly, preliminary data from our laboratory suggest that transcription from the BETA2 gene is repressed by HES-1 (unpublished results). Thus, NGF signaling could potentially result in the derepression of the BETA2 gene through the inhibition of HES-1 DNA-binding activity. However, any role NGF signaling plays in the transcriptional regulation of BETA2 or in pancreatic islet development *in vivo* is purely speculative at this time.

5. Phosphatases and Kinases

The transmission of signals from the cell surface to the nucleus is propagated by way of a cascade of kinases and phosphatases. Such molecules are intimately involved in cell proliferation, cell cycle, and gene regulation. It follows that there are several such molecules that are specifically expressed in the pancreas during pancreas development and thereafter in mature islet cells. The following section presents a sampling of phosphatase and kinase molecules believed to be important to pancreas development.

The MLK family members are MAPK Kinase Kinases that preferentially act on the JNK/SAPK signaling pathway. The factor MLK-1 has both kinase and leucine zipper motifs as well as a bHLH domain downstream of the

leucine zipper (DeAizpurua *et al.*, 1997). It is expressed primarily in immature β-cells in the developing embryo between E14 and E16 and is not
detectable at later stages of gestation or postnatally. Also, MLK-1 expression
has been reported to be associated with cell lines that display a more immature pre-β-cell phenotype (RIN-5AH) and not in cell lines displaying a more
mature cell type (RIN-A12 cells) (DeAizpurua *et al.*, 1997). Considering the
developmentally regulated expression pattern of MLK-1 in the pancreas, it
is likely that this factor plays a role in pancreas development. Furthermore,
another MLK family member, ZPK, which is expressed in a cell-specific
manner in the adult pancreas, activates JNK/SAPK by activating JNKK, a
MAPK kinase class of protein kinase (Nadeau *et al.*, 1997). Thus, there
appears to be a family of protein kinases that display pancreas-specific
expression. However, there is little information in regard to how these
protein kinases affect pancreas development, leaving open a very promising
avenue of research.

There is at least one member of the receptor tyrosine phosphatase family,
PTP-NP, expressed in the nervous system and in the pancreas during development. PTP-NP has now been implicated in the development of the endocrine
pancreas (Chiang and Flanagan, 1996). Strikingly, PTP-NP expression can
be detected in the region of the developing pancreas as early as E8.5. This
precedes pancreatic morphogenesis and the expression of most pancreas
markers. At E9.5, PTP-NP is expressed in the pancreatic bud, but its expression is somewhat restricted to cells that express either glucagon or insulin.
Additionally, PTP-NP is not expressed in the surrounding duodenum, demonstrating a highly pancreas-specific function. Throughout development and
into adulthood, PTP-NP is coexpressed with endocrine cell markers, but
not exocrine cell markers such as amylase. In addition, there is a small
population of cells expressing PTP-NP that do not express any hormones.
Chiang and Flanagan (1996) speculate that PTP-NP-expressing cells that
are hormone negative represent stem cells or undifferentiated progenitor
cells that have not yet begun to express hormone markers, and that PTP-
NP is a pancreatic progenitor cell marker.

IV. Summary and Perspective

We have presented a wide range of topics ranging from pancreatic gene
regulation to pancreatic islet development. Factors determined to influence
islet-cell differentiation have been described, and finally signaling factors
that may specify the pancreatic cell fate have been introduced. Pancreatic
disorders can and do arise from defects at all levels of pancreas development,
involving inappropriate transcriptional regulation, cellular differentiation,
and extracellular and intracellular signaling, as well as enzymatic dysfunction and organogenesis. Thus, the intention was not to be comprehensive,

but rather to provide a guide to pancreatic development from several different points of view.

In terms of gene regulation, much is now known about how pancreatic hormone genes are regulated. Many of the promoter elements that dictate islet-cell-specific gene expression have been characterized, and many of the protein factors that mediate these activation and repression events have been identified. Many factors and the elements to which they bind are redundant between genes that have differing expression patterns, creating more interesting questions about how genes such as the islet hormones are specifically expressed in a single cell type. As more tissue-specific factors are cloned and more elements are defined, we are beginning to see a second level of complexity, involving cooperative and synergistic functional interactions. For example, the interaction of homeobox factors such as PDX-1 or bHLH factors such as BETA2 with other factors appears to be important for cell-specific pancreatic hormone expression.

Promoter analysis has helped to identify and isolate several key pancreatic transcription regulators, and analysis of the corresponding loss-of-function mutants has helped to decipher the *in vivo* function of these factors. As a result, many of these factors have been determined to be crucial for aspects of the endocrine cell differentiation and for the maintenance of differentiated cell types. Clearly, knockout and transgenic animal technology is indispensable for the determination of gene function. Each of the seven knockout strains described herein demonstrates a block in pancreatic development at a different point, and the analysis of these mutants is beginning to add more detail to the endocrine cell lineage model. The β-cell branch of the endocrine lineage, for example, is already becoming more complex with the possibility that two separate β-cell populations are derived from endocrine progenitor cells. At the same time, the mutant analysis has demonstrated that β-cell development occurs in several stages. Accordingly, factors have been identified that can serve as stage-specific markers. Nkx2.2, for example, functions downstream of PAX4 but upstream of Nkx6.1 in β-cell differentiation. In this way, we can begin to trace the β-cell differentiation pathway.

Indeed, the greatest advances that have been made in pancreas and diabetes research have been toward a better understanding of endocrine cell differentiation and pancreas morphogenesis. From this, a model for the endocrine cell lineage has been developed. More is known about the mesenchymal epithelial interactions necessary for pancreas development. Significant steps have been made toward understanding how pancreas development is induced. The notochord has now been shown to be the source of inductive signals for the specification of the gut endoderm. In particular, notochord signaling molecules, activin-βB and FGF2, have now been implicated in the specification of gut endoderm to the pancreatic fate through the repression of Shh. However, many questions remain to be answered about how pancreas cell fate is specified—in particular, how the pancreatic epithelium responds to inductive signals emanating from the mesoderm. Receptors for several cell surface receptors and their ligands have been identified in the pancreas. In addition, members of the MLK family of MAPK Kinase Kinases and at least one phosphatase, PTP-NP, appear to be expressed highly in the developing pancreas. Moreover, the extremely early appearance of PTP-NP in the pancreatic anlage suggests that it is potentially part of the first response to environmental signals that specify gut endoderm to the pancreatic fate. Overall, however, little is known about the signaling mechanisms that instruct cells of the gut endoderm to proceed down the pancreas cell fate pathway.

It is important to consider the relevance of artificially generated pancreatic defects to the disease states that occur naturally in the human population. Several diabetic mouse models have been generated, but none of these actually recapitulate the etiology of type II diabetes, though they do provide invaluable information in regard to pancreatic development mechanisms. This chapter began with reference to several new tools for pancreatic research. These are the identification of a virtual menagerie of developmentally expressed pancreas-specific gene regulators in the past 4 to 6 years. Analyses of the corresponding knockout mouse strains have generated data that support Teitelman's endocrine lineage model and provide additional details. Further analysis of each knockout mouse strain should reveal the identity of upstream signaling factors that induce endocrine progenitor cell specification, as well as downstream target genes that influence pancreatic cell differentiation and maintenance. Without a doubt, the final outcome of these studies will be the identification of the pancreatic stem cells and the mechanism(s) that govern their differentiation to each of the different pancreatic cell types. This level of understanding will blaze a path toward discovering the causes of diabetes and ultimately better treatments for this debilitating disease.

Acknowledgments

We thank Dr. Roland Stein for his knowledgeable suggestions and kind assistance during the preparation of this manuscript.

References

Abmayr, S. M., and Keller, C. A. (1998). Drosophila myogenesis and insights into the role of nautilus [review; 296 refs]. *Curr. Top. Devel. Biol.* **38,** 35–80.

Ahlgren, U., Jonsson, J., and Edlund, H. (1996). The morphogenesis of the pancreatic mesenchyme is uncoupled from that of the pancreatic epithelium in IPF1/PDX1-deficient mice. *Development* **122,** 1409–1416.

Ahlgren, U., Pfaff, S. L., Jessell, T. M., Edlund, T., and Edlund, H. (1997). Independent requirement for ISL1 in formation of pancreatic mesenchyme and islet cells. *Nature* **385**, 257–260.

Alpert, S., Hanahan, D., and Teitelman, G. (1988). Hybrid insulin genes reveal a developmental lineage for pancreatic endocrine cells and imply a relationship with neurons. *Cell* **53**, 295–308.

Andrisani, O. M., Hayes, T. E., Roos, B., and Dixon, J. E. (1987). Identification of the promoter sequences involved in the cell specific expression of the rat somatostatin gene. *Nucl. Acids Res.* **15**, 5715–5728.

Apelquist, A., Ahlgren, U., and Edlund, H. (1997). Sonic hedgehog directs specialized mesoderm differentiation in the intestine and pancreas. *Curr. Biol.* **7**, 801–804.

Beck, T. C., and Madsen, O. D. (1989). Monoclonal antibodies as probes to the differentiated exocrine pancreas react to monoclonal islet tumor tissue. *Exp. Clin. Endocrinol.* **93**, 255–260.

Benezra, R., Davis, R. L., Lackshon, D., Turner, D. L., and Weintraub, H. (1990). The protein Id: A negative regulator of helix-loop-helix DNA bindig proteins. *Cell* **61**, 49–59.

Blume, N., Skouv, J., Larsson. L. I., Holst, J. J., and Madsen, O. D. (1995). Potent inhibitory effects of transplantable rat glucagonomas and insulinomas on the respective endogenous islet cells are associated with pancreatic apoptosis. *J. Clin. Invest.* **96**, 2227–2235.

Boam, D. S., Clark, A. R., and Docherty, K. (1990). Positive and negative regulation of the human insulin gene by multiple trans-acting factors. *J. Biol. Chem.* **265**, 8285–8296.

Bober, E., Baum, C., Braun, T., and Arnold, H. H. (1994). A novel NK-related mouse homeobox gene: Expression in central and peripheral nervous structures during embryonic development. *Devel. Biol.* **162**, 288–303.

Bock, P., Abdel-Moneim, M., Egerbacher, M. (1997). Development of pancreas. *Microsc. Res. Techn.* **37**, 374–383.

Boncinelli, E., and Mallamaci, A. (1995). Homeobox genes in vertebrate gastrulation. *Curr. Opin. Genet. Devel.* **5**, 619–627.

Bonner-Weir, S., Baxter, L. A., Schuppin, G. T., and Smith, F. E. (1993). A second pathway for regeneration of adult exocrine and endocrine pancreas. A possible recapitulation of embryonic development. *Diabetes* **42**, 1715–1720.

Bonny, C., Nicod, P., and Waeber, G. (1998). IB1, a JIP-1-related nuclear protein present in insulin-secreting cells. *J. Biol. Chem.* **273**, 1843–1846.

Botas, J. (1993). Control of morphogenesis and differentiation by HOM/Hox genes. *Curr. Opin. Cell Biol.* **5**, 1015–1022.

Bottinger, E. P., Jakubczak, J. L., Roberts, I. S., Mumy, M., Hemmati, P., Bagnall, K., Merlino, G., and Wakefield, L. M. (1997). Expression of a dominant-negative mutant TGF-beta type II receptor in transgenic mice reveals essential roles for TGF-beta in regulation of growth and differentiation in the exocrine pancreas. *EMBO J.* **16**, 2621–2633.

Bray, S. J. (1997). Expression and function of Enhancer of split bHLH proteins during *Drosophila* neurogenesis [review; 65 refs]. *Persp. Devel. Neurobiol.* **4**, 313–323.

Buchberger, A., Pabst, O., Brand, T., Seidl, K., and Arnold, H. H. (1996). Chick NKx-2.3 represents a novel family member of vertebrate homologues to the *Drosophila* homeobox gene tinman: differential expression of cNKx-2.3 and cNKx-2.5 during heart and gut development. *Mech. Devel.* **56**, 151–163.

Caudy, M., Vassin, H., Brand, M., Tuma, R., and Jan Y. N. (1988). Daughterless, a *Drosophila* gene essential for both neurogenesis and sex determination, has sequence similarities to myc and the acheate-scute complex. *Cell* **55**, 1061–1067.

Chalepakis, G., Wijnholds, J., Giese, P., Schachner, M., and Gruss, P. (1994). Characterization of Pax-6 and Hoxa-1 binding to the promoter region of the neural cell adhesion molecule L1. *DNA Cell Biol.* **13**, 891–900.

Chiang, M.-K., and Flanagan, J. G. (1996). PTP-NP a new member of the receptor protein tyrosine phosphatase family, implicated in development of nervous system and pancreatic endocrine cells. *Development* **122**, 2239–2250.

Cirulli, V., Baitens, D., Rutishauser, U., Halban, P. A., Orci, L., and Rouiller, D. G. (1994). Expression of neural cell adhesion molecule (N-CAM) in rat islets and its role in islet cell type aggregation. *J. Cell Sci.* **107**, 1429–1436.

Clark, J. T., Kalra, P. S., Crowely, W. R., and Kalra, S. P. (1984). Neuropeptide Y and human panreatic polypeptide stimulate feeding behaviour in rats. *Endocrinology* **115**, 427–429.

Clark, A. R., Wilson, M. E., Leibiger, I., Scott, V., and Docherty, K. (1995). A silencer and an adjacent positive element interact to modulate the activity of the human insulin promoter. *Eur. J. Biochem.* **232**, 627–632.

Cockell, M., Stolarczyk, D., Frutiger, S., Hughes, G. J., Hagenbuchle, O., and Wellauer, P. K. (1995). Binding sites for hepatocyte nuclear factor 3 beta or 3 gamma and pancreas transcription factor 1 are required for efficient expression of the gene encoding pancreatic alpha-amylase. *Mol. Cell. Biol.* **15**, 1933–1941.

Cordle, S. R., Henderson, E., Masuoka, H., Weil, P. A., and Stein, R. (1990). Pancreatic β-cell-specific transcription of the insulin gene is mediated by basic helix–loop–helix DNA-binding proteins. *Mol. Cell. Biol.* **11**, 1734–1738.

Crowe, D. T., and Tsai, M.-J. (1989). Mutagenesis of the rat insulin II 5′-flanking region defines sequences important for expression in HIT cells. *Mol. Cell. Biol.* **9**, 1784–1789.

Cvekl, A., Kashanchi, F., Sax, C. M., Brady, J. N., and Piatigorsky, J. (1995). Transcriptional regulation of the mouse alpha A-crystallin gene: activation dependent on a cyclic AMP-responsive element (DE1/CRE) and a Pax-6-binding site. [Abstract]. *Mol. Cell. Biol.* **15**(2), 653–660.

Dahl, U., Sjödin, A., and Semb, H. (1996). Cadherins regulate aggregation of pancreatic β-cells *in vivo*. *Development* **122**, 2895–2902.

DeAizpurua, H. J., Cram, D. S., Naselli, G., Devereux, L., and Dorow, D. S. (1997). Expression of mixed lineage kinase-1 in pancreatic β-cell lines at different stages of maturation and during embryonic pancreas development. *J. Biol. Chem.* **272**, 16364–16373.

Diedrich, T., Furstenau, U., and Knepel, W. (1997). Glucagon gene G3 enhancer: Evidence that activity depends on combination of an islet-specific factor and a winged helix protein. *Biol. Chem.* **378**, 89–98.

Docherty, K., and Clark, A. R. (1994). Nutrient regulation of insulin gene expression. *FASEB J.* **8**, 20–27.

Drews, U. (1995). "Color Atlas of Embryology," pp. 316–317. Thieme Medical Publishers, Inc., New York.

Drucker, D. J. (1998). Glucagon-like peptides. *Diabetes* **47**, 159–169.

Drucker, D. J., Philippe, J., Mojsov, S., Chick, W. I., and Habener, J. F. (1987). Glucagon-like peptide I stimulates insulin gene expression and increases cyclic AMP levels in a rat islet cell line. *Proc. Natl. Acad. Sci. USA* **84**, 3434–3438.

Dudek, R. W., Lawrence, I. E. J., Hill, R. S., and Johnson, R. C. (1991). Induction of islet cytodifferentiation by fetal mesenchyme in adult pancreatic ductal epithelium. *Diabetes* **40**, 1041–1048.

Dunn, N. R., Winnier, G. E., Hargett, L. K., Schrick, J. J., Fogo, A. B., and Hogan, B. L. (1997). Haploinsufficient phenotypes in Bmp4 heterozygous null mice and modification by mutations in Gli3 and Alx4. *Dev. Biol.* **188**, 235–247.

Durocher, D., Charron, F., Warren, R., Schwartz, R. J., and Nemer, M. (1997). The cardiac transcription factors Nkx2-5 and GATA-4 are mutual cofactors. *EMBO J.* **16**, 5687–5696.

Duvillie, B., Cordonnier, N., Deltour, L., Dandoy-Dron, F., Itier, J. M., Monthioux, E., Jami, J., Joshi, R. L., and Bucchini, D. (1997). Phenotypic alterations in insulin-deficient mutant mice. *Proc. Natl. Acad. Sci. USA* **94**, 5137–5140.

Echelard, Y., Epstein, D. J., St-Jacques, B., Shen, L., Mohler, J., McMahon, J. A., McMahon, A. P. (1993). Sonic hedgehog, a member of a family of putative signaling molecules, is implicated in the regulation of CNS polarity. *Cell* **75**, 1417–1430.

Edlund, T., Walker, M. D., Barr, P. J., and Rutter, W. J. (1985). Cell-specific expression of the rat insulin gene: Evidence for role of two distinct 5′ flanking elements. *Science* **230**, 912–916.

Emens, L. A., Landers, D. W., and Moss, L. G. (1992). Hepatocyte nuclear factor 1 alpha is expressed in a hamster insulinoma line and transactivates the rat insulin I gene. *Proc. Natl. Acad. Sci. USA* **89**, 7300–7304.

Ericson, J., Rashbass, P., Schedl, A., Brenner-Morton, S., Kawakami, A., van Heyningen, V., and Jessell, T. M. (1997). Pax6 controls progenitor cell identity and neuronal fate in response to graded Shh signalling [Abstract]. *Cell* **90**, 169–180.

Erlinge, D., Brunkwall, J., and Edvinsson, L. (1994). Neuropeptide Y stimulates proliferation of human vascular smooth muscle cells: Cooperation with noradrenaline and ATP. Regul. Pept. **50**, 259–265.

Fode, C., Gradwohl, G., Morin, X., Diedrich, A., LeMeur, M., Goridis, C., and Guillemot, F. (1998). [Abstract]. *Neuron* **20**(3), 483–494.

Furukawa, M., Eto, Y., and Kojima, I. (1995). Expression of immunoreactive activin A in fetal rat pancreas. *Endocrine J.* **42**, 63–68.

Furuta, M., Yano, H., Zhou, A., Rouille, Y., Holst, J. J., Carroll, R., Ravazzola, M., Orci, L., Furuta, H., and Steiner, D. F. (1997). Defective prohormone processing and altered pancreatic islet morphology in mice lacking active SPC2. *Proc. Natl. Acad. Sci. USA* **94**, 6646–6651.

Galsgaard, E. D., Gouilleux, F., Groner, B., Serup, P., Neilsen, J. H., and Bilestrup, N. (1996). Identification of a growth hormone-responsive STAT5-binding element in the rat insulin 1 gene. *Mol. Endocrinol.* **10**, 652–660.

Gamer, L. W., and Wright, C. V. (1995). Autonomous endodermal determination in Xenopus: Regulation of expression of the pancreatic gene X1Hbox 8. *Devel. Biol.* **171**, 240–251.

Gehring, W. J. (1987). Homeo boxes in the study of development. *Science* **236**, 1245–1252.

Gehring, W. J., Affolter, M., and Burglin, T. (1994). Homeodomain proteins. *Ann. Rev. Biochem.* **63**, 487–526.

German, M. S., Blaner, M. A., Nelson, C., Moss, J. B., and Rutter, W. J. (1991). Two related helix-loop-helix proteins participate in separate cell-specific complexes that bind the insulin enhancer. *Mol. Endocrinol.* **5**, 292–299.

German, M. S., Wang, J., Chadwick, R. B., and Rutter, W. J. (1992). Synergistic activation of the insulin gene by a LIM-homeo domain protein and a basic helix-loop-helix protein: Building a functional insulin minienhancer complex. *Genes Devel.* **6**, 2165–2176.

German, M., Ashcroft, S., Docherty, K., Edlund, H., Edlund, T., Goodison, S., Imura, H., Kennedy, G., Madsen, O., and Melloul, D. (1995). The insulin gene promoter. A simplified nomenclature [letter]. *Diabetes* **44**, 1002–1004.

Gittes, G. K., and Rutter, W. J. (1992). Onset of cell-specific gene expression in the developing mouse pancreas. *Proc. Natl. Acad. Sci. USA* **89**, 1128–1132.

Gittes, G. K., Galante, P. E., Hanahan, D., Rutter, W. J., and Debase, H. T. (1996). Lineage-specific morphogenesis in the developing pancreas: Role of mesenchymal factors. *Development* **122**, 439–447.

Glaser, T., Jepeal, L., Edwards, J. G., Yound, S. R., Favor, J., and Maas, R. L. (1994). Pax6 gene dosage effect in a family with congenital cataracts, aniridia, anophothal and central nervous system detects [published erratum appears in *Nature Genet.* 1994, 8(2):203]. *Nature Genet.* **7**, 463–471.

Golosow, N., and Grobstien, C. (1962). Epitheliomesenchymal interaction in pancreatic morphogenesis. *Devel. Biol.* **4**, 242–255.

Goodman, P. A., Medina-Martinez, O., and Fernandez-Mejia, C. (1996). Identification of the human insulin negative regulatory element as a negative glucocorticoid response element. *Mol. Cell. Endocrinol.* **120**, 139–146.

Gradwohl, G., Fode, C., and Guillemot, F. (1996). Restricted expression of a novel murine atonal-related bHLH protein in undifferentiated neural precursors. *Dev. Biol.* **180**, 227–241.

Gu, D., and Sarvetnick, N. (1993). Epithelial cell proliferation and islet neogenesis in IFN-g transgenic mice. *Development* **118**, 33–46.

Gu, D., Lee, M. S., Krahl, T., and Sarvetnick, N. (1994). Transitional cells in the regenerating pancreas. *Development* **120**, 1873–1881.

Guz, Y., Montminy, M. R., Stein, R., Leonard, J., Gamer, L. W., Wright, C. V., and Teitelman, G. (1995a). Expression of murine STF-1, a putative insulin gene transcription factor, in beta cells of pancreas, duodenal epithelium and pancreatic exocrine and endocrine progenitors during ontogeny. *Development* **121**, 11–18.

Halevy, O., Novitch, B. G., Spicer, D. B., Skapek, S. X., Rhee, J., Hannon, G. J., Beach, D., and Lasser, A. B. (1995). Correlation of terminal cell cycle arrest of skeletal muscle with induction of p21 by MyoD. *Science* **267**, 1018–1021.

Hebrok, M., Kim, S. K. K., and Melton, D. A. (1998). Notochord repression of endodermal Sonic hedgehog permits pancreas development [Abstract]. *Genes Devel.*, 1705–1713.

Henry, G. L., Brivanlou, I. H., Kessler, D. S., Hemmati-Brivanlou, A., and Melton, D. A. (1996). TGF-beta signals and a pattern in *Xenopus laevis* endodermal development. *Development* **122**, 1007–1115.

Herrera, P. L., Huarte, J., Sanvito, F., Meda, P., Orci, L., and Vassalli, J. D. (1991). Embryogenesis of the murine endocrine pancreas: Early expression of pancreatic polypeptide gene. *Development* **113**, 1257–1265.

Herrera, P. L., Huarte, J., Zufferey, R., Nichols, A., Mermillod, B., Philippe, J., Muniesa, P., Sanvito, F., Orci, L., and Vassalli, J. D. (1994). Ablation of islet endocrine cells by targeted expression of hormone-promoter-driven toxigenes. *Proc. Natl. Acad. Sci. USA* **91**, 12999–13003.

Herrera, P. L., Orci, L., and Vassalli, J. D. (1998). Two transgenic approaches to define the cell lineages in endocrine pancreas development [Abstract]. *Mol. Cell. Endocrinol.* **140**(1–2): 45–50.

Hill, R. E., Favor, J., Hogan, B. L., Ton, C. C., Saunders, G. F., Hanson, I. M., Prosser, J., Jordan, T., Hastie, N. D., and van Heyningen, V. (1991). Mouse small eye results from mutaions in a paired-like homeobox-containing gene [published erratum appears in *Nature* (1992) Feb 20;355 (6362);750]. *Nature* **354**, 522–525.

Hogan, B., Beddington, R., Costantini, F., and Lacy, E. (1994). "Manipulating the Mouse Embryo: A Laboratory Manual." Cold Spring Harbor Laboratory Press, pp. 21–113.

Holman, G. D., and Kasuga, M. (1997). From receptor to transporter insulin signalling to glucose transport. *Diabetologia* **40**, 991–1003.

Hutton, J. C., Christofori, G., Chi, W. Y., Edman, U., Guest, P. C., Hanahan, D., and Kelly, B. (1993). Molecular cloning of mouse pancreatic islet R-cadherin: Differential expression in endocrine and exocrine tissue. *Mol. Endocrinol.* **7**, 1151–1160.

Hwung, Y.-P., Wang, L.-H., Tsai, S. Y., and Tsai, M.-J. (1988). Differential binding of the chicken ovalbumin upstream promoter (COUP) transcription factor to two different promoters. *J. Biol. Chem.* **263**, 13470–13474.

Hwung, Y.-P., Gu, Y. Z., and Tsai, M.-J. (1990). Cooperativity of sequence elements mediates tissue-specificity of the rat insulin II gene. *Mol. Cell. Biol.* **10**, 1784–1788.

Jackerott, M., and Larsson, L. I. (1997). Immunocytochemical localization of the NPY/PYY Y1 receptor in enteric neurons, endothelial cells, and endocrine-like cells of the rat intestinal tract. *J. Histochem. Cytochem.* **45**, 1643–1650.

Jackerott, M., Oster, A., and Larsson, L. I. (1996). PYY in developing murine islet cells: Comparisons to development of islet hormones, NPY, and BrdU incorporation. *J. Histochem. Cytochem.* **44**, 809–817.

Jensen, J., Serup, P., Karlsen, C., Nielsen, T. F., and Madsen, O. D. (1996). mRNA profiling of rat islet tumors reveals nkx 6.1 as a beta-cell-specific homeodomain transcription factor. *J. Biol. Chem.* **271**, 18749–18758.

Johnson, L., Greenbaum, D., Cichowski, K., Mercer, K., Murphy, E., Schmitt, E., Bronson, R. T., Umanoff, H., Edelmann, W., Kucherlapati, R., Jacks, T. (1997). K-ras is an essential gene in the mouse with partial functional overlap with N-ras. *Genes Devel.* **11**, 2468–2481.

Jonsson, J., Carlsson, L., Edlund, J., and Edlund, H. (1994). Insulin-promoter-factor 1 is required for pancreas development in mice. *Nature* **371**, 606–609.

Jordan, T., Hanson, I. M., Zaletayev, D., Hodgson, S., Prosser, J., Seawright, A., Hastie, N. D., and van Heyningen, V. (1992). The human PAX6 gene is mutated in tow patients with aniridia [Abstract]. *Nature Genet.* **1**(5), 328–332.

Joyner, A. L. (1996). Engrailed, Wnt and Pax genes regulate midbrain–hindbrain development. *Trends Genet.* **12**, 15–20.

Kahn, B. B. (1998). Type 2 Diabetes: When insulin secreation fails to compensate for insulin resistance. *Cell* **92**, 593–596.

Kanaka-Gantenbein, C., Dicou, E., Czernichow, P., and Scharfmann, R. (1995a). Presence of nerve growth factor and its receptors in an *in Vitro* model of islet cell development: implication in normal islet morphogenesis. *Endocrinology* **136**, 3154–3162.

Kanaka-Gantenbein, C., Tazi, A., Czernichow, P., and Scharfmann, R. (1995b). In vivo presence of the high affinity nerve growth factor receptor Trk-A in the rat pancreas: differential localization during pancreatic development. *Endocrinology* **136**, 761–769.

Karlsson, D., Edlund, T., Moss, J. B., Rutter, W. J., and Walker, M. D. (1987). A mutational analysis of the insulin gene trascription control region: Expression in beta cells is dependent on two related sequences within the enhancer. *Proc. Natl. Acad. Sci. USA* **84**, 8819–8823.

Karlsson, O., Thor, S., Norberg, T., Ohlsson, H., and Edlund, T. (1990). Insulin gene enhancer binding protein Isl-1 is a member of a novel class of proteins containing both a homeo- and a Cys-His domain. *Nature* **344**, 879–882.

Kash, S. F., Johnson, R. S., Tecott, L. H., Noebels, J. L., Mayfield, R. D., Hanahan, D., and Baekkeskov, S. (1997). Epilepsy in mice deficient in the 65-kDa isoform of glutamic acid decarboxylase. *Proc. Natl. Acad. Sci. USA* **94**, 14060–14065.

Kazhdan, I., Rickard, D., and Leboy, P. S. (1997). HLH transcription factor activity in osteogenic cells. *J. Cell. Biochem.* **65**, 1–10.

Kennedy, G. C., and Rutter, W. J. (1992). Pur-1, a zinc-finger protein that binds to purine-rich sequences, transactivates an insulin promoter in heterologous cells. *Proc. Natl. Acad. Sci. USA* **89**, 11498–11502.

Kennedy, G. C., German, M. S., and Rutter, W. J. (1995). The minisatellite in the diabetes susceptibility locus IDDM2 regulates insulin transcription [see comments]. *Nature Genet.* **9**, 293–298.

Kent, G., Iles, R., Bear, C. E., Huan, L. J., Griesenbach, U., McKerlie, C., Frndova, H., Ackerley, C., Gosselin, D., Radzioch, D., O'Brodovich, H., Tsui, L. C., Buchwald, M., Tanswell, A. K. (1997). Lung disease in mice with cystic fibrosis. *J. Clin. Invest.* **100**, 3060–3069.

Kim, Y., and Nirenberg, M. (1989). Drosophila NK-homeobox genes. *Proc. Natl. Acad. Sci. USA* **86**, 7716–7720.

Kim, S. K., Matthias, H., and Melton, D. A. (1997). Notochord to endoderm signaling is required for pancreas development. *Development*, 4243–4252.

Knepel, W., Vallejo, M., Chafitz, J. A., and Habener, J. F. (1991). The pancreatic islet-specific glucagon G3 transcription factors recognize control elements in the rat somatostatin and insulin-I genes. *Mol. Endocrinol.* **5**, 1457–1466.

Laimins, L., Holmgren-König, M., and Khoury, G. (1986). Transcriptional "silencer" element in rat repetitive sequences associated with the rat insulin 1 gene locus. *Proc. Natl. Acad. Sci. USA* **83**, 3151–3155.

Laser, B., Meda, P., Constant, I., and Philippe, J. (1996). The caudal-related homeodomain protein Cdx-2/3 regulates glucagon gene expression in islet cells. *J. Biol. Chem.* **271**, 28984–28994.

Lee, J. E. (1997). Basic helix-loop-helix genes in neural development. [Review] [45 refs]. *Curr. Opin. Neurobiol.* **7**, 13–20.

Lee, J. K., Hollenberg, S. M., Snider, L., Turner, D. L., Lipnick, N., and Weintraub, H. (1995). Conversion of Xenopus ectoderm into neurons by neuroD, a basic helix-loop-helix transcription factor. *Science* **268**, 836–844.

Lemercier, C., To, R. Q., Swanson, B. J., Lysons, G. E., and Konieczny, S. F. (1997). Mist1: A novel basic helix-loop-helix transcription factor exhibits a developmentally regulated expression pattern. *Devel. Biol.* **182**, 101–113.

Leonard, J., Serup, P., Gonzalez, G., Edlund, T., and Montminy, M. (1992). The LIM family transcription factor Isl-1 requires cAMP response element binding protein to promote somatostatin expression in pancreatic islet cells. *Proc. Natl. Acad. Sci. USA* **89**, 6247–6251.

Leonard, J., Peers, B., Johnson, T., Ferreri, K., Lee, S., and Montminy, M. R. (1993). Characterization of somatostatin transactivating factor-1, a novel homeobox factor that stimulates somatostatin expression in pancreatic islet cells. *Mol. Endocrinol.* **7**, 1275–1283.

Leshkowitz, D., Aronheim, A., and Walker, M. D. (1992). Insulin-producing cells contain a cell-specific repressor activity that functions through multiple E-box sequences. *DNA Cell Biol.* **11**, 549–558.

Lu, M., Seufert, J., and Habener, J. F. (1997). Pancreatic beta-cell-specific repression of insulin gene transcription by CCAAT/enhancer-binding protein beta. Inhibitory interactions with basic helix-loop-helix transcription factor E47. *J. Biol. Chem.* **272**, 28349–28359.

Ludwig, T., Eggenschwiler, J., Fisher, P., D'Ercole, A. J., Davenport, M. L., and Efstratiadis, A. (1996). Mouse mutants lacking the type 2 IGF receptor (IGF2R) are rescued from perinatal lethality in Igf2 and Igf1r null backgrounds. *Devel. Biol.* **177**, 517–535.

Lyons, K. M., Hogan, B. L., and Robertson, E. J. (1995). Colocalization of BMP 7 and BMP 2 RNAs suggests that these factors cooperatively mediate tissue interactions during murine development. *Mech. Dev.* **50**, 71–83.

Ma, Q., Chen, Z., and Anderson, D. J. (1998). Neurogenin1 is essential for the determination of neuronal precursors for proximal cranial sensory ganglia [Abstract]. *Neuron* **20**(3), 469–482.

MacFarlane, W. M., Read, M. L., Gilligan, M., Bujalska, I., and Docherty, K. (1994). Glucose modulates the binding of the β-cell transcription factor IUF1 in a phosphorylation-dependent manner. *Biochem. J.* **303**, 526–631.

Madsen, O. D., Larsson, L. I., Rehfeld, J. F., Schwartz, T. W., Lernmark, A., Labrecque, A. D., and Steiner, D. F. (1986). Cloned cell lines from a transplantable islet cell tumor are heterogeneous and express cholecystokinin in addition to islet hormones. *J. Cell Biol.* **103**, 2025–2034.

Madsen, O. D., Andersen, L. C., Michelsen, B., Owerbach, D., Larsson, L. I., Lernmark, A., and Steiner, D. F. (1988). Tissue-specific expression of transfected human insulin genes in pluripotent clonal rat insulinoma lines induced during passage in vivo. *Proc. Natl. Acad. Sci. USA* **85**, 6652–6656.

Madsen, O. D., Nielsen, J. H., Michelsen, B., Westermark, P., Betsholtz, C., Nishi, M., and Steiner, D. F. (1991). Islet amyloid polypeptide and insulin expression are controlled differently in primary and transformed islet cells. *Mol. Endocrinol.* **5**, 143–148.

Madsen, O. D., Karlsen, C., Nielsen, E., Lund, K., Kofod, H., Welinder, B., Rehfeld, J. F., Larsson, L. I., Steiner, D. F., and Holst, J. J. (1993). The dissociation of tumor-induced weight loss from hypoglycemia in a transplantable pluripotent rat islet tumor results in the segregation of stable alpha- and beta-cell tumor phenotypes. *Endocrinology* **133**, 2022–2030.

Madsen, O. D., Jensen, J., Blume, N., Petersen, H. V., Lund, K., Karlsen, C., Andersen, F. G., Jensen, P. B., Larsson, L. I., and Serup, P. (1996). Pancreatic development and maturation of the islet B cell. Studies of pluripotent islet cultures. [Review] [125 refs]. *Eur. J. Biochem.* **242**, 435–445.

Madsen, O. D., Jensen, J., Petersen, H. V., Pedersen, E. E., Oster, A., Andersen, F. G., Jorgensen, M. C., Jensen, P. B., Larsson, L. I., and Serup, P. (1997). Transcription factors contributing to the pancreatic beta-cell phenotype. [Review] [60 refs]. *Horm. Metabol. Res.* **29**, 265–270.

Mansouri, A., Hallonet, M., and Gruss, P. (1996). Pax genes and their roles in cell differentiation and development [Abstract]. *Curr. Opin. Cell Biol.* 8(6), 851–857.

Massague, J. (1996). TGFbeta signaling: receptors, transducers, and Mad proteins. *Cell* 85, 947–950.

Matsuo, T., Osumi-Yamashita, N., Noji, S., Ohuchi, H., Koyama, E., Myokai. F., Matsuo, N., Taniguchi, S., Doi, H., Iseki, S., Ninomiya T., Fujwara M., Watanabe T., Eto K. (1993). A mutation in the Pax6 gene in rat small ey is associated with impaired migration of midbrain crest cells. [Abstract]. *Nature Genet.* 3(4), 299–304.

Mellentin, J. D., Smith, S. D., and Cleary, M. L. (1989). *lyl-1*, a novel gene altered by chromosomal translocation in T cell leukemia, codes for a protein with a helix-loop-helix DNA binding motif. *Cell* 58, 77–83.

Miller, C. P., McGehee, R. E., Jr., and Habener, J. F. (1994). IDX-1: a new homeodomain transcription factor expressed in rat pancreatic islets and duodenum that transactivates the somatostatin gene. *EMBO J.* 13, 1145–1156.

Miralles, F., Czernishow, P., and Scharfman, R. (1998). Follistatin regulates the relative proportions of endocrine versus exocine tissue during pancreatic development. *Development* 125, 1017–1024.

Moller, C. J., Christgau, S., Williamson, M. R., Madsen, O. D., Niu, Z.-P., Bock, E., and Baekkeskov, S. (1992). Differential expression of neural cell adhesion molecule and cadherins in pancreatic islets, glucagonomoas and insulinomas. *Mol. Endocrinol.* 6, 1332–1342.

Moltz, J. H., and McDonald, J. K. (1985). Neuropeptide Y: direct and indirect action on insulin scretion in the rat. *Peptides* 6, 1155–1159.

Montminy, M., Brindle, P., Arias, J., Ferreri, K., and Armstrong, R. (1996). Regulation of somatostatin gene transcription by cAMP. [Review] [20 refs]. *Adv. Pharmacol.* 36, 1–13.

Morel, C., Cordier-Bussat, M., and Philippe, J. (1995). The upstream promoter element of the glucagon gene, G1, confers pancreatic alpha cell-specific expression. *J. Biol. Chem.* 270, 3046–3055.

Motta, P. M. (1997). Histology of the Exocine Pancreas. *Microsc. Res. Techn.* 37, 384–398.

Mutoh, H., Fung, B. P., Naya, F. J., Tsai, M. J., Nishitani, J., and Leiter, A. B. (1997). The basic helix-loop-helix transcription factor BETA2/NeuroD is expressed in mammalian enteroendocrine cells and activates secretin gene expression. *Proc. Natl. Acad. Sci. USA* 94, 3560–3564.

Myers, M. G., and White, M. F. (1996). Insulin signal Transduction and the IRS proteins. *Annu. Rev. Pharmacol. Toxicol.* 36, 615–658.

Myrsén-Axcrona, U., Ekblad, E., and Sundler, F. (1997). Developmental expression of NPY, PYY and PP in the rat pancreas and their coexistance with islet hormones. *Regul. Pept.* 68, 165–167.

Nadeau, A., Grondin, G., and Blouin, R. (1997). In situ hybridization analysis of ZPK gene expression during murine embryogenesis. *J. Histochem. Cytochem.* 45, 107–118.

Naya, F. J., Stellrecht, C. M., and Tsai, M. J. (1995). Tissue-specific regulation of the insulin gene by a novel basic helix-loop-helix transcription factor. *Genes Devel.* 9, 1009–1019.

Naya, F. J., Huang, H. P., Qiu, Y., Mutoh, H., DeMayo, F. J., Leiter, A. B., and Tsai, M. J. (1997). Diabetes, defective pancreatic morphogenesis, and abnormal enteroendocrine differentiation in BETA2/neuroD-deficient mice. *Genes Devel.* 11, 2323–2334.

Nir, U., Walker, M. D., and Rutter, W. J. (1986). Regulation of rat insulin I gene expression: evidence for negative regulation in nonpancreatic cells. *Proc. Natl. Acad. Sci. USA* 83, 3180–3184.

Odagiri, H., Wang, J., and German, M. S. (1996). Function of the human insulin promoter in primary cultured islet cells. *J. Biol. Chem.* 271, 1909–1915.

Offield, M. F., Jetton, T. L., Labosky, P. A., Ray, M., Stein, R. W., Magnuson, M. A., Hogan, B. L., and Wright, C. V. (1996). PDX-1 is required for pancreatic outgrowth and differentiation of the rostral duodenum. *Development* 122, 983–995.

Ohlsson, H., Karlsson, K., and Edlund, T. (1993a). IPF-1, a homeodomain-containing transactivator of the insulin gene. *EMBO J.* **12,** 4251–4259.

Ohlsson, H., Karlsson, K., and Edlund, T. (1993b). IPF1, a homeodomain-containing transactivator of the insulin gene. *EMBO Journal* **12,** 4251–4259.

Oliver, G., Sosa-Pineda, B., Geisendorf, S., Spana, E. P., Doe, C. Q., and Gruss, P. (1993). Prox 1, a prospero-related homeobox gene expressed during mouse development. *Mech. Devel.* **44,** 3–16.

Olson, L. K., Sharma, A., Peshavaria, M., Wright, C. V., Towle, H. C., Rodertson, R. P., and Stein, R. (1995). Reduction of insulin gene transcription in HIT-T15 beta cells chronically exposed to a supraphysiologic glucose concentration is associated with loss of STF-1 transcription factor expression [published erratum appears in *Proc. Natl. Acad. Sci. USA* (1995) Nov 21;92(24):11322]. *Proc. Natl. Acad. Sci. USA* **92,** 9127–9131.

Pabst, O., Schneider, A., Brand, T., and Arnold, H. H. (1997). The mouse Nkx2-3 homeodomain gene is expressed in gut mesenchyme during pre- and postnatal mouse development. *Devel. Dynam.* **209,** 29–35.

Pang, K., Mukonoweshuro, C., and Wong, G. G. (1994). Beta cells arise from glucose transporter type 2 (Glut2)-expressing epithelial cells of the developing rat pancreas. *Proc. Natl. Acad. Sci. USA* **91,** 9559–9563.

Park, C. W., and Walker, M. D. (1992). Subunit structure of cell-specific E box-binding proteins analyzed by quantitation of electrophoretic mobility shift. *J. Biol. Chem.* **267,** 15642–15649.

Peers, B., Leonard, J., Sharma, S., Teitelman, G., and Montminy, M. R. (1994). Insulin expression in pancreatic islet cells relies on cooperative interactions between the helix loop helix factor E47 and the homeobox factor STF-1. *Mol. Endocrinol.* **8,** 1798–1806.

Peers, B., Sharma, S., Johnson, T., Kamps, M., and Montminy, M. (1995). The pancreatic islet factor STF-1 binds cooperatively with Pbx to a regulatory element in the somatostatin promoter: importance of the FPWMK motif and of the homeodomain. *Mol. Cell. Biol.* **15,** 7091–7097.

Peluso, J. J. (1997). Putative mechanism through N-cadherin-mediated cell contact maintains calcium homeostasis and thereby prevent ovarian cells from undergoing apoptosis [Abstract]. *Biochem. Pharmacol.* **54**(8), 847–853.

Petersen, H. V., Serup, P., Leonard, J., Michelsen, B. K., and Madsen, O. D. (1994). Transcriptional regulation of the human insulin gene is dependent on the homeodomain protein STF1/IPF1 acting through the CT boxes. *Proc. Natl. Acad. Sci. USA* **91,** 10465–10469.

Petterson, M., Ahrén, B., Lundquist, I., Böttcher, G., and Sundler, F. (1987). Neuropeptide Y: intrapancreatic neuronal localisation and effets on insulin secretion in the mouse. *Cell Transplant.* **248,** 43–48.

Peyton, M., Moss, L. G., and Tsai, M. J. (1994). Two distinct class A helix-loop-helix transcription factors, E2A and BETA1, form separate DNA binding complexes on the insulin gene E box. *J. Biol. Chem.* **269,** 25936–25941.

Peyton, M., Stellrecht, C. M., Naya, F. J., Huang, H. P., Samora, P. J., and Tsai, M. J. (1996). BETA3, a novel helix-loop-helix protein, can act as a negative regulator of BETA2 and MyoD-responsive genes. *Mol. Cell. Biol.* **16,** 626–633.

Pfaff, S. L., Mendelsohn, M., Stewart, C. L., Edlund, T., and Jessell, T. M. (1996). Requirement for LIM homeobox gene Isl1 in motor neuron generation reveals a motor neuron-dependent step in interneuron differentiation. *Cell* **84,** 309–320.

Philippe, J. (1995). Hepatocyte-nuclear factor 3 beta gene transcripts generate protein isoforms with different transactivation properties on the glucagon gene. *Mol. Endocrinol.* **9,** 368–374.

Philippe, J., and Missotten, M. (1990). Functional characterization of a cAMP-responsive element of the rat insulin I gene. *J. Biol. Chem.* **265,** 1465–1469.

Philippe, J., Powers, A. C., Mojsov, S., Drucker, D. J., Comi, R., and Habener, J. F. (1988). Expression of peptide hormone genes in human islet cell tumors. *Diabetes* **37,** 1647–1651.

Philippe, J., Morel, C., and Prezioso, V. R. (1994). Glucagon gene expression is negatively regulated by hepatocyte nuclear factor 3 beta. *Mol. Cell. Biol.* **14**, 3514–3523.

Philippe, J., Morel, C., and Cordier-Bussat, M. (1995). Islet-specific proteins interact with the insulin-response element of the glucagon gene. *J. Biol. Chem.* **270**, 3039–3045.

Pictet, R., and Rutter, W. J. (1972). Development of the embryonic endocrine pancreas. *In* "Handbook of Physiology" (Williams and Wilkins, eds.), Vol 1, pp. 25–66. American Physiological Society, Washington DC.

Placzek, M., and Furley, A. (1996). Patterning cascades in the neural tube. Neural development. [Review] [17 refs]. *Curr. Biol.* **6**, 526–529.

Placzek, M., Tessier-Lavigne, M., Yamada, T., Dodd, J., and Jessell, T. M. (1990). Guidance of developing axons by diffusible chemoattractants. *Cold Spring Harbor Symposia on Quantitative Biology* **55**, 279–289.

Placzek, M., Yamada, T., Tessier-Lavigne, M., Jessell, T., and Dodd, J. (1991). Control of dorsoventral pattern in vertebrate neural development: induction and polarizing properties of the floor plate. *Development (Suppl.* 2), 105–122.

Poitout, V., Olson, L. K., and Robertson, R. P. (1996). Chronic exposure of betaTC-6 cells to supraphysiologic concentrations of glucose decreases binding of the RIPE3b1 insulin gene transcription activator. *J. Clin. Invest.* **97**, 1041–1046.

Pontoglio, M., Sreenan, S., Roe, M., Pugh, W., Ostrega, D., Doyen, A., Pick, A. J., Baldwin, A., Velho, G., Froguel, P., Levisetti, M., Bonner-Weir, S., Bell, G. I., Yaniv, M., Polonsky, K. S. (1998). Defective insulin secretion in hepatocyte nuclear factor 1alpha-deficient mice. *J. Clin. Invest.* **101**, 2215–2222.

Poulin, G., Turgeon, B., and Drouin, J. (1997). NeuroD1/beta2 contributes to cell-specific transcription of the proopiomelanocortin gene. *Mol. Cell. Biol.* **17**, 6673–6682.

Powers, A. C., Tedeschi, F., Wright, K. E., Chan, J. S., and Habener, J. F. (1989). Somatostatin gene expression in pancreatic islet cells is directed by cell-specific DNA control elements and DNA-binding proteins. *J. Biol. Chem.* **264**, 10048–10056.

Price, M., Lazzaro, D., Pohl, T., Mattei, M. G., Ruther, U., Olivo, J. C., Duboule, D., and Di Lauro, R. (1992). Regional expression of the homeobox gene Nkx-2.2 in the developing mammalian forebrain. *Neuron* **8**, 241–255.

Quiring, R., Walldorf, U., Kloter, U., and Gehring, W. J. (1994). Homology of the eyeles gene of Drosophila to the Small eye gene in mice and Aniridia in humans [Abstract]. *Science* **265**(5173), 785–789.

Rauskolb, C., Peifer, M., and Wieschaus, E. (1993). *extradenticle* a regulator of homeotic gene activity, is a homolog of the homeobox-containing human proto-oncogene pbx1. *Cell* **74**, 1101–1112.

Reecy, J. M., Yamada, M., Cummings, K., Sosic, D., Chen, C. Y., Eichele, G., Olson, E. N., and Schwartz, R. J. (1997). Chicken Nkx-2.8: a novel homeobox gene expressed in early heart progenitor cells and pharyngeal pouch-2 and -3 endoderm. *Devel. Biol.* **188**, 295–311.

Rinkwitz-Brandt, S., Justus, M., Oldenettel, I., Arnold, H. H., and Bober, E. (1995). Distinct temporal expression of mouse Nkx-5.1 and Nkx-5.2 homeobox genes during brain and ear development. *Mech. Devel.* **52**, 371–381.

Roberts, D. J., Johnson, R. L., Burke, A. C., Nelson, C. E., and Morgan, B. A. (1995). Sonic hedgehog is an endodermal signal inducing Bmp-4 and Hox genes during induction and regionalization of the chick hindgut [Abstract]. *Development* **121**, 3163–3174.

Rosenberg, L., Vinik, A. I., Pittenger, G. L., Rafaeloff, R., and Duguid, W. P. (1996). Islet-cell regeneration in the diabetic hamster pancreas with restoration of normoglycaemia can be induced by a local growth factor(s). *Diabetologia* **39**, 256–262.

Rouiller, D. G., Cirulli, V., and Halban, P. A. (1991). Uvomorulin meidates calcium-dependent aggregationo f islet cells, whereas calcium-independent cell adhesion molecules distinguich between cell types. *Devel. Biol.* **148**, 233–242.

Sander, M., and German, M. S. (1997). The beta cell transcription factors and development of the pancreas. [Review] [129 refs]. *J. Mol. Med.* 75, 327–340.

Sander, M., Neubuser, A., Kalamaras, J., Ee, H. C., Martin, G. R., and German, M. S. (1997). Genetic analysis reveals that PAX6 is required for normal transcription of pancreatic hormone genes and islet development. *Genes Devel.* 11, 1662–1673.

Sander, M., Kalamaras, J., and German, M. S. (1998). The homeobox gene Nkx6.1 is essential for differentiation of insulin-producing β-cells in the mouse pancreas [Abstract]. Keystone Symposia on Molecular and Cellular Biology, 111.

Scaglia, L., Cahill, C. J., Finegood, D. T., and Bonner-Weir, S. (1997). Apoptosis participates in the remodeling of the endocrine pancreas in the neonatal rat. *Endocrinology* 138, 1736–1741.

Schwab, M. H., Druffel-Augustin, S., Gass, P., Jung, M., Klugmann, M., Bartholomae, A., Rossner, M. J., and Nave, K. A. (1998). Neuronal basic helix-loop-helix proteins (NEX, neuroD, NDRF): spatiotemporal expression and targeted disruption of the NEX gene in transgenic mice. *J. Neurosci.* 18, 1408–1418.

Sciavolino, P. J., Abrams, E. W., Yang, L., Austenberg, L. P., Shen, M. M., and Abate-Shen, C. (1997). Tissue-specific expression of murine Nkx3.1 in the male urogenital system. *Devel. Dynam.* 209, 127–138.

Semb, H., Cremer, H., Radice, G., and Esni, F. (1998). N-CAM and N-cadherin are required for different pancreatic morphogenetic events [Abstract]. Keystone Symposia on Molecular and Cellular Biology.

Serup, P., Petersen, H. V., Pedersen, E. E., Edlund, H., Leonard, J., Petersen, J. S., Larsson, L. I., and Madsen, O. D. (1995). The homeodomain protein IPF-1/STF-1 is expressed in a subset of islet cells and promotes rat insulin 1 gene expression dependent on an intact E1 helix-loop-helix factor binding site. *Biochem. J.* 310, 997–1003.

Sharma, A., Olson, L. K., Robertson, R. P., and Stein, R. (1995). The reduction of insulin gene transcription in HIT-T15 beta cells chronically exposed to high glucose concentration is associated with the loss of RIPE3b1 and STF-1 transcription factor expression. *Mol. Endocrinol.* 9, 1127–1134.

Sharma, S., Jhala, U. S., Johnson, T., Ferreri, K., Leonard, J., and Montminy, M. (1997). Hormonal regulation of an islet-specific enhancer in the pancreatic homeobox gene STF-1. *Mol. Cell. Biol.* 17, 2598–2604.

Shieh, S., and Tsai, M. J. (1991). Cell specific and ubiquitious factors are responsible for the enhancer activity of the rat insulin II gene. *J. Biol. Chem.* 266, 16708–16714.

Shiran, R., Aronheim, A., Rosen, A., Park, C. W., Leshkowitz, D., and Walker, M. D. (1993). Positive and negative regulation of insulin gene transcription. [Review] [36 refs]. *Biochem. Soc. Trans.* 21, 150–154.

Silva, A. J., Simpson, E. M., Takahashi, J. S., Lipp, H., Nakanishi, S., Wehner, J. M., Giese, K. P., Tully, T., Abel, T., Chapman, P. F., Fox, F., Grant, S., Itohara, S., Lathe, R., Mayford, M., McNamara, J. O., Morris, R. J., Picciotto, M., Roder, J., Shin, H.-S., Slesinger, P. A., Storm, D. R., Stryker, M. P., Tonegawa, S., Wang, Y., Wolfer, D. P. (1997). *Neuron* 19, 755–759.

Skoglund, G., Gross, R. A., Bertrand, G. R., Ahrén, B., and Loubatières-Mariani, M. M. (1991). Comparison of effects of neuropeptide Y and norepinephrine on insulin secretion and vascular resistance in perfused rat pancreas. *Diabetes* 40, 660–665.

Slack, J. M. (1995). Developmental biology of the pancreas. [Review] [80 refs]. *Development* 121, 1569–1580.

Smeyne, R. J., Klein, R., Schnapp, A., Long, L. K., Bryant, S., Lewin, A., Lira, S. A., and Barbacid, M. (1994). Severe sensory and sympathetic neuropathies in mice carrying a disrupted Trk/NGF receptor gene [see comments]. *Nature* 368, 246–249.

Soares, M. B., Schon, E., Henderson, A., Karathanasis, S. K., Cate, R., Zeitlin, S., Chirgwin, J., and Efstratiadis, A. (1985). RNA-mediated gene duplication: the rat preproinsulin I gene is a functional retroposon. *Mol. Cell. Biol.* 5, 2090–2103.

Sommer, L., and Ma, Q. A. D. J. (1998). Neurogenins, a novel family of atonal-related bHLH transcription factors, are putative mammalian neuronal determination genes that reveal progenitor cell heterogeneity in the developing CNA and PNS. *Mol. Cell. Neurosci.* **8**, 221–241.

Sosa-Pineda, B., Chowdhury, K., Torres, M., Oliver, G., and Gruss, P. (1997). The Pax4 gene is essential for differentiation of insulin-producing beta cells in the mammalian pancreas. *Nature* **386**, 399–402.

St-Onge, L., Sosa-Pineda, B., Chowdhury, K., Mansouri, A., and Gruss, P. (1997). Pax6 is required for differentiation of glucagon-producing alpha-cells in mouse pancreas. *Nature* **387**, 406–409.

Stellrecht, C. M., DeMayo, F. J., Finegold, M. J., and Tsai, M. J. (1997). Tissue-specific and developmental regulation of the rat insulin II gene enhancer, RIPE3, in transgenic mice. *J. Biol. Chem.* **272**, 3567–3572.

Stoffers, D. A., Zinkin, N. T., Stanojevic, V., Clarke, W. L., and Habener, J. F. (1997). Pancreatic agenesis attributable to a single nucleotide deletion in the human IPF1 gene coding sequence. *Nature Genet.* **15**, 106–110.

Stoykova, A., Gotz, M., Gruss, P., and Price, J. (1997). PAX6-dependedent regulation of adhesive patterning, R-cadherin expression and boundary formation indeveloping forebrain. *Development* **124**, 3765–3777.

Strom, A., Castella, P., Rockwood, J., Wagner, J., and Caudy, M. (1997). Mediation of NGF signaling by post-translational inhibition of HES-1, a basic helix-loop-helix repressor of neuronal differentiation. *Genes Dev.* **11**, 3168–3181.

Sussel, L., Kalamaras, J., Hartigan-O'Connor, D. J., Meneses, J. J., Pederson, R. A., Rubenstein, J. L. R., and German, M. S. (1998a). Mice lacking Nkx-2.2 have diabetes due to arrested differentiation of pancreatic β cells [Abstract]. Keystone Symposia on Molecular and Cellular Biology, 113.

Sussel, L., Kalamaras, J., Hartigan-O'Connor, D. J., Meneses, J. J., Pederson, R. A., Rubenstein, J. L. R., and German, M. S. (1998b). Mice lacking the homeodomain transcription factor Nkx2.2 have diabetes due to arrested differentiation of pancreatic beta cells. *Development* **125**, 2213–2221.

Teitelman, G. (1993). On the origin of pancreatic endocrine cells, proliferation and neoplastic transformation. [Review] [19 refs]. *Tumour Biol.* **14**, 167–173.

Teitelman, G. (1996). Induction of β-cell Neogenesis by Islet Injury [Abstract]. *Diabetes/Metab. Rev.* **12**(2), 91–102.

Teitelman, G., Alpert, S., Polak, J. M., Martinez, A., and Hanahan, D. (1993). Precursor cells of mouse endocrine pancreas coexpress insulin, glucagon and the neuronal proteins tyrosine hydroxylase and neuropeptide Y, but not pancreatic polypeptide. *Development* **118**, 1031–1039.

Tickle, C., and Eichele, G. (1996). Vertebrate limb development. *Annu. Rev. Physiol.* **10**, 121–152.

Tsutsui, H., Sakatsume, O., Itakura, K., and Yokoyama, K. K. (1996). Members of the MAZ family: a novel cDNA clone for MAZ from human pancreatic islet cells. *Biochem. Biophys. Res. Commun.* **226**, 801–809.

Turque, N., Plaza, S., Radvanyi, F., Carriere, C., and Saule, S. (1994). Pax-QNR/Pax6, a paired box- and homeobox-containing gene expressed in neurons, is also expressed in pancreatic endocrine cells. *Mol. Endocrinol.* **8**(7), 929–938.

Upchurch, B. H., Aponte, G. W., and Leiter, A. B. (1994). Expression of peptide YY in all four islet cell types in the developing mouse pancreas suggests a common peptide YY in all four islet cell types in the developing mouse pancreas suggests a common peptide YY-producing progenitor. *Development* **120**, 245–252.

Vallejo, M., Miller, C. P., and Habener, J. F. (1992a). Somatostatin gene transcription regulated by a bipartite pancreatic islet D-cell-specific enhancer coupled synergetically to a cAMP response element. *J. Biol. Chem.* **267**, 12868–12875.

Vallejo, M., Penchuk, L., and Habener, J. F. (1992b). Somatostatin gene upstream enhancer element activated by a protein complex consisting of CREB, Isl-1-like, and alpha-CBF-like transcription factors. *J. Biol. Chem.* **267**, 12876–12884.

Vallejo, M., Miller, C. P., Beckman, W., and Habener, J. F. (1995). Repression of somatostatin gene transcription mediated by two promoter silencer elements. *Mol. Cell. Endocrinol.* **113**, 61–72.

Vionnet, N., Stoffel, M., Takeda, J., Yasuda, K., Bell, G. I., Zouali, H., Lesage, S., Velho, G., Iris, F., and Passa, P. (1992). Nonsense mutation in the glucokinase gene causes early-onset non-insulin-dependent diabetes mellitus. *Nature* **356**, 721–722.

Waeber, G., Thompson, N., Nicod, P., and Bonny, C. (1996). Transcriptional activation of the GLUT2 gene by the IPF-1/STF-1/IDX-1 homeobox factor. *Mol. Endocrinol.* **10**, 1327–1334.

Walker, M. D., Edlund, T., Boulet, A. M., and Rutter, W. J. (1983). Cell-specific expression controlled by the 5′-flanking region of insulin and chymotrypsin genes. *Nature* **306**, 557–561.

Wang, M., and Drucker, D. J. (1995). The LIM domain homeobox gene isl-1 is a positive regulator of islet cell-specific proglucagon gene transcription. *J. Biol. Chem.* **270**, 12646–12652.

Wang, M., and Drucker, D. J. (1996). Activation of amylin gene transcription by LIM domain homeobox gene isl-1. *Mol. Endocrinol.* **10**, 243–251.

Wang, R. N., Kloppel, G., and Bouwens, L. (1995). Duct- to islet-cell differentiation and islet growth in the pancreas of duct-ligated adult rats. *Diabetologia* **38**, 1405–1411.

Watada, H., Kajimoto, Y., Miyagawa, J., Hanafusa, T., Hamaguchi, K., Matsuoka, T., Yamamoto, K., Matsuzawa, Y., Kawamori, R., and Yamasaki, Y. (1996a). PDX-1 induces insulin and glucokinase gene expressions in alphaTC1 clone 6 cells in the presence of betacellulin. *Diabetes* **45**, 1826–1831.

Watada, H., Kajimoto, Y., Umayahara, Y., Matsuoka, T., Kaneto, H., Fujitani, Y., Kamada, T., Kawamori, R., and Yamasaki, Y. (1996b). The human glucokinase gene beta-cell-type promoter: an essential role of insulin promoter factor 1/PDX-1 in its activation in HIT-T15 cells. *Diabetes* **45**, 1478–1488.

Weintraub, H. (1993). The MyoD family and myogenesis: Redundancy, networks and thresholds. *Cell* **75**, 1241–1244.

Wessells, N. K., and Cohen J. H. (1967). Early pancreas organogenesis:morphogenesis, tissue interactions and mass interations and mass effects. *Devel. Biol.* **15**, 237–270.

Whelan, J., Poon, D., Weil, P. A., and Stein, R. (1989). Pancreatic beta-cell-type specific expression of the rat insulin II gene is controlled by positive and negative cellular transcriptional elements. *Mol. Cell. Biol.* **9**, 3253–3259.

Withers, D. J., Gutierrez, J. S., Towery, H., Burks, D. J., Ren, J.-M., Previs, S., Zhang, Y., Bernal, D., Pons, S., Shulman, G., Bonner-Weir, S., and White, M. F. (1998). Disruption of IRS-2 causes type 2 diabetes in mice. *Nature* **391**, 900–904.

Wrege, A., Diedrich, T., Hochhuth, C., and Knepel, W. (1995a). Transcriptional activity of domain A of the rat glucagon G3 element conferred by an islet-specific nuclear protein that also binds to similar pancreatic islet cell-specific enhancer sequences (PISCES). *Gene Expr.* **4**, 205–216.

Wrege, A., Diedrich, T., Hochhuth, C., and Knepel, W. (1995b). Transcriptional activity of domain A of the rat glucagon G3 element conferred by an islet-specific nuclear protein that also binds to similar pancreatic islet cell-specific enhancer sequences (PISCES). *Gene Expr.* **4**, 205–216.

Wright, C. V. E., and Schnegelsberg, P. & D. R. E. M. (1988). XlHbox8: a novel *Xenopus* homeoprotein restricted to a narrow band of endoderm. *Development (Camb.)* **104**, 787–794.

Wu, K. L., Gannon, M., Peshavaria, M., Offield, M. F., Henderson, E., Ray, M., Marks, A., Gamer, L. W., Wright, C. V., and Stein, R. (1997). Hepatocyte nuclear factor 3beta is involved in pancreatic beta-cell-specific transcription of the pdx-1 gene. *Mol. Cell. Biol.* **17,** 6002–6013.

Xu, X. J., Hao, J. X., Andell-Jonsson, S., Poli, V., Bartfai, T., and Wiesenfeld-Hallin, Z. (1997). Nociceptive responses in interleukin-6-deficient mice to peripheral inflammation and peripheral nerve section. *Cytokine* **9,** 1028–1033.

Yamagata, K., Furuta, H., Oda, N., Kaisaki, P. J., Menzel, S., and Cox, N. J. (1996a). Mutations in the hepatocyte nuclear factor-4alpha gene in maturity-onset diabetes of the young (MODY1) [Abstract]. *Nature* **384**(6608), 458–460.

Yamagata, K., Oda, N., Kaisaki, P. J., Menzel, S., Furuta, H., Vaxillaire, M., Southam, L., Cox, R. D., Lathrop, G. M., Boriraj, V. V., Chen, X., Cox, N. J., Oda, Y., Yano, H., LeBeau, M. M., Yamada, S., Nishigori, H., Takeda, J., Fajans, S. S., Hattersley, A. T., Iwasaki, N., Hansen, T., Pedersen, O., Polonsky, K. S., Bell, G. I. (1996b). Mutations in the hepatocyte nuclear factor-1alphagene in maturity-onset diabetes of the young (MODY3). *Nature* **384,** 455–458.

Yamanaka, Y., Friess, H., Buchler, M., Beger, H. G., Gold, L. I., and Korc, M. (1993). Synthesis and expression of transforming growth factor beta-1, beta-2, and beta-3 in the endocrine and exocrine pancreas. *Diabetes* **42,** 746–756.

Yingling, J. M., Wang, X. F., and Bassing, C. H. (1995). Signaling by the transforming growth factor-beta receptors. [Review] [119 refs]. *Biochim. Biophys. Acta* **1242,** 115–436.

Zaret, K. S. (1996). Molecular genetics of early liver development. [Review] [132 refs]. *Annu. Rev. Physiol.* **58,** 231–251.

A. O. Brinkmann*
J. Trapman†

*Department of Endocrinology and Reproduction
Erasmus University Rotterdam
†Department of Pathology
Erasmus University Rotterdam

Genetic Analysis of Androgen Receptors in Development and Disease

I. Introduction

A. Androgens and Sexual Differentiation

Androgens are important steroid hormones for expression of the male phenotype. They have characteristic roles during male sexual differentiation, during development and maintenance of secondary male characteristics, and during the initiation and maintenance of spermatogenesis (George and Wilson, 1994). The two most important androgens in this respect are testosterone and 5α-dihydrotestosterone (Fig. 1). Each androgen has its own specific role during male sexual differentiation: Testosterone is directly involved in the development and differentiation of the Wolffian duct–derived structures (epididymis, vas deferens, seminal vesicles, and ejaculatory ducts), whereas 5α-dihydrotestosterone, a metabolite of testosterone, is the active ligand in a number of other androgen target tissues, such as the urogenital

Hormones and Signaling

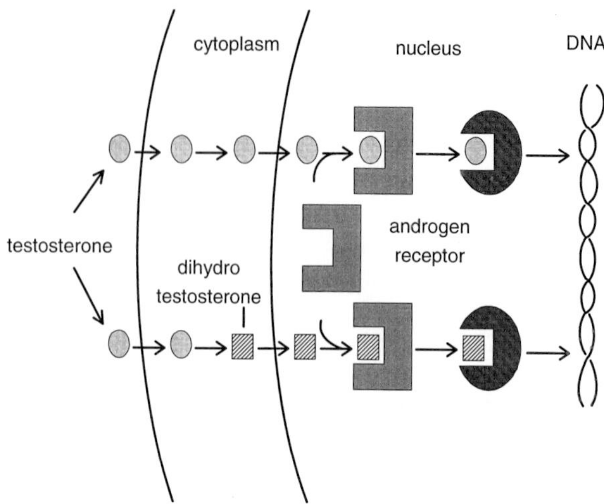

FIGURE I Simplified model for the mechanism of androgen action. Testosterone enters the androgen target cell by passive diffusion and either is metabolized in the cytoplasm to 5α-dihydrotestosterone or interacts directly with the androgen receptor. Both androgens can bind to the same androgen receptor protein.

sinus and tubercle and their derived structures (prostate gland, penis), and in the pilosebacious unit (hair follicles) (Wilson *et al.*, 1993; Randall, 1994). These findings are based on the phenotype of patients with 5α-reductase type II deficiency, the tissue-specific expression of this enzyme, and the measurement of steroid hormone levels in the different target tissues during early development (Imperato-McGinley *et al.*, 1982; Andersson *et al.*, 1991; Wilson *et al.*, 1993). The concept of two hormones for the different actions of androgens has been generally accepted for some time, but recent insights reveal a much more complex picture, involving also estrogens as modulators of androgen action in androgen target tissues (Sharpe, 1997). In that respect it may be more appropriate to speak of three hormones determining the male phenotype.

B. The Androgen Receptor and the Nuclear Receptor Family

The actions of androgens are mediated by the androgen receptor. This ligand-dependent transcription factor belongs to the superfamily of nuclear receptors, including those for the other steroid hormones, the retinoids, the thyroid hormones, and a still growing number of orphan receptors (Evans, 1988; Laudet *et al.*, 1992). In the last decade, since the cloning of the human androgen receptor cDNA, our insights in the mechanism of androgen action have been increased tremendously. Only one androgen receptor cDNA has been identified and cloned, despite the two different ligands (Chang *et al.*, 1988; Lubahn *et al.*, 1988; Trapman *et al.*, 1988; Tilly *et al.*, 1989). The tissue-specific actions of testosterone and 5α-dihydrotestosterone, mediated by the same androgen receptor, suggest a ligand-specific recruitment of transcription intermediary factors (TIFs). However, experimental evidence for ligand-specific TIFs for the androgen receptor has not been provided as yet. The androgen receptor protein displays a large homology in the DNA-binding domain and in the ligand-binding domain with the other members of the steroid hormone receptor subfamily (e.g., receptors for glucocorticoids, estradiol, progesterone and mineralocorticoids; Hollenberg *et al.*, 1985; Green *et al.*, 1986; Misrahi *et al.*, 1987; Arriza *et al.*, 1987; Trapman *et al.*, 1988).

C. The Androgen Receptor Gene

The androgen receptor gene is located on the X-chromosome at Xq11–12 and codes for a protein with a molecular mass of approx. 110 kDa (Brown *et al.*, 1989; Van Laar *et al.*, 1989) (Fig. 2). The gene consists of eight coding exons, and the structural organization is essentially identical to those of the genes coding for the other steroid hormone receptors (e.g., exon/intron boundaries are highly conserved; Kuiper *et al.*, 1989; Lubahn *et al.*, 1989; Keaveney *et al.*, 1991; Zong *et al.*, 1990). The promoter region of the androgen receptor gene contains two transcription initiation sites in a 13-bp region (Tilley *et al.*, 1990; Faber *et al.*, 1991). Homologous-down regulation of androgen receptor gene expression occurs in the prostate, but an androgen response element has not been identified in the promoter region. Also, TATA and CCAAT boxes are lacking. However, the presence and functional involvement of an SP1 site, a purine-rich region, and a cAMP-responsive element in the androgen receptor promoter region have been extensively documented (Faber *et al.*, 1993; Mizokami *et al.*, 1994). Despite this information, it is unknown in which way exactly androgen receptor gene expression is regulated during embryonic development.

As a result of differential splicing in the $3'$-untranslated region, two androgen receptor mRNA species (8.5 and 11 kb, respectively) have been identified in several cell lines (Faber *et al.*, 1991). In the human prostate and in genital skin fibroblasts, predominantly the 11 kb size mRNA is expressed.

D. Androgen Receptor Polymorphism

As indicated in Section I,B, the androgen receptor DNA- and ligand-binding domains have a high homology with the corresponding domains of the other members of the steroid receptor subfamily. Remarkable is the low homology of the NH_2-terminal domain with that of the other steroid receptors (Hollenberg *et al.*, 1985; Green *et al.*, 1986; Misrahi *et al.*, 1987;

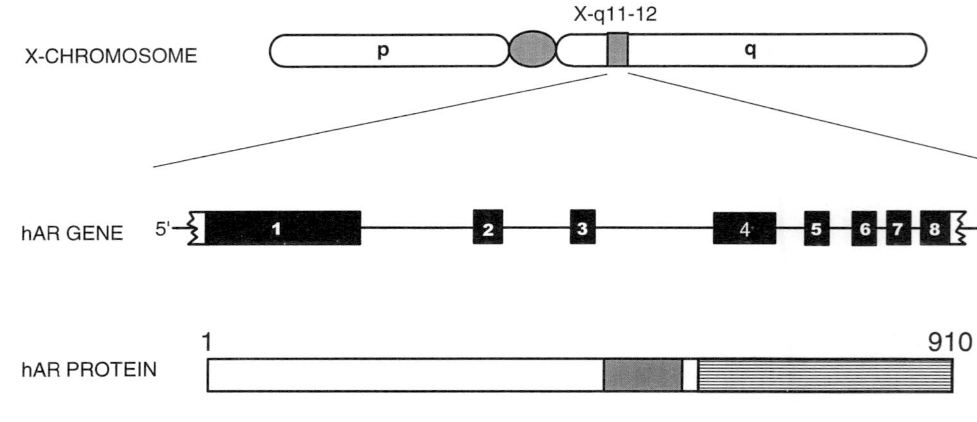

FIGURE 2 Human androgen receptor gene and protein. The androgen receptor gene has been mapped to the long arm of the X-chromosome (locus: Xq11–12). The androgen receptor protein is encoded by eight exons and consists of several distinct functional domains. (DBD = DNA-binding domain; LBD = ligand-binding domain.)

Arriza *et al.*, 1987; Faber *et al.*, 1989). A polyglutamine stretch, encoded by a polymorphic (CAG)$_n$CAA repeat, is present in the NH$_2$-terminal domain (Sleddens *et al.*, 1992) (Fig. 3). The variation of the length of this repeat can be used as a marker for the androgen receptor gene, and also as an X-chromosome marker (Boehmer *et al.*, 1997; Li *et al.*, 1998). Variation in length (9–33 glutamine residues) is observed in the normal population and has been suggested to be associated with a very mild modulation of androgen receptor activity (Nance, 1997). This assumption is based on *in vitro* experiments after transient transfection of androgen receptor cDNA's containing (CAG)$_n$CAA repeats of different lengths (Jenster *et al.*, 1994; Kazemi-Esfarjani *et al.*, 1995). Whether subtle differences in (CAG)$_n$CAA repeat lengths are important for modulation *in vivo* of androgen receptor activity is still a matter of debate.

E. Androgen Receptor Pathology

In general, mutated androgen receptors are directly involved in three pathological situations: (1) androgen insensitivity syndrome (AIS), (2) spinal bulbar muscular atrophy (SBMA), and (3) prostate cancer.

1. Androgen Insensitivity Syndrome

In the X-linked AIS, a large variety of mutations in the androgen receptor gene resulting in amino acid substitutions in the DNA- or ligand-binding domain, or resulting in premature stop codons, have been shown to prevent the normal development of both internal and external male structures in 46,XY individuals (Quigley *et al.*, 1995; Gottlieb *et al.*, 1998).

2. Spinal Bulbar Muscular Atrophy

Spinal and bulbar muscular atrophy is characterized by progressive muscle weakness and atrophy. Clinical symptoms usually manifest in the

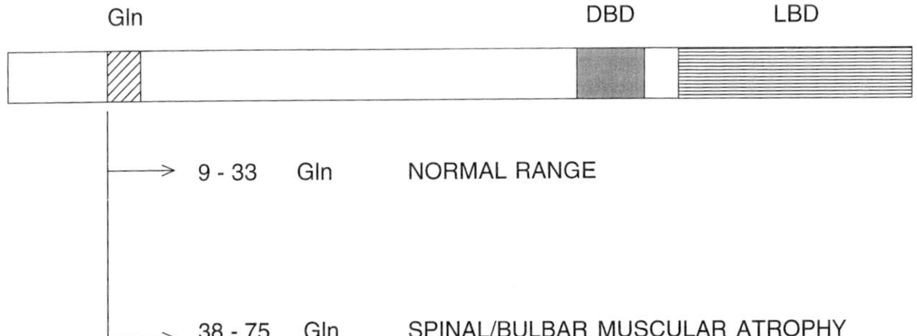

FIGURE 3 Position of the glutamine (Gln) stretch in the amino-terminal domain of the human androgen receptor, which is susceptible to variation in normal individuals (9–33 Glns). The abnormal length (38–75 Gln's) of the polyglutamine stretch is associated with X-linked spinal bulbar muscular atrophy.

third to fifth decade and result from severe depletion of lower motornuclei in spinal cord and brainstem (Kennedy *et al.*, 1968; Robitaille *et al.*, 1997; Nance, 1997). In addition, SBMA patients frequently exhibit endocrinological abnormalities including testicular atrophy, infertility, gynecomastia, and elevated LH, FSH, and estradiol levels. Sex differentiation proceeds normally and characteristics of mild androgen insensitivity appear later in life. The $(CAG)_n CAA$ repeat in exon 1 of the androgen receptor gene is expanded in all investigated SBMA patients and varies between 38 and 75 repeat units (La Spada *et al.*, 1991; Nance, 1997) (Fig. 3). From the clinical signs observed in SBMA, one can conclude that SBMA is likely to result from a combination of a gain-of-function mechanism in motor neurons and a loss-of-function mechanism, causing partial loss of receptor function in androgen target tissues. The observation that neurological symptoms of SBMA are not observed in cases with AIS, which is caused by inactivating mutations of the *AR* gene, is in accordance with this hypothesis. The severity of the motor-neurological symptoms appears to correlate roughly with the $(CAG)_n CAA$ repeat length (Doyu *et al.*, 1994; Igarashi *et al.*, 1992; La Spada *et al.*, 1992).

3. Prostate Cancer

In prostate cancer the initial tumor growth depends on the presence of an activated androgen receptor. In general prostate cancer patients benefit temporarily from androgen ablation therapies. However, the majority of prostate tumors develop into an androgen-independent stage. In these apparently hormone-independent tumors, high nuclear androgen receptor expression levels exist. In part of the hormone-refractory prostate cancers, androgen receptor gene amplification occurs or androgen receptor mutations are found, which can modify the ligand specificity (Visakorpi *et al.*, 1995; Koivisto *et al.*, 1997; Veldscholte *et al.*, 1990; Taplin *et al.*, 1995).

Some studies have indicated that the length of the $(CAG)_n CAA$ repeat in prostate cancer patients is related to the aggressiveness and the onset of prostate cancer. In this case the presence of a shorter repeat (fewer than 19 CAGs) in the androgen receptor gene predicts a more aggressive tumor, with a somewhat earlier onset (Hardy *et al.*, 1996; Giovannucci *et al.*, 1997; Stanford *et al.*, 1997) (Fig. 3). More studies with a larger number of patients are needed to get a more accurate picture of the role of the $(CAG)_n CAA$ repeat length in the etiology of prostate cancer and also in other androgen responses during lifetime.

F. Focus

The aim of the present review is to highlight some aspects in androgen action unravelled in the last 10 years. In this overview a short introduction will be given on functional domain structure of the human androgen receptor with emphasis on the recent findings of functional interactions between the NH$_2$-terminal domain and the ligand-binding domain. Posttranslational modifications (phosphorylation) of the androgen receptor protein in relation to function will be discussed next, followed by androgen-regulated gene expression. Finally, we will concentrate on ligand specificity in a special case: An androgen receptor mutation in prostate cancer. Throughout the text the numbering of the different codons is based on a total number of 910 amino acid residues of the androgen receptor (Brinkmann *et al.*, 1989). This number differs from the amino acid content published by others (Chang *et al.*, 1988; Lubahn *et al.*, 1998; Tilly *et al.*, 1989). These differences are caused by the variation in length of the polyglutamine and polyglycine stretches in the NH$_2$-terminal domain of the receptor.

II. Androgen Receptor Functional Domains ———————

A. Ligand Binding Domain

It is well established for nuclear receptors that the variable NH$_2$-terminal domain is involved in transcription activation and contains a transactivation function AF1 (Gronemeyer, 1992). The centrally located, highly conserved DNA-binding domain mediates the interaction with hormone-response elements on the DNA. The COOH-terminal 250 amino acid residues are involved in ligand binding. This ligand-binding domain is also involved in receptor dimerization and can functionally interact with TIFs (Parker, 1993; Horwitz *et al.*, 1996). Elucidation of the 3D crystallographic structure of the ligand binding domain of several nuclear receptors [e.g., 9-*cis*-retinoic acid receptor alpha (RXRα), all-*trans* retinoic acid receptor gamma (RARγ), thyroid hormone receptor alpha (TRα), estrogen receptor alpha (ERα) and progesterone receptor (PR)] has established that a variable number (10–12, depending on the type of receptor) of α-helices and an antiparallel β sheet arranged in a helical sandwich are involved in formation of the ligand binding pocket and the formation of an interaction surface for TIFs (Bourguet *et al.*, 1995; Renaud *et al.*, 1995; Wagner *et al.*, 1995; Brzozowski *et al.*, 1997; Williams and Sigler, 1998; Feng *et al.*, 1998). Interestingly, despite structural differences and the number of α-helices, a relatively high homology is observed in the 3D structure of the ligand-binding domains of the TRα, ERα, and PR. Helices 3, 5, 7, 11, and 12 and the β-turn are predominantly involved in the formation of the hydrophobic binding pocket, and amino acid residues in these helices are in contact with the ligand. Upon ligand binding, an interaction surface is formed that allows interactions with other proteins (e.g., TIFs). The ligand-binding domains of various nuclear receptors contain a ligand-dependent transactivation function AF2 (Danielian *et al.*, 1992; Barettino *et al.*, 1994; Durand *et al.*, 1994). An autonomous activating domain in this AF2 region, designated as AF-2 AD, is conserved among many nuclear receptors and is located in the COOH-terminal part of the ligand binding domain (Wurtz *et al.*, 1996). A core region in the AF2 AD, located in helix 12, appeared to be important for transcriptional activity and for the hormone-dependent interaction with TIFs (Danielian *et al.*, 1992; Barettino *et al.*, 1994; Lanz and Rusconi, 1994; Montano *et al.*, 1996; Feng *et al.*, 1998; Berrevoets *et al.*, 1998). These TIFs can modulate the transcriptional activity of a broad range of nuclear receptors (Cavailles *et al.*, 1995; LeDouarin *et al.*, 1995; Vom Baur *et al.*, 1996; Voegel *et al.*, 1996; Onate *et al.*, 1995; Yeh and Chang, 1996; Berrevoets *et al.*, 1998). Mutations in the AF2 AD core region abolish the *in vitro* association of the receptor with these TIFs. The recent finding that TIFs display histone acetyltransferase activity has provided further insights in the molecular events occuring at the chromatin level during transcription activation (Jenster *et al.*, 1997). Wurtz *et al.* (1996) proposed a general mechanism for nuclear receptor activation, in which the AF-2 AD core, present in helix 12, plays a central role in the generation of an interaction surface, allowing binding of TIFs to the ligand-binding domain. The residues involved in this surface have been investigated and identified in the TR ligand-binding domain, and it appears that the interaction surface contains a hydrophobic cleft (Feng *et al.*, 1998). The amino acid residues that form the surface cleft are located in helices 3, 4/5, 6, and 12. Formation of this surface occurs upon ligand binding when the COOH-terminal alpha helix 12 is folded against a scaffold of the three other helices (Feng *et al.*, 1998).

For the androgen receptor ligand-binding domain, no information is available with respect to 3D structure, but it can be predicted to resemble the progesterone receptor to a large extent, based on the high homologies of the ligand binding domains of these receptors (Trapman *et al.*, 1988; Tanenbaum *et al.*, 1998).

I. Conformational Changes Induced by Androgens and Antiandrogens

Binding of androgens by the androgen receptor results in two conformational changes of the receptor molecule (Kuil and Mulder, 1994; Kuil *et al.*, 1996). Initially, a fragment of 35 kDa, spanning the complete ligand-binding domain and part of the hinge region, is protected by the ligand, but after prolonged incubation times a second conformational change occurs resulting in protection of a smaller fragment of 29 kDa. In the presence of several antiandrogens (e.g., cyproterone acetate, hydroxyflutamide, and bicalutamide), only the 35-kDa fragment is protected, and no smaller fragments are detectable upon longer incubations. Obviously, the 35-kDa fragment is correlated with an inactive conformation, whereas the second conformational change, only inducible by agonists and considered as the necessary step for transcription activation, is lacking upon binding of anti-androgens. Further analyses with specific antibodies against different epitopes in the 35- and 29-kDa fragments reveal that only the most COOH-terminal end of the androgen receptor protein is represented by the 29-kDa fragment (Kuil *et al.*, 1995).

2. Transactivation Function in the Ligand Binding Domain

Deletion and mutation studies, as well as mutations found in patients with either the androgen insensitivity syndrome or prostate cancer, have given some insight in which amino acid residues are important for ligand binding (Jenster *et al.*, 1991; Quigley *et al.*, 1995; Gottlieb *et al.*, 1998). The overall picture is that large deletions (>10 amino acid residues) severely affect hormone binding, but interestingly deletion of the complete ligand binding domain results in a constitutive active molecule (Jenster *et al.*, 1991, 1995) (Fig. 4). Such COOH-terminal truncated androgen receptor proteins have escaped any hormonal control and could theoretically be prostate growth-promoting molecules in hormone-independent prostate cancer. Until now this type of truncated androgen receptor molecule has not been identified in prostate cancer tumor cell lines or prostate cancer specimens.

In the ligand-binding domain of the human androgen receptor, a transcription activation function (designated as AF2) has been identified, although it is very weak in comparison with that found in other steroid receptors (e.g., estrogen and glucocorticoid receptors; Voegel *et al.*, 1996; Hong *et al.*, 1996; Berrevoets *et al.*, 1998). The AF2 domain in the androgen receptor can be activated in a hormone-dependent way and is strongly enhanced in a promoter dependent-way by the co-activators TIF2 and GRIP1 (Voegel *et al.*, 1996; Hong *et al.*, 1996; Berrevoets *et al.*, 1998) (Fig. 5). The boundaries of the AF2 domain in the androgen receptor ligand-binding domain have not been determined as yet, but it contains the core region as defined in the ligand-binding domains of several members of the ligand-dependent nuclear receptor family. This AF2 activation domain (AD) core region contains the conserved sequence 884-Glu-Met-Met-Ala-Glu-888. Mutations in this region can result in a decrease in activation function without affecting the ligand-binding capability. This indicates that the amino acid residues of the AF2-AD core region are not directly involved in ligand binding, but are part of or are determining the interaction surface. Studies on mutations in this region and the interaction of coactivators have confirmed this presumption (Feng *et al.*, 1998; Berrevoets *et al.*, 1998). Interestingly, mutations have not been reported in the AF2-AD core region of either individuals with the androgen insensitivity syndrome or prostate cancer patients, which implies that none of the individual amino acids in the AF2-AD core region is essential in the full-length androgen receptor.

B. The NH₂-Terminal Domain

I. Transactivation Functions

The boundaries of the NH₂-terminal transactivation domain in the androgen receptor (designated as AF1) are not exactly defined, but generally

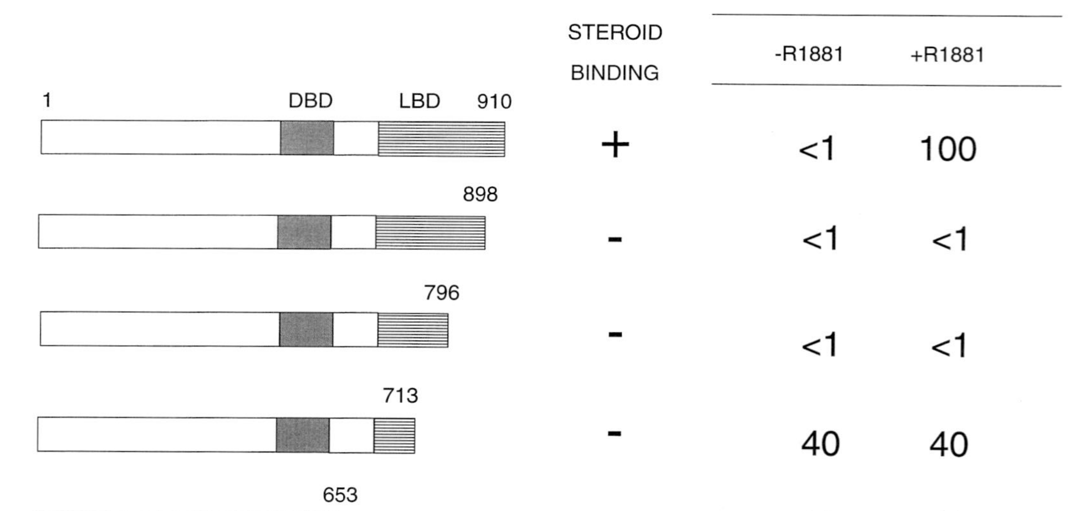

FIGURE 4 Steroid binding and CAT enzyme induction capacity of human androgen receptor COOH-terminal deletion mutants. Transcriptional activity was examined after cotransfection of androgen receptor expression plasmids and a (GRE)2-tk-CAT reporter plasmid in HeLa cells in the absence (−R1881) or presence (+R1881) of 10 nM R1881. Hormone binding was determined by whole cell labeling with ^{3}H-R1881. (DBD = DNA-binding domain; LBD = ligand-binding domain.)

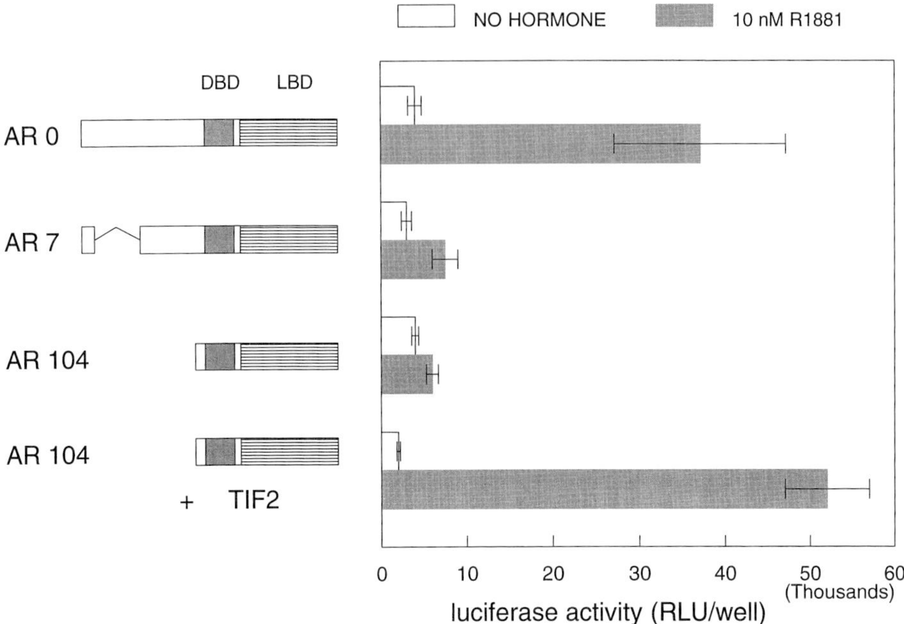

FIGURE 5 Effect of TIF2 on the activity of the transcription activation function (AF2) in the ligand-binding domain of the androgen receptor. CHO cells were transfected either with the wild-type androgen receptor (AR0) or with a COOH-terminal construct lacking the complete NH$_2$-terminal domain (AR104). AR104 activity was tested in the presence (+TIF2) or absence of TIF2. For comparison, a NH$_2$-terminal domain mutant lacking AF1 (AR7) was also transfected. MMTV-LUC was used as a reporter gene. Cells were incubated with vehicle (open bars) or 10 nM R1881 (gray bars). Each bar represents the mean (±SEM) luciferase activity of three experiments. TIF2 expression plasmid was kindly provided by Drs. Gronemeyer and Chambon.

it appears that the region between amino residues 51 and 211 is essential for transactivation activity in the full-length receptor (Jenster *et al.*, 1991). This region is not involved in the transactivation capacity of the COOH-terminal truncated androgen receptor, which displays constitutive activity (Jenster *et al.*, 1995). The most important activating region in the constitutive receptor molecule is located in the NH$_2$-terminal domain between residues 370 and 494. This region is designated as AF5 (Fig. 6).

So, the androgen receptor can use different transactivation domains (AF1 and AF5, respectively, in the NH$_2$-terminal domain and AF2-AD in the COOH-terminal domain) depending on the "form" of the receptor protein (Fig. 7). Two AF functions are ligand dependent (AF1 and AF2), whereas AF5 operates in a ligand-independent way. The ligand dependency of AF1 in the full-length androgen receptor and the switch to AF5 in the COOH-terminal truncated androgen receptor strongly suggest a functional inhibitory action of the ligand-binding domain on AF1 in the absence of

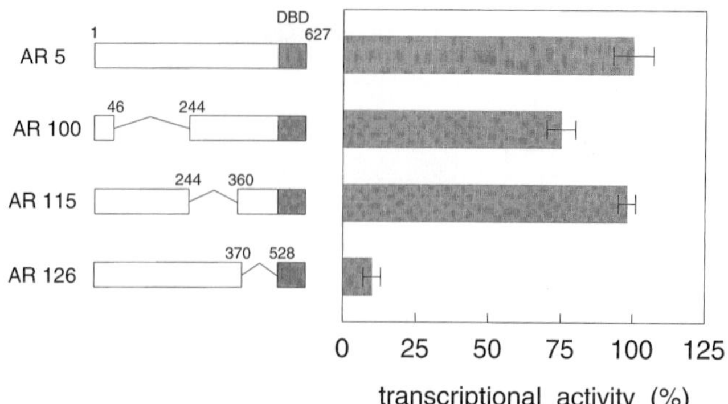

FIGURE 6 Functional analysis of NH$_2$-terminal deletion mutants of the constitutive active androgen receptor construct (AR5). Transcriptional activity was examined by cotransfection of androgen receptor expression plasmids and a (GRE)$_2$-tk-CAT reporter plasmid in HeLa cells cultured in the absence of androgens. Activities were expressed as the percentage relative to that of AR5 and as the mean (±SEM) of five independent assays.

ligand and on AF5 in the presence of ligand. The AF2 function is strongly dependent on the presence of ligand and androgen receptor coactivators.

C. Functional Interaction of the NH$_2$-Terminal Domain and the COOH-Terminal Domain

In the previous section (II,B,1) evidence is presented for a possible interaction between the ligand-binding domain and the AF functions in the NH$_2$-terminal domain. Investigating this NH$_2$-terminal domain–COOH-

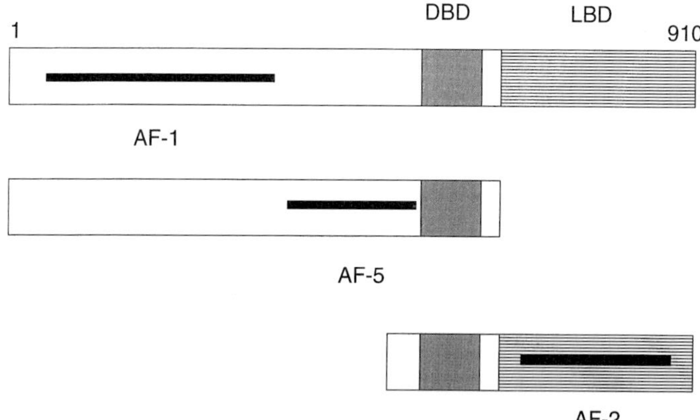

FIGURE 7 Overview of the regions in the NH$_2$-terminal domain and in the COOH-terminal domain of the androgen receptor responsible for the transactivating capacity of the wild-type androgen receptor (AF1), or the ligand-binding domain truncated constitutive active mutant (AF5), or the NH$_2$-terminal domain truncated mutant (AF2).

terminal domain (N/C) interaction in more detail reveals that only certain regions in the NH$_2$-terminal domain are involved in the interaction (Langley *et al.*, 1995, 1997; Doesburg *et al.*, 1997; Berrevoets *et al.*, 1998). Interesting is that the AF1 core region is not involved in this interaction; amino acid residues 3–36 as well as amino acid residues 370–494 are necessary for a proper functional interaction (Fig. 8). In the COOH-terminal domain the AF2-AD core region (amino acid residues: 884Glu-Met-Met-Ala-Glu888) is involved in the interaction as was established by substituting an essential amino acid residue (Glu 888) by a glutamine residue. A similar mutation also affects the functional interaction of the androgen receptor ligand-binding domain with TIF2, suggesting that both the NH$_2$-terminal domain and TIF2 are recognizing the same interaction surface of the ligand-binding domain upon hormone binding (Berrevoets *et al.*, 1998).

D. Dimerization of the Androgen Receptor via the Ligand Binding Domain

Structural analyses of several nuclear receptors have predicted dimerization surfaces for the ligand-binding domain. The size of the dimerization

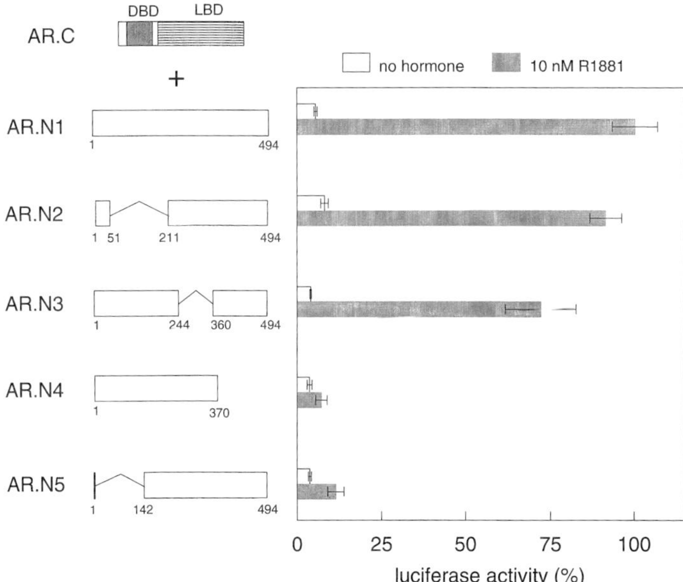

FIGURE 8 Transcriptional activities of androgen receptor NH$_2$-terminal domain deletion mutants AR.N1–AR.N5 cotransfected with a ligand-binding domain construct (AR.C) in CHO cells. Transcriptional activities were determined by cotransfection of the NH$_2$-terminal constructs together with a MMTV-LUC reporter plasmid. Cells were incubated either with vehicle (open bars) or with 10 nM R1881 (closed bars). Each bar represent the mean (±SEM) luciferase activity of three experiments.

surface in the progesterone receptor ligand-binding domain is different in comparison with that of the estrogen receptor (Tanenbaum *et al.*, 1998). Since the androgen receptor ligand-binding domain is more homologous with the progesterone receptor ligand-binding domain than with that of the estrogen receptor, it can be predicted that the dimerization surface of the androgen receptor ligand-binding domain resembles that of the progesterone receptor. Evidence for a dimerization via the ligand-binding domain of the androgen receptor was obtained in a yeast protein–protein interaction system (Doesburg *et al.*, 1997; Berrevoets *et al.*, 1998). The interaction is ligand dependent and relatively weak.

III. Functional Posttranslational Modifications of the Androgen Receptor

A. Hormone-Independent Phosphorylation and Function

The newly synthesized androgen receptor protein migrates as a 110-kDa protein during SDS–PAGE and becomes phosphorylated within 10 min upon synthesis, resulting in an additional protein band at 112 kDa (Kuiper *et al.*, 1991; Jenster *et al.*, 1994). This rapid posttranslational modification is important for the acquisition of the hormone-binding properties of the androgen receptor (Blok *et al.*, 1998). Evidence for this phosphorylation function was obtained from experiments in which dephosphorylation of the 112 kDa was associated with a decreased hormone binding capacity, which could not be explained by an altered K_d value or decreased androgen receptor protein levels (Table I). Dephosphorylation of the 112-kDa isoform of the endogenous androgen receptor in the prostate cancer cell line LNCaP or the transiently expressed androgen receptor in COS-1 cells was accomplished via activation of the protein kinase A pathway, by stimulation of adenylyl cyclase by forskolin (Blok *et al.*, 1998) (Fig. 9). In order to establish which amino acid residues in the androgen receptor were phosphorylated in control and forskolin-treated cells, trypsin-digested androgen receptors were sub-

TABLE I Binding Characteristics of Androgen Receptors in LNCaP Cells Cultured for 24 h in the Absence (Control) or Presence of 20 μM of Forskolin[a]

	Control	Forskolin
K_d (nM)	0.99 ± 0.4	0.72 ± 0.2
Bmax (fmol/mg P)	842 ± 97	459 ± 130

[a] Binding of R1881 was determined by Scatchard analysis.

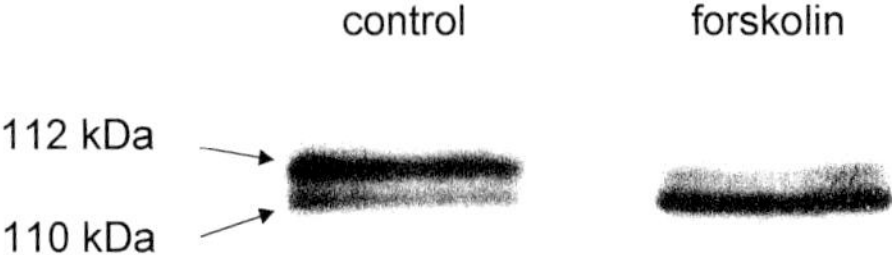

FIGURE 9 Forskolin-induced dephosphorylation of the androgen receptor. LNCaP cells were cultured for 4 h in the absence (control) or presence (forskolin) of 20 μM forskolin. The androgen receptor protein was immunopurified before separation on SDS–PAGE. The two androgen receptor isoforms are indicated by 110 kDa and 112 kDa, respectively.

jected to reverse-phase high pressure liquid chromatography and the isolated phosphorylated peptides to Edman degradation. It was observed that serine residues 506, 641, and 653 were potentially phosphorylated in control cells, whereas after forskolin treatment strong evidence was obtained that phosphorylation of serines 641 and 653 was significantly reduced (Fig. 10). The forskolin-induced dephosphorylation also had consequences for androgen-induced transcription regulation in these cells, as was illustrated by the decreased upregulation of PSA mRNA and the diminished downregulation of the β1-subunit of Na,K-ATPase (Blok *et al.*, 1998).

The mechanism by which forskolin induced the dephosphorylation of the androgen receptor is at present unknown. The swift nature of this process (dephosphorylation within 10 min) suggests that androgen receptor dephosphorylation is an active process that involves activation of phosphatases rather than inhibition of kinases. Indeed, the activity of some phosphatases is known to be regulated by stimulation of protein kinase A. For example, the nuclear protein phosphatase-1 (PP-1N) is activated by protein kinase A–induced phosphorylation of NIPP-1 (nuclear inhibitor of protein phosphatase-1) (Beullens *et al.*, 1993; Van Eynde *et al.*, 1994, 1995). There-

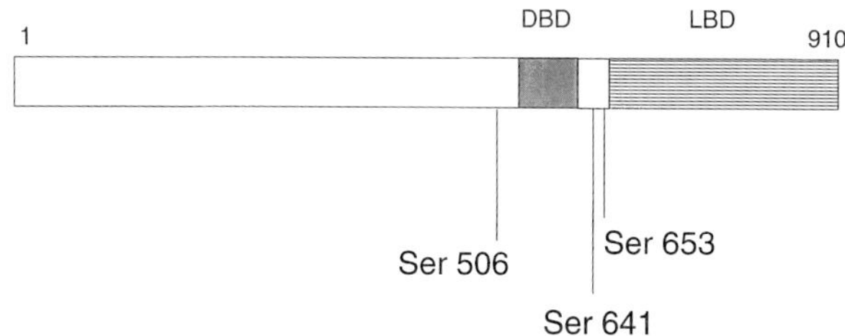

FIGURE 10 Potential serine phosphorylation sites in the androgen receptor protein, which most likely become phosphorylated in the newly synthesized androgen receptor. The casein kinase II consensus sites Ser641 and Ser653 are dephosphorylated by activation of the protein kinase A (PKA) pathway. A possible mechanism for this dephosphorylation event is presented in Fig. 11. The mitogen-activated protein kinase (MAPK) consensus site Ser506 remains phosphorylated upon PKA stimulation.

fore, a possible cascade of events could be the following (Fig. 11): forskolin activates adenylyl cyclase, which in turn activates protein kinase A. The protein kinase A either activates directly a protein phosphatase by a phosphorylation mechanism or inactivates by phosphorylation a protein phosphatase inhibitor, resulting also in activation of protein phosphatase, which ultimately dephosphorylates the androgen receptor. Dephosphorylation of a protein upon stimulation of protein kinase A is not new. For example, ribosomal protein S6 and the retinoblastoma gene product (Rb) are dephosphorylated as a result of stimulation of the protein kinase A pathway (Hara-Yokoyama *et al.*, 1994; Christoffersen *et al.*, 1994).

B. Hormone-Dependent Phosphorylation and Function

A second important phosphorylation step of the androgen receptor occurs upon hormone binding resulting in a third isoform migrating at 114 kDa during SDS–PAGE (Jenster *et al.*, 1994; Brüggenwirth *et al.*, 1997). All three isoforms (e.g., 110, 112, and 114 kDa) exist in several androgen-responsive cell lines in the presence of androgens and migrate as a triplet. The presence of the triplet correlates very well with DNA binding by the androgen receptor, because mutational analysis of certain amino acid residues in the DNA-binding domain that severely affects the DNA-binding properties of the androgen receptor simultaneously displays a defective hormone-induced phosphorylation (Jenster *et al.*, 1994).

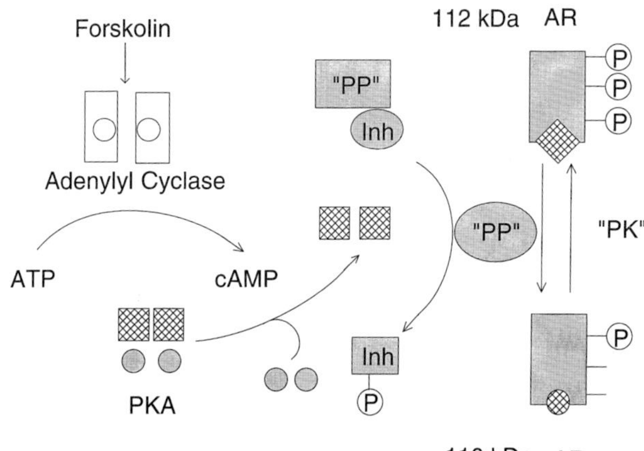

FIGURE 11   Possible cascade of events during forskolin induced dephosphorylation of the androgen receptor in LNCaP cells. The newly synthesized androgen receptor (110-kDa AR) is phosphorylated by a protein kinase ("PK") resulting in the acquisition of ligand binding capabilities and the formation of the 112-kDa form. Forskolin activates adenylyl cyclase, which in turn activates protein kinase A (PKA). Protein kinase A inactivates by phosphorylation a protein phosphatase inhibitor (Inh), resulting in activation of a protein phosphatase ("PP"), which ultimately dephosphorylates the 112-kDa androgen receptor form.

The absence of the androgen receptor triplet in genital skin fibroblasts from a patient with the androgen insensitivity syndrome has been used as indicator for an androgen receptor defective in DNA-binding (Brüggenwirth *et al.*, 1997). In the androgen receptor gene of this patient, a mutation was found in the splice acceptor site of intron 2, resulting in a defective splicing of the androgen receptor mRNA. The mature transcript contained an additional 69 nucleotides between exon 2 and exon 3 sequences. The translation of this altered splice product is a protein with an insertion of 23 amino acid residues between the first and the second zinc cluster of the DNA-binding domain. Additional protein analysis experiments revealed that in genital skin fibroblasts of the index patient, the extended protein was expressed in large quantities. Tight nuclear binding of this mutated receptor protein could not be observed, corresponding with the absence of the triplet isoforms normally seen for wild-type androgen receptors. Only a doublet of 110–112 kDa was expressed, indicating a defective DNA dependent phosphorylation of the human androgen receptor.

The experiments just described indicate that posttranslational modification (e.g., phosphorylation) of the androgen receptor protein might be important at two different steps of receptor activation: (1) during acquisition of ligand binding capabilities and (2) during transformation to the DNA-binding /transcription activation form.

IV. An Androgen Receptor Mutation in a Prostate Cancer Cell Line

Prostate cancer is initially hormone dependent and can be treated successfully with androgen ablation therapy or with anti-androgens. The main goal is either preventing endogenous ligand synthesis or blocking receptor binding with the antagonist. However, essentially all prostate cancers that were initially hormone responsive become hormone independent. The mechanisms underlying this change in hormone dependency is largely unknown. It is remarkable that high levels of androgen receptors are found in the androgen-independent tumors, suggesting some role of the androgen receptor in the androgen-independent growth of the tumor (Van der Kwast *et al.*, 1991; Ruizeveld de Winter *et al.*, 1994).

Several mechanisms have been suggested involving the androgen receptor in the androgen-independent growth of prostate tumors. One mechanism is the changed ligand specificity of the receptor due to mutations in the ligand-binding domain, causing the androgen receptor to become activated by other steroids, such as estrogens, progestagens, anti-androgens, and adrenal androgens. Threonine 868 appeared to be a hot spot in this respect, because it was found to be substituted by alanine in prostate tumors from several different patients (Gottlieb *et al.*, 1998). Initially this mutation was detected in the prostate cancer cell line LNCaP, derived from a lymph node carcinoma of the prostate (Veldscholte *et al.*, 1990). The position of the threonine residue in helix 11 of the ligand-binding domain is a very interesting one, because the corresponding residue in the progesterone receptor ligand-binding domain is cysteine 891, which is involved in interacting with the C_{20}-keto group in progesterone (Trapman *et al.*, 1988; Tanenbaum *et al.*, 1998) (Fig. 12). It can be hypothesized therefore that this position is very important in ligand-binding domains of steroid receptors to discriminate between C_{19} steroids (e.g., androgens) and C_{21} steroids (progesterone, cortisol, aldosterone). Furthermore, in the glucocorticoid and mineralocorticoid receptors (both receptors have C_{21} ligands) the amino acid residue at this position is also a cysteine. If the amino residue threonine 868 in the ligand-binding domain of the androgen receptor is changed into an alanine residue (as found in some prostate cancers), a cysteine residue (as in the progesterone, glucocorticoid, and mineralocorticoid receptors), or a serine residue (as in the estrogen receptor and in a prostate cancer), the androgen receptor ligand-binding domain binding pocket can no longer discriminate between C_{19} steroids and C_{21} steroids.

Another interesting observation with respect to the 868 position is the interaction with the partial androgen agonist/antagonist RU486. This compound is supposed to bind weakly to the androgen receptor. Binding of RU486 to the mutated androgen receptor from LNCaP cells is not different from that to the wild-type receptor, which is strikingly different from other antagonists or partial agonists such as hydroxyflutamide and cyproterone acetate (Kuil *et al.*, 1995). The explanation is most likely that RU486 is interacting in a different way with the androgen receptor ligand-binding domain. This is supported by proteolytic degradation studies of RU486-liganded androgen receptor molecules in which different fragments (e.g., 30 and 25 kDa) are protected by RU486 as compared with normal agonists (e.g.,

         868

hAR: His Gln Phe *Thr* Phe Asp Leu (865 - 871)

hPR: . Leu Tyr *Cys* Leu Asn Thr (888 - 894)

hGR: Leu Asn Tyr *Cys* . Gln Thr (733 - 739)

hER: . Leu Tyr *Ser* Met Lys Cys (524 - 530)

FIGURE 12   Amino acid homologies of parts of the ligand-binding domains of four steroid hormone receptors. The amino acid sequence of the human androgen receptor residues 865–871 (hAR, Brinkmann *et al.*, 1989) was aligned with the sequences of the human progesterone receptor (hPR, Misrahi *et al.*, 1987), the human glucocorticoid receptor (hGR, Hollenberg *et al.*, 1985) and the human estrogen receptor (hER, Green *et al.*, 1986). Identical amino acid residues are indicated by points.

35 and 29 kDa), indicating that this ligand induces different conformational changes in the ligand-binding domain of the androgen receptor. It seems that RU486 does not need the most COOH-terminal part of the receptor (helices 11 and 12) for optimal binding and for partial agonistic/antagonistic action. In this respect the interaction of RU486 with the androgen receptor resemble that of RU486 with the progesterone receptor ligand binding domain (Allan *et al.*, 1992).

V. Concluding Remarks

Androgen action is mediated by the androgen receptor, a ligand-dependent transcription factor belonging to the superfamily of nuclear receptors. The two most important androgens are testosterone and 5α-dihydrotestosterone, and their tissue-specific actions are mediated by the same androgen receptor protein.

Binding of androgens by the androgen receptor results in two consecutive conformational changes, which are different from those induced by anti-androgens.

The androgen receptor can use different transactivation domains (AF1 and AF5, respectively, in the NH_2-terminal domain and AF2-AD in the COOH-terminal domain) depending on the "form" of the receptor protein. The AF2 function is strongly dependent on the presence of nuclear receptor coactivators. Two AF functions are ligand dependent (AF1 and AF2), whereas AF5 operates in a ligand-independent way. The ligand dependency of AF1 in the full-length androgen receptor and the switch to AF5 in the COOH-terminal truncated androgen receptor strongly suggests a functional inhibitory action of the ligand-binding domain on AF1 in the absence of ligand and on AF5 in the presence of ligand. In vivo experiments favour a ligand dependent functional interaction between the AF2 AD core region in the ligand-binding domain with the NH2-terminal domain. This interaction might be either direct or indirect, requiring additional factors, and results in androgen receptor–driven transcription activation.

The androgen receptor protein can undergo two posttranslational modifications during receptor activation: First, upon synthesis the protein is rapidly phosphorylated to acquire hormone binding capacities, and second, upon hormone binding an additional phosphorylation occurs during transformation to the DNA-binding transcription activation form.

Mutated androgen receptors are directly involved in three pathological situations: Androgen insensitivity syndrome, spinal bulbar muscular atrophy, and prostate cancer. In the X-linked androgen insensitivity syndrome, a large variety of mutations in the androgen receptor gene resulting in amino acid substitutions in the DNA- or ligand-binding domain, or resulting in premature stop codons, have prevented the normal development of both internal and external male structures in 46,XY individuals. Spinal and bulbar muscular atrophy is characterized by progressive muscle weakness and atrophy. The $(CAG)_n CAA$ repeat in exon 1 of the androgen receptor gene is expanded in all investigated SBMA patients and varies between 38 and 75 repeat units. From the clinical observations one can conclude that SBMA is likely to result from a combination of a gain-of-function mechanism in motor neurons and a loss-of-function mechanism causing partial loss of receptor function in androgen target tissues. In prostate cancer one particular mutation in helix 11 of the ligand-binding domain has been found to occur most frequently. The mutation (Thr868Ala) has changed the ligand-binding specificity in such a way that the ligand-binding pocket in the androgen receptor ligand-binding domain can no longer discriminate between C_{19} steroids and C_{21} steroids.

Acknowledgments

The authors are grateful for the valuable contributions of Dr. J. H. van Laar, Dr. P. Faber, Dr. J. Veldscholte, Dr. C. Ris-Stalpers, Dr. G. Jenster, Dr. G. G. J. M. Kuiper, Dr. C. Kuil, Dr. H. T. Brüggenwirth, Dr. A. L. M. Boehmer, Dr. K. Steketee, and Dr. L. J. Blok. The excellent technical assistance during the investigations by C. Berrevoets, P. E. de Ruiter, J. A. G. M. van der Korput, P. Doesburg, and M. C. T. Verleun-Mooiman is gratefully acknowledged. This research was supported by the Dutch Cancer Society, The Netherlands Organization for Scientific Research through GB-MW, and the Sophia Foundation for Scientific Research.

References

Allan, G. F., Leng, X., Tsai, S. Y., Weigel, N. L., Edwards, D. P., Tsai, M-J., and O'Malley, B. W. (1992). Hormone and antihormone induce distinct conformational changes which are central to steroid receptor activation. *J. Biol. Chem.* **267**, 19513–19520.

Andersson, S., Berman, D. M., Jenkins, E. P., and Russell, D. W. (1991). Deletion of steroid 5α-reductase 2 gene in male pseudohermaphroditism. *Nature* **354**, 159–161.

Arriza, J. L., Weinberger, C., Cerelli, G., Glaser, T. M., Handelin, B. L., Housman, D. E., and Evans, R. M. (1987). Cloning of human mineralocorticoid receptor complementary DNA: structural and functional kinship with the glucocorticoid receptor. *Science* **237**, 268–275.

Barettino, D., Vivanco Ruiz, M. M., and Stunnenberg, H. G. (1994). Characterization of the ligand-dependent transactivation domain of thyroid hormone receptor. *EMBO J.* **13**, 3039–3049.

Berrevoets, C. A., Doesburg, P., Steketee, K., Trapman, J., and Brinkmann, A. O. (1998). Functional interactions of the AF-2 activation domain core region of the human androgen receptor with the amino-terminal domain and with the transcriptional coactivator TIF(-transcriptional intermediary factor 2). *Mol. Endocrinol.* **12**, 1172–1183.

Beullens, M., Van Eynde, A., Bollen, M., and Stalmans, W. (1993). Inactivation of nuclear inhibitory polypeptides of protein phosphatase-1 (NIPP-1) by protein kinase A. *J. Biol. Chem.* **268**, 13172–13177.

Blok, L. J., De Ruiter, P. E., and Brinkmann, A. O. (1998). Forskolin-induced dephosphorylation of the androgen receptor impairs ligand binding. *Biochemistry* **37**, 3850–3857.

Boehmer, A. L. M., Brinkmann, A. O., Niermeijer, M. F., Bakker, L., Halley, D. J. J., and Drop, S. L. S. (1997). Germ-line and somatic mosaicism in the androgen insensitivity syndrome: Implications for genetic counseling. *Am. J. Hum. Genet.* **60**, 1003–1006.

Bourguet, W., Ruff, M., Chambon, P., Gronemeyer, H., and Moras, D. (1995). Crystal structure of the ligand-binding domain of the nuclear receptor RXR-α. *Nature* **375**, 377–382.

Brinkmann, A. O., Faber, P. W., van Rooy, H. C. J., Kuiper, G. G. J. M., Ris, C., Klaassen, P., van der Korput, J. A. G. M., Voorhorst, M. M., van Laar, J. H., Mulder, E., and Trapman, J. (1989). The human androgen receptor: Domain structure, genomic organization and regulation of expression. *J. Steroid Biochem.* **34**, 307–310.

Brown, C. J., Goss, S. J., Lubahn, D. B., Joseph, D. R., Wilson, E. M., French, F. S., and Willard, H. F. (1989). Androgen receptor locus on the X chromosome: Regional localization to Xq11-12 and description of a DNA polymorphism. *Am. J. Hum. Genet.* **44**, 264–269.

Brüggenwirth, H. T., Boehmer, A. L. M., Verleun-Mooijman, M. C. T., Hoogenboezem, T., Kleijer, W. J., Otten, B. J., Trapman, J., and Brinkmann, A. O. (1996). Molecular basis of androgen insensitivity. *J. Steroid Biochem. Mol. Biol.* **58**, 569–575.

Brüggenwirth, H, T., Boehmer, A. L. M., Ramnarain, S., Verleun-Mooijman, M. C. T., Satijn, D. P. E., Trapman, J., Grootegoed, J. A., and Brinkmann, A.O. (1997). Molecular analysis of the androgen receptor gene in a family with receptor-positive partial androgen insensitivity: An unusual type of intronic mutation. *Am. J. Hum. Genet.* **61**, 1067–1077.

Brzozowski, A. M., Pike, A. C. W., Dauter, Z., Hubbard, R. E., Bonn, T., Engström, O., Öhman, L., Greene, G. L., Gustafsson, J-A., and Carlquist, M. (1997). Molecular basis of agonism and antagonism in the oestrogen receptor. *Nature* **389**, 753–758.

Cavailles, V., Dauvois, S., L'Horset, F., Lopez, G., Hoare, S., Kushner, P. J., and Parker, M. G. (1995). Nuclear factor RIP140 modulates transcriptional activation by the estrogen receptor. *EMBO J.* **14**, 3741–3751.

Chang, C., Kokontis, J., and Liao, S. (1988). Molecular cloning of human and rat complementary DNA encoding androgen receptors. *Science* **240**, 324–326.

Christoffersen, J., Smeland, E. B., Stokke, T., Tasken, K., Andersson, K. B., and Blomhoff, H. K. (1994). Retinoblastoma protein is rapidly dephosphorylated by elevated cyclic adenosine monophosphate levels in human B-lymphoid cells. *Cancer Res.* **54**, 2245–2250.

Danielian, P. S., White, R., Lees, J. A., and Parker, M. G. (1992). Identification of a conserved region required for hormone dependent transcriptional activation by steroid hormone receptors. *EMBO J.* **11**, 1025–1033.

Doesburg, P., Kuil, C. W., Berrevoets, C. A., Steketee, K., Faber, P. W., Mulder, E., Brinkmann, A. O., and Trapman, J. (1997). Functional *in vivo* interaction between the amino-terminal, transactivation domain and the ligand binding domain of the androgen receptor. *Biochemistry* **36**, 1052–1064.

Doyu, M., Sobue, G., Kimata, K., Yamamoto, K., and Mitsuma, T. (1994). Androgen receptor mRNA with increased size of tandem CAG repeat is widely expressed in the neural and non-neural tissues of X-linked recessive bulbospinal neuronopathy. *J. Neurol. Sci.* **127**, 43–47.

Durand, B., Saunders, M., Gaudon, C., Roy, B., Losson, R., and Chambon, P. (1994). Activation function 2 (AF-2) of retinoic acid receptor and 9-cis retinoic acid receptor: presence of a conserved autonomous constitutive activating domain and influence of the nature of the response element on AF-2 activity. *EMBO J.* **13**, 5370–5382.

Evans, R. (1988). The steroid and thyroid hormone receptor superfamily. *Science* **240**, 889–894.

Faber, P. W., Kuiper, G. G. J. M., van Rooij, H. C. J., van der Korput, J. A. G. M., Brinkmann, A. O., and Trapman, J. (1989). The N-terminal domain of the human androgen receptor is encoded by one, large exon. *Mol. Cell. Endocrinol.* **61**, 257–262.

Faber, P. W., van Rooij, H. C. J., van der Korput, H. A. G. M., Baarends, W. M., Brinkmann, A. O., Grootegoed, J. A., and Trapman, J. (1991). Characterization of the human androgen receptor transcription unit. *J. Biol. Chem.* **266**, 10743–10749.

Faber, P. W., van Rooij, H. C. J., Schipper, H. J., Brinkmann, A. O., and Trapman, J. (1993). Two different, overlapping pathways of transcription initiation are active on the TATA-less human androgen receptor promoter: the role of Sp1. *J. Biol. Chem.* **268**, 9296–9301.

Feng, W., Ribeiro, R. C. J., Wagner, R. L., Nguyen, H., Apriletti, J. W., Fletterick, R. J., Baxter, J. D., Kushner, P. J., and West, B. L. (1998). Hormone-dependent coactivator binding to a hydrophobic cleft on nuclear receptors. *Science* **280**, 1747–1749.

George, F. W., and Wilson, J. D. (1994). Sex determination and differentiation. *In* "The Physiology of Reproduction" (E. Knobil and J. D. Neill, eds.), Chapter 1. Raven Press Ltd., New York.

Giovannucci, E., Stampfer, M. J., Krithivas, K., Brown, M., Dahl, D., Brufsky, A., Talcott, J., Hennekens, C. H., and Kantoff, P. W. (1997). The CAG repeat within the androgen receptor gene and its relationship to prostate cancer. *Proc. Natl. Acad. Sci. USA* **94**, 3320–3323.

Gottlieb, B., Lehvaslaiho, H., Beitel, L. K., Lumbroso, R., Pinsky, L., and Trifiro, M. (1998). The androgen receptor gene mutations database. *Nucleic Acids Res.* **26**, 151–154.

Green, S., Walter, P., Kumar, V., Krust, A., Bornert, J. M., Argos, P., and Chambon P. (1986). Human oestrogen receptor cDNA: sequence, expression and homology to v-*erb*-A. *Nature* **320**, 134–139.

Gronemeyer, H. (1992). Control of transcription activation by steroid hormone receptors. *FASEB J.* **6**, 2524–2529.

Hara-Yokoyama, M., Sugiya, H., Furuyama, S., Wang, J. H., and Yokoyama, N. (1994). Dephosphorylation of ribosomal protein S6 phosphorylated via the cAMP-mediated signaling pathway in rat parotid gland: Effect of okadaic acid and Zn2+. *Biochem. Mol. Biol. Int.* **34**, 1177–1187.

Hardy, D. O., Scher, H. I., Bogenreider, T., Sabbatini, P., Zhang, Z-F., Nanus, D. M., and Catterall, J. F. (1996). Androgen receptor CAG repeat lengths in prostate cancer: Correlation with age of onset. *J. Clin. Endocrinol. Metab.* **81**, 4400–4405.

Hollenberg, S. M., Weinberger, C., Ong, E. S., Cerelli, G., Oro, A., Lebo, R., Thompson, E. B., Rosenfeld, M. G., and Evans, R. M. (1985). Primary structure and expression of a functional human glucocorticoid receptor cDNA. *Nature* **318**, 635–641.

Hong, H., Kohli, K., Trivedi, A., Johnson, D. L., and Stallcup, M. R. (1996). GRIP-1, a novel mouse protein that serves as a transcriptional coactivator in yeast for the hormone binding domains of steroid receptors. *Proc. Natl. Acad. Sci. USA* **93**, 4948–4952.

Horwitz, K. B., Jackson, T. A., Bain, D. L., Richer, J. K., Takimoto, G. S., and Tung, L. (1996). Nuclear receptor coactivators and corepressors. *Mol. Endocrinol.* **10**, 1167–1177.

Igarashi, S., Tanno, Y., Onodera, O., Yamazaki, M., Sato, S., Ishikawa, A., Miyatani, N., Nagashima, M., Ishikawa, Y., Sahashi, K., Ibi, T., Miyatake, T., and Tsuji, S. (1992). Strong correlation between the number of CAG repeats in androgen receptor genes and the clinical onset of features of spinal and bulbar muscular atrophy. *Neurology* **42**, 2300–2302.

Imperato-McGinley, J., Peterson, R. E., Gautier, T., Cooper, G., Danner, R., Arthur, A., Morris, P. L., Sweeney, W. J., and Shackleton, C. (1982). Hormonal evaluation of a large kindred with complete androgen insensitivity: Evidence for secondary 5α-reductase deficiency. *J. Clin. Endocrinol. Metab.* **54**, 931–941.

Jenster, G., van der Korput, H. A. G. M., van Vroonhoven, C., van der Kwast, Th. H., Trapman, J., and Brinkmann, A. O. (1991). Domains of the human androgen receptor involved in steroid binding, transcriptional activation and subcellular localization. *Mol. Endocrinol.* **5**, 1396–1404.

Jenster, G., de Ruiter, P. E., van der Korput, H. A. G. M., Kuiper, G. G. J. M., Trapman, J., and Brinkmann, A. O. (1994). Changes in abundance of androgen receptor isotypes: Effects of ligand treatment, glutamine-stretch variation and mutation of putative phosphorylation sites. *Biochemistry* **33**, 14064–14072.

Jenster, G., van der Korput, J. A. G. M., Trapman, J., and Brinkmann, A. O. (1995). Identification of two transcription activation units in the N-terminal domain of the human androgen receptor. *J. Biol. Chem.* **270**, 7341–7346.

Jenster, G., Spencer, T. E., Burcin, M. M., Tsai, S. Y., Tsai, M-J., and O'Malley, B. W. (1997). Steroid receptor induction of gene transcription: A two-step model. *Proc. Natl. Acad. Sci. USA* **94**, 7879–7884.

Kazemi-Esfarjani, P., Trifiro, M. A., and Pinsky, L. (1995). Evidence for a repressive function of the long polyglutamine tract in the human androgen receptor: Possible pathogenic relevance for the (CAG)n-expanded neuronopathies. *Hum. Mol. Genet.* **4**, 523–527.

Keaveney, M., Klug, J., Dawson, M. T., Nester, P. V., Neilan, J. G., Forde, R. C., and Gannon, F. (1991). Evidence for a previously unidentified upstream exon in the human oestrogen receptor gene. *J. Mol. Endocrinol.* **6**, 111–115.

Kennedy, W. R., Alter, M., and Sung, J. H. (1968). Progressive proximal spinal and bulbar muscular atrophy of late onset: a sex-linked recessive trait. *Neurology* **18**, 671–680.

Koivisto, P., Kononen, J., Palmberg, C., Tammela, T., Hyytinen, E., Isola, J., Trapman, J., Cleutjens, K., Noordzij, A., Visakorpi, T., and Kallioniemi, O-P. (1997). Androgen receptor gene amplification: A possible molecular mechanism for androgen deprivation therapy failure in prostate cancer. *Cancer Res.* **57**, 314–319.

Kuil, C. W., and Mulder, E. (1994). Mechanism of antiandrogen action: Conformational changes of the receptor. *Mol. Cell. Endocrinol.* **102**, R1–R5.

Kuil, C. W., Berrevoets, C. A., and Mulder, E. (1995). Ligand-induced conformational alterations of the androgen receptor analyzed by limited trypsinization—Studies on the mechanism of antiandrogen action. *J. Biol. Chem.* **270**, 27569–27576.

Kuiper, G. G. J. M., Faber, P. W., van Rooij, H. C. J., van der Korput, J. A. G. M., Ris-Stalpers, C., Klaassen, P., Trapman, J., and Brinkmann, A. O. (1989). Structural organization of the human androgen receptor gene. *J. Mol. Endocrinol.* **2**, R1–R4.

Kuiper, G. G. J. M., de Ruiter, P. E., Grootegoed, J. A., and Brinkmann, A. O. (1991). Synthesis and post-translational modification of the androgen receptor in LNCaP cells. *Mol. Cell. Endocrinol.* **80**, 65–73.

Langley, E., Zhou, Z.-X., and Wilson, E. M. (1995). Evidence for an anti-parallel orientation of the ligand-activated human androgen receptor dimer. *J. Biol. Chem.* **270**, 29983–29990.

Langley, E., Kemppainen, J. A., and Wilson, E. M. (1998). Intermolecular NH_2-/carboxy-terminal interactions in androgen receptor dimerization revealed by mutations that cause androgen insensitivity. *J. Biol. Chem.* **273**, 92–101.

Lanz, R. B., and Rusconi, S. (1994). A conserved carboxy-terminal subdomain is important for ligand interpretation and transactivation by nuclear receptors. *Endocrinology* **135**, 2183–2195.

La Spada, A. R., Wilson, E. M., Lubahn, D. B., Harding, A. E., and Fischbeck, K. H. (1991). Androgen receptor gene mutations in X-linked spinal and bulbar muscular atrophy. *Nature* **352**, 77–79.

La Spada, A. R., Roling, D. B., Harding, A. E., Warner, C. L., Spiegel, R., Hausmanowa Petrusewicz, I., Yee, W. C., and Fischbeck, K. (1992). Meiotic instability and genotype–phenotype correlation of the trinucleotide repeat in X-linked spinal and bulbar muscular atrophy. *Nat. Genet.* **2**, 301–304.

Laudet, V., Hänni, C., Coll, J., Catzeflis, F., and Stéhelin, D. (1992). Evolution of the nuclear receptor gene superfamily. *EMBO J.* **11**, 1003–1013.

LeDouarin, B., Zechel, C., Garnier, J. M., Lutz, Y., Tora, L., Pierrat, P., Heery, D., Gronemeyer, H., Chambon, P., and Losson, R. (1995). The N-terminal part of TIF1, a putative mediator of the ligand-dependent activation function (AF-2) of nuclear receptors, is fused to B-raf in the oncogenic protein T18. *EMBO J.* **14**, 2020–2033.

Li, S. L., Ting, S. S., Lindeman, R., French, R., and Ziegler, J. B. (1998). Carrier identification in X-linked immunodeficiency diseases. *J. Paediatr. Child Health* **34**, 273–279.

Lubahn, D. B., Joseph, D. R., Sullivan, P. M., Willard, H. F., French, F. S., and Wilson, E. M. (1988). Cloning of human androgen receptor complementary DNA and localization to the X chromosome. *Science* **240**, 327–330.

Lubahn, D. B., Brown, T. R., Simental, J. A., Higgs, H. N., Migeon, C. J., Wilson, E. M., and French, F. S. (1989). Sequence of the intron/exon junctions of the coding region of the human androgen receptor gene and identification of a point mutation in a family with complete androgen insensitivity. *Proc. Natl. Acad. Sci. USA* **86**, 9534–9538.

Misrahi, M., Atger, M., d'Auriol, L., Loosfelt, H., Meriel, C., Fridlansky, F., Guiochon-Mantel, A., Galibert, F., and Milgrom, E. (1987). Complete amino acid sequence of the human progesterone receptor deduced from cloned cDNA. *Biochem. Biophys. Res. Commun.* **143**, 740–748.

Mizokami, A., Yeh, S-Y., and Chang, C. (1994). Identification of $3'5'$-cyclic adenosine monophosphate response element and other *cis*-acting elements in the human androgen receptor gene promoter. *Mol. Endocrinol.* **8**, 77–88.

Montano, M. M., Ekena, K., Krueger, K. D., Keller, A. L., and Katzenellenbogen, B. S. (1996). Human estrogen receptor ligand activity inversion mutants: receptors that interpret antiestrogens as estrogens and discriminate among different antiestrogens. *Mol. Endocrinol.* **10**, 230–242.

Nance, M. A. (1997). Clinical aspects of CAG repeat diseases. *Brain Pathol.* **7**, 881–900.

Onate, S. A., Tsai, S. Y., Tsai, M-J., and O'Malley, B. W. (1995). Sequence and characterization of a coactivator for the steroid hormone receptor superfamily. *Science* **270**, 1354–1357.

Parker, M. G. (1993). Steroid and related receptors. *Curr. Opin. Cell Biol.* **5**, 499–504.

Quigley, C. A., De Bellis, A., Marschke, K. B., El-Awady, M. K., Wilson, E. M., and French, F. S. (1995). Androgen receptor defects: historical, clinical and molecular perspectives. *Endocr. Rev.* **16**, 271–321.

Randall, V. A. (1994). Androgens and human hair growth hair growth. *Clin. Endocrinol.* **40**, 439–457.

Renaud, J.-P., Rochel, N., Ruff, M., Vivat, V., Chambon, P., Gronemeyer, H., and Moras, D. (1995). Crystal structure of the RAR-γ ligand-binding domain bound to all-*trans* retinoic acid. *Nature* **378**, 681–689.

Robitaille, Y., Lopes-Cendes, I., Becher, M., Rouleau, G., and Clark, A. W. (1997). The neuropathology of CAG-repeat diseases:review and update of genetic and molecular features. *Brain Pathol.* **7**, 901–626.

Ruizeveld de Winter, J. A., Janssen, P. J. A., Sleddens, H. M. E. B., Verleun-Mooijman, M. C. T., Trapman, J., Brinkmann, A. O., Santerse, A. B., Schroder, F. H., and van der Kwast, Th. H. (1994). Androgen receptor status in localized and locally progressive hormone refractory human prostate cancer. *Am. J. Pathol.* **144**, 735–746.

Sharpe, R. M. (1997). Do males rely on female hormones? *Nature* **390**, 447–448.

Sleddens, H. F. B. M., Oostra, B. A., Brinkmann, A. O., and Trapman, J. (1992). Trinucleotide repeat polymorphism in the androgen receptor gene (AR). *Nucleic Acids Res.* **20**, 1427.

Stanford, J. L., Just, J. J., Gibbs, M., Wicklund, K. G., Neal, C. L., Blumenstein, B. A., and Ostrander, E. A. (1997). Polymorphic repeats in the androgen receptor gene: Molecular markers of prostate cancer risk. *Cancer Res.* **57**, 1194–1198.

Tanenbaum, D. M., Wang, Y., Williams, S. P., and Sigler, P. B. (1998). Crystallographic comparison of the estrogen and progesterone receptor's ligand binding domains. *Proc. Natl. Acad. Sci. USA* **95**, 5998–6003.

Taplin, M. E., Bubley, G. J., Shuster, T. D., Frantz, M. E., Spooner, A. E., Ogata, G. K., Keer, H. N., and Balk, S. P. (1995). Mutation of the androgen receptor gene in metastatic androgen-independent prostate cancer. *N. Engl. J. Med.* **332**, 1393–1398.

Tilley, W. D., Marcelli, M., Wilson, J. D., and McPhaul, M. J. (1989). Characterization and expression of a cDNA encoding the human androgen receptor. *Proc. Natl. Acad. Sci. USA* **86**, 327–331.

Tilley, W. D., Marcelli, M., and McPhaul, M. J. (1990). Expression of the human androgen receptor gene utilizes a common promoter in diverse human tissues and cell lines. *J. Biol. Chem.* **265**, 13776–13781.

Trapman, J., Klaassen, P., Kuiper, G. G. J. M., van der Korput, J. A. G. M., Faber, P. W., van Rooij, H. C. J., Geurts van Kessel, A., Voorhorst, M. M., Mulder, E., and Brinkmann, A. O. (1988). Cloning, structure and expression of a cDNA encoding the human androgen receptor. *Biochem. Biophys. Res. Commun.* **153**, 241–248.

Van der Kwast, Th. H., Schalken, J., Ruizeveld de Winter, J. A., van Vroonhoven, C. C. J., Mulder, E., Boersma, W., and Trapman, J. (1991). Androgen receptors in endocrine-therapy-resistant human prostate cancer. *Int. J. Cancer* **48**, 189–193.

Van Eynde, A., Beullens, M., Stalmans, W., and Bollen, M. (1994). Full activation of a nuclear species of proetein phosphatase-1 by phosphorylation with protein kinase A and casein kinase-2. *Biochem. J.* **297**, 447–449.

Van Eynde, A., Wera, S., Beullens, M., Torrekens, S., van Leuven, F., Stalmans, W., and Bollen, M. (1995). Molecular cloning of NIPP-1, a nuclear inhibitor of protein phosphatase-1, reveals homology with polypeptides involved in RNA processing. *J. Biol. Chem.* **270**, 28068–28074.

Van Laar, J. H., Bolt-de Vries, J., Voorhorst-Ogink, M. M., and Brinkmann, A. O. (1989). The human androgen receptor is a 110 kDa protein. *Mol. Cell. Endocrinol.* **63**, 39–44.

Veldscholte, J., Ris-Stalpers, C., Kuiper, G. G. J. M., Jenster, G., Berrevoets, C., Claassen, E., van Rooij, H. C. J., Trapman, J., Brinkmann, A. O., and Mulder, E. (1990). A mutation in the ligand-binding domain of the androgen receptor of human LNCaP cells affects steroid binding characteristics and response to antiandrogens. *Biochem. Biophys. Res. Commun.* **173**, 534–540.

Visakorpi, T., Hyytinen, E., Koivisto, P., Tanner, M., Keinanen, R., Palmberg, C., Palotie, A., Tammela, T., Isola, J., and Kallioniemi, O-P. (1995). *In vivo* amplification of the androgen receptor gene and progression of human prostate cancer. *Nat. Genet.* **9**, 401–406.

Voegel, J. J., Heine, M. J., Zechel, C., Chambon, P., and Gronemeyer, H. (1996). TIF2, a 160 kDa transcriptional mediator for the ligand-dependent activation function AF-2 of nuclear receptors. *EMBO J.* **15**, 3667–3675.

Vom Baur, E., Zechel, C., Heery, D., Heine, M. J., Garnier, J. M., Vivat, V., Le Douarin, B., Gronemeyer, H., Chambon, P., and Losson, R. (1996). Differential ligand-dependent interactions between the AF-2 activating domain of nuclear receptors and the putative transcriptional intermediary factors mSUG1 and TIF1. *EMBO J.* **15**, 110–124.

Wagner, R. L., Apriletti, J. W., McGrath, M. E., West, B. L., Baxter, J. D., and Fletterick, R. J. (1995). A structural role for hormone in the thyroid hormone receptor. *Nature* **378**, 690–697.

Williams, S. P., and Sigler, P. B. (1998). Atomic structure of progesterone complexed with its receptor. *Nature* **393**, 392–396.

Wilson, J. D., Griffin, J. E., and Russell, D. W. (1993). Steroid 5α-reductase 2 deficiency. *Endocr. Rev.* **14**, 577–593.

Wurtz, J. M., Bourguet, W., Renaud, J. P., Vivat, V., Chambon, P., and Moras, D. (1996). A canonical structure for the ligand-binding domain of nuclear receptors. *Nature Struct. Biol.* **3**, 87–94.

Yeh, S., and Chang, C. (1996). Cloning and characterization of a specific coactivator, ARA70, for the androgen receptor in human prostate cells. *Proc. Natl. Acad. Sci. USA* **93**, 5517–5521.

Zong, J., Ashraf, J., and Thompson, E. B. (1990). The promoter and first, untranslated exon of the human glucocorticoid receptor gene are GC-rich but lack consensus glucocorticoid receptor element sites. *Mol. Cell. Biol.* **10**, 5580–5585.

Yaolin Wang*
Sophia Y. Tsai
Bert W. O'Malley

Department of Cell Biology
Baylor College of Medicine
Houston, TX 77030

An Antiprogestin Regulable Gene Switch for Induction of Gene Expression *in Vivo*

I. Introduction

In an organism, coordinate expression of various genes is critical for development as well as growth and survival. Disregulation of the timing or the level of gene expression often results in pathological changes leading to the development of pathology. To better regulate individual genes, a "gene switch" that can control the timing and expression level of genes in a particular cell or tissue would be advantageous.

Eukaryotic genes respond to various stimuli, including metabolites, growth factors, and hormonal or environmental agents. Constitutive expression of foreign genes may result in cytotoxicity as well as undesired immune responses. The quest to develop a tight inducible and regulatable system for

* Present address: Department of Oncology, Schering-Plough Research Institute, 2015 Galloping Hill Road, Kenilworth, NJ 07033.

controlling gene expression has been initiated over the past decade in a number of laboratories around the world.

We believe that any successful and applicable gene switch should fulfill the following criteria: (1) It should only induce expression of newly introduced target genes and should not affect endogenous genes; (2) the target gene could be activated only by an exogenous signal, preferably a small molecule that distributes throughout all bodily tissues; (3) the target induction should be reversible; (4) the exogenous signal molecule should be biologically safe and easily administered, preferably orally; and (5) the target gene should have low basal activity and a high level of inducibility.

Several inducible systems for regulating gene expression have been established. They include the use of the metal response promoter (Searle *et al.*, 1985), the heat shock promoter (Fuqua *et al.*, 1989), the glucocorticoid-inducible (MMTV-LTR) promoter (Hirt *et al.*, 1994), the *lac* repressor/operator system using IPTG as an inducer (Figge *et al.*, 1988; Baim *et al.*, 1991), and the Tet repressor/operator system (tTA) with tetracycline as an inducer (Gossen and Bujard, 1992; Freudlieb *et al.*, 1997; Furth *et al.*, 1994). More recently, gene activator sequences and selective DNA binding sequences have been fused to the hormone-binding domains (HBDs) of steroid receptors, rendering the fusion protein responsive to steroid (Picard, 1994; Walker and Enrietto, 1995). For example, a GAL4-HBD fusion protein capable of transactivating a target gene by binding to GAL4 binding sites (17 mer) located upstream of the target gene has been constructed (Braselmann *et al.*, 1993).

II. Construction of RU486 Inducible Gene Expression System

An interesting discovery was made several years ago in our laboratory while we were studying the structural conformation of steroid hormone receptors and the role of the hormone binding domain in transcriptional activation of steroid-responsive target genes. We found that deletion of 42 amino acids at the C terminus of the human progesterone receptor (hPR) resulted in a loss of agonist binding and response to progesterone, so that the mutated hPR could no longer activate progesterone responsive genes (Vegeto *et al.*, 1992). Surprisingly, this mutated hPR still bound the progesterone antagonist RU486, an exogenous synthetic antiprogestin, and activated transcription of target genes containing binding sites for progesterone receptor. Further analysis of steroid hormone binding confirmed that this mutated hPR binds to no endogenous steroid hormones such as progesterone, glucocorticoid, estrogen, androgen, secosteroids, and retinoids.

Utilizing the unique features of the mutated human progesterone receptor, we set out to develop a recombinant steroid hormone receptor that could regulate target gene expression in the presence of only exogenous synthetic antiprogestin ligands. We constructed a chimeric molecule (GL) by fusing the HBD of the mutated hPR (residues 640–891), as a ligand-dependent regulatory domain, to a segment of the yeast transcriptional activator GAL4 (residues 1–94). This region of GAL4 contains a DNA binding function (residues 1–65), a dimerization function (residues 65–94), and a nuclear localization signal (residues 1–29). By replacing the DNA binding domain of PR with that of GAL4, we eliminated the possibility of simultaneous activation of any endogenous progesterone-responsive genes. To enhance the transcriptional activation function of the chimeric protein, we fused the C-terminal fragment of the herpes simplex virus protein VP16 (residues 411–487) to the N terminus of GAL4 in the chimera GL, creating a chimeric gene regulator (gene switch) GLVP (Fig. 1).

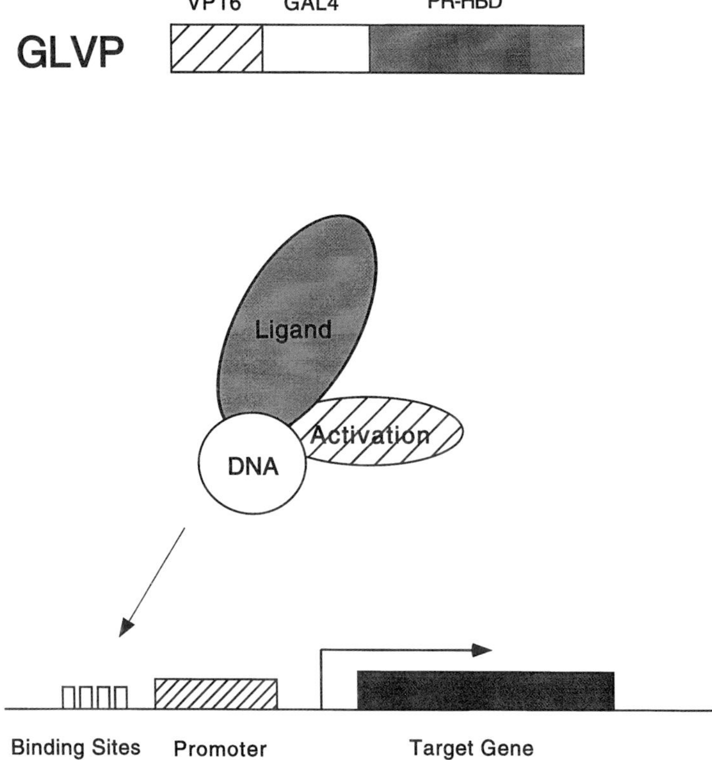

FIGURE I   Schematic diagram of the inducible system. The recombinant steroid receptor GLVP consists of the hormone binding domain of human progesterone receptor (residues 640–891), the DNA binding domain of GAL4, and the transcription activation domain of VP16. The reporter plasmid contains four copies of 17 mer (GAL4 binding sites) linked to the promoter and target gene of interest.

The expression of GLVP can be driven by any viral promoter (RSV, CMV) or tissue-specific promoter of interest. We have demonstrated that this novel regulator GLVP, when expressed in the cells through transient transfection, could activate a target gene containing the GAL4 DNA binding sites (17 mer). More importantly, the target gene expression only occurs in the presence of exogenous RU486 but not endogenous steroid hormones (Wang *et al.*, 1994). This tight regulation of target gene expression was confirmed in several mammalian cell lines. We also found that in addition to RU486, the GLVP can be activated by other known synthetic progesterone antagonists at low concentration (Wang *et al.*, 1994). Various target genes encoding intracellular proteins (chloramphenicol acetyl transferase, tyrosine hydroxylase) or secretory protein (human growth hormone) were evaluated and shown to be induced effectively at very low concentration (~1 nM) of RU486. In addition, this inducible system has been validated through the construction of RU486 inducible stable cell lines. Finally, we have confirmed the effectiveness of this inducible system using an *ex vivo* transplantation approach in which cells containing the stably integrated chimeric regulator GLVP and a target gene were grafted to rats. In this case, the expression of the target gene (tyrosine hydroxylase) was again tightly regulated by the administration of RU486. The dosage of RU486 used is significantly lower than that required for antagonizing *in vivo* progesterone actions.

III. Summary of Characteristics of the RU486 Inducible System

1. It can be regulated by an exogenous signal, in this case, RU486.
2. RU486 is a small synthetic molecule (molecular weight ~430 Da) and has been used safely as an oral drug for other medical purposes (Brogden *et al.*, 1993; Grunberg *et al.*, 1991).
3. Since the yeast GAL4 protein has no mammalian homologue, it is unlikely that the target gene driven by the 17 mer binding sites and promoter would be activated or repressed by endogenous proteins.
4. It does not affect endogenous gene expression, since the regulator GL-VP only efficiently activates the target gene bearing multiple copies of the 17 mer sequences juxtaposed to the promoter of the target gene (Lin *et al.*, 1988).
5. The regulator can be activated by a very low dose of RU486 (1 nM). At this low concentration, RU486 does not affect the normal function of endogenous progesterone and glucocorticoid receptors, since it only antagonizes the activity of these receptors at much higher doses. Thus, the high binding affinity of antiprogestin for the GLVP gene switch ensures its specificity for regulation of only target gene expression.

IV. Optimization of Inducible Gene Expression

Researchers working with the various inducible systems (RU486, tetracycline, IPTG) sometimes encounter a problem of leaky expression of the target gene. In certain instances, this could significantly affect the outcome of an experiment. To address this question, we can choose to use a minimal promoter (E1B TATA box) instead of a stronger thymidine kinase (tk) promoter, thereby reducing the basal level of target gene expression significantly (Fig. 2). In cases where a strong promoter is required to maximally induce the expression of a target gene, we found it necessary to optimize the ratio of the GLVP plasmid vs the reporter plasmid in transient transfection assays. In general we use a ratio of 1 : 10 to 1 : 5 of GLVP vs reporter. To ensure the successful application of the inducible system, it is important that the individual researcher empirically determine the amount and the ratio of the regulator and the reporter to be used in the investigation. When constructing a stable cell line harboring the inducible system, it is useful to

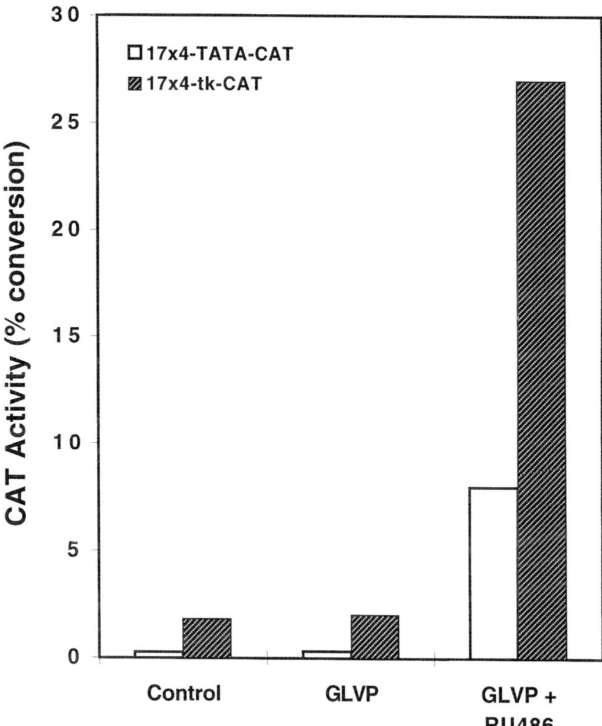

FIGURE 2 Optimization of target gene expression. CAT assay showing cotransfection of 5 μg of reporter plasmid (17 × 4-TATA-CAT or 17 × 4-tk-CAT) with 2 μg of GLVP expression plasmid (driven by RSV promoter). CV-1 cells were then incubated with RU486 (10^{-7} M) as indicated.

screen for multiple clones until a desired one has been identified. In most cases, some residual basal expression would be detected (depending on the sensitivity of assay) as do many endogenous genes, and should not be of general concern. We found that clones with more than 10-fold induction will likely be sufficient to present the researcher with the desired biological function or phenotype.

V. Protocol for Transient Transfection of Cell Cultures

The RU486-inducible system has been used successfully in various cell cultures such as HeLa, CV1, and HepG2. There may be some background variation in other cell types, and we suggest that one of the aforementioned cell lines be employed as a control. The following protocol describes the polybrene-mediated transfection of HeLa cells. When using different cell lines, appropriate media should be employed according to the supplier's specification. Other transfection methods such as calcium precipitation and liposome-mediated transfection also can be used.

1. Cells are grown in Dulbecco's modified Eagle's medium (DMEM) with 10% fetal bovine serum (FBS), 100 U/ml penicillin G, and 100 mg/ml streptomycin (GIBCO-BRL, Gaithersburg, MD).

2. Split exponentially growing cells into 10 cm tissue culture dishes the day before transfection at 1×10^6 per dish. The optimal cell density before transfection should be around 30–50% confluency.

3. To a 15 ml sterile Falcon tube, add 1 ml of sterile transfection buffer (prepared by mixing 0.2 g $MgCl_2 \cdot 6H_2O$, 0.132 g $CaCl_2 \cdot 2H_2O$, 8 g NaCl, 0.38 g KCl, 0.19 g $Na_2HPO_4 \cdot 7H_2O$, and 23.5 ml 1 M HEPES to 1 L and titrate pH to 7.05).

4. Add 10–50 μg DNA of choice to each tube followed by gentle vortexing. We normally use 0.5–2 μg of transactivator plasmid (pGLVP) and 10 μg of reporter plasmid (17 $\times$ 4-tk-CAT) for each transfection (dish) and add calf thymus DNA to balance the amount of DNA in each tube to 20 μg. We emphasize that to achieve optimal induction and low basal activity of reporter gene expression, it is critical to titrate the amount of transactivator and reporter plasmid used in the initial experiment by varying the ratio between transactivator and reporter.

5. To each tube, add 5 μl of polybrene solution (10 mg/ml, prepared in transfection buffer) while vortexing the tube gently. Polybrene solution can be stored as frozen aliquots in the freezer, and fresh aliquot should be used each time.

6. Let the mixture sit at room temperature for 30 min.

7. While incubating the DNA mixture, wash the cells twice with 5 ml of Hanks' Balanced Salt Solutions (HBSS) (Gibco-BRL) to remove serum from the media. Then add 10 ml of serum-free DMEM to each plate.

8. Gently vortex the tube containing the DNA mixture and add the mixture drop by drop to the cell culture plate. Swirl the plate so that the media and the DNA mixture are well mixed.

9. Incubate the cells for 4–5 h in a cell culture incubator.

10. Remove the plates, aspirate the media, and immediately add 3 ml of 25% glycerol (prepared in DMEM).

11. Remove the glycerol after 30 s. Do not incubate longer than 1 min. Wash the cells with 5 ml DMEM, twice, to remove glycerol.

12. Feed the cells with 10 ml DMEM (with 10% FBS).

13. At this point, 10 μl of RU486 (10^{-6} or 10^{-5} M) can be added to each plate to make a final concentration of 10^{-9} M (or 10^{-8} M). A stock solution of RU486 can be prepared by dissolving RU486 in 80% ethanol and can be stored in a freezer for 12 months.

14. Incubate the cells for 36 h and harvest the cells for reporter assay using the specified lysis buffer for the designated assay method.

VI. Combining the Inducible Transactivator and Reporter into One Vector

One way to increase the efficiency of simultaneous uptake of the inducible chimeric regulator GLVP and reporter DNA by cells is to construct a plasmid vector containing both gene products. For this purpose, we first constructed a plasmid vector of the desired target gene and GLVP driven by a tissue-specific promoter. In the reporter plasmid, we fused the human growth hormone gene (genomic fragment) with the adenoviral E1B minimal TATA promoter linked to GAL4 binding sites (four copies of a 17 mer; Wang *et al.*, 1997). For the expression of GLVP, we chose the transthyretin enhancer/promoter previously shown to confer liver-specific expression (Yan *et al.*, 1990). These two plasmids were constructed and transiently transfected into a liver-specific cell line, HepG2. In the presence of RU486, approximately 30- to 90-fold induction of human growth hormone expression was observed (Wang *et al.*, 1997a, 1997b). We then fused the two genes by standard cloning procedure, generating a plasmid containing both the regulator and target genes as illustrated in Fig. 3a. When this plasmid was introduced into HepG2 cells, approximately 120-fold induction of hGH expression was observed. This study demonstrated that a single vector incorporating both the regulator and reporter gene works as efficiently as two individual plasmid vectors. This single vector inducible system should be particularly suitable for application to gene therapy where viral vectors are utilized.

VII. Generation of a More Potent Gene Switch

We employed mutagenesis to generate a gene switch regulator that would be more potent in the presence of RU486 while maintaining the same

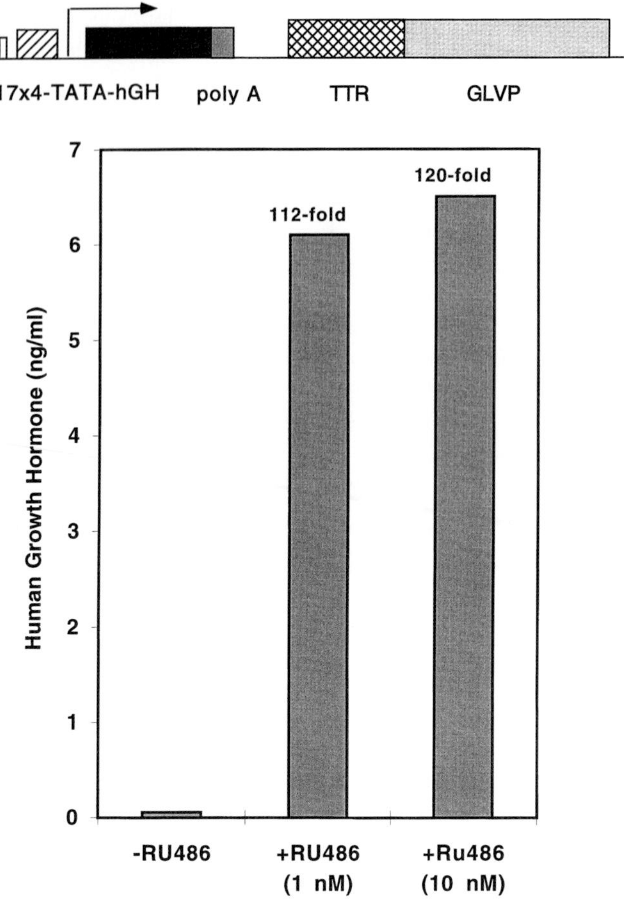

FIGURE 3   Inducible system in one vector. (a) Diagram of a single plasmid vector carrying both the regulator GLVP, driven by the liver-specific transthyretin enhancer/promoter (TTR) and reporter gene, human growth hormone (hGH). (b): Two microgram (0.3 pmol) of the plasmid DNA was transfected into HepG2 cells and hGH expression in response to various concentration of RU486 was measured using RIA (Wang *et al.*, 1997a). Numbers above the bar indicate fold of induction in the presence of RU486.

low basal level of target gene expression in the absence of the ligand (Wang *et al.*, 1997c). We found that deletion of fewer amino acids from the C terminus of the human progesterone receptor hormone binding domain, 19AAs instead of 42AAs, results in a more potent RU486 dependent regulator (GL$_{914}$VP) that activates target gene expression more strongly because the expanded sequence allows greater dimerization of the regulator (data not shown). Interestingly, when we fused the VP16 activation domain to the C terminus of the hPR hormone binding domain (640–914), we observed that the new chimeric regulator (GL$_{914}$VP$_{C'}$) was even more potent than the

GL$_{914}$VP in which the VP16 activation domain is located at the N terminus of the chimeric molecule. More importantly, the GL$_{914}$VP$_{C'}$ activates target gene expression at a lower concentration of RU486 (0.01 nM) (Fig. 4). It is likely that this new fusion protein adopts a slightly different conformation, allowing better binding to RU486 as well as better interaction with the general transcription machinery.

VIII. Construction of an Inducible Repressor

Although most current inducible systems involve activation of target genes, it is obvious that development of an inducible transcriptional repression system would be advantageous in certain instances. To explore the possibility of converting the chimeric transactivator GLVP into a regulatable repressor, we replaced the VP16 activation domain with a transcriptional repression domain, the Krüppel-associated box-A (Krab), from the zinc finger protein *Kid-1*. *Kid-1* was identified as a kidney-specific transcription factor that is regulated during renal ontogeny and injury (Wirzgall *et al.*, 1994). The Krab domain (residues 1–70) was inserted at the C-terminus of GL$_{914}$ generating GL$_{914}$Krab (Fig. 5). The reporter plasmid 17 × 4-tk-CAT was cotransfected with GL$_{914}$Krab into HeLa cells and repression by RU486 was analyzed. Since we are looking at repression of the basal promoter and

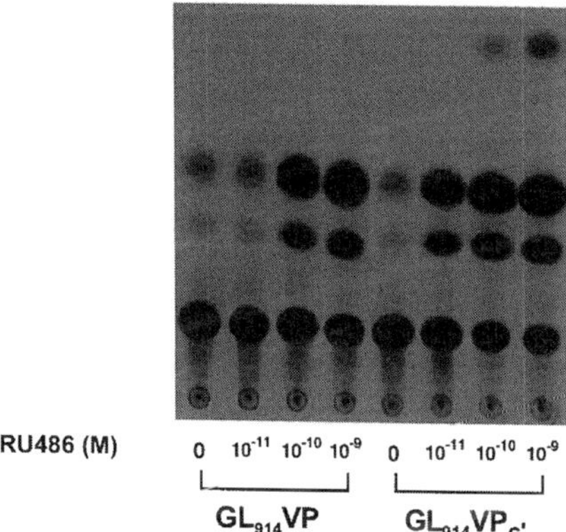

FIGURE 4 Comparison of transactivation activity of the original gene switch GLVP with the new and potent version GL$_{914}$VP and GL$_{914}$VP$_{C'}$ in response to RU486. HeLa cells were transfected with 4 μg of chimeric regulator (driven by RSV promoter) as indicated and 10 μg of reporter 17 × 4-TATA-CAT. After transfection, cells were incubated with either solvent control (−) (85% ethanol, final concentration in the medium diluted 1000-fold or 0.085%), or RU486 (RU) at various concentrations. GL$_{914}$VP$_{C'}$ activation on reporter gene (CAT) expression is stronger than that of the original GL$_{914}$VP. Dose–response analysis of chimeric regulators demonstrates that GL$_{914}$VP$_{C'}$ responds to very low concentrations of RU486 (0.01 nM) and exhibit maximal induction of target gene expression at 1 nM RU486.

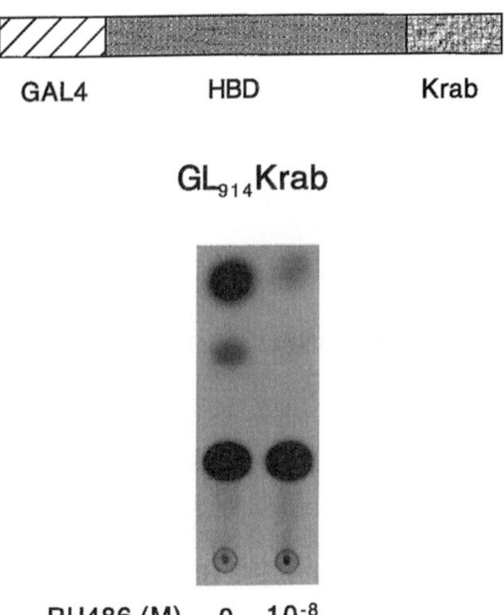

FIGURE 5 Inducible repression of target gene expression. HeLa cells were transfected with 4 μg of expression plasmid pGL$_{914}$Krab (driven by CMV promoter) and 10 μg of reporter plasmid 17 × 4-tk-CAT. CAT activity was measured in the absence or presence of RU486 (10 nM).

enhancer (tk) activity from the reporter construct, we used a higher amount of reporter plasmid (10 μg) and 4 μg of regulator plasmid (driven by CMV promoter). As shown in Fig. 5, the chimeric regulator Gl$_{914}$Krab strongly repressed (88% reduction) expression of the target gene (CAT) in an RU486-dependent manner. It should be noted that achieving maximal repression of a target promoter is more difficult than activation of a target gene, since it usually requires almost maximal occupation (competitive binding) of the promoter site by the regulator. A 10-fold repression would correspond to ~90% reduction of the basal activity. These results suggest that the Krab domain contains a potent transcriptional repression function and the chimeric repressor molecule GL$_{914}$Krab binds to DNA efficiently in the presence of RU486 to repress gene expression.

IX. Regulable and Tissue-Specific Gene Expression in Transgenic Mice

One of the most challenging aspects of research in gene regulation is to achieve regulable gene expression *in vivo* in an intact animal. The major problem for biologists is the fact that integration of foreign gene into the chromosome occurs at random sites. For example, integration into a heterochromatin region may result in complete silence of the foreign gene. Not surprisingly, multiple transgenic mouse lines need to be screened before one with satisfactory transgene expression can be obtained. To express the chimeric transactivator GLVP in a tissue-specific fashion we again chose the enhancer/promoter of the (TTR) gene, known to target transgene expression specifically to the liver in transgenic mouse (Yan *et al.*, 1990).

Our initial attempts to express the TTR-GLVP transgene in the liver of transgenic mice failed to produce a GLVP-expressing line in 10 founders. This could have resulted from the integration of the GLVP transgene into a heterochromatin region or a site influenced by a neighboring negative regulatory element. To minimize the effect associated with site of integration, a tandem repeat of a chromosomal insulator sequence obtained from the 5′ region of the chicken β-globin gene was placed upstream of the TTR-GLVP construct (Chung *et al.*, 1993). This insulator sequence was shown to possess no intrinsic enhancing or silencing activity (Chung *et al.*, 1993; Y. W., unpublished data). Four founder lines bearing the insulator-TTR-GLVP transgene were established and three showed liver-specific expression of GLVP.

A target mouse line (17 × 4-tk-hGH) was generated and its serum was examined for hGH expression. No hGH was detected above the assay background in the serum of these mice, indicating that endogenous proteins do not activate this target gene, supporting the notion that no mammalian homologue of the yeast transcriptional activator GAL4 exists in higher eukaryotes (Brand and Perrimon, 1993) and vertebrates (Spitz and Bardin, 1993).

Individual lines of mice containing the regulator GLVP were crossed to mice containing the target gene, and bigenic mice harboring both transgenes (TTR-GLVP/hGH) were selected for further analysis. No, or very little, expression of hGH was detected in the serum of bigenic mice, demonstrating that the inducible system was not activated by endogenous ligand. In contrast, intraperitoneal administration of a single dose of RU486 (250 μg/kg), resulted in a 1500-fold induction of hGH expression in serum (Fig. 6a). High hGH expression (~200 ng/ml) was achieved by 12 hours. For comparison, mouse endogenous GH level peaks at ~10 ng/ml depending on the time of the day, age, and sex of the mouse. Transgene hGH expression peaked at ~12 h after RU486 administration and returned to background levels ~100 h postinjection. RU486-inducible expression of hGH transgene was dose dependent and liver specific (Wang *et al.*, 1997b). Repeated injection of RU486 (one every 2 days) to bigenic mice resulted in a significant increase (30–37%) in weight (within 10 days) compare to the control group (Fig. 6b).

These studies represented the first kinetic study of inducible and dose-dependent transgene activation *in vivo* accompanied by a gross phenotypic change in response to a synthetic inducer. The use of such bigenic mice should allow the creation of an array of novel transgenic mouse models mimicking human diseases that might otherwise be impossible because of

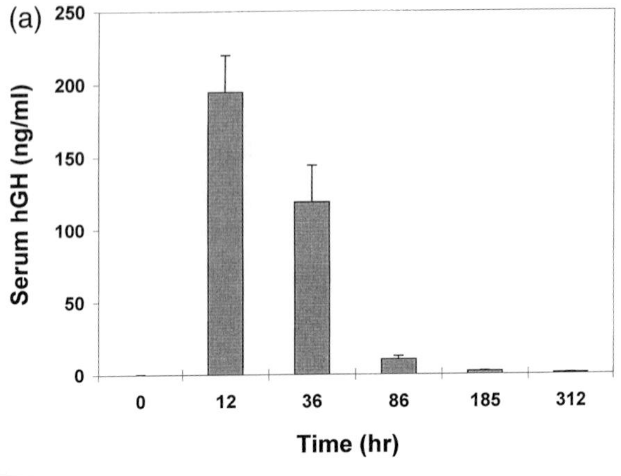

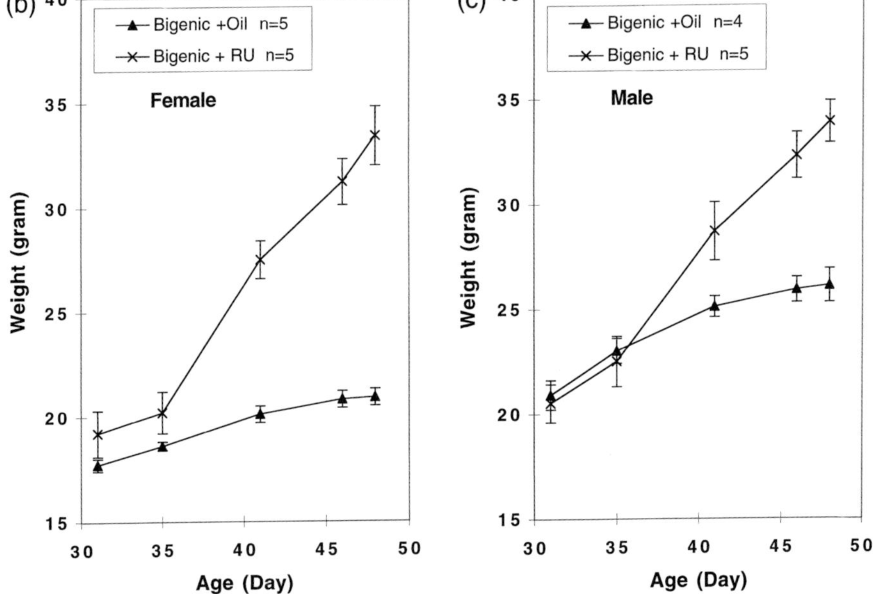

FIGURE 6   RU486 inducible gene switch system in transgenic mice. (a) Kinetics of RU486-inducible hGH expression in transgenic mice. Bigenic mice ($n = 14$) were given a single injection (i.p.) of 250 μg/kg dose of RU486 (dissolved in sesame oil) and serum hGH was measured (Wang *et al.*, 1997b) at different time points as indicated. (b) Phenotype of bigenic mice in response to RU486 treatment. Bigenic mice (grouped in female and male respectively) were given either sesame oil or RU486 (250 μg/kg) at day 31 after birth and continued to receive the same dose every two days. The mouse weight was recorded at the indicated points. The number of mice used in each study group is shown in the legend box.

the deleterious effects of constitutive expression, such as genes involved in apoptosis and tumorigenesis. Since RU486 (Mifeprestone) can be administered orally and has an established safety record (Spitz and Bardin, 1993), we envision that this inducible system should have broad application both in regulation of target gene expression in transgenic mice and in the future development of methodology for human gene therapy.

Acknowledgments

We thank Dr. Ming-Jer Tsai and members of our laboratory for helpful discussions throughout the development of the RU486-inducible gene expression system.

References

Baim, S. B., Labow, M. A., Levine, A. J., and Shenk, T. (1991). *Proc. Natl. Acad. Sci. USA* **88**, 5072.

Brand, A. H., and Perrimon, N. (1993). *Development* **118**, 401.

Braselmann S., Graninger, P., and Busslinger, M., (1993). *Proc. Natl. Acad. Sci. USA* **90**, 1657.

Brogden, R. N., Goa, K. L., and Faulds, D. (1993). *Drugs* **45**, 384.

Chung, J., Whiteley, M., and Felsenfeld, G. (1993). *Cell* **74**, 505.

Figge, J., Wright, C., Collins, C. J., Roberts, T. M., and Livingston, D. M. (1988). *Cell* **52**, 713.

Freudlieb, S., Baron, U., Bonin, A. L., Gossen, M., and Bujard, H. (1997). *Methods Enzymol.* **283**, 159.

Fuqua, S. A., Blum-Salingaros, M., and McGuire, W. L. (1989). *Cancer Res.* **49**, 4126.

Furth, P. A., Onge, L. T., Boger, H., Gruss, P., Gossen, M., Kistner, A., Bujard, H., and Hennighausen, L. (1994). *Proc. Natl. Acad. Sci. USA* **91**, 9302.

Gossen, M., and Bujard, H. (1992). *Proc. Natl. Acad. Sci. USA* **89**, 5547.

Grunberg, S. M., Weiss, M. H., Spitz, I. M., Ahmadi, J., Sadun, A., Russell, C. A., Lucci, L., and Stevenson, L. L. (1991). *J. Neurosurg.* **74**, 861.

Hirt, R., Fasel, N., and Kraehenbuhl, J. P. (1994). *Meth. Cell Biol.* **43**, 247.

Lin, Y.-S., Carey, M. F., Ptashne, M., and Green, M. R. (1988). *Cell* **52**, 713.

Picard, D. (1994). *Curr. Opin. Biotechnol.* **5**, 511.

Searle, P. F., Stuart, G. W., and Palmiter, R. (1985). *Mol. Cell. Biol.* **5**, 1480.

Spitz, I. M., and Bardin, C. W. (1993). *N. Engl. J. Med.* **329**, 404.

Vegeto, E., Allan, G. F., Schrader, W. T., Tsai, M.-J., McDonnell, D. P., and O'Malley, B. W. (1992). *Cell* **69**, 703.

Walker, A. K., and Enrietto, P. J. (1995). *Methods Enzymol.* **254**, 469.

Wang, Y., O'Malley, B. W., Jr., Tsai, S. Y., and O'Malley, B. W. (1994). *Proc. Natl. Acad. Sci. USA* **91**, 8180.

Wang, Y., O'Malley, B. W., and Tsai, S. Y. (1997a). *Meth. Mol. Biol.* **63**, 401.

Wang, Y., DeMayo, F. J., Tsai, S. Y., and O'Malley, B. W. (1997b). *Nature Biotechnol.* **15**, 239.

Wang, Y., Xu, J., Pierson, T., O'Malley, B. W., and Tsai, S. Y. (1997c). *Gene Ther.* **4**, 432.

Witzgall, R., O'Leary, E., Leaf, A., Onaldi, D., and Bonventre, J. V. (1994). *Proc. Natl. Acad. Sci. USA* **91**, 4514.

Yan, C., Costa, R. H., Darnell, J. E., Jr., Chen, J., and Van Dyke, T. A. (1990). *EMBO J.* **9**, 869.

Sylvia Hewitt Curtis
Kenneth S. Korach

LRDT, Receptor Biology Section
National Institute of Environmental Health Sciences, NIH

Steroid Receptor Knockout Models: Phenotypes and Responses Illustrate Interactions between Receptor Signaling Pathways *in Vivo*

I. Introduction

Steroid receptors have been characterized and analyzed since they were first identified and purified from different target tissues of a variety of species. Hormones such as estradiol and progesterone exert their effects on tissues containing receptors, which are nuclear transcription factors that bind the hormone and regulate transcription of hormone-responsive genes. Steroid receptors bind to specific DNA sequences in the target gene and also interact with transcriptional coactivators or co-repressors in response to ligand binding. It is the interaction of these cofactors with the transcriptional machinery of the cell that regulates the activity of RNA polymerase. This mechanism is more completely described and reviewed elsewhere (Chen and Li, 1998; Evans, 1988; Carson Jurica *et al.*, 1990; O'Malley, 1990; Tsai and O'Malley, 1994; Vegeto *et al.*, 1996).

Gonadal steroids have numerous sex- and tissue-specific effects on development and reproduction (George and Wilson, 1986; Wilson *et al.*, 1995; Werner *et al.*, 1996). In the physiological context, nuclear receptor–mediated signaling pathways interact with each other as well as with other signaling pathways. The complexity of this cross-talk is difficult to dissect *in vivo* or to reproduce *in vitro*; thus, the development of the transgenic mouse has provided useful models that facilitate study of steroid receptor mechanisms and interactions. The aim of this chapter is to review some of the interactions between steroid receptor pathways as well as cross-talk with other signaling pathways, and to discuss how these interactions have been clarified using transgenic knockout mouse models.

II. Mammalian Reproduction

The reproductive cycle of the female mouse involves a complex interplay of several organs and hormones. Ultimately, the uterus is the target organ for implantation of fertilized oocytes, but the proper hormonal environment must be obtained and maintained for reproduction to be successful. The major steroid hormones involved in the preparation of the uterine endometrium for implantation are progesterone and estradiol, both produced by the ovary. Estrogen is important for the proliferation of the uterine epithelium (Galand *et al.*, 1971) and enhances progesterone action via induction of the progesterone receptor (PR) gene (Graham and Clarke, 1997). Progesterone plays a role in proliferation, differentiation, and maintenance of the uterine stroma and endometrium (Weitlauf, 1988).

Estrogen and progesterone are also important for mammary gland development and differentiation associated with pregnancy. Estrogen induces growth factors in the mammary gland and acts systemically to elevate pituitary prolactin secretion, which also stimulates growth of mammary tissues. Estrogen is also required for ductal morphogenesis and accentuates progesterone action by inducing PR. Progesterone is required for branching and lobuloalveolar development (Daniel *et al.*, 1987; Silberstein *et al.*, 1994). The engineering of the PR knockout (PRKO) and ERα knockout (ERKO) mice has allowed the dissection of the roles of progesterone and estrogen in reproduction and development (Lubahn *et al.*, 1993; Korach, 1994; Couse *et al.*, 1995a, 1995b; Lydon *et al.*, 1995, 1996; Eddy *et al.*, 1996; Korach *et al.*, 1996; Chappell *et al.*, 1997; Couse and Korach, 1999). For example, the PRKO mammary gland has epithelial ducts that lack side branches and do not develop lobuloalveolar structures (Lydon *et al.*, 1995). In contrast, the ERKO mammary gland contains an epithelial duct rudiment that does not grow into the fat pad (Korach *et al.*, 1996). This indicates the importance of estrogen in ductal morphogenesis and progesterone in ductal branching and alveolar development. In this review, the relative roles of the sex steroids as revealed by genetically altered rodent models will be described.

III. Phenotypes of ERKO Illustrate the Normal Roles of Estradiol

The development of the ERαKO mouse has allowed the further definition of estrogen action in target tissues and the role of estrogen in growth and development. The phenotypes of the ERKO female are reviewed elsewhere (Korach, 1994; Couse and Korach, 1999; Korach *et al.*, 1996; Couse and Korach, 1998) and are summarized in Table I. The uterus is underdeveloped (Fig. 1) and refractory to estradiol in terms of a lack of cell proliferation, wet weight increase, and induction of estrogen responsive genes. The phenotype of the ERKO uterus indicates that ERα does have a role in uterine physiology but is not required for development of a uterus. Studies using tissue recombination have indicated that ERα need only be present in the stromal compartment for epithelial mitogenesis to occur, as stromal cells from wild-type mice cultured together with epithelial cells from an ERKO uterus resulted in estrogen-dependent induction of DNA synthesis in epithelial cells (Cooke *et al.*, 1997).

The ERKO ovaries develop normally but become increasingly enlarged and filled with hemorrhagic cysts following puberty (Fig. 1) (Lubahn *et al.*, 1993). They produce dramatically elevated levels of both estradiol and testosterone (Couse *et al.*, 1995) (see Table II), but are not functional in terms of ovulation. Follicles appear to become atretic following the preantral stage. The abnormal appearance and functioning of the ERKO ovaries ap-

TABLE I Phenotypes of Estrogen and Progesterone Target Tissues in the ERαKO Female

Tissue	Observation	References
Uterus	Underdeveloped, unresponsive to estradiol	Lubahn *et al.*, 1993; Couse and Korach, 1999
Ovary	Enlarged, hemhorrhagic cysts, follicles arrested at preantral stage, no corpora lutea, no ovulation, elevated serum E and T levels	Lubahn *et al.*, 1993; Couse and Korach, 1999
Mammary	Ducts do not develop beyond epithelial rudiment at nipple, no alveolar development	Korach, 1994; Korach *et al.*, 1996
Pituitary	FSHβ, LHβ αGSU, mRNAs all elevated, prolactin mRNA reduced	Scully *et al.*, 1997
Brain	No mating behaviors	Ogawa *et al.*, 1996, 1998

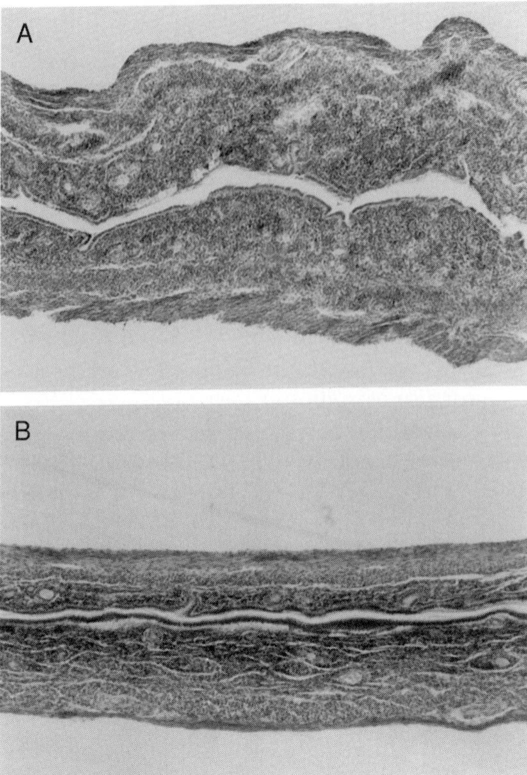

FIGURE 1   Phenotypes of reproductive tissues of the ERKO female. Reproductive tissues from wild-type and ERKO females. (A, B) Uterine tissue cross sections, H&E stained, showing the wild-type (A) and ERKO uterus (B). The ERKO uterus is small and underdeveloped compared to the wild type. Both uteri are from ovariectomized mice. Note that the luminal epithelium does not develop invaginations and the stromal layer is thinner in the ERKO (B). (C, D) Cross-sections from wild type (C) and ERKO (D) ovaries. Note the numerous hemorrhagic cysts and the lack of corpora lutea, indicating that ovulation is not occurring in the ERKO (D). Reproduced from Couse *et al.* (1995b). (E, F) Mammary gland whole mounts showing wild-type (E) and ERKO (F) mammary glands. Note the underdeveloped epithelial rudiment only at the nipple (arrow) in the ERKO (F), while the wild-type epithelium fills the gland (E).

pears to be due to chronic LH stimulation. LH levels are elevated due to the lack of ERα to mediate estrogen feedback on the pituitary and hypothalamus and thus the lack of downregulation of LH production and secretion (Korach *et al.*, 1996; Scully *et al.*, 1997). Interestingly, ovariectomy results in a decrease in uterine weight in the ERKO. This effect can be attributed to the elevated serum testosterone acting through the androgen receptor in the uterus, as DHT treatment following ovariectomy results in recovery of uterine weight (Lindzey *et al.*, 1996). This observation illustrates the usefulness of a transgenic knockout model in unmasking other unrecognized sig-

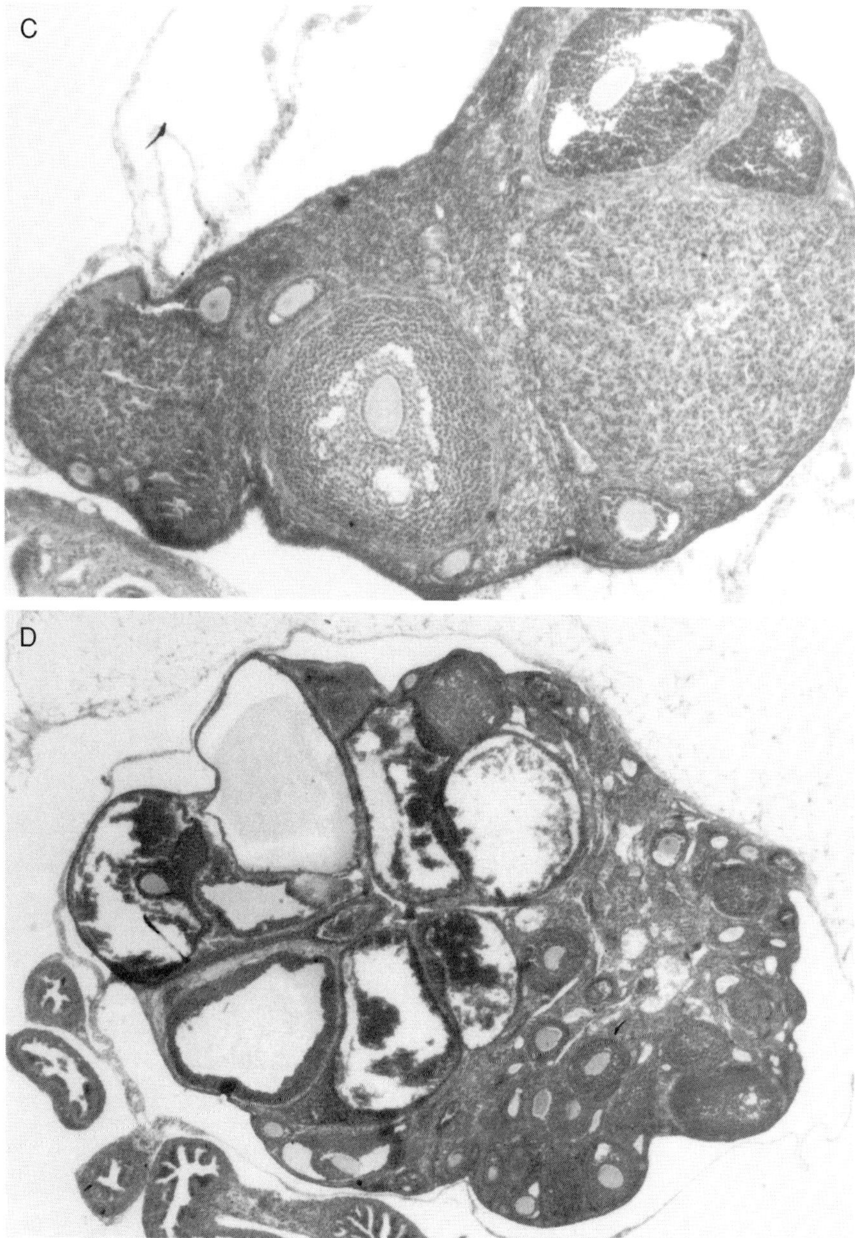

FIGURE 1. (*Continued*)

naling mechanisms. Androgen receptor signaling would not normally be associated with female reproductive tract function, but use of the ERKO has allowed observation of this ER-independent testosterone effect.

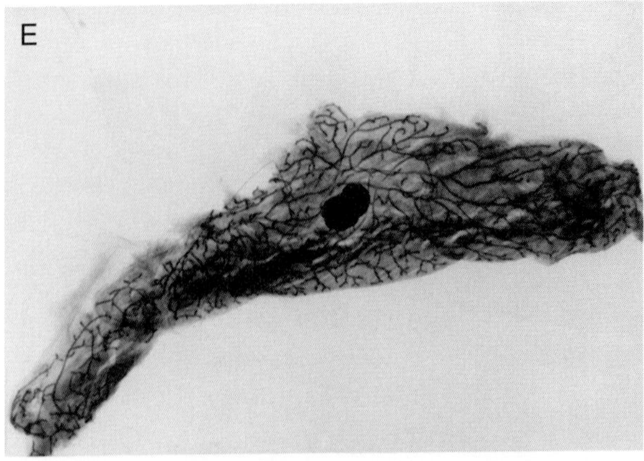

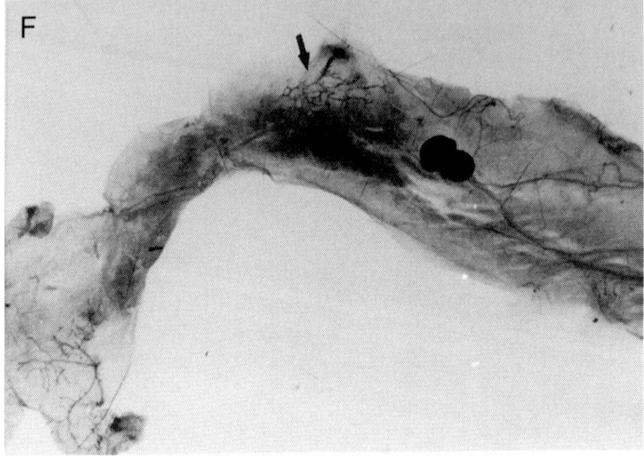

FIGURE 1. (*Continued*)

TABLE II   Serum Hormone Levels in Wild-Type and ERKO Females

Hormone	Wild type	ERKO
Estradiol[a], pg/ml	29.5 ± 2.5	84.3 ± 12.5[b]
Testosterone[a], ng/ml	0.36 ± 0.4	3.2 ± 0.6
Progesterone[a], ng/ml	2.3 ± 0.6	4.0 ± 1.1

[a] (Couse *et al.* 1995) The estradiol values in the ERKO differ from those previously reported (Couse *et al.* 1995b). The initial data was obtained from pooled serum samples, whereas the above values are the means from assays on individual samples and are therefore more likely to reflect the true levels.

[b] Range: 45–197.

The mammary glands in the ERKO female lack epithelial tissue except for a rudiment at the nipple that does not develop despite the high circulating estradiol levels, indicating the importance of estradiol and ERα action for epithelial proliferation in the mammary gland (Korach *et al.*, 1996) (Fig. 1).

A. Progesterone in Female Reproduction and the PRKO Phenotype

The progesterone receptor is an essential component in female reproduction, and its importance is illustrated by the phenotypes of normally progesterone responsive tissues of PRKO mice (Lydon *et al.*, 1995, 1996; Chappell *et al.*, 1997) (Table III). Progesterone functions in many organs involved in normal cycling as well as maintenance of pregnancy (Graham and Clarke, 1997). The primary targets of progesterone include the uterus, where it is involved in proliferation of the stroma, implantation of embryos, and maintenance of pregnancy. In response to apposition of the conceptus, the stromal cells of the uterus undergo a massive transformation to decidual cells, resulting in a dramatic increase in size and weight of the uterus (Weitlauf, 1988). This is called a decidual reaction, and it can be experimentally induced using a combination of hormonal stimulation and physical trauma to the uterine lumen to mimic early pregnancy. Progesterone is essential for this reaction to occur, as illustrated by hormonal manipulation experiments (Finn, 1966; Milligan and Mirembe, 1985; Finn and Pope, 1986) and by the fact that the PRKO uterus is not able to undergo decidual transformation in response to artificial stimulus (Lydon *et al.*, 1995).

In the mammary gland, progesterone is important for lobular alveolar development during pregnancy. The role of progesterone is again illustrated by the PRKO, which lacks lobular alveolar development and shows no alveolar buds after administration of progesterone (Lydon *et al.*, 1995) (Fig. 2).

In the ovary, follicles develop normally in PRKO mice to the terminal stage of the ovulatory follicle, but no rupture and ovulation occurs. This

TABLE III   Phenotypes of Estrogen and Progesterone Responsive Tissues in PRKO Female

Tissue	Phenotype	References
Uterus	Exaggerated inflammatory response to estradiol, no decidual response	Lydon *et al.*, 1995
Ovary	No ovulation, no corpora luteum, many preovulatory follicles but no rupture	Lydon *et al.*, 1995
Mammary	No alveolar buds	Lydon *et al.*, 1995
Pituitary	LH elevated, no LH or FSH surges	Chappell *et al.*, 1997
Brain	No lordosis response	Lydon *et al.*, 1995

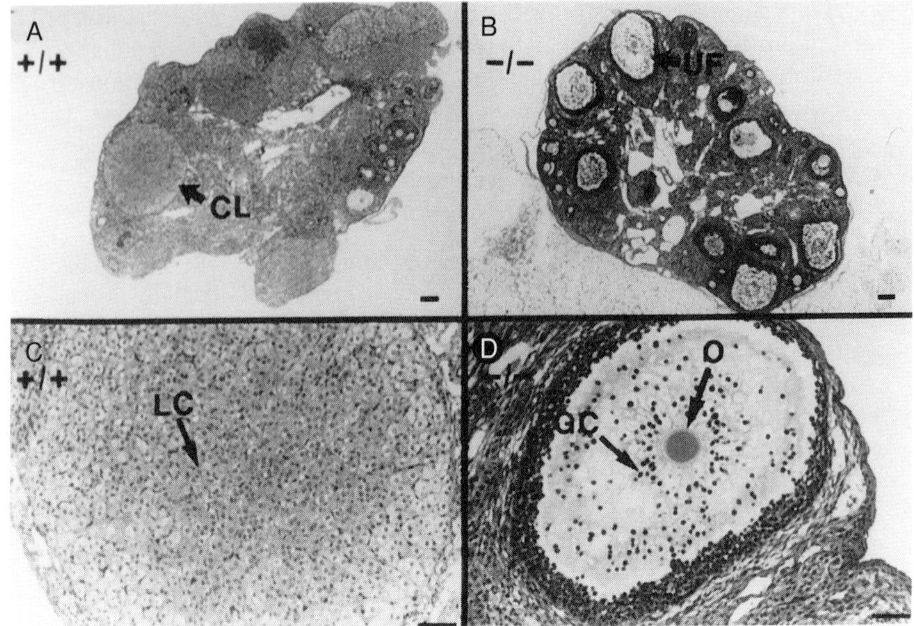

FIGURE 2   Phenotypes of PRKO reproductive tissues. I. Comparison of wild-type (+/+; A and C) and PRKO (−/−; B and D) ovaries. Note the numerous follicles but lack of corpora luteum (CL) in the PRKO. Reproduced from Lydon *et al.* (1995). II. Comparison of wild-type (+/+; A, C, E) and PRKO (−/−; B, D, F) mammary glands from females treated with progesterone (1 mg/day) and estradiol (1 μg/day) for 21 days. Note the lack of lobular alveolar structures in the PRKO. Reproduced from Lydon *et al.* (1995).

indicates that progesterone action through the PR is important for release of the oocyte by rupture of the follicle (Lydon *et al.*, 1995) (Fig. 2).

Progesterone action in the brain is important for normal mating behavior to occur, and disruption of progesterone signaling results in an inability to elicit a lordosis response (Lydon *et al.*, 1995). Interestingly, the lordosis response can by initiated in wild-type female mice by using the neurotransmitter dopamine in the absence of progesterone (Power *et al.*, 1991; Mani *et al.*, 1994a, 1994b). The importance of the PR protein in this mechanism was illustrated by showing that dopamine does not induce the lordosis response in the PRKO (Mani *et al.*, 1996). This study illustrates how useful the transgenic model can be for determining the role of the steroid receptor in nontraditional mechanisms.

B. Interaction between Estrogen and Progesterone Function

It has been shown that ER- and PR-mediated processes occur in overlapping tissues and that the responses often affect one another. For example,

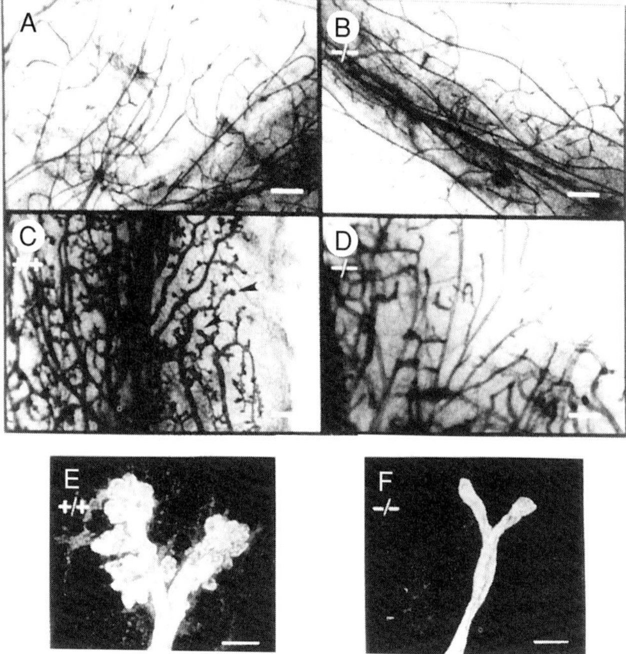

FIGURE 2. (*Continued*)

estradiol induces both the mRNA and protein level of PR, resulting in greater responsiveness to progesterone in target tissues (Graham and Clarke, 1997). Progesterone, through unknown mechanisms, is also known to dampen or oppose some of the effects of estradiol, as illustrated by a smaller increase in uterine weight following coadministration of estradiol and progesterone as compared to estradiol alone. Similarly, lactoferrin is induced in the uterus following estradiol treatment, but this induction is attenuated when estradiol is combined with progesterone (Tibbetts *et al.*, 1998). The interaction between these two signaling pathways is illustrated by the exaggerated and abnormal response of the PRKO uterus to progesterone and estradiol treatment (Lydon *et al.*, 1995). Removal of PR and thus the opposing effects of progesterone caused the uterus to become abnormally enlarged and fluid-filled. Cellular and histological analyses indicate that an inflammatory response occurred and that the luminal epithelium was hyperplastic and disorganized. The endometrial glands were enlarged and the epithelia were hypertrophied.

C. Progesterone Signaling in the ERKO Uterus

Because estradiol and progesterone effects in the uterus are so intertwined, the functioning of progesterone in the ERKO uterus was studied. Previous characterization has shown the ERKO exhibits profound phenotypes in PR responsive tissues (Table I): the rudimentary ductal structure of the mammary gland, the arrested development of ovarian follicles, and the underdeveloped uterus. Additionally, ERKO females show no mating behaviors (Ogawa *et al.*, 1996, 1998). In the uterus, the actions of estradiol and progesterone balance one another, as illustrated earlier by the exaggerated response of the PRKO uterus to estradiol. Since PR is induced by estradiol, removal of ER might also disrupt progesterone signaling in these tissues, and the additional loss of progesterone responsiveness could be contributing to the phenotypes of these tissues. Thus, the progesterone-signaling pathway was examined in the ERKO uterus.

D. Biochemical Characteristics of PR in the ERKO Uterus

Serum progesterone levels in ERKO females were within normal range, but did not achieve the postovulatory levels seen in normal-cycling wild-type animals (Couse and Korach, 1999) (Table II). Northern blot analysis of total uterine RNA showed that ERKO PR mRNA is present at a constitutive level similar to that of an ovariectomized wild-type animal and that the mRNA is not induced by estradiol in the ERKO (Couse and Korach, 1999) (Fig. 3).

PR levels were measured by ^{3}H-R5020 binding in uterine tissue extracts from wild-type and ERKO mice. The level of labeled R5020 bound to extracted uterine protein from the wild type and heterozygous (not shown) animals was the same (Curtis *et al.*, 1999) (Table IV), although heterozygotes do have approximately half the level of ERα (Couse *et al.*, 1995). The total (nuclear + cytosolic) ^{3}H-R5020 binding in the ERKO uterus is approximately 60% of the wild-type level, and a greater proportion of binding is in the nuclear compartment compared to wild-type or heterozygote samples (24% in ERKO vs 4.5% in wild type). The amount of binding was decreased in uterine tissue from ovariectomized wild-type and ERKO mice. However, ovariectomized ERKO mice retained 80% of the R5020 binding level of the ovariectomized wild type (Table IV). No R5020 binding was detected in the nuclear fraction of uterine tissue samples from ovexed mice of either genotype.

The PR gene encodes two protein isoforms (PR-A and PR-B) that are the products of different promoters (Graham and Clarke, 1997). PR-B contains additional N-terminal sequences and can be antagonized by PR-A. These PR isoforms do not function identically, and the mechanism regulating their relative expression levels is not known. Since the R5020 binding assay indicated that uterine PR levels were reduced in the absence of ERα, the relative expression levels of the PR isoforms were also analyzed to determine whether their expression ratio was altered. Western blot analysis of uterine cytosolic protein shows that PR-A and PR-B isoforms are present in the same relative amounts in the ERKO and wild-type uterine extracts (Curtis

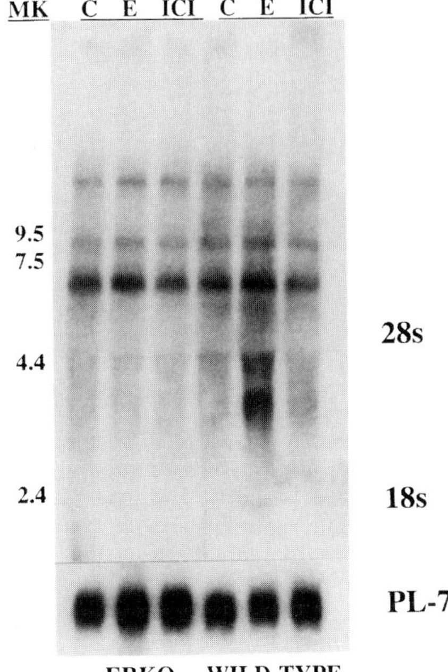

FIGURE 3   PR mRNA is not induced by estradiol in the ERKO. Total uterine RNA from wild-type and ERKO animals was analyzed by Northern blot. Ovariectomized mice were treated for 24 h with saline vehicle (C) or estradiol (E), or pretreated with the antiestrogen ICI182780 and then with estradiol (ICI). Blots were probed with a mouse PR cDNA, then stripped and re-probed with ribosomal protein L7 cDNA (PL7) to normalize for loading differences. The positions and sizes of the RNA markers are indicated in kilobases, as are the positions of 28s and 18s rRNA. Reproduced from Couse *et al.* (1995b).

TABLE IV   Progesterone Binding Is Decreased in ERKO Uteri[a]

	Intact	*Ovex*
Wild-type	509 ± 23 (4.5%)	153 ± 18
ERKO	338 ± 18 (24%)	119 ± 26

[a] The combined nuclear and cytosol binding of ^{3}H-R5020 was measured in wild-type and ERKO uteri from intact or ovex animals (Curtis *et al.*, 1999). Binding is expressed as specific (total-nonspecific) moles bound $\times$ 10^{-15} per mg total cytosolic protein. Parentheses indicate the percent of binding detected in the nuclear fraction.

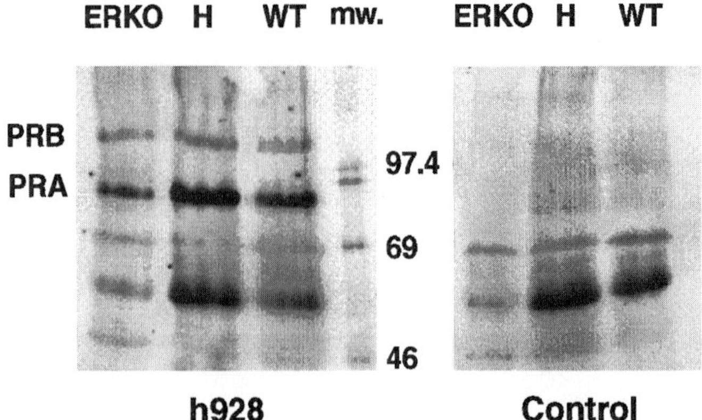

FIGURE 4   PR A and B isoforms are present in all three ERα genotypes. Cytosol from ERKO, heterozygous (H), and wild-type (WT) animals was analyzed by SDS–PAGE/Western blot. Two PR isoforms, PR-A and PR-B, are detected with anti-PR antibody h928, but not in the absence of primary antibody (control). Size (in kilodaltons) of ^{14}C-labeled molecular weight markers (mw) is indicated. Reproduced from Curtis *et al.* (1999).

et al., 1999) (Fig. 4). This result indicates that estrogen action is not required to regulate the relative amounts of PR-A and PR-B isoforms in the uterus.

E. Analysis of Progesterone-Responsive Genes in the ERKO

To determine if the reduced PR levels in the ERKO uterus were sufficient to mediate genomic responses, induction of progesterone responsive genes was analyzed. Calcitonin is a hormone that prevents bone resorption and increases calcium excretion in the kidney. In the rat uterus, calcitonin mRNA was shown to be induced at the time of implantation and in response to progesterone treatment (Ding *et al.,* 1994, 1995). Estrogen "priming" or pretreatment, which increases PR levels, is not required for calcitonin induction, but does enhance it approximately two-fold (Ding *et al.,* 1994). Estradiol priming for 1 day followed by 3 days of progesterone treatment caused a robust increase in calcitonin mRNA levels (Curtis *et al.,* 1999) (Table V) in both WT and ERKO ovariectomized mice. When the level of calcitonin mRNA was normalized to cyclophilin mRNA, which is not regulated by estrogen or progesterone, the average fold induction of calcitonin mRNA in progesterone treated ERKO mice (24.6-fold) was lower than that of the wild type (78.7-fold; Table II). Although PR levels are lower in the ERKO and are not induced by estrogen "priming" (Couse *et al.,* 1995b), the reduced level of PR is sufficient to mediate calcitonin mRNA induction.

Since induction of calcitonin expression occurred after 3 days of progesterone treatment, a more rapid progesterone-induced gene response was

TABLE V Calcitonin and Amphiregulin Are Induced by Progesterone in the ERKO Uterus[a]

	Wild-type	ERKO
Calcitonin	78.7 ± 48	24.6 ± 1.7
Amphiregulin (P)	1.9 ± 0.6	2.55 ± 0.8
Amphiregulin (P + E)	0.42 ± 0.1	2.28 ± 1.5

[a] Uterine RNA was analyzed by Northern blot and quantified using a Molecular Dynamics Storm phospho-imager and Image Quant software. Calcitonin or amphiregulin signals were normalized to cyclophilin and expressed as fold induction over vehicle control ± range of duplicate samples (Curtis *et al.*, 1999). Calcitonin was induced with 1 day of estradiol priming followed by 3 days of progesterone treatment. Amphiregulin was induced with one injection of progesterone for 4 h.

also examined. Amphiregulin is an EGF-like hormone that activates the EGF receptor and was shown to be induced at the time of implantation in the mouse uterus (Das *et al.*, 1995). Amphiregulin mRNA can also be induced beginning 2 h after progesterone treatment of ovariectomized mice (Das *et al.*, 1995). Therefore, wild-type and ERKO mice were ovariectomized and treated with progesterone, and after 4 h uterine RNA was isolated and analyzed by Northern blot. Amphiregulin mRNA levels were induced in both genotypes (Curtis *et al.*, 1999) (Table V), indicating that the PR levels in the ERKO were sufficient to mediate an acute genomic response.

It has been shown previously that amphiregulin mRNA induction by progesterone in the uterus is inhibited by cotreatment with estrogen (Das *et al.*, 1995). However, estradiol does not repress amphiregulin induction by progesterone in the ERKO, indicating that ERα was involved in the mechanism of inhibition (Curtis *et al.*, 1999) (Table V). This result illustrates how useful a transgenic model can be for dissection of the mechanism of interaction between two steroid hormonal signaling pathways *in vivo*. Although pharmacological studies can be useful in defining a mechanism, in some cases, removal of a pathway by ablation of the receptor gene allows definitive conclusions to be drawn.

F. Decidual Response in the ERKO Uterus

The ERKO mice are infertile because of ovarian dysfunction and a lack of uterine mitogenic response to estrogen (Lubahn *et al.*, 1993; Korach, 1994). Since the preceding biochemical studies indicate that the PR is present and functional in the ERKO uterus, we wanted to determine whether the ERKO could undergo the progesterone-dependent stromal decidualization that normally occurs in response to implantation. Because of the diminished size of the ERKO uteri (Lubahn *et al.*, 1993; Couse and Korach, 1999), physical trauma of the uterine lumen to experimentally induce a decidual reaction was not possible using established techniques that involve intraluminal scratching or oil injection (Ledford *et al.*, 1976; Lydon *et al.*, 1995). Instead, sesame oil was forced into the uterine horns through the cervix using a medicine dropper attached to a 3-cc syringe. ERKO mice were estrogen primed and then treated with progesterone to mimic pregnancy and implantation (Ledford *et al.*, 1976; Lydon *et al.*, 1995) (Fig. 5). After trauma 7/10 wild type and 5/7 ERKO uteri responded with deciduoma formation (Curtis *et al.*, 1999) (Table VI, Fig. 6). The ERKO animals averaged a 21-fold increase in uterine weight over nontraumatized controls, whereas the wild-type uteri showed only a 6.4-fold increase in uterine weight (Curtis *et al.*, 1999).

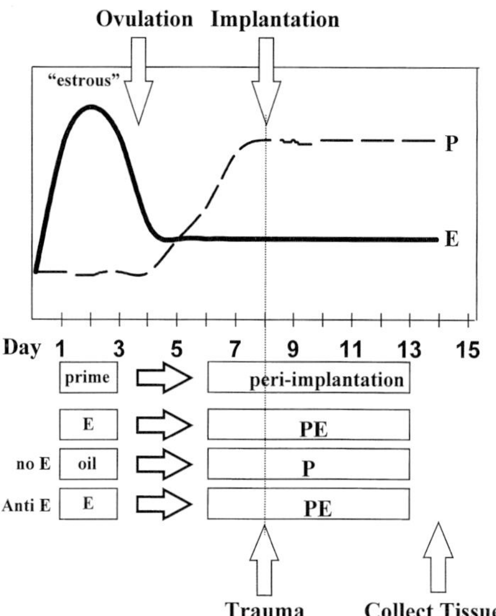

FIGURE 5   Schematic diagram of hormonal and surgical regimen used to induce decidual response. The top portion of the figure is a schematic representation of progesterone (P) and estradiol (E) levels in the mouse during the estrous cycle and early pregnancy. Estradiol levels peak just prior to ovulation during the estrous phase and then drop, but remain above the basal level throughout the peri-implantation period. The progesterone level begins to rise after ovulation and is maintained at a plateau if implantation occurs. To mimic this cycle, animals are given a priming dose of estradiol on days 1, 2, and 3 (E). After 2 days without injection, mice are treated with progesterone and a low dose of estradiol (PE) on days 6–13. Six hours following the third P injection, the uterine lumen is traumatized to mimic implantation. The tissue is collected and analyzed on the day following the final injection. The hormonal regimen was altered to determine the role of estrogen by excluding estrogen from all injections (no E), or cotreating with anti-estrogen ICI 182,780 (anti E) with all injections. Reproduced from Curtis *et al.* (1999).

TABLE VI Decidualization Response Is Estrogen
Independent in the ERKO[a]

Treatment	Wild-type	ERKO
E, PE	70% (7/10)	71% (5/7)
ICI + E. ICI + PE	10% (1/10)	100% (5/5)
Oil, P only	0% (0/3)	100% (2/2)

[a] Uteri collected after the hormonal treatments described in Fig. 5
were inspected for appearance of decidual nodules, and percent
uteri responding was calculated for each treatment group. Number responding/total number in each group are indicated in parentheses (Curtis *et al.,* 1999).

G. The Role of Estradiol in the Decidual Reaction

The decidual response in the ERKO uterus was unexpected, since estrogen priming was thought to be essential for a decidual transformation. To
define the role of estrogen in the mechanism of decidualization, mice were
given the anti-estrogen, ICI 182,780, with the daily estradiol and PE treatments as described previously. Only 1 of 10 wild-type uteri displayed a
decidual reaction, whereas 5 of 5 ERKO uteri responded (Curtis *et al.,* 1999)
(Table VI). The increase in wet weight following the decidualization response
was more robust in the ERKO group (15-fold increase) than in the wild-
type group (2.4-fold increase, Fig. 6) (Curtis *et al.,* 1999). When no estrogen
was used either at priming or with the progesterone, all the ERKO uteri
responded with a 22-fold increase in weight (Table VI), while none of the
wild-type uteri responded (Curtis *et al.,* 1999). Thus, estrogen action through
ERα is not necessary for decidualization in the ERKO. The ERKO uterus is
fully progesterone-responsive in terms of gene regulation and morphological
changes of the tissues despite the absence of ERα.

The function of progesterone during implantation and reproduction is
strongly associated with estrogen and ERα signaling. Most interesting is
our observation that the decidualization reaction is estrogen-dependent in
the wild-type, but not in the ERKO uterus. Others have demonstrated the
necessity for estrogen priming (at a time mimicking estrus) as well as a low
dose of estrogen (at a time mimicking the peri-implantation period) to induce
decidualization with oil (see Fig. 5) (Finn, 1966; Milligan and Mirembe,
1985; Finn and Pope, 1986). However, more traumatic stimuli, such as
crushing the uterine horns, can produce decidualization without estrogen
priming in the mouse (Finn, 1965). This indicates that the mechanism of
decidualization does not absolutely require estrogen. The threshold level of
stimulus required for initiation of decidualization might be lowered by
estrogen treatment in a wild type. Similarly, the estrogen-independent decidualization in the ERKO may indicate that the ovexed ERKO uterine tissue

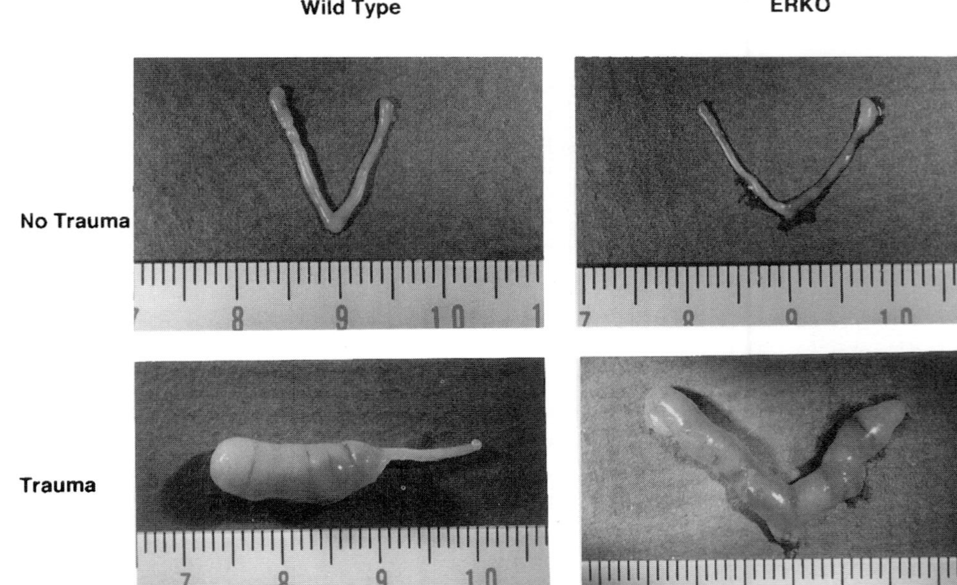

FIGURE 6   Decidualized wild-type and ERKO uteri. Wild-type and ERKO mice were treated as described in Fig. 5. Uteri in the bottom panels were traumatized on the third day of PE treatment as described. The wild-type uterus pictured in the bottom panel shows a decidual reaction in one horn, indicating that oil may only have entered one horn. Reproduced from Curtis *et al.* (1999).

is more sensitive to a decidualization stimulus so that the threshold level of trauma required to initiate decidualization is as low as that of an estrogen-primed wild type.

Although the role of estrogen in the mechanism of decidualization has not been defined, it is likely that estrogen induces genes required for decidual transformation. However, the gene targets of estrogen in initiating and maintaining the deciduoma have not been identified. There is evidence that estrogen can induce several physiologic responses that might be involved in implantation and decidual transformation. These responses include histamine and prostaglandin release (Weitlauf, 1988), increased vascular permeability (Milligan and Mirembe, 1985), and induction of various growth factor receptor ligands (Wang *et al.*, 1994; Das *et al.*, 1995; Lim *et al.*, 1997).

Induction of PR, especially in the priming phase of the experiment, has been proposed to be a key component of the mechanism. Although PR is increased in both the stroma and epithelium, ERα-mediated induction of PR seems to occur mainly in the stroma, since antiestrogen treatment blocks stromal but not epithelial induction of PR mRNA (Das *et al.*, 1998). There is also evidence that ERα-independent estrogen induction of PR might occur in the stroma, as increase in PR protein has been reported in the stroma following estradiol treatment of both wild-type and ERKO uteri (Tibbetts *et al.*, 1998; Kurita *et al.*, 1998). The discrepancy between these results may indicate a posttranscriptional effect of estradiol on stromal ER. Clearly the amount of PR expressed in the ERKO uterus is sufficient to mediate decidualization with or without estradiol induction. Interestingly, the stromal PR has been shown, through tissue recombination, to be the mediator of inhibition of estrogen-induced mitosis (Kurita *et al.*, 1998).

These observations further illustrate the usefulness and unique role of transgenic "knockout" animals. In this case, all evidence gathered from wild-type animals indicates a requirement for estradiol and ER for decidualization to occur. However, the lack of estrogen dependence in the ERKO indicates that the decidual reaction can occur in the absence of ER, shedding new light on the role of ER in the process. Observations such as this must be interpreted with some caution, as the ERKO uterus is not merely a wild-type uterus without ER. Further analysis of the effect of development and functioning in the estrogen-insensitive environment and within the abnormal ovarian and gonadatrophic hormonal milieu are necessary before final conclusions on the role of ER in this process can be drawn.

IV. Nuclear/Membrane Receptor Cross-Talk

Estradiol, epidermal growth factor (EGF), and insulin-like growth factors (IGF) are known mitogens in the rodent reproductive tract (Gannon *et al.*, 1976; Green and Chambon, 1991; Das *et al.*, 1994). Estrogen has been shown to increase the uterine levels of both EGF and its receptor (EGF-R) (Mukku and Stancel, 1985; Stancel *et al.*, 1987, 1990; DiAugustine *et al.*, 1988; Lingham *et al.*, 1988; Gardner *et al.*, 1989; Huet Hudson *et al.*, 1990; Das *et al.*, 1994), suggesting a link between the mitogenic effects of estrogens and growth factors. Furthermore, EGF has been shown to mimic the effects of estrogen in the mouse reproductive tract in terms of increased DNA synthesis and cornification of the vaginal epithelium (Nelson *et al.*, 1991), as well as increased phosphorylation and nuclear retention of the estrogen receptor (ER) (Ignar Trowbridge *et al.*, 1992). When estradiol is administered in conjunction with an EGF-specific antibody, a 60–70% reduction in the hormone-induced proliferation of the epithelium is observed in the mouse uterus and vagina (Nelson *et al.*, 1991). These data indicate a possible role for EGF as a mediator of estrogen action. Further evidence of EGF/estrogen cross-talk was provided by experiments showing that pretreatment of mice with the pure anti-estrogen ICI 164,384 greatly diminished the uterine response to EGF (Ignar Trowbridge *et al.*, 1992). Since ICI 164,384 significantly reduces the level of uterine ER (Gibson *et al.*, 1991), these studies suggest the necessity for the ER in the mitogenic actions of EGF. This was supported by studies in Ishikawa cells, a human endometrial carcinoma cell line devoid of ER, in which an estrogen-responsive chloramphenicol acetyl transferase (CAT) reporter gene could only be activated by EGF after cotransfection with an ER-expression plasmid (Ignar Trowbridge *et al.*, 1993).

These studies have led to a model in which EGF plays a role in ER mediated events in a ligand-independent manner. This model is illustrated in Fig. 7. Two independent but interacting pathways are present in the uterus: The membrane-bound EGFR pathway and the nuclear ER pathway. Ligands that bind to and activate the EGFR induce autophosphorylation and activation of a cascade of cellular proteins, ultimately leading to regulation of nuclear transcription factors and altered expression of EGF-responsive genes. Estrogenic ligands, alternatively, are able to penetrate the cell membrane and bind the nuclear ER protein, resulting in regulation of estrogen-responsive genes. The data described earlier suggest interaction between these two pathways, as depicted by the striped arrow. This indicates that activators of the membrane-bound EGFR can regulate activity of estrogen responses in the absence of estrogenic hormones. To show the validity of this model of EGFR–ER cross-talk, studies were carried out in the ERKO mice. Since the model depends on ER protein for EGF-initiated activity, removal of the ER protein should decouple the pathway, resulting in a lack of ER-mediated responses to EGF, but EGFR-mediated responses should be retained.

A. Characterization of the EGFR Pathway in the ERKO Uterus

EGF and EGFR are both known to be induced by estradiol in the uterus. To address the concern that lack of ER might disrupt the expression and

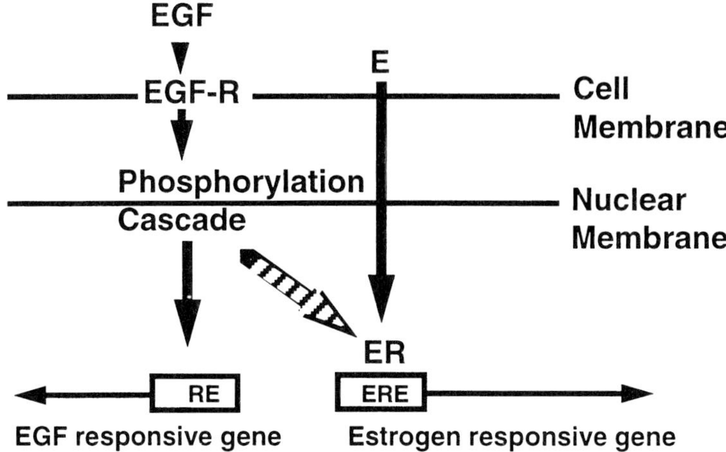

FIGURE 7 Model of cross-talk between EGF-R/ER signaling pathways. The model depicts the separate EGF- and estrogen-inducible pathways. EGF binds to its membrane-bound receptor, which initiates a phosphorylation cascade modulating the activities of cellular kinases and nuclear transcription factors (depicted as the phosphorylation cascade in the model). The activation of transcription factors results in regulation of responsive genes. In contrast, estrogen (E) diffuses to the nucleus where it binds and activates the nuclear ER, thereby modulating transcription of estrogen responsive genes. Cross-talk between the pathways (depicted by the dashed arrow) occurs when EGF initiates the phosphorylation cascade, which activates the ER, presumably via ER phosphorylation, regardless of the presence of estrogen. Reproduced from Curtis *et al.* (1996).

functioning of the EGF-R signaling pathway, EGF-R levels were determined and EGFR autophosphorylation was assayed. Both parameters were found to be similar to those for wild-type animals (Curtis *et al.*, 1996) (Table VII). In addition, the induction of c-*fos*, an EGF-regulated gene, was measured

TABLE VII EGF Signaling in ERKO

Response	Wild-type	ERKO
EGFR, CPM ^{125}I EGF bound/mg protein[a]	16.3×10^3	10.6×10^3
EGFR autophosphorylation[b]	Yes	Yes
c-fos mRNA fold induction[c]	18	13

[a] [125]I EGF binding to membrane fractions was determined as previously described (Sewall *et al.*, 1995; Curtis *et al.*, 1996).

[b] Solubilized membrane from uterus and liver was subjected to an *in vitro* autophosphorylation assay by incubation in the presence (+) or absence (−) of EGF (Rubin and Earp, 1983) and analysis by SDS–PAGE (Curtis *et al.*, 1996).

[c] RNase protection assay was used to measure the induction of c-fos mRNA after treatment with EGF for 30 min compared to vehicle treated controls (Curtis *et al.*, 1996).

TABLE VIII   Responses to EGF in ERKO and
Wild-Type Uteri

Response	Wild-type	ERKO
DNA Synthesis[a]	Yes	No
PR mRNA induction	Yes	No

[a] Curtis *et al.* (1996).

and again found to be comparable to that of wild-type animals, indicating
that the EGFR pathway is fully functional in the ERKO uterus (Curtis *et
al.*, 1996).

B. ER-Mediated Responses to EGF in the ERKO Uterus

ERKO animals, as well as wild-type controls, were treated with EGF,
and induction of DNA synthesis and the progesterone receptor (PR) gene
was measured to confirm whether ER is required for the estrogenic activities
of EGF. Both responses occurred only in the wild-type animals, showing
the dependence upon ER for the EGF induction of these responses (Curtis
et al., 1996) (Table VIII). Thus, removal of the ER in the ERKO has removed
the estradiol independent estrogen-like responses to EGF, indicating that
these responses require the ER protein.

V. Conclusions

Clearly, transgenic steroid receptor knockout mice are powerful tools
that enable dissection of complex interactions between signaling pathways
in a physiological setting. Some tissues and processes are altered in such a
way that results are complicated and difficult to interpret, as in the case of
decidualization in the ERKO. Some processes, for example, the phenotype
of the ERKO ovary, are altered via peripheral effects of steroid receptor
ablation such as altered gonadotropin levels. Importantly, other processes
are unmasked, such as testosterone-dependent uterotropic effects. In addition, these transgenic animals provide convincing evidence for cross-talk
with membrane receptor–mediated signaling.

References

Carson Jurica, M. A., Schrader, W. T., and O'Malley, B. W. (1990). Steroid receptor family:
    structure and functions. *Endocr. Rev.* **11**(2), 201–220.

Chappell, P. E., Lydon, J. P., Conneely, O. M., O'Malley, B. W., and Levine, J. E. (1997). Endocrine defects in mice carrying a null mutation for the progesterone receptor gene. *Endocrinology* 138(10), 4147–4152.

Chen, J. D., and Li, H. (1998). Coactivation and corepression in transcriptional regulation by steroid/nuclear hormone receptors. *Crit. Rev. Eukaryot. Gene Expr.* 8(2), 169–190.

Cooke, P. S., Buchanan, D. L., Young, P., Setiawan, T., Brody, J., Korach, K. S., Taylor, J., Lubahn, D. B., and Cunha, G. R. (1997). Stromal estrogen receptors mediate mitogenic effects of estradiol on uterine epithelium. *Proc. Natl. Acad. Sci. USA* 94(12), 6535–6540.

Couse, J. F., and Korach, K. S. (1999). Estrogen Receptor Null Mice: What have we learned and where will they lead us. Endocrine Reviews 20, 358–417.

Couse, J. F., and Korach, K. S. (1998). Exploring the role of sex steroids through studies of receptor deficient mice. *J. Mol. Med.* 76(7), 497–511.

Couse, J. F., Curtis, S. W., Washburn, T. F., Eddy, E. M., Schomberg, D. W., and Korach, K. S. (1995a). Disruption of the mouse oestrogen receptor gene: resulting phenotypes and experimental findings. *Biochem. Soc. Trans.* 23(4), 929–935.

Couse, J. F., Curtis, S. W., Washburn, T. F., Lindzey, J., Golding, T. S., Lubahn, D. B., Smithies, O., and Korach, K. S. (1995b). Analysis of transcription and estrogen insensitivity in the female mouse after targeted disruption of the estrogen receptor gene. *Mol. Endocrinol.* 9, 1441–1454.

Curtis, S. W., Washburn, T., Sewall, C., DiAugustine, R., Lindzey, J., Couse, J. F., and Korach, K. S. (1996). Physiological coupling of growth factor and steroid receptor signaling pathways: Estrogen receptor knockout mice lack estrogen-like response to epidermal growth factor. *Proc. Natl. Acad. Sci. USA* 93(22), 12626–12630.

Curtis, S. W., Clark, J., Myers, P., and Korach, K. S. (1999). Disruption of estrogen signaling does not prevent progesterone action in the estrogen receptor knockout mouse uterus. *Proc. Natl. Acad. Sci. USA,* 96, 3646–3651.

Daniel, C. W., Silberstein, G. B., and Strickland, P. (1987). Direct action of 17 beta-estradiol on mouse mammary ducts analyzed by sustained release implants and steroid autoradiography. *Cancer Res.* 47(22), 6052–6057.Das, S. K., Tsukamura, H., Paria, B. C., Andrews, G. K., and Dey, S. K. (1994). Differential expression of epidermal growth factor receptor (EGF-R) gene and regulation of EGF-R bioactivity by progesterone and estrogen in the adult mouse uterus. *Endocrinology* 134(2), 971–981.

Das, S. K., Chakraborty, I., Paria, B. C., Wang, X. N., Plowman, G. and Dey, S. K. (1995). Amphiregulin is an implantation-specific and progesterone-regulated gene in the mouse uterus. *Mol. Endocrinol.* 9, 691–705.

Das, S. K., Tan, J., Johnson, D. C. and Dey, S. K. (1998). Differential spatiotemporal regulation of lactoferrin and progesterone receptor genes in the mouse uterus by primary estrogen, catechol estrogen, and xenoestrogen. *Endocrinology* 139, 2905–2915.

DiAugustine, R. P., Petrusz, P., Bell, G. I., Brown, C. F., Korach, K. S., McLachlan, J. A., and Teng, C. T. (1988). Influence of estrogens on mouse uterine epidermal growth factor precursor protein and messenger ribonucleic acid. *Endocrinology* 122(6), 2355–2363.

Ding, Y. Q., Zhu, L. J., Bagchi, M. K., and Bagchi, I. C. (1994). Progesterone stimulates calcitonin gene expression in the uterus during implantation. *Endocrinology* 135, 2265–2274.

Ding, Y. Q., Bagchi, M. K., Bardin, C. W. and Bagchi, I. C. (1995). Calcitonin gene expression in the rat uterus during pregnancy. *Rec. Prog. Hormone Res.* 50, 373–378.

Eddy, E. M., Washburn, T. F., Bunch, D. O., Goulding, E. H., Gladen, B. C., Lubahn, D. B., and Korach, K. S. (1996). Targeted disruption of the estrogen receptor gene in male mice causes alteration of spermatogenesis and infertility. *Endocrinology* 137(11), 4796–4805.

Evans, R. M. (1988). The steroid and thyroid hormone receptor superfamily. *Science* 240(4854), 889–895.

Finn, C. A. (1965). Oestrogen and the decidual cell reaction of implantation in mice. *J. Endocrinol.* 32, 223–229.

Finn, C. A. (1966). Endocrine control of endometrial sensitivity during the induction of the decidual cell reaction in the mouse. *J. Endocrinol.* **36**, 239–248.

Finn, C. A. and Pope, M. (1986). Control of leucocyte infiltration into the decidualized mouse uterus. *J. Endocrinol.* **110**, 93–96.

Galand, P., Leroy, F. and Chretien, J. (1971). Effect of oestradiol on cell proliferation and histological changes in the uterus and vagina of mice. *J. Endocrinol.* **49**(2), 243–252.

Gannon, F., Katzenellenbogen, B., Stancel, G., and Gorski, J. (1976). Estrogen-receptor movement to the nucleus: Discussion of a cytoplasmic-exclusion hypothesis. *Symp. Soc. Dev. Biol.* **34**, 137–149.

Gardner, R. M., Verner, G., Kirkland, J. L., and Stancel, G. M. (1989). Regulation of uterine epidermal growth factor (EGF) receptors by estrogen in the mature rat and during the estrous cycle. *J. Steroid Biochem.* **32**(3), 339–343.

George, F. W., and Wilson, J. D. (1986). Hormonal control of sexual development. *Vitam. Horm.* **43**, 145–196.

Gibson, M. K., Nemmers, L. A., Beckman, W. C., Jr., Davis, V. L., Curtis, S. W., and Korach, K. S. (1991). The mechanism of ICI 164,384 antiestrogenicity involves rapid loss of estrogen receptor in uterine tissue. *Endocrinology* **129**(4), 2000–2010.

Graham, J. D., and Clarke, C. L. (1997). Physiological action of progesterone in target tissues. *Endocrine Rev.* **18**, 502–519.

Green, S., and Chambon, P. (1991). The oestrogen receptor: From perception to mechanism. *In* "Nuclear Hormone Receptors Molecular Mechanisms, Cellular Functions, Clinical Abnormalities" (M. G. Parker, ed.). Academic Press, San Diego, pp. 15–38.

Huet Hudson, Y. M., Chakraborty, C., De, S. K., Suzuki, Y., Andrews, G. K. and Dey, S. K. (1990). Estrogen regulates the synthesis of epidermal growth factor in mouse uterine epithelial cells. *Mol. Endocrinol.* **4**(3), 510–523.

Ignar Trowbridge, D. M., Nelson, K. G., Bidwell, M. C., Curtis, S. W., Washburn, T. F., McLachlan, J. A., and Korach, K. S. (1992). Coupling of dual signaling pathways: epidermal growth factor action involves the estrogen receptor. *Proc. Natl. Acad. Sci. USA* **89**(10), 4658–4662.

Ignar Trowbridge, D. M., Teng, C. T., Ross, K. A., Parker, M. G., Korach, K. S., and McLachlan, J. A. (1993). Peptide growth factors elicit estrogen receptor-dependent transcriptional activation of an estrogen-responsive element. *Mol. Endocrinol.* **7**(8), 992–998.

Korach, K. S. (1994). Insights from the study of animals lacking functional estrogen receptor. *Science* **266**(5190), 1524–1527.

Korach, K. S., Couse, J. F., Curtis, S. W., Washburn, T. F., Lindzey, J., Kimbro, K. S., Eddy, E. M., Migiaccio, S., Snedeker, S. M., Lubahn, D. B., Schomberg, D. W., and Smith, E. P. (1996). Estrogen receptor gene disruption: Molecular characterization and experimental and clinical phenotypes. *In* "Recent Progress in Hormone Research (P. Conn, ed.), Vol. 51, pp. 159–188. The Endocrine Society, Bethesda, MD.

Kurita, T., Young, P., Brody, J. R., Lydon, J. P., O'Malley, B. W. and Cunha, G. R. (1998). Stromal progesterone receptors mediate the inhibitory effects of progesterone on estrogen-induced uterine epithelial cell deoxyribonucleic acid synthesis. *Endocrinology* **139**(11), 4708–4713.

Ledford, B. E., Rankin, J. C., Markwald, R. R., and Baggett, B. (1976). Biochemical and morphological changes following artificially stimulated decidualization in the mouse uterus. *Biol. Reprod.* **15**, 529–535.

Lim, H., Dey, S. K., and Das, S. K. (1997). Differential expression of the erbB2 gene in the periimplantation mouse uterus: Potential mediator of signaling by epidermal growth factor-like growth factors. *Endocrinology* **138**, 1328–1337.

Lindzey, J., Curtis, S., Washburn, T., and Korach, K. (1996). Uterotropic effects of dihydrotestosterone in estrogen receptor knockout and wild-type mice. 10th International Congress of Endocrinology, San Francisco, CA, The Endocrine Society.

Lingham, R. B., Stancel, G. M., and Loose Mitchell, D. S. (1988). Estrogen regulation of epidermal growth factor receptor messenger ribonucleic acid. *Mol. Endocrinol.* 2(3), 230–235.

Lubahn, D. B., Moyer, J. S., Golding, T. S., Couse, J. F., Korach, K. S., and Smithies, O. (1993). Alteration of reproductive function but not prenatal sexual development after insertional disruption of the mouse estrogen receptor gene. *Proc. Natl. Acad. Sci. USA* 90(23), 11162–11166.

Lydon, J. P., DeMayo, F. J., Funk, C. R., Mani, S. K., Hughes, A. R., Montgomery, C. A., Shyamala, G., Conneely, O. M. and O'Malley, B. W. (1995). Mice lacking progesterone receptor exhibit pleiotropic reproductive abnormalities. *Genes Devel.* 9, 2266–2278.

Lydon, J. P., DeMayo, F. J., Conneely, O. M., and O'Malley, B. W. (1996). Reproductive phenotpes of the progesterone receptor null mutant mouse. *J. Steroid Biochem. Mol. Biol.* 56, 67–77.

Mani, S. K., Allen, J. M., Clark, J. H., Blaustein, J. D., and O'Malley, B. W. (1994a). Convergent pathways for steroid hormone- and neurotransmitter-induced rat sexual behavior. Science 265(5176), 1246–1249.

Mani, S. K., Blaustein, J. D., Allen, J. M., Law, S. W., O'Malley, B. W. and Clark, J. H. (1994b). Inhibition of rat sexual behavior by antisense oligonucleotides to the progesterone receptor. *Endocrinology* 135(4), 1409–1414.

Mani, S. K., Allen, J. M., Lydon, J. P., Mulac-Jericevic, B., Blaustein, J. D., DeMayo, F. J., Conneely, O., and O'Malley, B. W. (1996). Dopamine requires the unoccupied progesterone receptor to induce sexual behavior in mice. Mol. Endocrinol 10, 1728–1737.

Milligan, S. R., and Mirembe, F. M. (1985). Intraluminally injected oil induces changes in vascular permeability in the "sensitized" and "non-sensitized" uterus of the mouse. *J. Reprod. Fertil.* 74, 95–104.

Mukku, V. R., and Stancel, G. M. (1985). Regulation of epidermal growth factor receptor by estrogen. *J. Biol. Chem.* 260(17), 9820–9824.

Nelson, K. G., Takahashi, T., Bossert, N. L., Walmer, D. K., and McLachlan, J. A. (1991). Epidermal growth factor replaces estrogen in the stimulation of female genital-tract growth and differentiation. *Proc. Natl. Acad. Sci. USA* 88(1), 21–25.

Ogawa, S., Taylor, J. A., Lubahn, D. B., Korach, K. S., and Pfaff, D. W. (1996). Reversal of sex roles in genetic female mice by disruption of estrogen receptor gene. *Neuroendocrinology* 64(6), 467–470.

Ogawa, S., Eng, V., Taylor, J., Lubahn, D. B., Korach, K. S., and Pfaff, D. W. (1998). Roles of estrogen receptor-alpha gene expression in reproduction-related behaviors in female mice. *Endocrinology* 139(12), 5070–5081.

O'Malley, B. (1990). The steroid receptor superfamily: more excitement predicted for the future. *Mol. Endocrinol.* 4(3), 363–369.

Power, R. F., Mani, S. K., Codina, J., Conneely, O. M., and O'Malley, B. W. (1991). Dopaminergic and ligand-independent activation of steroid hormone receptors. *Science* 254(5038), 1636–1639.

Rubin, R. A., and Earp, H. S. (1983). Solubilization of EGF receptor with Triton X-100 alters stimulation of tyrosine residue phosphorylation by EGF and dimethyl sulfoxide. *J. Biol. Chem.* 258(8), 5177–5182.

Scully, K. M., Gleiberman, A. S., Lindzey, J., Lubahn, D. B., Korach, K. S., and Rosenfeld, M. G. (1997). Role of estrogen receptor-alpha in the anterior pituitary gland. *Mol. Endocrinol.* 11(6), 674–681.

Sewall, C. H., Clark, G. C., and Lucier, G. W. (1995). TCDD reduces rat hepatic epidermal growth factor receptor: Comparison of binding, immunodetection, and autophosphorylation. *Toxicol. Appl. Pharmacol.* 132(2), 263–272.

Silberstein, G. B., Van Horn, K., Shyamala, G., and Daniel, C. W. (1994). Essential role of endogenous estrogen in directly stimulating mammary growth demonstrated by implants containing pure antiestrogens. *Endocrinol.* 134(1), 84–90.

Stancel, G. M., Gardner, R. M., Kirkland, J. L., Lin, T. H., Lingham, R. B., Loose Mitchell, D. S., Mukku, V. R., Orengo, C. A. and Verner, G. (1987). Interactions between estrogen and EGF in uterine growth and function. *Adv. Exp. Med. Biol.* **230**, 99–118.

Stancel, G. M., Chiapetta, C., Gardner, R. M., Kirkland, J. L., Lin, T. H., Lingham, R. B., Loose Mitchell, D. S., Mukku, V. R. and Orengo, C. A. (1990). Regulation of the uterine epidermal growth factor receptor by estrogen. *Prog. Clin. Biol. Res.* **322**, 213–226.

Tibbetts, T. A., Mendoza-Meneses, M., O'Malley, B. W., and Conneely, O. M. (1998). Mutual and intercompartmental regulation of estrogen receptor and progesterone receptor expression in the mouse uterus. *Biol. Reprod.* **59**(5), 1143–1152.

Tsai, M. J. and O'Malley, B. W. (1994). Molecular mechanisms of action of steroid/thyroid receptor superfamily members. *Annu. Rev. Biochem.* **63**, 451–486.

Vegeto, E., Wagner, B. L., Imhof, M. O., and McDonnell, D. P. (1996). The molecular pharmacology of ovarian steroid receptors. *Vitam. Horm.* **52**, 99–128.

Wang, X. N., Das, S. K., Damm, D., Klagsbrun, M., Abraham, J. A., and Dey, S. K. (1994). Differential regulation of heparin-binding epidermal growth factor-like growth factor in the adult ovariectomized mouse uterus by progesterone and estrogen. *Endocrinology* **135**(3), 1264–1271.

Weitlauf, H. M. (1988). Biology of implantation. *In* "The Physiology of Reproduction" (E. Knobil and J. Neill, eds.), pp. 231–262. Raven Press, New York.

Werner, M. H., Huth, J. R., Gronenborn, A. M., and Clore, G. M. (1996). Molecular determinants of mammmalian sex. *Trends Biochem. Sci.* **21**(8), 302–308.

Wilson, J. D., George, F. W., and Renfree, M. B. (1995). The endocrine role in mammalian sexual differentiation. *In* "Recent Progress in Hormone Research" (C. W. Bardin, ed.), Vol. 50, pp. 349–364. Academic Press, San Diego.

Index

Contents of Previous Volumes

Unnatural Nucleotide Sequences in Biopharmaceutics
Lawrence A. Loeb

Pharmacology of the Neurotransmitter Release Enhancer
Linopirdine (DuP 996), and Insights into Its Mechanism
of Action
Simon P. Aiken, Robert Zaczek, and Barry S. Brown

Volume 36

Regulation of Somatostatin Gene Transcription by cAMP
M. Montminy, P. Brindle, J. Arias, K. Ferreri, and R. Armstrong

Dissection of Protein Kinase Cascades That Mediate Cellular
Response to Cytokines and Cellular Stress
Philip Cohen

Cyclic Nucleotide Phosphodiesterases: Gene Complexity,
Regulation by Phosphorylation, and Physiological Implications
Fiona Burns, Allan Z. Zhao, and Joseph A. Beavo

Structural Analysis of the MAP Kinase ERK2 and Studies of MAP
Kinase Regulatory Pathways
Melanie H. Cobb, Shuichan Xu, Mangeng Cheng, Doug Ebert, David Robbins,
Elizabeth Goldsmith, and Megan Robinson

Novel Protein Phosphatases That May Participate in
Cell Signaling
Patricia T. W. Cohen, Mao Xiang Chen, and Christopher G. Armstrong

Protein Tyrosine Phosphatases and the Control of Cellular
Signaling Responses
N. K. Tonks

Roles of the MAP Kinase Cascade in Vertebrates
Tetsuo Moriguchi, Yukiko Gotoh, and Eisuke Nishida

Signal Transductions of SH2/SH3: Ash/Grb-2
Downstream Signaling
Tadaomi Takenawa, Kenji Miura, Hiroaki Miki, and Kazutada Watanabe

Sphingolipid-Dependent Protein Kinases
Sen-itiroh Hakomori

Volume 37

Volume 40

Volume 41

Role of *p53* in Apoptosis
Christine E. Canman and Michael B. Kastan

Chemotherapy-Induced Apoptosis
Peter W. Mesner, Jr., I. Imawati Budihardjo, and Scott H. Kaufmann

Bcl-2 Family Proteins: Strategies for Overcoming Chemoresistance in Cancer
John C. Reed

Role of Bcr-Abl Kinase in Resistance to Apoptosis
Afshin Samali, Adrienne M. Gorman, and Thomas G. Cotter

Apoptosis in Hormone-Responsive Malignancies
Samuel R. Denmeade, Diane E. McCloskey, Ingrid B. J. K. Joseph, Hillary A. Hahm, John T. Isaacs, and Nancy E. Davidson

Volume 42

Catecholamine: Bridging Basic Science
Edited by David S. Goldstein, Graeme Eisenhofer, and Richard McCarty

Part A. Catecholamine Synthesis and Release

Part B. Catecholamine Reuptake and Storage

Part C. Catecholamine Metabolism

Part D. Catecholamine Receptors and Signal Transduction

Part E. Catecholamine in the Periphery

Part F. Catecholamine in the Central Nervous System

Part G. Novel Catecholaminergic Systems

Part H. Development and Plasticity

Part I. Drug Abuse and Alcoholism

Volume 43

Overview: Pharmacokinetic Drug–Drug Interactions
Albert P. Li and Malle Jurima-Romet

Role of Cytochrome P450 Enzymes in Drug–Drug Interactions
F. Peter Guengerich

The Liver as a Target for Chemical–Chemical Interactions
John-Michael Sauer, Eric R. Stine, Lhanoo Gunawardhana, Dwayne A. Hill, and I. Glenn Sipes

Application of Human Liver Microsomes in Metabolism-Based Drug–Drug Interactions: *In Vitro–in Vivo* Correlations and the Abbott Laboratories Experience
A. David Rodrigues and Shekman L. Wong

Primary Hepatocyte Cultures as an *in Vitro* Experimental Model for the Evaluation of Pharmacokinetic Drug–Drug Interactions
Albert P. Li

Liver Slices as a Model in Drug Metabolism
James L. Ferrero and Klaus Brendel

Use of cDNA-Expressed Human Cytochrome P450 Enzymes to Study Potential Drug–Drug Interactions
Charles L. Crespi and Bruce W. Penman

Pharmacokinetics of Drug Interactions
Gregory L. Kedderis

Experimental Models for Evaluating Enzyme Induction Potential of New Drug Candidates in Animals and Humans and a Strategy for Their Use
Thomas N. Thompson

Metabolic Drug–Drug Interactions: Perspective from FDA Medical and Clinical Pharmacology Reviewers
John Dikran Balian and Atiqur Rahman

Drug Interactions: Perspectives of the Canadian Drugs Directorate
Malle Jurima-Romet

Overview of Experimental Approaches for Study of Drug Metabolism and Drug–Drug Interactions
Frank J. Gonzalez

Volume 44

Drug Therapy: The Impact of Managed Care
Joseph Hopkins, Shirley Siu, Maureen Cawley, and Peter Rudd

The Role of Phosphodiesterase Enzymes in Allergy and Asthma
D. Spina, L. J. Landells, and C. P. Page

Modulating Protein Kinase C Signal Transduction
Daria Mochly-Rosen and Lawrence M. Kauvar

Preventive Role of Renal Kallikrein—Kinin System in the Early
Phase of Hypertension and Development of New Antihypertensive
Drugs
Makoto Kartori and Masataka Majima

The Multienzyme PDE4 Cyclic Adenosine Monophosphate-Specific
Phosphodiesterase Family: Intracellular Targeting, Regulation, and
Selective Inhibition by Compounds Exerting Anti-inflammatory
and Antidepressant Actions
Miles D. Houslay, Michael Sullivan, and Graeme B. Bolger

Clinical Pharmacology of Systemic Antifungal Agents: A
Comprehensive Review of Agents in Clinical Use, Current
Investigational Compounds, and Putative Targets for Antifungal
Drug Development
Andreas H. Groll, Stephen C. Piscitelli, and Thomas J. Walsh

Volume 45

Cumulative Subject Index
Volumes 25–44

Volume 46

Therapeutic Strategies Involving the Multidrug Resistance
Phenotype: The *MDR1* Gene as Target, Chemoprotectant, and
Selectable Marker in Gene Therapy
Josep M. Aran, Ira Pastan, and Michael M. Gottesman

The Diversity of Calcium Channels and Their Regulation in
Epithelial Cells
Min I. N. Zhang and Roger G. O'Neil

Gene Therapy and Vascular Disease
Melina Kibbe, Timothy Billiar, and Edith Tzeng

Heparin in Inflammation: Potential Therapeutic Applications beyond Anticoagulation
David J. Tyrrell, Angela P. Horne, Kevin R. Holme, Janet M. H. Preuss, and Clive P. Page

The Regulation of Epithelial Cell cAMP- and Calcium-Dependent Chloride Channels
Andrew P. Morris

Calcium Channel Blockers: Current Controversies and Basic Mechanisms of Action
William T. Clusin and Mark E. Anderson

Mechanisms of Antithrombotic Drugs
Perumal Thiagarajan and Kenneth K. Wu